HYPERSPECTRAL REMOTE SENSING OF VEGETATION
SECOND EDITION
VOLUME III

Biophysical and Biochemical Characterization and Plant Species Studies

Edited by
Prasad S. Thenkabail
John G. Lyon
Alfredo Huete

CRC Press
Taylor & Francis Group
Boca Raton London New York

CRC Press is an imprint of the
Taylor & Francis Group, an **informa** business

CRC Press
Taylor & Francis Group
6000 Broken Sound Parkway NW, Suite 300
Boca Raton, FL 33487-2742

First issued in paperback 2022

© 2019 by Taylor & Francis Group, LLC
CRC Press is an imprint of Taylor & Francis Group, an Informa business

No claim to original U.S. Government works

ISBN 13: 978-1-03-247586-8 (pbk)
ISBN 13: 978-1-138-36471-4 (hbk)

DOI: 10.1201/9780429431180

**Visit the Taylor & Francis Web site at
http://www.taylorandfrancis.com**

**and the CRC Press Web site at
http://www.crcpress.com**

Dr. Prasad S. Thenkabail, *Editor-in-Chief of these four volumes would like to dedicate the four volumes to three of his professors at the Ohio State University during his PhD days:*

1. Late Prof. Andrew D. Ward, *former professor of The Department of Food, Agricultural, and Biological Engineering (FABE) at The Ohio State University,*

2. Prof. John G. Lyon, *former professor of the Department of Civil, Environmental and Geodetic Engineering at the Ohio State University, and*

3. Late Prof. Carolyn Merry, *former Professor Emerita and former Chair of the Department of Civil, Environmental and Geodetic Engineering at the Ohio State University.*

Contents

SECTION I Vegetation Biophysical and Biochemical Properties

SECTION II Plant Species Identification and Discrimination

SECTION III   Conclusions

Foreword to the First Edition

The publication of this book, *Hyperspectral Remote Sensing of Vegetation*, marks a milestone in the application of imaging spectrometry to study 70% of the Earth's landmass which is vegetated. This book shows not only the breadth of international involvement in the use of hyperspectral data but also in the breadth of innovative application of mathematical techniques to extract information from the image data.

Imaging spectrometry evolved from the combination of insights from the vast heterogeneity of reflectance signatures from the Earth's surface seen in the ERTS-1 (Landsat-1) 4-band images and the field spectra that were acquired to help fully understand the causes of the signatures. It wasn't until 1979 when the first hybrid area-array detectors, mercury-cadmium-telluride on silicon CCD's, became available that it was possible to build an imaging spectrometer capable of operating at wavelengths beyond 1.0 μm. The AIS (airborne imaging spectrometer), developed at NASA/JPL, had only 32 cross-track pixels but that was enough for the geologists clamoring for this development to see *between* the bushes to determine the mineralogy of the substrate. In those early years, vegetation cover was just a nuisance!

In the early 1980s, spectroscopic analysis was driven by the interest to identify mineralogical composition by exploiting absorptions found in the SWIR region from overtone and combination bands of fundamental vibrations found in the mid-IR region beyond 3 μm and the electronic transitions in transition elements appearing, primarily, short of 1.0 μm. The interests of the geologists had been incorporated in the Landsat TM sensor in the form of the add-on, band 7 in the 2.2 μm region based on field spectroscopic measurements. However, one band, even in combination with the other six, did not meet the needs for mineral identification. A summary of mineralogical analyses is presented by Vaughan et al. in this volume. A summary of the historical development of hyperspectral imaging can be found in Goetz (2009).

At the time of the first major publication of the AIS results (Goetz et al., 1985), very little work on vegetation analysis using imaging spectroscopy had been undertaken. The primary interest was in identifying the relationship of the chlorophyll absorption red-edge to stress and substrate composition that had been seen in airborne profiling and in field spectral reflectance measurements. Most of the published literature concerned analyzing NDVI, which only required two spectral bands.

In the time leading up to the 1985 publication, we had only an inkling of the potential information content in the hundreds of contiguous spectral bands that would be available to us with the advent of AVIRIS (airborne visible and infrared imaging spectrometer). One of the authors, Jerry Solomon, presciently added the term "hyperspectral" to the text of the paper to describe the "...multidimensional character of the spectral data set," or, in other words, the mathematically, over-determined nature of hyperspectral data sets. The term hyperspectral as opposed to multispectral data moved into the remote sensing vernacular and was additionally popularized by the military and intelligence community.

In the early 1990s, as higher quality AVIRIS data became available, and the first analyses of vegetation using statistical techniques borrowed from chemometrics, also known as NIRS analysis used in the food and grain industry, were undertaken by John Aber and Mary Martin of the University of New Hampshire. Here, nitrogen contents of tree canopies were predicted from reflectance spectra by regression techniques using reference measurements from laboratory wet chemical analyses of needle and leaf samples acquired by shooting down branches. At the same time, the remote sensing community began to recognize the value of "too many" spectral bands and the concomitant wealth of spatial information that was amenable to information extraction by statistical techniques. One of them was Eyal Ben-Dor who pioneered soil analyses using hyperspectral imaging and who is one of the contributors to this volume.

As the quality of AVIRIS data grew, manifested in increasing SNR, an ever-increasing amount of information could be extracted from the data. This quality was reflected in the increasing number of nearly noiseless principal components that could be obtained from the data or, in other words, its dimensionality. The explosive advances in desktop computing made possible the application of image processing and statistical analyses that revolutionized the uses of hyperspectral imaging. Joe Boardman and others at the University of Colorado developed what has become the ENVI software package to make possible the routine analysis of hyperspectral image data using "unmixing techniques" to derive the relative abundance of surface materials on a pixel-by-pixel basis.

Many of the analysis techniques discussed in this volume, such as band selection and various indices, are rooted in principal components analysis. The eigenvector loadings or factors indicate which spectral bands are the most heavily weighted allowing others to be discarded to reduce the noise contribution. As sensors become better, more information will be extractable and fewer bands will be discarded. This is the beauty of hyperspectral imaging, allowing the choice of the number of eigenvectors to be used for a particular problem. Computing power has reached such a high level that it is no longer necessary to choose a subset of bands just to minimize the computational time.

As regression techniques such as PLS (partial least squares) become increasingly adopted to relate a particular vegetation parameter to reflectance spectra, it must be remembered that the quality of the calibration model is a function of both the spectra and the reference measurement. With spectral measurements of organic and inorganic compounds under laboratory conditions, we have found that a poor model with a low coefficient of determination (r^2) is most often associated with inaccurate reference measurements, leading to the previously intuitive conclusion that "spectra don't lie."

Up to this point, AVIRIS has provided the bulk of high-quality hyperspectral image data but on an infrequent basis. Although Hyperion has provided some time series data, there is no hyperspectral imager yet in orbit that is capable of providing routine, high-quality images of the whole Earth on a consistent basis. The hope is that in the next decade, HyspIRI will be providing VNIR and SWIR hyperspectral images every 3 weeks and multispectral thermal data every week. This resource will revolutionize the field of vegetation remote sensing since so much of the useful information is bound up in the seasonal growth cycle. The combination of the spectral, spatial, and temporal dimensions will be ripe for the application of statistical techniques and the results will be extraordinary.

Dr. Alexander F. H. Goetz PhD
Former Chairman and Chief Scientist
ASD Inc.
2555 55th St. #100
Boulder, CO 80301, USA
303-444-6522 ext. 108
Fax 303-444-6825
www.asdi.com

REFERENCES

Goetz, A. F. H., 2009, Three decades of hyperspectral imaging of the Earth: A personal view, *Remote Sensing of Environment*, 113, S5–S16.
Goetz, A.F.H., G. Vane, J. Solomon and B.N. Rock, 1985, Imaging spectrometry for Earth remote sensing, *Science*, 228, 1147–1153.

BIOGRAPHICAL SKETCH

Dr. Goetz is one of the pioneers in hyperspectral remote sensing and certainly needs no introduction. Dr. Goetz started his career working on spectroscopic reflectance and emittance studies of the Moon and Mars. He was a principal investigator of Apollo-8 and Apollo-12 multispectral photography studies. Later, he turned his attention to remote sensing of Planet Earth working in collaboration with Dr. Gene Shoemaker to map geology of Coconino County (Arizona) using Landsat-1 data and went on to be an investigator in further Landsat, Skylab, Shuttle, and EO-1 missions. At NASA/JPL he pioneered field spectral measurements and initiated the development of hyperspectral imaging. He spent 21 years on the faculty of the University of Colorado, Boulder, and retired in 2006 as an Emeritus Professor of Geological Sciences and an Emeritus Director of Center for the Study of Earth from Space. Since then, he has been Chairman and Chief Scientist of ASD Inc. a company that has provided more than 850 research laboratories in over 60 countries with field spectrometers. Dr. Goetz is now retired. His foreword was written for the first edition and I have retained it in consultation with him to get a good perspective on the development of hyperspectral remote sensing.

Foreword to the Second Edition

The publication of the four-volume set, *Hyperspectral Remote Sensing of Vegetation*, second edition, is a landmark effort in providing an important, valuable, and timely contribution that summarizes the state of spectroscopy-based understanding of the Earth's terrestrial and near shore environments. Imaging spectroscopy has had 35 years of development in data processing and analysis methods. Today's researchers are eager to use data produced by hyperspectral imagers and address important scientific issues from agricultural management to global environmental stewardship. The field started with development of the Jet Propulsion Lab's Airborne Imaging Spectrometer in 1983 that measured across the reflected solar infrared spectrum with 128 spectral bands. This technology was quickly followed in 1987 by the more capable Advanced Visible Infrared Imaging Spectrometer (AVIRIS), which has flown continuously since this time (albeit with multiple upgrades). It has 224 spectral bands covering the 400–2500 nm range with 10 nm wavelength bands and represents the "gold standard" of this technology. In the years since then, progress toward a hyperspectral satellite has been disappointingly slow. Nonetheless, important and significant progress in understanding how to analyze and understand spectral data has been achieved, with researchers focused on developing the concepts, analytical methods, and spectroscopic understanding, as described throughout these four volumes. Much of the work up to the present has been based on theoretical analysis or from experimental studies at the leaf level from spectrometer measurements and at the canopy level from airborne hyperspectral imagers.

Although a few hyperspectral satellites have operated over various periods in the 2000s, none have provided systematic continuous coverage required for global mapping and time series analysis. An EnMap document compiled the past and near-term future hyperspectral satellites and those on International Space Station missions (EnMap and GRSS Technical Committee 2017). Of the hyperspectral imagers that have been flown, the European Space Agency's CHRIS (Compact High Resolution Imaging Spectrometer) instrument on the PROBA-1 (Project for On-Board Autonomy) satellite and the Hyperion sensor on the NASA technology demonstrator, Earth Observing-1 platform (terminated in 2017). Each has operated for 17 years and have received the most attention from the science community. Both collect a limited number of images per day, and have low data quality relative to today's capability, but both have open data availability. Other hyperspectral satellites with more limited access and duration include missions from China, Russia, India, and the United States.

We are at a threshold in the availability of hyperspectral imagery. There are many hyperspectral missions planned for launch in the next 5 years from China, Italy, Germany, India, Japan, Israel, and the United States, some with open data access. The analysis of the data volumes from this proliferation of hyperspectral imagers requires a comprehensive reference resource for professionals and students to turn to in order to understand and correctly and efficiently use these data. This four-volume set is unique in compiling in-depth understanding of calibration, visualization, and analysis of data from hyperspectral sensors. The interest in this technology is now widespread, thus, applications of hyperspectral imaging cross many disciplines, which are truly international, as is evident by the list of authors of the chapters in these volumes, and the number of countries planning to operate a hyperspectral satellite. At least some of the hyperspectral satellites announced and expected to be launched in this decade (such as the HyspIRI-like satellite approved for development by NASA with a launch in the 2023 period) will provide high-fidelity narrow-wavelength bands, covering the full reflected solar spectrum, at moderate (30 m pixels) to high spatial resolution. These instruments will have greater radiometric range, better SNR, pointing accuracy, and reflectance calibration than past instruments, and will collect data from many countries and parts of the world that have not previously been available. Together, these satellites will produce an unprecedented flow of information about the physiological functioning (net primary production, evapotranspiration, and even direct measurements related to respiration), biochemical characteristics (from spectral indices and from radiative transfer first principle methods), and direct measurements of the distributions of plant and soil biodiversity of the terrestrial and coastal environments of the Earth.

This four-volume set presents an unprecedented range and scope of information on hyperspectral data analysis and applications written by leading authors in the field. Topics range from sensor characteristics from ground-based platforms to satellites, methods of data analysis to characterize plant functional properties related to exchange of gases CO_2, H_2O, O_2, and biochemistry for pigments, N cycle, and other molecules. How these data are used in applications range from precision agriculture to global change research. Because the hundreds of bands in the full spectrum includes information to drive detection of these properties, the data is useful at scales from field applications to global studies.

Volume I has three sections and starts with an introduction to hyperspectral sensor systems. Section II focuses on sensor characteristics from ground-based platforms to satellites, and how these data are used in global change research, particularly in relation to agricultural crop monitoring and health of natural vegetation. Section III provides five chapters that deal with the concept of spectral libraries to identify crops and spectral traits, and for phenotyping for plant breeding. It addresses the development of spectral libraries, especially for agricultural crops and one for soils.

Volume II expands on the first volume, focusing on use of hyperspectral indices and image classification. The volume begins with an explanation of how narrow-band hyperspectral indices are determined, often from individual spectral absorption bands but also from correlation matrices and from derivative spectra. These are followed by chapters on statistical approaches to image classification and a chapter on methods for dealing with "big data." The last half of this volume provides five chapters focused on use of vegetation indices for quantifying and characterizing photosynthetic pigments, leaf nitrogen concentrations or contents, and foliar water content measurements. These chapters are particularly focused on applications for agriculture, although a chapter addresses more heterogeneous forest conditions and how these patterns relate to monitoring health and production.

The first half of Volume III focuses on biophysical and biochemical characterization of vegetation properties that are derived from hyperspectral data. Topics include ecophysiological functioning and biomass estimates of crops and grasses, indicators of photosynthetic efficiency, and stress detection. The chapter addresses biophysical characteristics across different spatial scales while another chapter examines spectral and spatial methods for retrieving biochemical and biophysical properties of crops. The chapters in the second half of this volume are focused on identification and discrimination of species from hyperspectral data and use of these methods for rapid phenotyping of plant breeding trials. Lastly, two chapters evaluate tree species identification, and another provides examples of mapping invasive species.

Volume IV focuses on six areas of advanced applications in agricultural crops. The first considers detection of plant stressors including nitrogen deficiency and excess heavy metals and crop disease detection in precision farming. The second addresses global patterns of crop water productivity and quantifying litter and invasive species in arid ecosystems. Phenological patterns are examined while others focus on multitemporal data for mapping patterns of phenology. The third area is focused on applications of land cover mapping in different forest, wetland, and urban applications. The fourth topic addresses hyperspectral measurements of wildfires, and the fifth evaluates use of continuity vegetation index data in global change applications. And lastly, the sixth area examines use of hyperspectral data to understand the geologic surfaces of other planets.

Susan L. Ustin
Professor and Vice Chair, Dept. Land, Air and Water Resources
Associate Director, John Muir Institute
University of California
Davis California, USA

REFERENCE

EnMap Ground Segment Team and GSIS GRSS Technical committee, December, 2017. *Spaceborne Imaging Spectroscopy Mission Compilation*. DLR Space Administration and the German Federal Ministry of Economic Affairs and Technology. http://www.enmap.org/sites/default/files/pdf/Hyperspectral_EO_Missions_2017_12_21_FINAL4.pdf

BIOGRAPHICAL SKETCH

Dr. Susan L. Ustin is currently a Distinguished Professor of Environmental and Resource Sciences in the Department of Land, Air, and Water Resources, University of California Davis, Associate Director of the John Muir Institute, and is Head of the Center for Spatial Technologies and Remote Sensing (CSTARS) at the same university. She was trained as a plant physiological ecologist but began working with hyperspectral imagery as a post-doc in 1983 with JPL's AIS program. She became one of the early adopters of hyperspectral remote sensing which has now extended over her entire academic career. She was a pioneer in the development of vegetation analysis using imaging spectrometery, and is an expert on ecological applications of this data. She has served on numerous NASA, NSF, DOE, and the National Research Council committees related to spectroscopy and remote sensing. Among recognitions for her work, she is a Fellow of the American Geophysical Union and received an honorary doctorate from the University of Zurich. She has published more than 200 scientific papers related to ecological remote sensing and has worked with most of the Earth-observing U.S. airborne and spaceborne systems.

Preface

This seminal book on *Hyperspectral Remote Sensing of Vegetation* (Second Edition, 4 Volume Set), published by Taylor and Francis Inc.\CRC Press is an outcome of over 2 years of effort by the editors and authors. In 2011, the first edition of *Hyperspectral Remote Sensing of Vegetation* was published. The book became a standard reference on hyperspectral remote sensing of vegetation amongst the remote sensing community across the world. This need and resulting popularity demanded a second edition with more recent as well as more comprehensive coverage of the subject. Many advances have taken place since the first edition. Further, the first edition was limited in scope in the sense it covered some very important topics and missed equally important topics (e.g., hyperspectral library of agricultural crops, hyperspectral pre-processing steps and algorithms, and many others). As a result, a second edition that brings us up-to-date advances in hyperspectral remote sensing of vegetation was required. Equally important was the need to make the book more comprehensive, covering an array of subjects not covered in the first edition. So, my coeditors and myself did a careful research on what should go into the second edition. Quickly, the scope of the second edition expanded resulting in an increasing number of chapters. All of this led to developing the seminal book: *Hyperspectral Remote Sensing of Vegetation*, Second Edition, 4 Volume Set. The four volumes are:

Volume I: Fundamentals, Sensor Systems, Spectral Libraries, and Data Mining for Vegetation
Volume II: Hyperspectral Indices and Image Classifications for Agriculture and Vegetation
Volume III: Biophysical and Biochemical Characterization and Plant Species Studies
Volume IV: Advanced Applications in Remote Sensing of Agricultural Crops and Natural Vegetation

The goal of the book was to bring in one place collective knowledge of the last 50 years of advances in hyperspectral remote sensing of vegetation with a target audience of wide spectrum of scientific community, students, and professional application practitioners. The book documents knowledge advances made in applying hyperspectral remote sensing technology in the study of terrestrial vegetation that include agricultural crops, forests, rangelands, and wetlands. This is a very practical offering about a complex subject that is rapidly advancing its knowledge-base. In a very practical way, the book demonstrates the experience, utility, methods, and models used in studying terrestrial vegetation using hyperspectral data. The four volumes, with a total of 48 chapters, are divided into distinct themes.

- **Volume I**: There are 14 chapters focusing on hyperspectral instruments, spectral libraries, and methods and approaches of data handling. The chapters extensively address various preprocessing steps and data mining issues such as the Hughes phenomenon and overcoming the "curse of high dimensionality" of hyperspectral data. Developing spectral libraries of crops, vegetation, and soils with data gathered from hyperspectral data from various platforms (ground-based, airborne, spaceborne), study of spectral traits of crops, and proximal sensing at field for phenotyping are extensively discussed. Strengths and limitations of hyperspectral data of agricultural crops and vegetation acquired from different platforms are discussed. It is evident from these chapters that the hyperspectral data provides opportunities for great advances in study of agricultural crops and vegetation. However, it is also clear from these chapters that hyperspectral data should not be treated as panacea to every limitation of multispectral broadband data such as from Landsat or Sentinel series of satellites. The hundreds or thousands of hyperspectral narrowbands (HNBs) as well as carefully selected hyperspectral vegetation indices (HVIs) will help us make significant advances in characterizing, modeling, mapping, and monitoring vegetation biophysical, biochemical, and structural quantities. However, it is also important to properly understand hyperspectral data and eliminate redundant bands that exist for every application and to optimize computing as well as human resources to enable seamless and efficient handling enormous volumes of hyperspectral data. Special emphasis is also put on preprocessing and processing of Earth Observing-1 (EO-1) Hyperion, the first publicly available hyperspectral data from space. These methods, approaches, and algorithms, and protocols set the stage for upcoming satellite hyperspectral sensors such as NASA's HyspIRI and Germany's EnMAP.

- **Volume II**: There are 10 chapters focusing on hyperspectral vegetation indices (HVIs) and image classification methods and techniques. The HVIs are of several types such as: (i) two-band derived, (ii) multi-band-derived, and (iii) derivative indices derived. The strength of the HVIs lies in the fact that specific indices can be derived for specific biophysical, biochemical, and plant structural quantities. For example, you have carotenoid HVI, anthocyanin HVI, moisture or water HVI, lignin HVI, cellulose HVI, biomass or LAI or other biophysical HVIs, red-edge based HVIs, and so on. Further, since these are narrowband indices, they are better targeted and centered at specific sensitive wavelength portions of the spectrum. The strengths and limitations of HVIs in a wide array of applications such as leaf nitrogen content (LNC), vegetation water content, nitrogen content in vegetation, leaf and plant pigments, anthocyanin's, carotenoids, and chlorophyll are thoroughly studied. Image classification using hyperspectral data provides great strengths in deriving more classes (e.g., crop species within a crop as opposed to just crop types) and increasing classification accuracies. In earlier years and decades, hyperspectral data classification and analysis was a challenge due to computing and data handling issues. However, with the availability of machine learning algorithms on cloud computing (e.g., Google Earth Engine) platforms, these challenges have been overcome in the last 2–3 years. Pixel-based supervised machine learning algorithms like the random forest, and support vector machines as well as object-based algorithms like the recursive hierarchical segmentation, and numerous others methods (e.g., unsupervised approaches) are extensively discussed. The ability to process petabyte volume data of the planet takes us to a new level of sophistication and makes use of data such as from hyperspectral sensors feasible over large areas. The cloud computing architecture involved with handling massively large petabyte-scale data volumes are presented and discussed.

- **Volume III**: There are 11 chapters focusing on biophysical and biochemical characterization and plant species studies. A number of chapters in this volume are focused on separating and discriminating agricultural crops and vegetation of various types or species using hyperspectral data. Plant species discrimination and classification to separate them are the focus of study using vegetation such as forests, invasive species in different ecosystems, and agricultural crops. Performance of hyperspectral narrowbands (HNBs) and hyperspectral vegetation indices (HVIs) when compared with multispectral broadbands (MBBs) and multispectral broadband vegetation indices (BVIs) are presented and discussed. The vegetation and agricultural crops are studied at various scales, and their vegetation functional properties diagnosed. The value of digital surface models in study of plant traits as complementary\supplementary to hyperspectral data has been highlighted. Hyperspectral bio-indicators to study photosynthetic efficiency and vegetation stress are presented and discussed. Studies are conducted using hyperspectral data across wavelengths (e.g., visible, near-infrared, shortwave-infrared, mid-infrared, and thermal-infrared).

- **Volume IV**: There are 15 chapters focusing on specific advanced applications of hyperspectral data in study of agricultural crops and natural vegetation. Specific agricultural crop applications include crop management practices, crop stress, crop disease, nitrogen application, and presence of heavy metals in soils and related stress factors. These studies discuss biophysical and biochemical quantities modeled and mapped for precision farming, hyperspectral narrowbands (HNBs), and hyperspectral vegetation indices (HVIs) involved in assessing nitrogen in plants, and the study of the impact of heavy metals on crop health and stress. Vegetation functional studies using hyperspectral data presented and discussed include crop water use (actual evapotranspiration), net primary productivity (NPP), gross primary productivity (GPP), phenological applications, and light use efficiency (LUE). Specific applications discussed under vegetation functional studies using hyperspectral data include agricultural crop classifications, machine learning, forest management studies, pasture studies, and wetland studies. Applications in fire assessment, modeling, and mapping using hyperspectral data in the optical and thermal portions of the spectrum are presented and discussed. Hyperspectral data in global change studies as well as in outer planet studies have also been discussed. Much of the outer planet remote sensing is conducted using imaging spectrometer and hence the data preprocessing and processing methods of Earth and that of outer planets have much in common and needs further examination.

The chapters are written by leading experts in the global arena with each chapter: (a) focusing on specific applications, (b) reviewing existing "state-of-art" knowledge, (c) highlighting the advances made, and (d) providing guidance for appropriate use of hyperspectral data in study of vegetation and its numerous applications such as crop yield modeling, crop biophysical and biochemical property characterization, and crop moisture assessment.

The four-volume book is specifically targeted on hyperspectral remote sensing as applied to terrestrial vegetation applications. This is a big market area that includes agricultural croplands, study of crop moisture, forests, and numerous applications such as droughts, crop stress, crop productivity, and water productivity. To the knowledge of the editors, there is no comparable book, source, and/or organization that can bring this body of knowledge together in one place, making this a "must buy" for professionals. This is clearly a unique contribution whose time is now. The book highlights include:

1. Best global expertise on hyperspectral remote sensing of vegetation, agricultural crops, crop water use, plant species detection, crop productivity and water productivity mapping, and modeling;
2. Clear articulation of methods to conduct the work. Very practical;
3. Comprehensive review of the existing technology and clear guidance on how best to use hyperspectral data for various applications;
4. Case studies from a variety of continents with their own subtle requirements; and
5. Complete solutions from methods to applications inventory and modeling.

Hyperspectral narrowband spectral data, as discussed in various chapters of this book, are fast emerging as practical most advanced solutions in modeling and mapping vegetation. Recent research has demonstrated the advances and great value made by hyperspectral data, as discussed in various chapters in: (a) quantifying agricultural crops as to their biophysical and harvest yield characteristics, (b) modeling forest canopy biochemical properties, (c) establishing plant and soil moisture conditions, (d) detecting crop stress and disease, (e) mapping leaf chlorophyll content as it influences crop production, (f) identifying plants affected by contaminants such as arsenic, and (g) demonstrating sensitivity to plant nitrogen content, and (h) invasive species mapping. The ability to significantly better quantify, model, and map plant chemical, physical, and water properties is well established and has great utility.

Even though these accomplishments and capabilities have been reported in various places, the need for a collective "knowledge bank" that links these various advances in one place is missing. Further, most scientific papers address specific aspects of research, failing to provide a comprehensive assessment of advances that have been made nor how the professional can bring those advances to their work. For example, deep scientific journals report practical applications of hyperspectral narrowbands yet one has to canvass the literature broadly to obtain the pertinent facts. Since several papers report this, there is a need to synthesize these findings so that the reader gets the correct picture of the best wavebands for their practical applications. Also, studies do differ in exact methods most suited for detecting parameters such as crop moisture variability, chlorophyll content, and stress levels. The professional needs this sort of synthesis and detail to adopt best practices for their own work.

In years and decades past, use of hyperspectral data had its challenges especially in handling large data volumes. That limitation is now overcome through cloud-computing, machine learning, deep learning, artificial intelligence, and advances in knowledge in processing and applying hyperspectral data.

This book can be used by anyone interested in hyperspectral remote sensing that includes advanced research and applications, such as graduate students, undergraduates, professors, practicing professionals, policy makers, governments, and research organizations.

Dr. Prasad S. Thenkabail, PhD
Editor-in-Chief
Hyperspectral Remote Sensing of Vegetation, Second Edition, Four Volume Set

Acknowledgments

This four-volume *Hyperspectral Remote Sensing of Vegetation* book (second edition) was made possible by sterling contributions from leading professionals from around the world in the area of hyperspectral remote sensing of vegetation and agricultural crops. As you will see from list of authors and coauthors, we have an assembly of "**who is who**" in hyperspectral remote sensing of vegetation who have contributed to this book. They wrote insightful chapters, that are an outcome of years of careful research and dedication, to make the book appealing to a broad section of readers dealing with remote sensing. My gratitude goes to (mentioned in no particular order; names of lead authors of the chapters are shown in bold): **Drs. Fred Ortenberg** (Technion–Israel Institute of Technology, Israel), **Jiaguo Qi** (Michigan State University, USA), **Angela Lausch** (Helmholtz Centre for Environmental Research, Leipzig, Germany), **Andries B. Potgieter** (University of Queensland, Australia), **Muhammad Al-Amin Hoque** (University of Queensland, Australia), **Andreas Hueni** (University of Zurich, Switzerland), **Eyal Ben-Dor** (Tel Aviv University, Israel), **Itiya Aneece** (United States Geological Survey, USA), **Sreekala Bajwa** (University of Arkansas, USA), **Antonio Plaza** (University of Extremadura, Spain), **Jessica J. Mitchell** (Appalachian State University, USA), **Dar Roberts** (University of California at Santa Barbara, USA), **Quan Wang** (Shizuoka University, Japan), **Edoardo Pasolli** (University of Trento, Italy), (Nanjing University of Science and Technology, China), **Anatoly Gitelson** (University of Nebraska- Lincoln, USA), **Tao Cheng** (Nanjing Agricultural University, China), **Roberto Colombo** (University of Milan-Bicocca, Italy), **Daniela Stroppiana** (Institute for Electromagnetic Sensing of the Environment, Italy), **Yongqin Zhang** (Delta State University, USA), **Yoshio Inoue** (National Institute for Agro-Environmental Sciences, Japan), Yafit Cohen (Institute of Agricultural Engineering, Israel), **Helge Aasen** (Institute of Agricultural Sciences, ETH Zurich), **Elizabeth M. Middleton** (NASA, USA), **Yongqin Zhang** (University of Toronto, Canada), **Yan Zhu** (Nanjing Agricultural University, China), **Lênio Soares Galvão** (Instituto Nacional de Pesquisas Espaciais [INPE], Brazil), **Matthew L. Clark** (Sonoma State University, USA), **Matheus Pinheiro Ferreira** (University of Paraná, Curitiba, Brazil), **Ruiliang Pu** (University of South Florida, USA), **Scott C. Chapman** (CSIRO, Australia), **Haibo Yao** (Mississippi State University, USA), **Jianlong Li** (Nanjing University, China), **Terry Slonecker** (USGS, USA), **Tobias Landmann** (International Centre of Insect Physiology and Ecology, Kenya), **Michael Marshall** (University of Twente, Netherlands), **Pamela Nagler** (USGS, USA), **Alfredo Huete** (University of Technology Sydney, Australia), **Prem Chandra Pandey** (Banaras Hindu University, India), **Valerie Thomas** (Virginia Tech., USA), **Izaya Numata** (South Dakota State University, USA), **Elijah W. Ramsey III** (USGS, USA), **Sander Veraverbeke** (Vrije Universiteit Amsterdam and University of California, Irvine), **Tomoaki Miura** (University of Hawaii, USA), **R. G. Vaughan** (U.S. Geological Survey, USA), Victor Alchanatis (Agricultural research Organization, Volcani Center, Israel), Dr. Narumon Wiangwang (Royal Thai Government, Thailand), Pedro J. Leitão (Humboldt University of Berlin, Department of Geography, Berlin, Germany), James Watson (University of Queensland, Australia), Barbara George-Jaeggli (ETH Zuerich, Switzerland), Gregory McLean (University of Queensland, Australia), Mark Eldridge (University of Queensland, Australia), Scott C. Chapman (University of Queensland, Australia), Kenneth Laws (University of Queensland, Australia), Jack Christopher (University of Queensland, Australia), Karine Chenu (University of Queensland, Australia), Andrew Borrell (University of Queensland, Australia), Graeme L. Hammer (University of Queensland, Australia), David R. Jordan (University of Queensland, Australia), Stuart Phinn (University of Queensland, Australia), Lola Suarez (University of Melbourne, Australia), Laurie A. Chisholm (University of Wollongong, Australia), Alex Held (CSIRO, Australia), S. Chabrillant (GFZ German Research Center for Geosciences, Germany), José A. M. Demattê (University of São Paulo, Brazil), Yu Zhang (North Dakota State University, USA), Ali Shirzadifar (North Dakota State University, USA), Nancy F. Glenn (Boise State University, USA), Kyla M. Dahlin (Michigan State

University, USA), Nayani Ilangakoon (Boise State University, USA), Hamid Dashti (Boise State University, USA), Megan C. Maloney (Appalachian State University, USA), Subodh Kulkarni (University of Arkansas, USA), Javier Plaza (University of Extremadura, Spain), Gabriel Martin (University of Extremadura, Spain), Segio Sánchez (University of Extremadura, Spain), Wei Wang (Nanjing Agricultural University, China), Xia Yao (Nanjing Agricultural University, China), Busetto Lorenzo (Università Milano-Bicocca), Meroni Michele (Università Milano-Bicocca), Rossini Micol (Università Milano-Bicocca), Panigada Cinzia (Università Milano-Bicocca), F. Fava (Università degli Studi di Sassari, Italy), M. Boschetti (Institute for Electromagnetic Sensing of the Environment, Italy), P. A. Brivio (Institute for Electromagnetic Sensing of the Environment, Italy), K. Fred Huemmrich (University of Maryland, Baltimore County, USA), Yen-Ben Cheng (Earth Resources Technology, Inc., USA), Hank A. Margolis (Centre d'Études de la Forêt, Canada), Yafit Cohen (Agricultural research Organization, Volcani Center, Israel), Kelly Roth (University of California at Santa Barbara, USA), Ryan Perroy (University of Wisconsin-La Crosse, USA), Ms. Wei Wang (Nanjing Agricultural University, China), Dr. Xia Yao (Nanjing Agricultural University, China), Keely L. Roth (University of California, Santa Barbara, USA), Erin B. Wetherley (University of California at Santa Barbara, USA), Susan K. Meerdink (University of California at Santa Barbara, USA), Ryan L. Perroy (University of Wisconsin-La Crosse, USA), B. B. Marithi Sridhar (Bowling Green University, USA), Aaryan Dyami Olsson (Northern Arizona University, USA), Willem Van Leeuwen (University of Arizona, USA), Edward Glenn (University of Arizona, USA), José Carlos Neves Epiphanio (Instituto Nacional de Pesquisas Espaciais [INPE], Brazil), Fábio Marcelo Breunig (Instituto Nacional de Pesquisas Espaciais [INPE], Brazil), Antônio Roberto Formaggio (Instituto Nacional de Pesquisas Espaciais [INPE], Brazil), Amina Rangoonwala (IAP World Services, Lafayette, LA), Cheryl Li (Nanjing University, China), Deghua Zhao (Nanjing University, China), Chengcheng Gang (Nanjing University, China), Lie Tang (Mississippi State University, USA), Lei Tian (Mississippi State University, USA), Robert Brown (Mississippi State University, USA), Deepak Bhatnagar (Mississippi State University, USA), Thomas Cleveland (Mississippi State University, USA), Hiroki Yoshioka (Aichi Prefectural University, Japan), T. N. Titus (U.S. Geological Survey, USA), J. R. Johnson (U.S. Geological Survey, USA), J. J. Hagerty (U.S. Geological Survey, USA), L. Gaddis (U.S. Geological Survey, USA), L. A. Soderblom (U.S. Geological Survey, USA), and P. Geissler (U.S. Geological Survey, USA), Jua Jin (Shizuoka University, Japan), Rei Sonobe (Shizuoka University, Japan), Jin Ming Chen (Shizuoka University, Japan), Saurabh Prasad (University of Houston, USA), Melba M. Crawford (Purdue University, USA), James C. Tilton (NASA Goddard Space Flight Center, USA), Jin Sun (Nanjing University of Science and Technology, China), Yi Zhang (Nanjing University of Science and Technology, China), Alexei Solovchenko (Moscow State University, Moscow), Yan Zhu, (Nanjing Agricultural University, China), Dong Li (Nanjing Agricultural University, China), Kai Zhou (Nanjing Agricultural University, China), Roshanak Darvishzadeh (University of Twente, Enschede, The Netherlands), Andrew Skidmore (University of Twente, Enschede, The Netherlands), Victor Alchanatis (Institute of Agricultural Engineering, The Netherlands), Georg Bareth (University of Cologne, Germany), Qingyuan Zhang (Universities Space Research Association, USA), Petya K. E. Campbell (University of Maryland Baltimore County, USA), and David R. Landis (Global Science & Technology, Inc., USA), José Carlos Neves Epiphanio (Instituto Nacional de Pesquisas Espaciais [INPE], Brazil), Fábio Marcelo Breunig (Universidade Federal de Santa Maria [UFSM], Brazil), and Antônio Roberto Formaggio (Instituto Nacional de Pesquisas Espaciais [INPE], Brazil), Cibele Hummel do Amaral (Federal University of Viçosa, in Brazil), Gaia Vaglio Laurin (Tuscia University, Italy), Raymond Kokaly (U.S. Geological Survey, USA), Carlos Roberto de Souza Filho (University of Ouro Preto, Brazil), Yosio Edemir Shimabukuro (Federal Rural University of Rio de Janeiro, Brazil), Bangyou Zheng (CSIRO, Australia), Wei Guo (The University of Tokyo, Japan), Frederic Baret (INRA, France), Shouyang Liu (INRA, France), Simon Madec (INRA, France), Benoit Solan (ARVALIS, France), Barbara George-Jaeggli (University of Queensland, Australia), Graeme L. Hammer (University of Queensland, Australia), David R. Jordan (University of Queensland, Australia), Yanbo Huang (USDA, USA), Lie Tang (Iowa State

University, USA), Lei Tian (University of Illinois. USA), Deepak Bhatnagar (USDA, USA), Thomas E. Cleveland (USDA, USA), Dehua ZHAO (Nanjing University, USA), Hannes Feilhauer (University of Erlangen-Nuremberg, Germany), Miaogen Shen (Institute of Tibetan Plateau Research, Chinese Academy of Sciences, Beijing, China), Jin Chen (College of Remote Sensing Science and Engineering, Faculty of Geographical Science, Beijing Normal University, Beijing, China), Suresh Raina (International Centre of Insect Physiology and Ecology, Kenya and Pollination services, India), Danny Foley (Northern Arizona University, USA), Cai Xueliang (UNESCO-IHE, Netherlands), Trent Biggs (San Diego State University, USA), Werapong Koedsin (Prince of Songkla University, Thailand), Jin Wu (University of Hong Kong, China), Kiril Manevski (Aarhus University, Denmark), Prashant K. Srivastava (Banaras Hindu University, India), George P. Petropoulos (Technical University of Crete, Greece), Philip Dennison (University of Utah, USA), Ioannis Gitas (University of Thessaloniki, Greece), Glynn Hulley (NASA Jet Propulsion Laboratory, California Institute of Technology, USA), Olga Kalashnikova, (NASA Jet Propulsion Laboratory, California Institute of Technology, USA), Thomas Katagis (University of Thessaloniki, Greece), Le Kuai (University of California, USA), Ran Meng (Brookhaven National Laboratory, USA), Natasha Stavros (California Institute of Technology, USA).

Hiroki Yoshioka (Aichi Prefectural University, Japan), My two coeditors, **Professor John G. Lyon** and **Professor Alfredo Huete**, have made outstanding contribution to this four-volume *Hyperspectral Remote Sensing of Vegetation* book (second edition). Their knowledge of hyperspectral remote sensing is enormous. Vastness and depth of their understanding of remote sensing in general and hyperspectral remote sensing in particular made my job that much easier. I have learnt a lot from them and continue to do so. Both of them edited some or all of the 48 chapters of the book and also helped structure chapters for a flawless reading. They also significantly contributed to the synthesis chapter of each volume. I am indebted to their insights, guidance, support, motivation, and encouragement throughout the book project.

My coeditors and myself are grateful to **Dr. Alexander F. H. Goetz** and **Prof. Susan L. Ustin** for writing the foreword for the book. Please refer to their biographical sketch under the respective foreword written by these two leaders of Hyperspectral Remote Sensing.

Both the forewords are a must read to anyone studying this four-volume *Hyperspectral Remote Sensing of Vegetation* book (second edition). They are written by two giants who have made immense contribution to the subject and I highly recommend that the readers read them.

I am blessed to have had the support and encouragement (professional and personal) of my U.S. Geological Survey and other colleagues. In particular, I would like to mention Mr. Edwin Pfeifer (late), Dr. Susan Benjamin, Dr. Dennis Dye, and Mr. Larry Gaffney. Special thanks to Dr. Terrence Slonecker, Dr. Michael Marshall, Dr. Isabella Mariotto, and Dr. Itiya Aneece who have worked closely with me on hyperspectral research over the years. Special thanks are also due to Dr. Pardhasaradhi Teluguntla, Mr. Adam Oliphant, and Dr. Muralikrishna Gumma who have contributed to my various research efforts and have helped me during this book project directly or indirectly. I am grateful to Prof. Ronald B. Smith, professor at Yale University who was instrumental in supporting my early hyperspectral research at the Yale Center for Earth Observation (YCEO), Yale University. Opportunities and guidance I received in my early years of remote sensing from Prof. Andrew D. Ward, professor at the Ohio State University, Prof. John G. Lyon, former professor at the Ohio State University, and Mr. Thiruvengadachari, former Scientist at the National Remote Sensing Center (NRSC), Indian Space Research Organization, India, is gratefully acknowledged.

My wife (Sharmila Prasad) and daughter (Spandana Thenkabail) are two great pillars of my life. I am always indebted to their patience, support, and love.

Finally, kindly bear with me for sharing a personal story. When I started editing the first edition in the year 2010, I was diagnosed with colon cancer. I was not even sure what the future was and how long I would be here. I edited much of the first edition soon after the colon cancer surgery and during and after the 6 months of chemotherapy—one way of keeping my mind off the negative thoughts. When you are hit by such news, there is nothing one can do, but to be positive, trust your doctors, be thankful to support and love of the family, and have firm belief in the higher spiritual being (whatever your beliefs are). I am so very grateful to some extraordinary people who helped me through this difficult life event: Dr. Parvasthu Ramanujam (surgeon), Dr. Paramjeet K. Bangar (Oncologist), Dr. Harnath Sigh (my primary doctor), Dr. Ram Krishna (Orthopedic Surgeon and family friend), three great nurses (Ms. Irene, Becky, Maryam) at Banner Boswell Hospital (Sun City, Arizona, USA), courage-love-patience-prayers from my wife, daughter, and several family members, friends, and colleagues, and support from numerous others that I have not named here. During this phase, I learnt a lot about cancer and it gave me an enlightened perspective of life. My prayers were answered by the higher power. I learnt a great deal about life—good and bad. I pray for all those with cancer and other patients that diseases one day will become history or, in the least, always curable without suffering and pain. Now, after 8 years, I am fully free of colon cancer and was able to edit the four-volume *Hyperspectral Remote Sensing of Vegetation* book (second edition) without the pain and suffering that I went through when editing the first edition. What a blessing. These blessings help us give back in our own little ways. To realize that it is indeed profound to see the beautiful sunrise every day, the day go by with every little event (each with a story of their own), see the beauty of the sunset, look up to the infinite universe and imagine on its many wonders, and just to breathe fresh air every day and enjoy the breeze. These are all many wonders of life that we need to enjoy, cherish, and contemplate.

Dr. Prasad S. Thenkabail, PhD
Editor-in-Chief
Hyperspectral Remote Sensing of Vegetation

Editors

Prasad S. Thenkabail, Research Geographer-15, U.S. Geological Survey (USGS), is a world-recognized expert in remote sensing science with multiple major contributions in the field sustained over more than 30 years. He obtained his PhD from the Ohio State University in 1992 and has over 140+ peer-reviewed scientific publications, mostly in major international journals.

Dr. Thenkabail has conducted pioneering research in the area of hyperspectral remote sensing of vegetation and in that of global croplands and their water use in the context of food security. In hyperspectral remote sensing he has done cutting-edge research with wide implications in advancing remote sensing science in application to agriculture and vegetation. This body of work led to more than ten peer-reviewed research publications with high impact. For example, a single paper [1] has received 1000+ citations as at the time of writing (October 4, 2018). Numerous other papers, book chapters, and books (as we will learn below) are also related to this work, with two other papers [2,3] having 350+ to 425+ citations each.

In studies of global croplands in the context of food and water security, he has led the release of the world's first Landsat 30-m derived global cropland extent product. This work demonstrates a "paradigm shift" in how remote sensing science is conducted. The product can be viewed in full resolution at the web location www.croplands.org. The data is already widely used worldwide and is downloadable from the NASA\USGS LP DAAC site [4]. There are numerous major publication in this area (e.g. [5,6]).

Dr. Thenkabail's contributions to series of leading edited books on remote sensing science places him as a world leader in remote sensing science advances. He edited three-volume *Remote Sensing Handbook* published by Taylor and Francis, with 82 chapters and more than 2000 pages, widely considered a "magnus opus" standard reference for students, scholars, practitioners, and major experts in remote sensing science. Links to these volumes along with endorsements from leading global remote sensing scientists can be found at the location give in note [7]. He has recently completed editing *Hyperspectral Remote Sensing of Vegetation* published by Taylor and Francis in four volumes with 50 chapters. This is the second edition is a follow-up on the earlier single-volume *Hyperspectral Remote Sensing of Vegetation* [8]. He has also edited a book on *Remote Sensing of Global Croplands for Food Security* (Taylor and Francis) [9]. These books are widely used and widely referenced in institutions worldwide.

Dr. Thenkabail's service to remote sensing community is second to none. He is currently an editor-in-chief of the *Remote Sensing* open access journal published by MDPI; an associate editor of the journal *Photogrammetric Engineering and Remote Sensing* (PERS) of the American Society of Photogrammetry and Remote Sensing (ASPRS); and an editorial advisory board member of the International Society of Photogrammetry and Remote Sensing (ISPRS) *Journal of Photogrammetry and Remote Sensing*. Earlier, he served on the editorial board of *Remote Sensing of Environment* for many years (2007–2017). As an editor-in-chief of the open access *Remote Sensing* MDPI journal from 2013 to date he has been instrumental in providing leadership for an online publication that did not even have a impact factor when he took over but is now one of the five leading remote sensing international journals, with an impact factor of 3.244.

Dr. Thenkabail has led remote sensing programs in three international organizations: International Water Management Institute (IWMI), 2003–2008; International Center for Integrated Mountain Development (ICIMOD), 1995–1997; and International Institute of Tropical Agriculture (IITA),

1992–1995. He has worked in more than 25+ countries on several continents, including East Asia (China), S-E Asia (Cambodia, Indonesia, Myanmar, Thailand, Vietnam), Middle East (Israel, Syria), North America (United States, Canada), South America (Brazil), Central Asia (Uzbekistan), South Asia (Bangladesh, India, Nepal, and Sri Lanka), West Africa (Republic of Benin, Burkina Faso, Cameroon, Central African Republic, Cote d'Ivoire, Gambia, Ghana, Mali, Nigeria, Senegal, and Togo), and Southern Africa (Mozambique, South Africa). During this period he has made major contributions and written seminal papers on remote sensing of agriculture, water resources, inland valley wetlands, global irrigated and rain-fed croplands, characterization of African rainforests and savannas, and drought monitoring systems.

The quality of Dr. Thenkabail's research is evidenced in the many awards, which include, in 2015, the American Society of Photogrammetry and Remote Sensing (ASPRS) ERDAS award for best scientific paper in remote sensing (Marshall and Thenkabail); in 2008, the ASPRS President's Award for practical papers, second place (Thenkabail and coauthors); and in 1994, the ASPRS Autometric Award for outstanding paper (Thenkabail and coauthors). His team was recognized by the Environmental System Research Institute (ESRI) for "special achievement in GIS" (SAG award) for their Indian Ocean tsunami work. The USGS and NASA selected him to be on the Landsat Science Team for a period of five years (2007–2011).

Dr. Thenkabail is regularly invited as keynote speaker or invited speaker at major international conferences and at other important national and international forums every year. He has been principal investigator and/or has had lead roles of many pathfinding projects, including the ~5 million over five years (2014–2018) for the global food security support analysis data in the 30-m (GFSAD) project (https://geography.wr.usgs.gov/science/croplands/) funded by NASA MEaSUREs (Making Earth System Data Records for Use in Research Environments), and projects such as Sustain and Manage America's Resources for Tomorrow (waterSMART) and characterization of Eco-Regions in Africa (CERA).

REFERENCES

1. Thenkabail, P.S., Smith, R.B., and De-Pauw, E. 2000b. Hyperspectral vegetation indices for determining agricultural crop characteristics. *Remote Sensing of Environment*, 71:158–182.
2. Thenkabail, P.S., Enclona, E.A., Ashton, M.S., Legg, C., and Jean De Dieu, M. 2004. Hyperion, IKONOS, ALI, and ETM+ sensors in the study of African rainforests. *Remote Sensing of Environment*, 90:23–43.
3. Thenkabail, P.S., Enclona, E.A., Ashton, M.S., and Van Der Meer, V. 2004. Accuracy assessments of hyperspectral waveband performance for vegetation analysis applications. *Remote Sensing of Environment*, 91(2–3):354–376.
4. https://lpdaac.usgs.gov/about/news_archive/release_gfsad_30_meter_cropland_extent_products
5. Thenkabail, P.S. 2012. Guest Editor for Global Croplands Special Issue. *Photogrammetric Engineering and Remote Sensing*, 78(8).
6. Thenkabail, P.S., Knox, J.W., Ozdogan, M., Gumma, M.K., Congalton, R.G., Wu, Z., Milesi, C., Finkral, A., Marshall, M., Mariotto, I., You, S. Giri, C. and Nagler, P. 2012. Assessing future risks to agricultural productivity, water resources and food security: how can remote sensing help? *Photogrammetric Engineering and Remote Sensing*, August 2012 Special Issue on Global Croplands: Highlight Article. 78(8):773–782. IP-035587.
7. https://www.crcpress.com/Remote-Sensing-Handbook---Three-Volume-Set/Thenkabail/p/book/9781482218015
8. https://www.crcpress.com/Hyperspectral-Remote-Sensing-of-Vegetation/Thenkabail-Lyon/p/book/9781439845370
9. https://www.crcpress.com/Remote-Sensing-of-Global-Croplands-for-Food-Security/Thenkabail-Lyon-Turral-Biradar/p/book/9781138116559

John G. Lyon, educated at Reed College in Portland, OR and the University of Michigan in Ann Arbor, has conducted scientific and engineering research and carried out administrative functions throughout his career. He was formerly the Senior Physical Scientist (ST) in the US Environmental Protection Agency's Office of Research and Development (ORD) and Office of the Science Advisor in Washington, DC, where he co-led work on the Group on Earth Observations and the USGEO subcommittee of the Committee on Environment and Natural Resources and research on geospatial issues in the agency. For approximately eight years, he was director of ORD's Environmental Sciences Division, which conducted research on remote sensing and geographical information system (GIS) technologies as applied to environmental issues including landscape characterization and ecology, as well as analytical chemistry of hazardous wastes, sediments, and ground water. He previously served as professor of civil engineering and natural resources at Ohio State University (1981–1999). Professor Lyon's own research has led to authorship or editorship of a number of books on wetlands, watershed, and environmental applications of GIS, and accuracy assessment of remote sensor technologies.

Alfredo Huete leads the Ecosystem Dynamics Health and Resilience research program within the Climate Change Cluster (C3) at the University of Technology Sydney, Australia. His main research interest is in using remote sensing to study and analyze vegetation processes, health, and functioning, and he uses satellite data to observe land surface responses and interactions with climate, land use activities, and extreme events. He has more than 200 peer-reviewed journal articles, including publication in such prestigious journals as *Science* and *Nature*. He has over 25 years' experience working on NASA and JAXA mission teams, including the NASA-EOS MODIS Science Team, the EO-1 Hyperion Team, the JAXA GCOM-SGLI Science Team, and the NPOESS-VIIRS advisory group. Some of his past research involved the development of the soil-adjusted vegetation index (SAVI) and the enhanced vegetation index (EVI), which became operational satellite products on MODIS and VIIRS sensors. He has also studied tropical forest phenology and Amazon forest greening in the dry season, and his work was featured in a *National Geographic* television special entitled "The Big Picture." Currently, he is involved with the Australian Terrestrial Ecosystem Research Network (TERN), helping to produce national operational phenology products; as well as the AusPollen network, which couples satellite sensing to better understand and predict pollen phenology from allergenic grasses and trees.

Contributors

Helge Aasen
Crop Science Group, Institute of
  Agricultural Sciences, Federal Institute
  of Technology
Department of Environmental Systems
  Science, ETH Zurich
Zurich, Switzerland

Victor Alchanatis
Institute of Agricultural Engineering
Agricultural Research Organization
Bet-Dagan, Israel

Frederic Baret
French National Institute for Agricultural
  Research INRA
Manosque, Provence-Alpes-Côte d'Azur
Paris, France

Georg Bareth
GIS & RS Group, Institute of
  Geography, University of Koln
Koln, Germany

Fábio Marcelo Breunig
Department of Forestry
Federal University of Santa Maria
Santa Maria, RS, Brazil

Petya K. E. Campbell
University of Maryland, Baltimore
  County/Joint Center for Earth Systems
  Technology
Baltimore and Greenbelt, Maryland

Scott C. Chapman
CSIRO Agriculture and Food, Queensland
  Bioscience Precinct
St Lucia, Queensland, Australia

and

School of Agriculture and Food Sciences
The University of Queensland
Gatton, Queensland, Australia

Matthew L. Clark
Department of Geography, Environment and
  Planning
Sonoma State University
Rohnert Park, California

Yafit Cohen
Institute of Agricultural Engineering
Agricultural Research Organization
Bet-Dagan, Israel

Roshanak Darvishzadeh
Geo-Information Science and Earth
  Observation (ITC), Department of Natural
  Resources
University of Twente
Enschede, The Netherlands

Carlos Roberto de Souza Filho
The Instituto de Geociencias Department at
  Universidade Estadual de Campinas
Barão Geraldo, São Paulo, Brazil

Cibele Hummel do Amaral
Institute of Geosciences, University of São
  Paulo
São Paulo, Brazil

José Carlos Neves Epiphanio
The National Institute for Space Research
São José dos Campos, SP, Brazil

Matheus Pinheiro Ferreira
Military Institute of Engineering
Rio de Janeiro, Brazil

Antônio Roberto Formaggio
The National Institute for Space Research
São José dos Campos, SP, Brazil

K. Fred Huemmrich
University of Maryland, Baltimore
  County/Joint Center for Earth Systems
  Technology
Baltimore and Greenbelt, Maryland

Lênio Soares Galvão
The National Institute for Space Research
São José dos Campos, SP, Brazil

Barbara George-Jaeggli
Queensland Alliance for Agriculture and Food
    Innovation
The University of Queensland
St Lucia QLD, Australia

Anatoly A. Gitelson
University of Nebraska
Lincoln, Nebraska

and

Israel Institute of Technology
Haifa, Israel

Wei Guo
Institute for Sustainable Agro-ecosystem Services
Graduate School of Agricultural and Life
    Sciences
University of Tokyo
Tokyo, Japan

Graeme L. Hammer
The University of Queensland
St Lucia QLD, Australia

Alfredo Huete
School of Life Sciences
University of Technology Sydney
Ultimo, NSW, Australia

Yoshio Inoue
National Institute for Agro-Environmental
    Sciences, NARO
Tsukuba, Ibaraki, Japan

David R. Jordan
Queensland Alliance for Agriculture and Food
    Innovation
The University of Queensland
St Lucia QLD, Australia

Raymond Kokaly
Geology, Geophysics and Geochemistry
    Science Center
United States Geological Survey
Denver, Colorado

David R. Landis
Global Science & Technology, Inc.
Greenbelt, Maryland

Gaia Vaglio Laurin
Euro-Mediterranean Center for Climate
    Change (CMCC), IAFENT Division
and
Department for Innovation in Biological, Agro-
    Food and Forest Systems (DIBAF)
University of Tuscia
Viterbo. Tuscia, Italy

Shouyang Liu
French National Institute for Agricultural
    Research INRA
Manosque, Provence-Alpes-Côte d'Azur
Paris, France

John G. Lyon
Department of Civil, Environmental and
    Geodetic Engineering of Ohio State
    University
Columbus, Ohio

and

Fellow Member American Society for
    Photogrammetry and Remote Sensing
Chantilly, Virginia

Simon Madec
French National Institute for Agricultural
    Research INRA
Paris, France

Elizabeth M. Middleton
Biospheric Sciences Laboratory, National
    Aeronautics and Space Administration/
    Goddard Space Flight Center
Greenbelt, Maryland

Andries B. Potgieter
Queensland Alliance for Agriculture and Food
    Innovation
The University of Queensland
St Lucia QLD, Australia

Ruiliang Pu
School of Geosciences
University of South Florida
Tampa, Florida

Yosio Edemir Shimabukuro
National Institute for Space Research
São José dos Campos, SP, Brazil

Andrew Skidmore
Department of Environmental Sciences
Macquarie University
Sydney, Australia

and

Faculty of Geo-Information Science and Earth
    Observation (ITC), Department of Natural
    Resources
University of Twente
Enschede, The Netherlands

Benoit Solan
Department of Information Systems and
    Methodology
ARVALIS – Vegetable Institute for
    Technological Agriculture
Paris, France

Prasad S. Thenkabail
Research Geographer, Western Geographic
    Science Center
United States Geological Survey
Flagstaff, Arizona

Qingyuan Zhang
Universities Space Research Association
Columbia, Maryland

Bangyou Zheng
CSIRO Agriculture and Food, Queensland
    Biosciences Precinct
St Lucia, Queensland, Australia

Acronyms and Abbreviations

2D	Two-dimensional
3D	Three-dimensional
ALS	Airborne laser scanning
ANN	Artificial Neural Network
ARI1	Anthocyanin reflectance index 1
ARI2	Anthocyanin reflectance index 2
ARI	Anthocyanin Reflectance Index
ARVI	Atmospherically Resistant Vegetation Index
ASTER	Advanced Spaceborne Thermal Emission and Reflection Radiometer
AVHRR/NOAA-17	Advanced Very High Resolution Radiometer/National Oceanic and Atmospheric Administration-17
AVIRIS	Airborne Visible/Infrared Imaging Spectrometer
BB-PACs	Biophysical and biochemical properties of agricultural crops
BD-RDP	Beamlet-decorated recursive dyadic-partitioning
BRM	Biomass regression models
CALMIT	Center for Advanced Land Management and Information Technologies
CCD/CBERS-2	Charge-Coupled Device/China-Brazil Earth Resources Satellite-2
CFS	Correlation-based Feature Selection
CHRIS/PROBA	Compact High Resolution Imaging Spectrometer/Project for On-Board Autonomy
CI	Chlorophyll index
CNN	Convolutional neural network
CNPq	Conselho Nacional de Desenvolvimento Científico e Tecnológico (The Brazilian National Council for Scientific and Technological Development)
COVER%	Crop cover percentage
CRI1	Carotenoid reflectance index 1
CRI2	Carotenoid reflectance index 2
CRI	Carotenoid reflectance index
CSIRO	Commonwealth Scientific and Industrial Research Organisation (Australia)
CSM	Crop surface model
CWSI	Crop water stress index
D	Absorption band depth
DEM	Digital elevation model
DHP	Digital hemispherical photography
DSM	Digital surface mode
DSM	Digital surface model
DT	Decision Trees
DTM	Digital terrain model
DWSI	Disease Water Stress Index
ECHO	Extraction and classification of homogenous objects
EnMAP	Environmental Mapping and Analysis Program
EO-1	Earth Observing-One
ETM+/Landsat-7	Enhanced Thematic Mapper plus/Landsat-7
EV	Enhanced vegetation index
EVI	Enhanced Vegetation Index
fAPAR	Fraction of absorbed photosynthetically active radiation
fAPAR$_{green}$	Fraction of PAR absorbed by photosynthetically active vegetation

FAPESP	Fundação de Amparo à Pesquisa do Estado de São Paulo (São Paulo Research Foundation)
FBP	Field-based phenotyping
FIPAR	Fraction intercepted of photosynthetically active radiation
FLAASH	Fast Line-of-sight Atmospheric Analysis of Spectral Hypercubes
FVC	Fractional vegetation cover
GCP	Ground control point
GIS	Geographic Information System
GLCM	Gray levels co-occurrence matrix
GLT	Geographic look-up table
GPP	Gross primary productivity
GPS	Global Positioning Systems
GPU	Graphics processing unit
GRI	Red-green ratio index
GV	Green vegetation
GVF	Goodness variance of fit
HRG/SPOT-5	High Geometric Resolution Instrument/Système Pour l'Observation de la Terre-5
HS	Hyperspectral
HTPP	High-throughput plant phenotyping
HyspIRI	Hyperspectral Infrared Imager
INS	Inertial navigation system
iPLSR	Interval partial least squares regression
ITC	Individual tree crown
J-M	Jeffries–Matusita
K-T	Kaufman–Tanré aerosol retrieval
K	Potassium
LAI	Green leaf area index
LAI	Leaf Area Index
LiDAR	Light detection and ranging
LUE	Light use efficiency
LWP	Leaf water potential
LWVI-2	Leaf Water Vegetation Index -2
MCARI/OSAVI	Modified chlorophyll absorption ratio index/optimized soil-adjusted vegetation index
MCARI1	Modified chlorophyll absorption ratio index 1
MDA	Multiple Discriminant Analysis
MERIS	Medium Resolution Imaging Spectrometer
MESMA	Multiple end-member spectral mixture analysis
ML	Maximum likelihood
MLR	Multiple linear regression
MLR	Multivariate linear regression
MNB	Mean normalized bias
MNF	Minimum noise fraction
MODIS	Moderate Resolution Imaging Spectroradiometer
MODTRAN	MODerate resolution atmospheric TRANsmittance and radiance
MRF	Markov random field
MS	Multispectral
MSAVI	Moderate Resolution Imaging Spectrometer terrestrial chlorophyll index/improved OSAVI
MSI	Moisture Stress Index

MSI	Multi Spectral Instrument
MTCI	Terrestrial chlorophyll index
MTVI	Modified triangular vegetation index
N	Nitrogen
NDI or NDSI	Normalized difference spectral indices
NDII	Normalized difference infrared index
NDRE	Normalized difference red edge index
NDVI	Normalized difference vegetation index
NDWI	Normalized difference water index
NE	Noise equivalent
NIR	Near-infrared
NN	Neural network
NPLD	Non-pioneer light demanding
NPV	Nonphotosynthetic vegetation
NRCT	Normalized relative canopy temperature
NRI	Normalized ratio index
NRMSE	Normalized root mean square error
NSI	Nitrogen sufficiency index
NVI	Narrow-band vegetation index
OLI	Operational Land Imager
OSAVI	Optimized soil-adjusted vegetation index
P	Phosphorus
PAR	Photosynthetically active radiation
PAR$_{in}$	Incoming photosynthetically active radiation
PAR$_{pot}$	Potential photosynthetically active radiation
PCA	Principal component analysis
PCR	Principle components regression
PEM	Production efficiency model
PH	Plant height
PLS	Partial least squares
PLSR	Partial least squares regression
PRI	Photochemical Reflectance Index
PRISM	Processing Routines in IDL for Spectroscopic Measurements
PROSAIL	PROSPECT and Scattering by Arbitrary Inclined Leaves
PSRI	Plant Senescence Reflectance Index
R1	Reproductive stage 1 (beginning bloom)
R3	Reproductive stage 3 (beginning pod)
RENDVI	Red Edge Normalized Difference Vegetation Index
REP	Red Edge Position
RF	Random Forests
RGB	Red-green-blue
RMSE	Root mean square error
ROIs	Regions of interest
RUE	Radiation use efficiency
RWC	Relative water content
S2	Sentinel-2
SAIL	Scattering by Arbitrary Inclined Leaves
SAM	Spectral Angle Mapper
SB	Shade-bearer
SCR	Spatially coherent regions
SDSM	Spectral digital surface models

SfM	Structure from motion
SGI	Sum Green Index
SI	Simple ratio
SIPI	Structure-Insensitive Pigment Index
SMA	Spectral mixture analysis
SNR	Signal-to-Noise Ratio
SR	Simple Ratio
SRF	Spectral response function
SRI	Simple ratio index
STD	Standard deviation
SVM	Support Vector Machine
SWIR	Short-Wave InfraRed
SWP	Stem water potential
T_{canopy}	Canopy temperature
T_{dry}	Temperature of a nontranspiring leaf
T_{wet}	Temperature of a fully transpiring leaf
TCARI	Transformed chlorophyll absorption reflectance index
TIR	Thermal infrared
TLS	Terrestrial laser scanning
UAV	Unmanned Aerial Vehicle
UAV	Unmanned airborne vehicle
UAV	Unmanned autonomous vehicle
UnECHO	A developed algorithm based on ECHO
UNL	University of Nebraska-Lincoln
UQ	University of Queensland
UVE PLS	Uninformative variable elimination PLS
VARI	Visible Atmospherically Resistant Index
VF	Vegetation fraction
VI	Vegetation index
VIg	VIsible Green index
VIS-NIR	Visible and near infrared
VIS	Visible
VNIR	Visible and near-infrared
VOG-1	Vogelmann red edge index-1
VPD	Vapor pressure deficit
VReI	Vogelmann red edge index
VSWIR	Visible to short-wave infrared
WBI	Water Band Index
WDRVI	Wide dynamic range vegetation index
WEKA	Waikato Environment for Knowledge Analysis
WV-3	WorldView-3

Section I

Vegetation Biophysical and Biochemical Properties

1 Recent Developments in Remote Estimation of Crop Biophysical and Biochemical Properties at Various Scales

Anatoly A. Gitelson

CONTENTS

1.1 INTRODUCTION

Remote sensing has provided valuable insights into agronomic management over the past few decades. Use of remote sensing for determining crop physiological and phenological status has its roots in the pioneering work by William Allen, Harold Gausman, and Joseph Woolley [1–3], who provided much of the basic theory relating morphological characteristics of crop plants to their optical properties. These pioneering works have led to the understanding of how leaf reflectance changes in response to leaf thickness, species, canopy architecture, leaf age, nutrient and water status. Leaf chlorophyll content and its absorption in the visible spectrum provide the basis for utilizing reflectance as a tool either with broad-band radiometers or hyperspectral sensors that measure reflectance at narrow bands. The basic understanding of leaf reflectance has led to the development of various vegetation indices that have been extended to crop canopies and have been used to quantify various agronomic parameters (e.g., leaf area, crop cover, biomass, crop type, nutrient status, and yield). These tools are still being developed as we learn more about how to use the information contained in reflectance measurements from a range of different sensors.

A summary of the progress in applying remote sensing to agriculture has been published in a collection of articles in *Photogrammetric Engineering and Remote Sensing* [4–8]. Other recent reviews of the application of remote sensing methods to crops were developed by Hatfield et al. [9,10]. These articles provide a summary of the multispectral and hyperspectral remote sensing efforts in more detail and the reader is referred to these articles for a more thorough understanding.

Since first edition of this book was published [11], researchers at Center for Advanced Land Management and Information Technologies (CALMIT) at the University of Nebraska-Lincoln (UNL) have further developed and evaluated remote sensing techniques and tested them at close range and satellite levels. This chapter contains a summary of the experiences and advances made at UNL since 2010.

1.2 VEGETATION FRACTION

One of the principal variables in the growth of crops is the fraction of the solar radiation intercepted by foliage. The productivity of crops may be analyzed as the product of the solar energy intercepted over a season and the efficiency with which that energy is converted to biomass. In many crops, the relationship between radiation interception and green foliage cover/fractional vegetation cover is sufficiently close for the latter to be used as a substitute for more elaborate measurements of light interception [11]. Thus, vegetation fraction (VF) is an important trait that helps determine crop productivity.

Different vegetation indices (VIs) for the remote estimation of VF at close range in two crop types, maize and soybean, with contrasting canopy architectures and leaf structures were evaluated [12]. To determine the accuracy of VF estimation, the noise equivalent (NE) of VF was used:

$$NE\,\Delta VF = RMSE\,(VI\ vs.\ VF)/[d(VI)/d(VF)]$$

where RMSE (VI vs. VF) and $d(VI)/d(VF)$ are the root mean square error and the first derivative of the VI vs. VF relationship, respectively. The NE ΔVF provides a measure of how well the VI responds to VF across its entire range of its variation. NE ΔVF not only takes into account the RMSE of the VF estimation but also accounts for the sensitivity of the VI to VF, thus, providing a metric accounting for both scattering of the points from the best fit function and the slope of the best fit function.

Among the indices tested [12], the enhanced vegetation indices EVI, EVI2, wide dynamic range vegetation index (WDRVI), normalized difference vegetation index $NDVI_{green}$, and $NDVI_{red\,edge}$ were found to be the most accurate in estimating vegetation fraction (Figure 1.1). The algorithm for estimating VF by WDRVI $= (\alpha\rho_{NIR} - \rho_{red})/(\alpha\rho_{NIR} + \rho_{red})$ with $\alpha = 0.3$ was

$$VF = 80.84 * WDRVI_{\alpha=0.3} + 34$$

It was generic, not requiring parameterization for two crops studied with RMSE below 6% and mean normalized bias (MNB) below 2% (Figure 1.2). It was followed by red edge NDVI for both crops. EVI2 was accurate for soybeans (NE = 7.8%) and less accurate for maize (NE = 8.9%). Both $VARI_{green}$ and $VARI_{red\,edge}$, which were superior in estimating VF in wheat [11], were also quite accurate estimating VF in maize and soybeans; however, they were not the best among VF tested.

1.3 FRACTION OF ABSORBED PHOTOSYNTHETICALLY ACTIVE RADIATION

The fraction of absorbed photosynthetically active radiation (fAPAR) is one of the main traits used in production efficiency models (PEMs). It also plays tremendous role in accurate retrieval of light use efficiency, which is essential for assessing vegetation health. NDVI is the most-used VI for estimating fAPAR. In [13] relationships were established between fraction of PAR absorbed by photosynthetically active vegetation ($fAPAR_{green}$), and NDVI for two crops with contrasting leaf structures, photosynthetic pathways (C3 vs. C4), and canopy architectures, using *in situ* radiometric data and daily MODIS data over irrigated and rain-fed maize and soybean sites during eight years. Through the use of high temporal resolution *in situ* and MOSIS data, it was possible to identify specific phases in the growing season that aid in the interpretation of observations collected with coarser temporal resolution (or even single scenes). MODIS data are adequate for resolving distinct phases in the $fAPAR_{green}$/NDVI relationships within the growing season. The identification of these different phases has important implications for the interpretation of remotely sensed observations of crops, such as the estimation of light use efficiency (LUE) and productivity.

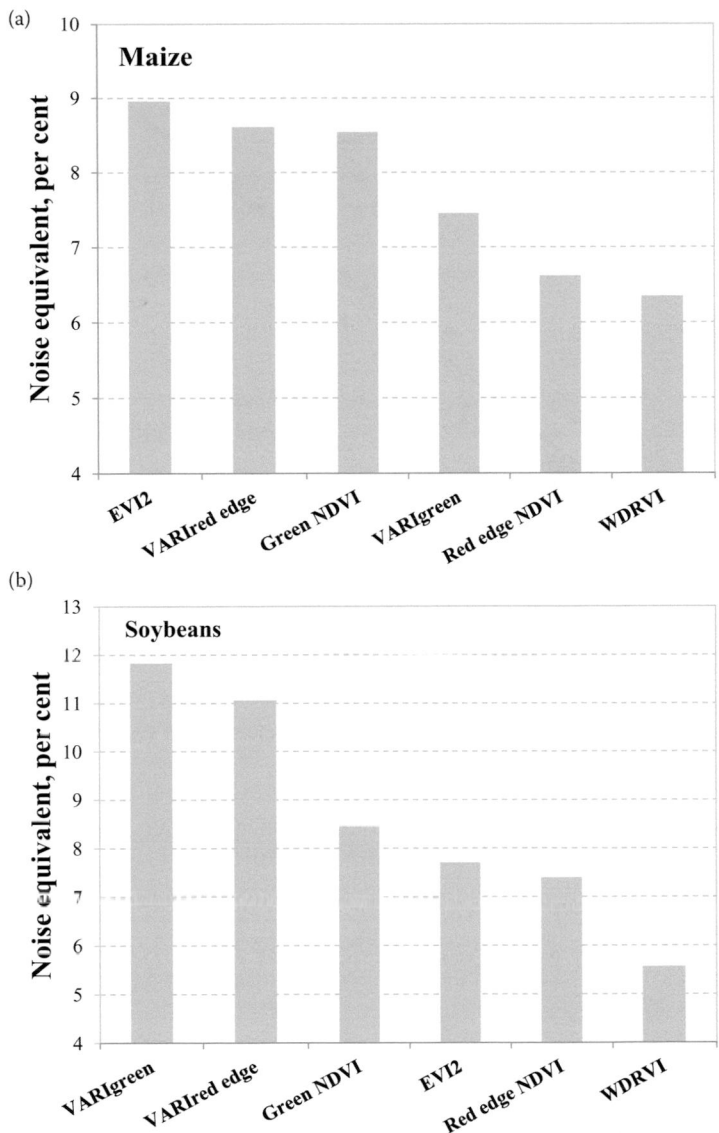

FIGURE 1.1 Noise equivalent of vegetation fraction estimation by vegetation indices tested for (a) maize, five years, nine irrigated and rain-fed sites; and (b) soybeans, three years, six irrigated and rain-fed sites.

Significantly, in [13] was shown that established relationships of $fAPAR_{green}$ vs. *in situ* NDVI were very close to that of $fAPAR_{green}$ vs. MODIS-retrieved NDVI. In vegetative stages, when $fAPAR_{green}$ was below 0.65, the $fAPAR_{green}$/NDVI relationships for crops with contrasting leaf structures and canopy architecture were close and almost linear, allowing accurate estimation of $fAPAR_{green}$ with RMSE = 5.8% (Figure 1.3). However, $fAPAR_{green}$/NDVI relationships in reproductive stages were very different for both crops (Figure 1.4), showing that canopy architecture and leaf structure greatly affect the relationship as leaf chlorophyll (Chl) content changes and vertical distribution of Chl content and green leaf area index (LAI) inside the canopy becomes heterogeneous.

The study [13] revealed fine details of the $fAPAR_{green}$/NDVI relationships, specifically two types of hysteresis that prevent accurate $fAPAR_{green}$ estimation using NDVI during the whole growing season. SAIL (Scattering by Arbitrary Inclined Leaves) model simulations of the $fAPAR_{green}$/NDVI

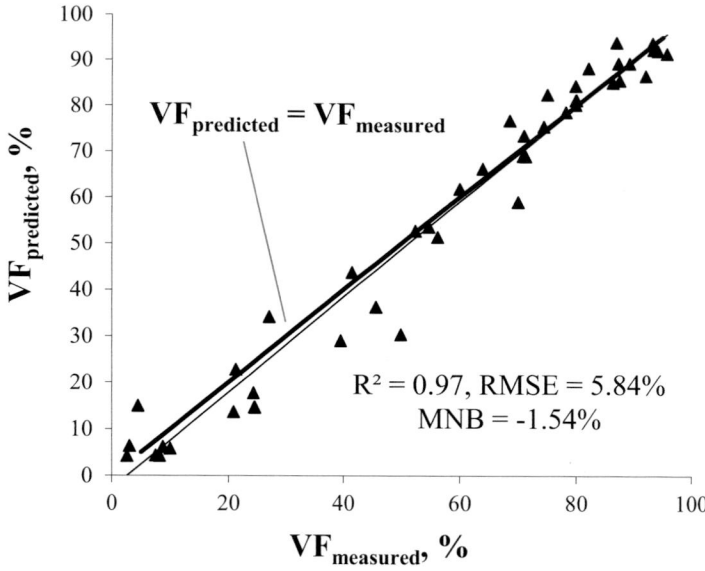

FIGURE 1.2   Vegetation fraction VF predicted by WDRVI with $\alpha = 0.3$ plotted versus vegetation fraction measured in irrigated and rain-fed maize and soybeans sites.

relationship for maize clearly displayed the existence of hysteresis in the relationship as revealed by empirical data.

It was also found that the $fAPAR_{green}$/NDVI relationships, established for vegetative stages in maize and soybean, are very different from other empirical studies at close range and satellite levels as well as from radiative transfer simulations [13]. This shows need for extensive research in

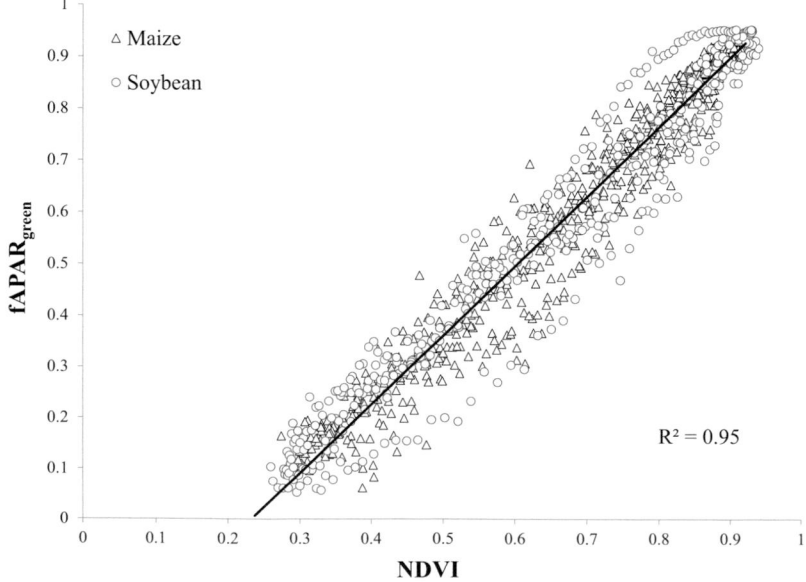

FIGURE 1.3   The MODIS-retrieved $fAPAR_{green}$/NDVI relationships for maize (collected in 2001 through 2008 in irrigated and rain-fed sites) and soybean (collected in 2002, 2004, 2006, and 2008 over two irrigated and rain-fed sites each year) in vegetative stage only (700 observations). Solid line is best-fit function.

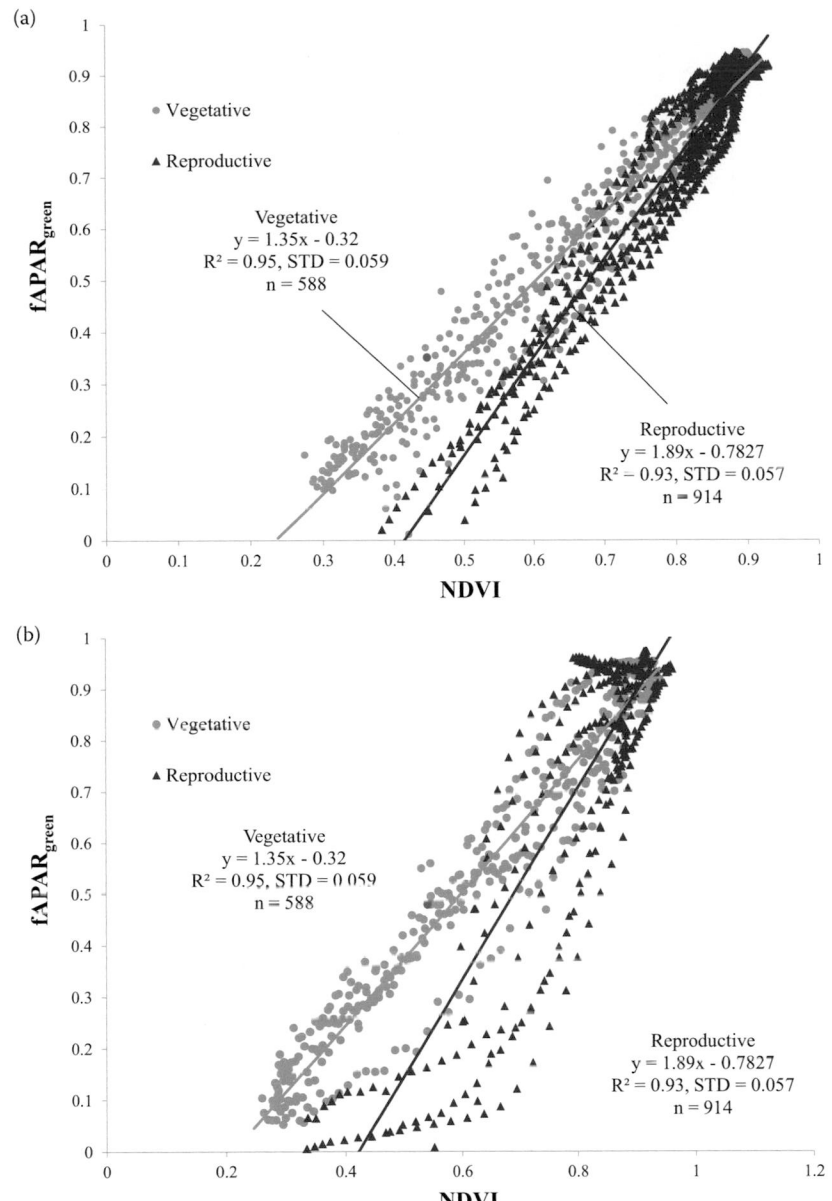

FIGURE 1.4 MODIS-retrieved NDVI vs. fAPAR$_{green}$ relationships for (a) maize in 2001 through 2008 (16 site years) and (b) soybean in 2002, 2004, 2006, and 2008 (8 site years).

remote sensing techniques for fAPAR$_{green}$ estimation. The issues of canopy vertical heterogeneity (in terms of leaf Chl content, leaf area, and leaf angle distribution), studied in [13], also affect other remote sensing problems such as estimating leaf and canopy Chl content, light use efficiency, and productivity. However, there has been little work addressing the issue of the effects of vertical variability in canopy structure and the paper [13] shows the importance of this.

In order to develop generic algorithms for fAPAR$_{green}$ estimation, VIs previously used for fAPAR estimation were tested. The reflectance spectra collected at close range [11] were resampled to spectral bands of the Moderate Resolution Imaging Spectroradiometer (MODIS) (green 545–565 nm, red 620–670 nm, and NIR 841–876 nm) using MODIS spectral response function and SR (simple ratio),

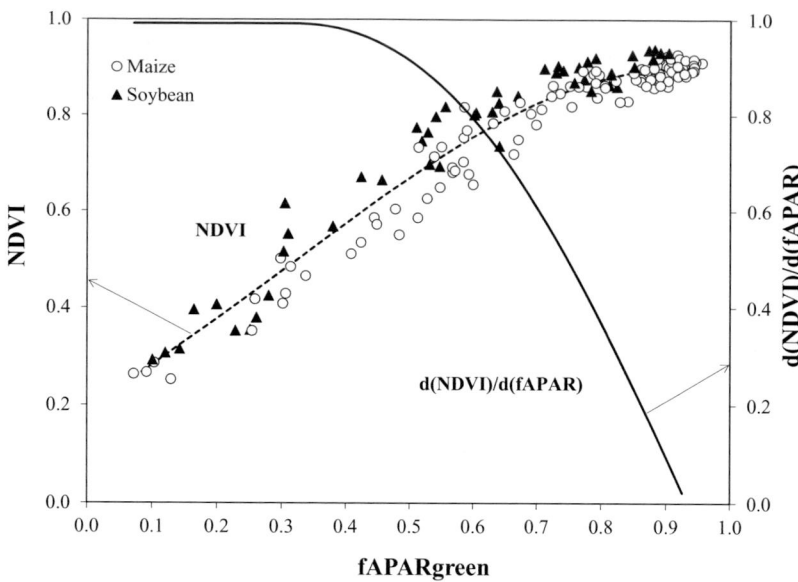

FIGURE 1.5   NDVI calculated with reflectance in spectral bands of MODIS and first derivative of NDVI with respect to fAPAR$_{green}$ plotted vs. fAPAR$_{green}$ in maize and soybean. Dashed line is best-fit function for NDVI vs. fAPAR$_{green}$ relationship.

NDVI, EVI2, TVI, MTVI1, MTVI2, VARI$_{green}$, OSAVI, WDRVI with $\alpha = 0.5$, and Green NDVI were calculated. The reflectance spectra were also resampled to spectral bands of the Multi Spectral Instrument (MSI) on the Sentinel-2 satellite system (green: 550–580 nm, red: 660–670 nm, red edge 1: 693–712 nm, red edge 2: 732–748 nm and NIR: 773–793 nm) using MSI spectral response function and MTCI, VARI$_{erd\ edge}$, red edge NDVI were calculated.

The relationship between NDVI and fAPAR$_{green}$ was almost non-species-specific (Figure 1.5). Thus, NDVI is supposed to be a proxy of fAPAR$_{green}$. However, the NDVI/fAPAR$_{green}$ relationship is asymptotic, with a decrease in the slope as fAPAR$_{green}$ exceeds 0.7 (first derivative dNDVI/dfAPAR$_{green}$ in Figure 1.5). Thus, NDVI exhibits limitations at moderate-to-high vegetation density. As fAPAR$_{green} > 0.7$, RMSE of fAPAR$_{green}$ estimation by NDVI grows exponentially, reaching 0.25 for fAPAR$_{green} = 0.8$. This means that in crop studied for more than two months during the growing season NDVI does not yield reliable information about fAPAR$_{green}$ [11].

EVI and EVI2 were closely related to fAPAR$_{green} < 0.7$, but further increase of EVI from 0.6 to 0.9 did not relate to fAPAR$_{green}$ (Figure 1.6). For the crops studied, the slope of fAPAR$_{green}$/LAI and fAPAR$_{green}$/[Chl] relationships increased gradually until LAI reached 3–4 and then it dropped due to decrease in depth of light penetration inside the canopy and decrease of Chl efficiency in light absorption [14]. Thus, for LAI > 3, EVI2 did follow the increase in LAI while fAPAR$_{green}$ increased a little, which disturbs the close fAPAR$_{green}$/EVI relationship.

Among VIs tested, only three had close linear non-species-specific relationships with fAPAR$_{green}$: WDRVI with $\alpha = 0.5$, green NDVI, and red edge NDVI. All three VIs were developed to avoid NDVI's limitation of estimating biophysical characteristics of dense vegetation. The main reasons for decreasing NDVI sensitivity to high-density vegetation are (i) a high ρ_{NIR}/ρ_{red} ratio that reaches 7–10 for moderate-to-high density vegetation, and (ii) saturation of red reflectance. WDRVI is a modification of NDVI that attenuates the effect of near infrared (NIR) reflectance by $\alpha < 1$. It makes the magnitudes of $\alpha\rho_{NIR}$ and ρ_{red} comparable and increases the sensitivity of WDRVI to such traits of dense vegetation as vegetation fraction and LAI. The fAPAR$_{green}$/WDRVI$_{\alpha=0.5}$ relationship was not species specific, with $R^2 = 0.92$ ($p < 0.001$) and RMSE $= 0.069$ (Figure 1.7, Table 1.1). Interestingly that close relationship between WDRVI and fAPAR$_{green}$ for Soil-Canopy Observation

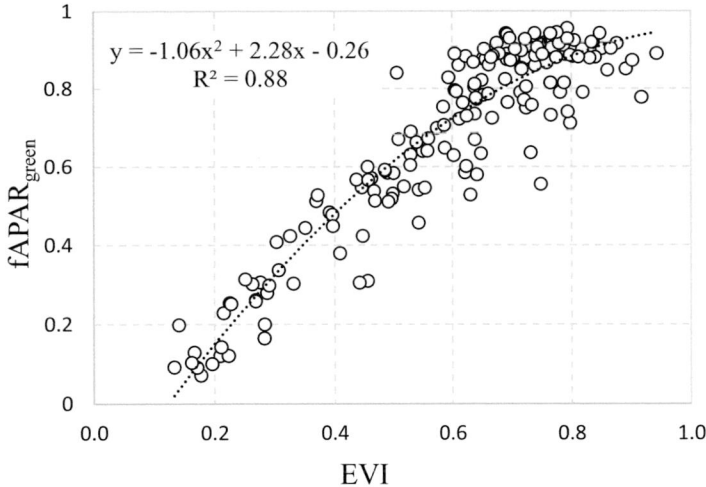

$$y = -1.06x^2 + 2.28x - 0.26$$
$$R^2 = 0.88$$

FIGURE 1.6 Relationship between $fAPAR_{green}$ and EVI2 for maize and soybean.

of Photosynthesis and Energy (SCOPE) simulations with LAI varying from 1 to 4, leaf chlorophyll content (20–80 $\mu g/cm^2$), solar zenith angle 20–60°, and three typical leaf inclination distribution functions (planophile, plagiophile, and spherical) was recently found [15].

The use of green and red edge spectral bands instead of red in NDVI is another way to increase the sensitivity of NDVI-like vegetation indices to traits of high-density vegetation. The absorption

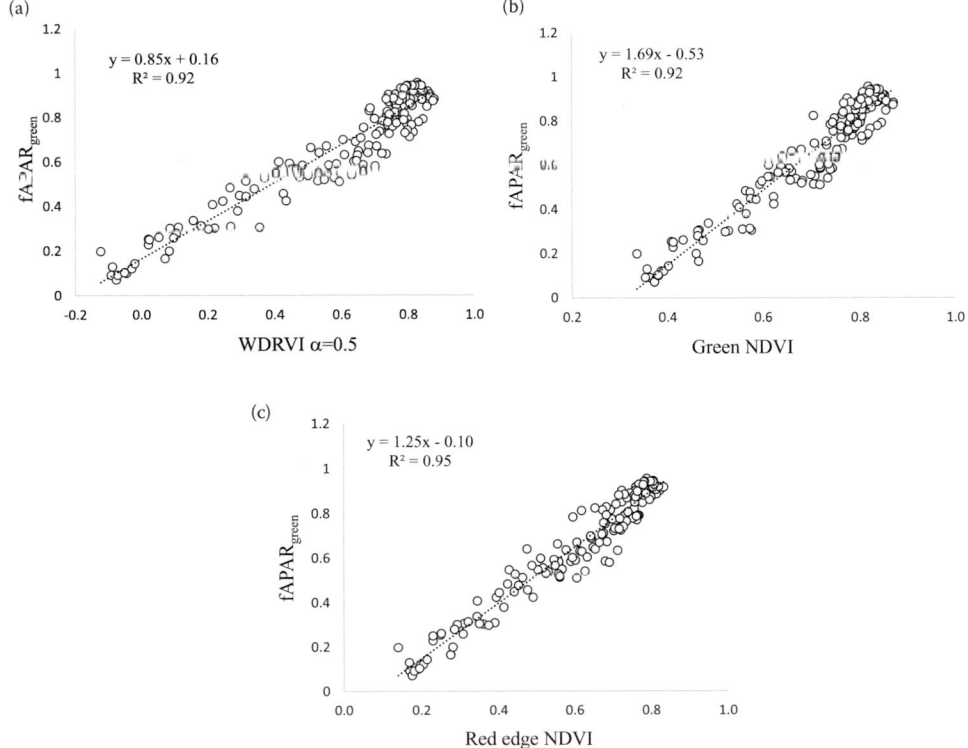

FIGURE 1.7 $WDRVI_{\alpha=0.5}$ (a), green NDVI (b), and red edge NDVI (c) plotted versus $fAPAR_{green}$ for maize and soybean.

TABLE 1.1

Algorithms, Determination Coefficients (R^2), and RMSE of fAPAR$_{green}$ Estimation in Maize and Soybean by Vegetation Indices

VI	fAPAR$_{green}$ vs. VI	R^2	RMSE
EVI	$y = -1.06x^2 + 2.28x - 0.26$	0.88	0.096
NDVI	$y = 0.07\exp(2.81x)$	0.92	0.075
WDRVI, $\alpha = 0.5$	$y = 0.85x + 0.16$	0.92	0.069
Green NDVI	$y = 1.6891x - 0.5271$	0.92	0.067
Red edge NDVI	$y = 1.2531x - 0.1035$	0.95	0.057

Note: The vegetation index names are given in full in the text. fAPAR, fraction of absorbed photosynthetically active radiation. RMSE, root mean square error.

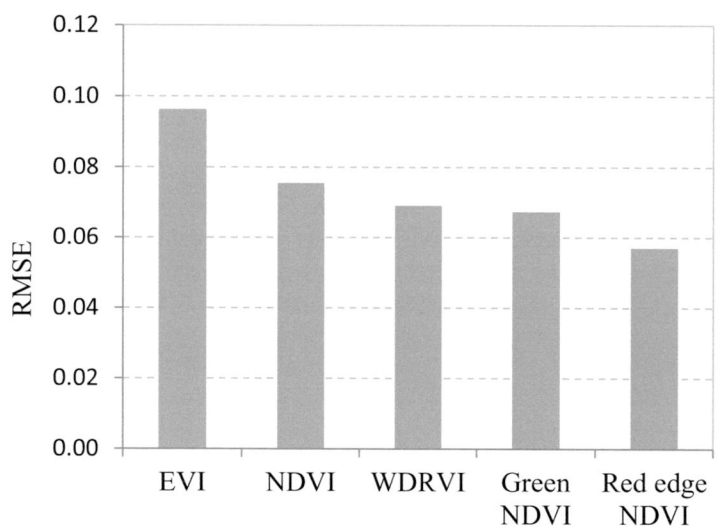

FIGURE 1.8   Root mean square error of fAPAR$_{green}$ estimation using different vegetation indices.

coefficient of Chl in the green and red edge spectral regions, located far from the main red absorption band of Chl (*in situ* around 670 nm), is not higher than 1%–2% of that in the red and the pathway of light inside a leaf and canopy is much larger than in the red. So, with increase in vegetation density, absorbance in these spectral regions continues to increase, enhancing sensitivity of the green and red edge reflectance to fAPAR$_{green}$.

The relationships of fAPAR$_{green}$ vs. green NDVI and fAPAR$_{green}$ vs. red edge NDVI were found to be very close ($p < 0.001$) with $R^2 = 0.92$ and 0.95, respectively (Figure 1.7). Red edge NDVI appears to be the best index for fAPAR$_{green}$ estimation in the whole range of its variation (Figure 1.8). The algorithms presented in Table 1.1 are not species specific for maize and soybean and do not require parameterization for these crops.

1.4   CHLOROPHYLL AND NITROGEN CONTENT

Canopy chlorophyll content (Chl) relates closely to plant photosynthetic capacity, nitrogen status, and productivity, and the necessity of remote Chl estimation in crops is recognized (e.g., [14,16]). Seasonal changes in the pigment pool and structural canopy properties greatly influence the light climate inside the canopy and modulate the extinction coefficient. For realistic modeling of reflectance, radiative transfer models should be fed by known vertical LAI and pigment content distributions as well as their changes during the season. It is important for monitoring crops with different structural properties when the spatial resolution of the sensor is low (e.g., comparable to field size or larger) and in regions with mixed-use cropping practices (e.g., maize/soybean rotation) requiring generic algorithms that do not need reparameterization for different crops.

Development of generic algorithms for Chl estimation that could be applied with no reparameterization for two contrasting crop species, maize and soybean, during the entire growing season was a goal of Peng et al. [16]. These two crops represent different biochemical mechanisms of photosynthesis, leaf structure, and canopy architecture. The relationships between canopy Chl and reflectance, collected at close range and resampled to bands of the Multi Spectral Instrument (MSI) aboard Sentinel-2, were analyzed in samples taken across the entire growing seasons in irrigated and rain-fed sites located in eastern Nebraska between 2001 and 2005.

Crop phenology was found to be a strong factor influencing canopy reflectance in two contrasting crops. Phenology caused a substantial species-specific difference (hysteresis) in the reflectance vs. canopy Chl relationships between the vegetative and reproductive stages. The reasons for the hysteresis were seasonal changes in canopy architecture, leaf structure, and foliar Chl, as well as seasonal changes in the influence of the soil/residue background. The effect of the hysteresis on vegetation indices applied for canopy Chl estimation depended on the bands selected in their formulation. For widely used VIs using NIR and red reflectance, NDVI, SR and EVI, there were significant differences in the VI vs. canopy Chl relationships between the vegetative and reproductive stages and between species, limiting their application for accurate canopy Chl estimation over the entire growing season. VIs with red edge and NIR bands, using reflectance simulated for the MSI sensor, included the chlorphyll index $CI_{740} = (\rho_{NIR}/\rho_{740}) - 1$, MERIS terrestrial chlorophyll index $MTCI = (\rho_{NIR} - \rho_{705})/(\rho_{705} - \rho_{670})$, and red edge $NDVI_{740} - (\rho_{NIR} - \rho_{740})/(\rho_{NIR} - \rho_{740})$. These VIs were accurate in estimating canopy Chl in maize and soybean with RMSE values below 0.38 g m^{-2} (Figure 1.9, Table 1.2). Algorithms utilizing these VIs require neither parameterization for each crop nor for each phenological stage (Table 1.2).

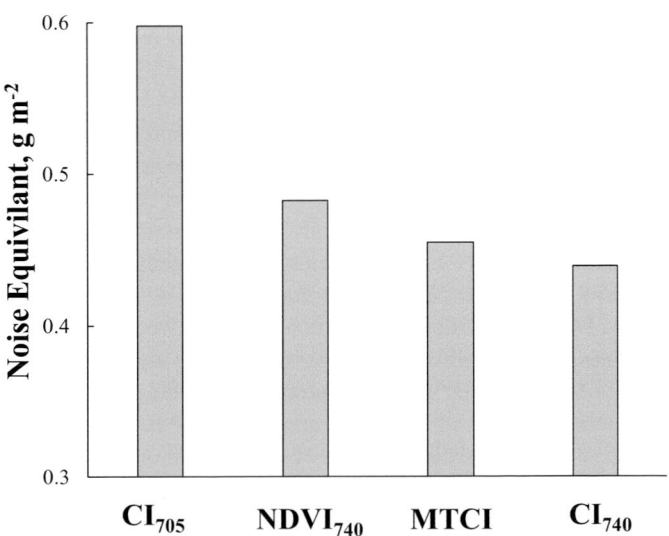

FIGURE 1.9 Noise equivalent of canopy Chl content estimation in maize and soybean combined by vegetation indices calculated using reflectance simulated at MSI spectral bands for chlorophyll index with red edge band at 705 nm (CI_{705}), red edge NDVI with red edge band at 740 nm ($NDVI_{740}$), MTCI, and chlorophyll index with red edge band at 740 nm (CI_{740}). Data taken from 2001 through 2005 over three irrigated and rain-fed sites (11 maize sites years and 4 soybeans sites year) with Chl content varied from 0 to 4.5 g m^{-2}.

TABLE 1.2

Algorithms for Canopy Chlorophyll Content Estimation, Generic for Maize and Soybean, Determination Coefficients (R^2), and RMSE in g m^{-2} for Three Vegetation Indices

Canopy Chl in g m^{-2}	R^2	RMSE
Chl = 0.241 × MTCI − 0.618	0.90	0.38
Chl = 18.509 × (red edge NDVI$_{740}$) − 0.999	0.91	0.37
Chl = 6.645 × CI$_{740}$ − 0.649	0.91	0.36

Note: The vegetation index names are given in full in the text. Chl, chlorophyl content. RMSE, root mean square error.

Development of generic algorithms for canopy Chl estimation in rice, wheat, corn, soybean, sugar beet, and natural grass using red edge and NIR spectral bands was presented in Inoue et al. [17]. The ratio of reflectances at 815 and 704 nm (ρ_{815}/ρ_{704}) was found to be superior to all other models in overall predictive ability of canopy Chl content. The soundness of the model was supported by simulation analyses using a radiative transfer model under various canopy conditions including plant types (canopy geometry), leaf Chl content, LAI, and soil background. Importantly, the partial least squares regression (PLSR) and interval partial least squares regression (iPLSR) models using much larger number of wavebands proved to be inferior to the VI-based models, especially in versatility.

The authors of Ciganda et al. [18] addressed a very important question: how deep into the maize canopy is Chl content sensed by the red-edge chlorophyll index, CI$_{red\ edge}$? Statistical techniques, a hierarchical regression, and three Aikaike Information Criterion, were used to determine how many leaf layers are sensed by the CI$_{red\ edge}$. The hierarchical regression procedure made it possible to assess the importance of each leaf Chl content in defining total Chl content in a maize canopy and documented very close relationships between CI$_{red\ edge}$ and total canopy Chl content when 8 to 10 top leaf layers were included in the model. Such deep sensing inside the maize canopy is the reason for the high accuracy in estimating maize canopy Chl content by CI$_{red\ edge}$, which employed the NIR and the red-edge (720–730 nm) spectral bands.

A strong correlation between foliar nitrogen (N) and Chl contents has been found for various plant species [19–21]. Since Chl is the main plant constituent determining the reflectance in the visible region of the spectrum, optical remote sensing techniques have great potential in providing information on canopy Chl and N content. Baret et al. [19] suggested that canopy Chl content is well suited for quantifying canopy level N content. Canopy Chl content is a physically sound quantity that represents the optical path in the canopy where absorption by Chl dominates the radiometric signal. Thus, absorption by Chl provides the necessary link between remote sensing observations and canopy-state variables that are used as indicators of N status and photosynthetic capacity.

PROSAIL (PROSPECT and Scattering by Arbitrary Inclined Leaves) simulations showed that the CI$_{red\ edge}$ is linearly related to the canopy Chl content over the full range of potential Chl values [22]. In that paper, the best results in estimating either canopy Chl or N content were obtained using CI$_{red\ edge}$ and CI$_{green}$. It was also shown that the precise position of the spectral bands in the CI$_{red\ edge}$ is not very critical. In [23], this was further elaborated by studying the spectral bands to be used in the CI$_{red\ edge}$ in order to get the minimum RMSE in estimating canopy Chl and N content for three crop species (potato, maize, and soybean) and grass. Although results varied for the various experiments, optimal results were obtained using a spectral band around 800 nm in the numerator of the CI$_{red\ edge}$ and a spectral band in the range 705–740 nm in the denominator. The choice of the denominator waveband was more critical than the choice of the numerator: for maize and grass (erectophile canopies) this was in a wide range of 720–740 nm, whereas for soybean and potato (planophile

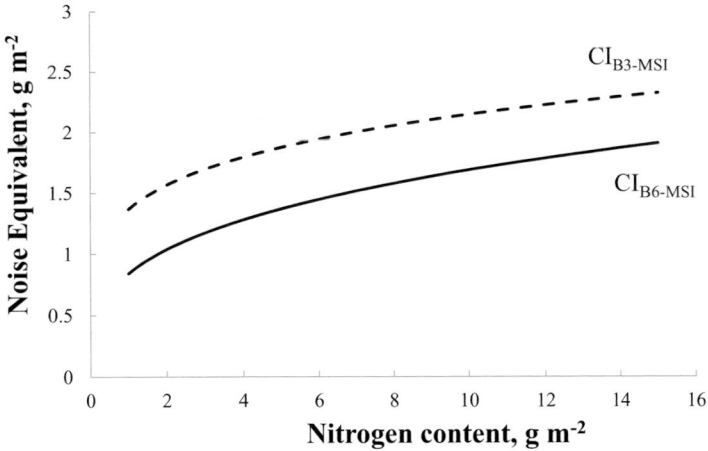

FIGURE 1.10 Noise equivalent of total nitrogen content estimation in maize by green and red edge chlorophyll indices (CI_{green} and $CI_{red\ edge}$), calculated using green B3 (543–577 nm) and red edge B6 (732–747 nm) spectral bands of MSI sensor onboard Sentinel-2. (From Schlemmer, M. et al., *International Journal of Applied Earth Observation and Geoinformation*, 25, 47–54, 2013. [24])

canopies) this band was around 705–710 nm. Subsequently, the Sentinel-2 spectral bands have been simulated using the data of the four experiments. In all experiments, the best results in estimating canopy Chl and N content were obtained using the $CI_{red\ edge}$, CI_{green}, and MTCI. Moreover, results using the Sentinel-2 band positions were quite similar to the optimal band positions for the $CI_{red\ edge}$. This confirms the importance of the red edge bands on Sentinel-2. However, CI_{green} with green band presented in MODIS and Landsat also seems very promising and requires further research.

In Schlemmer et al. [24] was shown that Chl and N content in maize can be estimated by the same remote sensing techniques, confirming a paradigm that absorption by Chl provides the necessary link between remote sensing observations and canopy-state variables that are used as indicators of N status. The study [24] presented the significance of the green (560 nm) and long wave red edge (740 nm) bands of the MSI sensor on Sentinel-2 for estimating Chl and N contents in maize (Figure 1.10). Also notable, $CI_{red\ edge}$ with quite a wide spectral band around 740 nm was optimal for N and Chl estimation. $CI_{red\ edge}$ with red edge band 720–730 nm allowed accurate non-species-specific estimation of gross primary production in maize and soybean [11,16] and the same range was found to be optimal for N estimation in rice [25]. Thus, it is likely that presented techniques for N and Chl estimation in maize could accurately estimate the same characteristics in other crops.

Despite encouraging results, thoughtful studies of reflectance vs. canopy Chl relationships are still required for different types of crops with contrasting biochemical and structural properties. The robustness of generic algorithms for different crops and their varieties should be confirmed in further studies. These algorithms should also be examined for their sensitivity to a range of typical soil backgrounds.

1.5 GREEN LEAF AREA INDEX

One of the key traits affecting primary production is the green leaf area index (green LAI), which is the ratio of the one-sided green leaf area to the ground area underneath. NDVI is widely used for estimating green LAI (see for review [10,26]). However, the relationship between NDVI and green LAI is essentially nonlinear and exhibits significant variations among various vegetation types. When green LAI > 2, NDVI is generally insensitive to green LAI. Thus, the main requirements for remote sensing techniques estimating green LAI are (i) increase of sensitivity to moderate-to-high vegetation density and (ii) decrease of their sensitivity to leaf structure and canopy architecture. Both issues are addressed below.

Remote sensing techniques for estimating green LAI in two crop types (maize and soybean) with contrasting canopy architectures and leaf structures were evaluated to develop algorithms not requiring reparameterization for each crop [26]. Among the VIs tested, the CI_{green}, the $CI_{red\,edge}$ and the MTCI exhibited strong and significant linear relationships with green LAI ranging from 0 to more than 6 m^2/m^2. The $CI_{red\,edge}$ was the only index insensitive to crop type and produced the most accurate estimations of green LAI in both crops (RMSE < 0.58 m^2/m^2). These results were obtained using data acquired at close range (i.e., field spectrometers mounted 6 m above the canopy) and from an aircraft-mounted Airborne Imaging Spectrometer for Applications (AISA). As the $CI_{red\,edge}$ also exhibited low sensitivity to soil background effects, it constitutes a simple yet robust tool for the remote and synoptic estimation of green LAI.

In [27] the results of the development of generic algorithms for green LAI estimation in four different crops, maize, soybean, wheat, and potato, were presented. Spectral measurements and green LAI data of wheat and potato were obtained in Israel and of maize and soybean in the United States. Among the VIs examined, two variants of the chlorophyll index (CI) and WDRVI with the green and red edge bands were the most accurate in estimating green LAI in all four crops. Hyperspectral reflectance data were used to determine optimal diagnostic bands for estimating green LAI in four crops using a universal algorithm. The green (530–570 nm) and red edge (700–730 nm) regions were identified as having the lowest errors in estimating green LAI. Since the Landsat 8 Operational Land Imager (OLI) has a green spectral band and the Sentinel-2, Sentinel-3, and VENμS have both green and red edge bands, it is expected that these VIs can be used to monitor green LAI in multiple crops using a single algorithm.

VIs that are maximally sensitive to green LAI along its entire range of variability were presented in [28]. In order to benefit from the different sensitivities of VIs along the entire green LAI range, combining of VIs was suggested. For sensors with spectral bands in the red and NIR regions, the best combination was NDVI and SR (maize normalized root mean square error (NRMSE) = 10%; soybean NRMSE = 11.5%). However, this combined index was species specific. For sensors with bands in the red edge and NIR regions, the best combination was red edge NDVI and $CI_{red\,edge}$, not requiring reparameterization, and was capable of accurately estimating green LAI in both crops (i.e., maize and soybean) with a NRMSE below 10%.

Informative spectral bands for estimating green LAI in maize (a C4 species) and soybean (a C3 species) retained in three types of methods—neural networks (NN), partial least squares (PLS) regression, and vegetation indices (VI)—were found in [29]. Hyperspectral reflectance and green LAI of irrigated and rain-fed maize and soybean were taken during eight years of observations (altogether 24 site-years) in very different weather conditions. The red edge and the NIR bands were selected by all methods and were found to be the most informative. The best results were obtained with NN using four spectral bands—two on red edge (700–710 and 720–740 nm), NIR (beyond 770 nm), and red (around 670 nm)—with NRMSE < 7.7%. These were followed by $CI_{red\,edge}$, using red edge and NIR bands, and PLS using three bands, both with NRMSE < 8.5%.

The validity of these bands was further confirmed via the uninformative variable elimination PLS technique, UVE PLS [30,31]. This technique assists in reducing the data dimension by eliminating spectral data that are uninformative or redundant and identifying the most informative spectral regions of the hyperspectral data. A smaller absolute value of the reliability parameter indicates that the data are less informative and can be removed at the user's discretion. Centner et al. [31] cautioned that this approach is not for band selection, but it is a way to eliminate variables that are useless. The most informative spectral bands for green LAI estimation were found by UVE PLS in the NIR, red edge, and green spectral ranges (Figure 1.11).

Informative spectral bands for green LAI estimation in maize and soybean using spectral data taken at close range [29] were tested in [32–34] using Aqua and Terra MODIS, Landsat TM

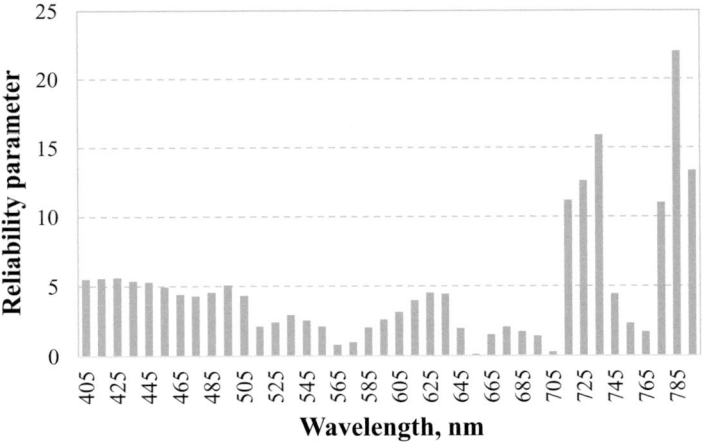

FIGURE 1.11   Spectra of reliability parameter calculated using hyperspectral data taken 6 m above the top of canopy and green LAI in maize and soybean combined by uninformative variable elimination partial least-squares (UVE PLS) technique. The magnitude of the reliability parameter is an indicator of useful information contained in reflectance spectra.

and ETM+, ENVISAT MERIS surface reflectance products, and simulated data of the recently launched Sentinel-2 MSI and Sentinel 3 OLCI (Ocean and Land Colour Instrument). Special emphasis was placed on testing generic algorithms that not require reparameterization for these species. Four techniques were investigated in [32]: support vector machines (SVM), neural network (NN), multiple linear regression (MLR), and vegetation indices (VI). All models tested provided a robust and consistent selection of spectral bands related to green LAI in crops representing a wide range of biochemical and structural traits. For TM/ETM+ Landsat, when only two spectral bands were allowed, all four techniques selected green and NIR bands. Among the nonparametric regression techniques, NN and SVM were the best with NRMSE below 14.4%. Addition of a third band (in the blue region) decreased the NRMSE only slightly (to 14%). When four bands were used (the fourth band was in the red region), NRMSE increased, due likely to overfitting at the training stage. WDRVI with two bands, green and NIR, was able to estimate green LAI with NRMSE below 13%.

The smallest NRMSEs of LAI estimation around 11.8% for all three techniques (MLR, SVM, and NN) were obtained using MERIS data. To achieve this accuracy, SVM used only three bands and addition of a fourth band decreased accuracy. In contrast, MLR reached maximal accuracy using five bands and NN six bands. However, when the fifth and sixth bands were added, the reduction in NRMSE was very small (0.1%–0.25%). WDRVI with two bands, red edge and NIR, achieved NRMSE < 12% and explained more than 83% of LAI variation in the two crops taken together (Figure 1.12). Sentinel-2 MSI and Sentinel 3 OLCI estimates based on simulated data had NRMSE below 8%. However the accuracy of these models with actual MSI and OLCI surface reflectance products remains to be determined.

These findings lay a strong foundation for the development of generic algorithms that are crucial for remote sensing of vegetation biophysical parameters. The bands retained by SVM, NN, PLS, and VI were in close agreement and were confirmed in [35] by Gaussian processes regression where top performances were found with between four and nine bands, and all of them relied on a band in the red edge and other bands in relevant absorption regions. Identifying informative spectral bands across all four techniques provided insight into spectral features of reflectance specific for each species as well as those that are common to species with different leaf structures, canopy architectures, and photosynthetic pathways.

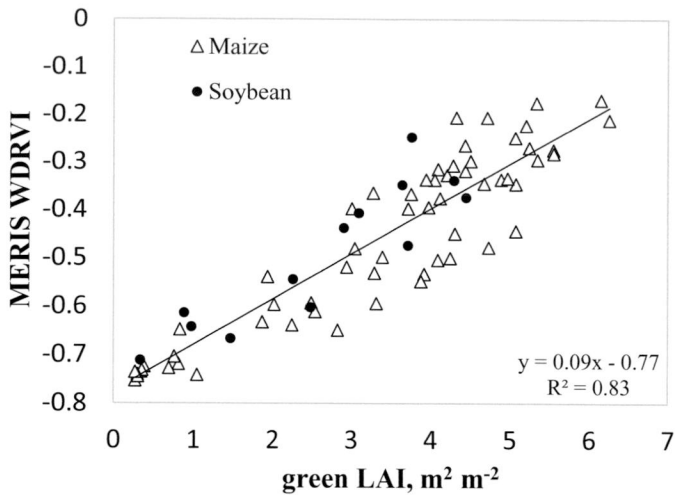

FIGURE 1.12   Relationships between MERIS WDRVI and green LAI in maize and soybean. WDRVI was calculated using MERIS (red edge and NIR) surface reflectance products. MERIS data were collected in 2003–2011 (61 images over maize and 14 over soybean). NRMSE < 12%. (From Kira, O. et al., *Remote Sensing*, 9, 318, 2017, doi: 10.3390/rs9040318. [32])

1.6   GROSS PRIMARY PRODUCTION

Vegetation productivity is the basis of all the biospheric functions on the land surface and is defined as the production of organic matter through photosynthesis. The total amount of organic matter produced through photosynthesis is termed the gross photosynthesis, and if expressed as the integral of the organic matter produced by all the individual plants in a defined area per unit of time, is termed the gross primary productivity (GPP). Given that the vegetation productivity is directly related to the interaction of solar radiation with the plant canopy [9–11], remote sensing techniques are used to measure vegetation productivity.

In the first CALMIT/UNL publications on remote estimation of crop GPP it was hypothesized that crop photosynthesis and GPP relate closely to total canopy/stand Chl content and thus that GPP can be estimated remotely using Chl-related models [11]. Using limited data sets it was shown that GPP could be estimated accurately by vegetation indices closely related to Chl content ($CI_{red\ edge}$, MTCI and CI_{green}). Rational for the hypothesis was (Figure 1.13): (i) fAPAR vs. Chl relationship was essentially not linear with significant (more than 5-fold) decrease of slope as Chl > 2 g m^{-2} and the slope was close to zero for Chl > 3 g m^{-2} (Figure 1.13a); (ii) in contrast to fAPAR, with increase in Chl above 2 g m^{-2} GPP steadily increased (Figure 1.13b), so GPP was sensitive to Chl content despite substantial decrease of fAPAR sensitivity to Chl; (iii) light use efficiency was found to be related to Chl content (Figure 1.13c) and it explained high sensitivity of GPP to moderate-to-high Chl (see [36–38] for detail).

This new paradigm based on total Chl content was elaborated using multiyear data taken over maize and soybean at three irrigated and rain-fed AmeriFlux sites in Nebraska, USA [36–38]. A model was suggested relating crop GPP to a product of total canopy Chl content and incoming photosynthetically active radiation, PAR_{in} [37–38]. Canopy Chl content was estimated by VIs closely related to Chl content. It was shown that the Chl–PAR_{in} model was able to accurately estimate GPP using VIs retrieved from reflectance data taken at close range over maize, soybean, and wheat [11,37–38] as well as grassland [39].

For application of the model for estimating GPP in C3 and C4 crops with no parameterization of algorithms, two questions were addressed: (i) Are the algorithms developed for maize and soybean

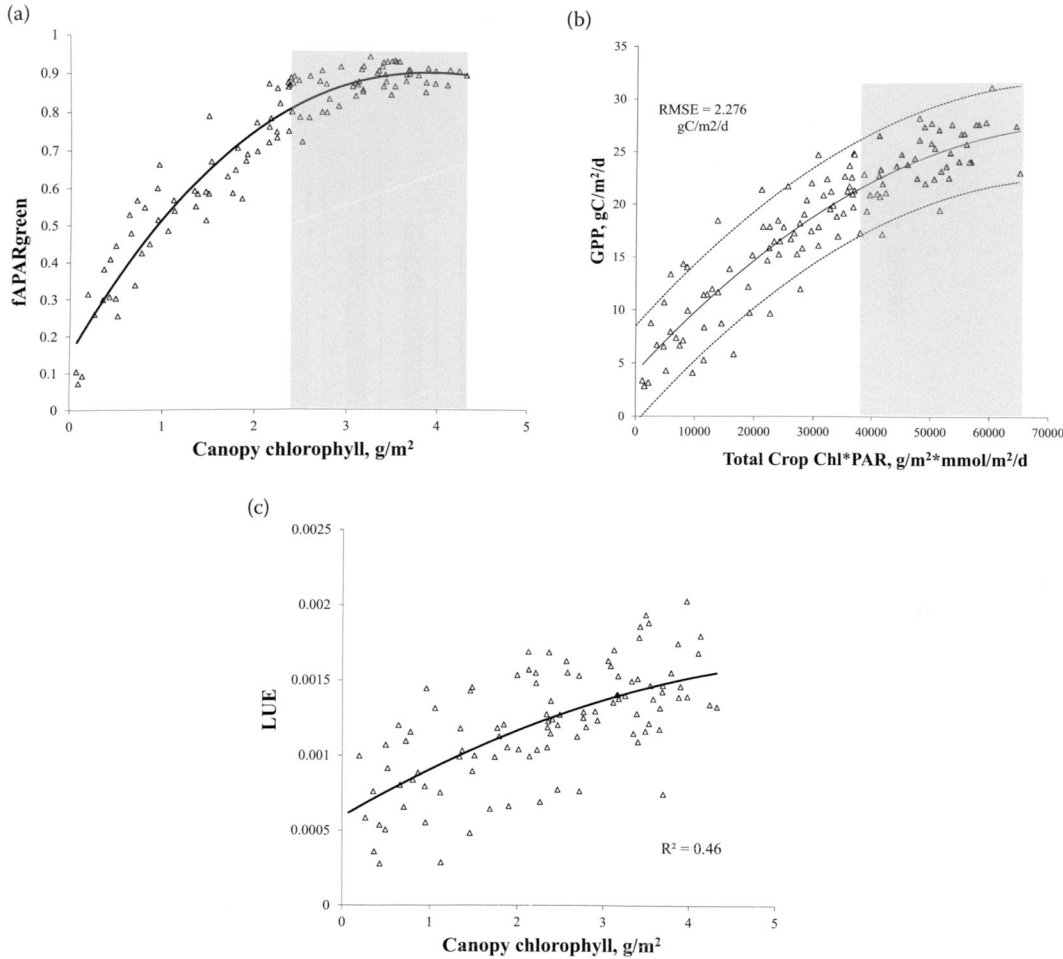

FIGURE 1.13 Fraction of PAR absorbed by photosynthetically active vegetation (fAPAR$_{green}$) (a), gross primary production (GPP) (b), and light use efficiency (LUE) (c) plotted versus canopy Chl content in maize. Hyperspectral reflectance data (110 observations) were taken at close range and GPP was measured at three AmeriFlux irrigated and rain-fed sites in Nebraska in 2001 through 2005.

different? (ii) Is it possible to develop a unified generic algorithm for GPP estimation in both maize and soybean? It was shown that several VIs may be used for generic GPP assessment—Figure 1.14 [37–38]. The use of red edge NDVI, red edge WDRVI, CI$_{green}$ and CI$_{red\ edge}$ allowed for estimation of GPP in both crops with no parameterization, with NRMSE < 10%. However, *only CI$_{red\ edge}$ and red edge NDVI were not species specific for maize and soybean.*

To apply the model for estimating crop GPP to satellite data, three approaches were used with respect to PAR: (i) incident PAR (PAR$_{in}$), (ii) PAR retrieved from short-wave radiation data [40], and (iii) potential photosynthetically active radiation (PAR$_{pot}$) [41–42]. PAR$_{pot}$ is the PAR$_{in}$ value under conditions of minimal aerosol loading; it represents the seasonal changes in hours of sunshine (i.e., day length). It was shown that the use of a product of Chl-related VI and PAR$_{pot}$ gave significantly decreased uncertainties of GPP estimation compared with other approaches [41–43].

Concurrent GPP and TM/ETM+ Landsat observations during 2001–2008 over the three Nebraska AmeriFlux sites represented a wide range of GPP variation (maize GPP ranging from 0 to 31 gC/m²/d; soybean GPP ranging from 0 to 18 gC/m²/d) [41]. The GPP vs. NDVI × PAR$_{pot}$

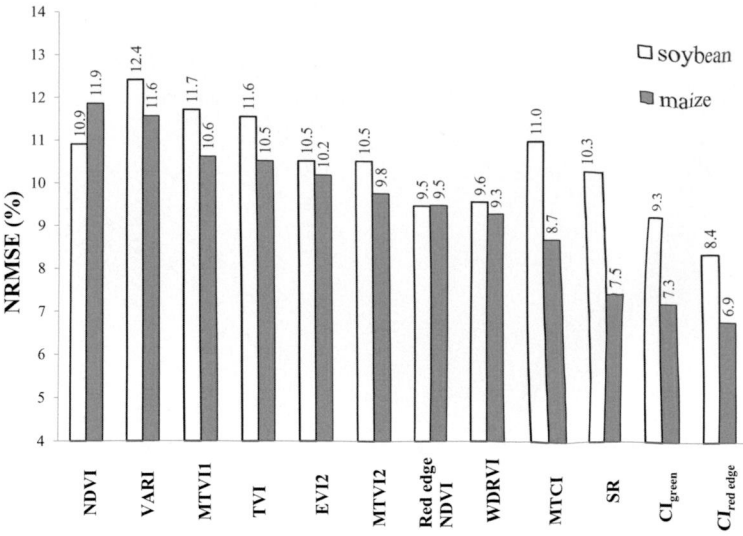

FIGURE 1.14 Normalized root mean square error (NRMSE) of GPP estimation in maize and soybean by vegetation indices. Hyperspectral reflectance data were taken at close range and GPP was measured at three AmeriFlux irrigated and rain-fed sites in Nebraska in 2001 through 2008.

relationship was nonlinear, with slope decreasing as GPP increased. MODIS and Landsat-derived NDVI was a good indicator of low-to-moderate GPP, but it was less accurate in detecting GPP when it exceeded 20 gC/m²/d. The normalized difference VIs (NDVI, green NDVI, and green WDRVI) performed better than ratio-based VIs (SR and CI_{green}): the NRMSE were >12% for ratio VIs but <8.5% for normalized difference VIs (Table 1.3). Except for green NDVI, all VIs tested in [41] were species specific for maize and soybean. For the same GPP, the value of VI × PAR_{pot} in soybean was consistently higher than that in maize with VIs calculated from reflectance in NIR and red bands. This result is due to contrasting leaf structures and canopy architectures of maize and soybean [11]. Thus, prior information about crop types is required when VIs with red and NIR bands are used for GPP estimation. However, when green band was used in NDVI, the green NDVI vs. GPP

TABLE 1.3

Determination Coefficients (R^2), RMSE, and NRMSE for Relationships GPP vs. VI × PAR_{pot} with VIs Retrieved from TM/ETM+ Landsat Atmospherically Corrected Images Taken over AmeriFlux Maize Sites from 2001 through 2008 and Soybean Sites in 2002, 2004, 2006, and 2008

Vegetation Index	Maize			Soybean		
	R^2	RMSE (gC/m²/d)	NRMSE (%)	R^2	RMSE (gC/m²/d)	NRMSE (%)
Green WDRVI	0.95	1.90	6.1	0.90	1.54	8.1
EVI2	0.95	1.92	6.2	0.87	1.79	9.5
Green NDVI	0.94	2.20	7	0.92	1.40	7.5
NDVI	0.93	2.22	7.1	0.89	1.65	8.7
CI_{green}	0.91	2.67	8.5	0.76	2.42	12.3
SR	0.84	3.49	11.0	0.67	2.79	14.2

Note: The vegetation index names are given in full in the text. PAR_{pot}, potential photosynthetically active radiation. RMSE, root mean square error. NRMSE, normalized root mean square error.

relationships for maize and soybean were close, allowing accurate GPP estimation in both crops using the same algorithm (Figure 1.15, Table 1.3).

MODIS 250-m data bring high temporal resolution. To test performance of the model for estimating GPP in irrigated and rain-fed maize and soybean croplands, eight years of MODIS 250-m data were collected and analyzed [40,42]. NDVI, EVI2, WDRVI, and SR were tested in the models. The best performance was found for WDRVI- and EVI2-based models [42] (Figure 1.16).

Thus, our results suggest that about 90% of GPP variation in crops is explained by total canopy Chl content [43–46]. It was also confirmed for wheat [47] and grasslands [48]. Assuming an invariant LUE_{green}, GPP in crops and grasslands may be accurately retrieved from close-range and satellite data [44–45,47–53]. LUE_{green} is affected by many factors, specifically cloudiness coefficient [54] as well as daytime temperature, vapor pressure deficit, and phenology [55]. It was shown that the effect of all these factors on LUE_{green} resulted in normalized standard deviations of LUE_{green} for irrigated and rain-fed maize of 11.9% and for irrigated and rain-fed soybean of 13.3%, thus demonstrating convergence of LUE_{green} to a narrow range [56]. Recently was found that the maximum daily LUE based on PAR absorption by canopy Chl, unlike other expressions of LUE, tends to converge across biome types [57]. Thus, taking into account conservative behavior of LUE_{green}, high accuracy of GPP estimation based on Chl-related vegetation indices and incident or potential PAR it is not surprising.

The question remains: Is a situation of limited resource availability and high resource acquisition costs a reason for efficient resource use and convergence of LUE_{green}? Such a scenario results in an optimization of resource allocation, which then results in a maximization of carbon gains and a convergence on a narrow range of LUE_{green} as was suggested in [58,59]. In this case, the plant response to stress is a decrease in radiation absorbed by photosynthetically active green vegetation such that LUE_{green} remains relatively invariant.

Our results have important implications for remote estimating of primary production in crops. Convergence of LUE_{green} allows the use of simple robust gross primary production models and also a better understanding of the role and constraints of LUE_{green} in process-based models. Assuming invariant LUE_{green}, the models based on either the canopy/stand/community Chl content or green LAI may facilitate accurate assessments of primary production and plant optimization patterns at multiple scales, from leaves to canopies and entire regions.

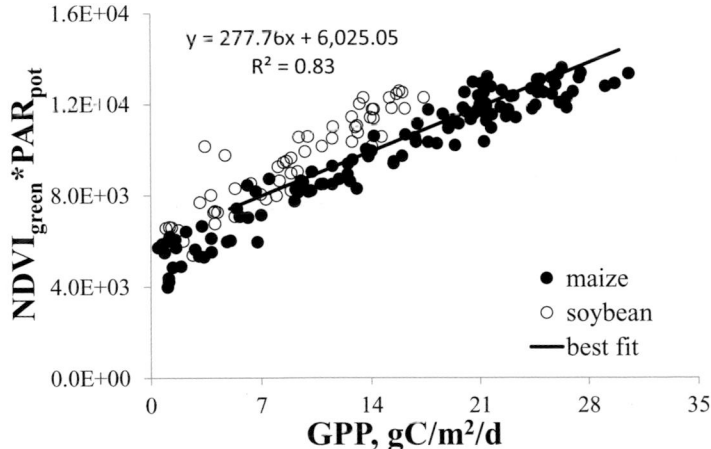

FIGURE 1.15 The relationships between GPP and the product of $NDVI_{green}$ and potential PAR (PAR_{pot}) for maize and soybean. Green NDVI was derived from TM and ETM+ Landsat data taken over irrigated and rain-fed AmeriFlux sites (120 observations over maize and 55 over soybean) from 2001 through 2008.

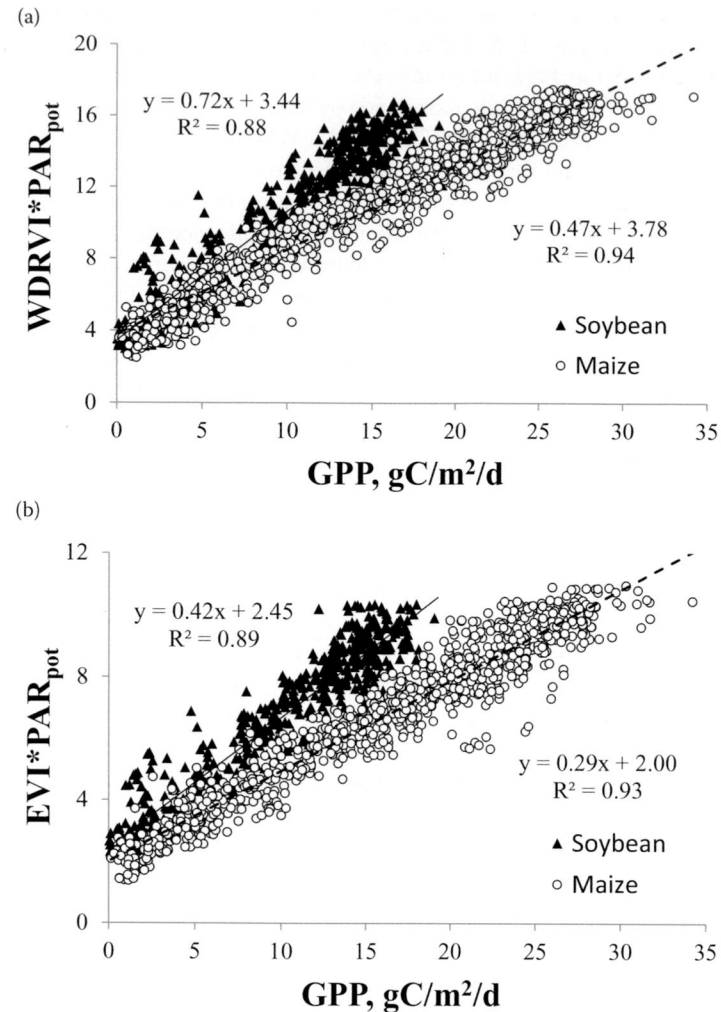

FIGURE 1.16 The product of (a) WDRVI $\times$ PAR$_{pot}$, and (b) EVI $\times$ PAR$_{pot}$ plotted versus GPP for 1058 observations over maize and 493 over soybean during growing seasons 2001 through 2008 in Nebraska when (PAR$_{po}$–PAR$_{in}$)/PAR$_{pot}$ was below 20%. Vegetation indices were retrieved from MODIS 250-m data.

1.7 CONCLUSIONS

Remote sensing techniques for estimating six crop biophysical and biochemical characteristics—vegetation fraction, fraction of PAR absorbed by photosynthetically active vegetation, chlorophyll content, nitrogen content, green leaf area index, and gross primary production—have been presented in this chapter. All techniques were tested using reflectances acquired from close range (6 meters above the top of the canopy) as well as TM/ETM+ Landsat, MODIS, and MERIS satellite data. It was shown that the aforementioned characteristics can be estimated accurately using remotely sensed data. Moreover, generic algorithms were developed that do not require parameterization for crops studied.

Tables 1.1 through 1.3 summarized the accuracy of estimating biophysical characteristics. The choice of index depends on the spectral characteristics of the radiometer or the satellite sensor being used. The indices employing red edge spectral bands, namely, CI$_{red\ edge}$, MTCI, VARI$_{red\ edge}$, NDVI$_{red\ edge}$, and WDRVI$_{red\ edge}$, can be used for satellite systems with spectral bands in the red edge region (Sentinel-2, Sentinel-3, and Venμs). The indices employing green spectral band, namely,

$NDVI_{green}$, $VARI_{green}$, CI_{green}, and $WDRVI_{green}$, can be used for satellite systems with spectral bands in the green region (e.g., Landsat, MODIS 500 m and 1 km spatial resolution, Venμs, Sentinel-2, and Sentinel-3). The indices using only red and NIR spectral bands, namely, NDVI, EVI2, and WDRVI, can be used for crop monitoring by satellite systems such as AVHRR, Landsat, MODIS (250 m spatial resolution) and Venμs.

The implications of these findings are far-reaching since the techniques described open a new possibility for accurate estimation of crop biochemical and biophysical characteristics at different scales, from close range to satellite altitudes. Some of the techniques based on the red, green, and NIR bands allow use of the extensive archive of Landsat and AVHRR imagery acquired since the early 1970s and the 250 m spatial resolution MODIS imagery acquired since 2001.

With these techniques, it is now possible to obtain global synoptic estimates of crop biochemical and biophysical characteristics at 20 and 30 m spatial resolution (Sentinel-2, Landsat TM/ETM+) and at 250 m/300 m resolution (MODIS and Sentinel-3). The performances of the algorithms were tested for maize, soybean, potato, rice, and wheat. These crops have very different canopy architectures and leaf structures. Still, the techniques developed yielded accurate estimations, which indicates that these techniques are likely applicable to other crops as well.

ACKNOWLEDGMENTS

I was honored to work with and learned a lot from Donald C. Rundquist, Shashi Verma, Timothy Arkebauer, and James Schepers. Their help at different stages of this study was invaluable and is greatly appreciated. I acknowledge contributions of my former PhD students Drs. Veronica Ciganda, Anthony Nguy-Robertson, Yi Peng, Andres Vina, and Arthur Zygielbaum. They contributed enormously in development and testing methods and techniques presented here. The support from the Center for Advanced Land Management and Information Technologies and Carbon Sequestration Program at University of Nebraska, USA is appreciated.

REFERENCES

1. Allen, W.A., Gausman, H.W., and Richardson, A.J., Willstatter-Stoll theory of leaf reflectance evaluated by ray tracing, *Applied Optics*, 12, 10, 2448–2453, 1973.
2. Gausman, H.W., Allen, W.A., Myers, V.I., and Cardenas, R., Reflectance and internal structure of cotton leaves Gossypium hirsutum L, *Agron. J.*, 61, 3, 374–376, 1969.
3. Woolley, J.T., Reflectance and transmittance of light by leaves, *Plant Physiology*, 47, 656–662, 1971.
4. Barnes, E.M., Sudduth, K.A., Hummel, J.W., Leach, S.M., Corwin, D.L., Yeng, C.C., Daughtry, S.T., and Bausch, W.C., Remote- and ground-based sensor techniques to map soil properties, *Photogrammetric Engineering and Remote Sensing*, 69, 619–630, 2003.
5. Kustas, W.P., French, A.N., Hatfield, J.L., Jackson, T.J., Moran, M.S., Rango, A., Ritchie, J.C., and Schmugge, T.J., Remote sensing research in hydrometeorology, *Photogrammetric Engineering and Remote Sensing*, 69, 631–646, 2003.
6. Pinter, P.J., Hatfield, J.L., Schepers, J.S., Barnes, E.M., Moran, M.S., Daughtry, C.S., and Upchurch, D.R., Remote sensing for crop management, *Photogrammetric Engineering and Remote Sensing*, 69, 647–664, 2003.
7. Doraiswamy, P.C., Moulin, S., and Cook, P.W., Crop yield assessment from remote sensing, *Photogrammetric Engineering and Remote Sensing*, 69, 665–674, 2003.
8. Moran, S., Fitzgerald, G., Rango, A., Walthall, C., Barnes, E., Bausch, W., Clarke, T. et al., Sensor development and radiometric correction for agricultural applications, *Photogrammetric Engineering and Remote Sensing*, 69, 705–718, 2003.
9. Hatfield, J.L., Prueger, J.H., and Kustas, W.P., Remote sensing of dryland crops. In: S.L. Ustin (ed.), *Remote Sensing for Natural Resource Management and Environmental Monitoring: Manual of Remote Sensing*, 3rd Edition (vol. 4, pp. 531–568). Hoboken, NJ, USA: Wiley, Chapter 10, 2004.
10. Hatfield, J.L., Gitelson, A.A., Schepers, J.S., and Walthall, C.L., Application of spectral remote sensing for agronomic decisions, *Agronomy Journal*, 100, 117–131, 2008, doi: 10.2134/agronj2006.0370c.

11. Gitelson, A.A., Remote sensing estimation of crop biophysical characteristics at various scales. In: P.S. Thenkabail, J.G. Lyon, and A. Huete (eds), *Hyperspectral Remote Sensing of Vegetation* (vol. 1, pp. 329–358). Boca Raton, FL, USA: Taylor & Francis/CRC Press, Chapter 15, 2011.

12. Gitelson, A.A., Remote estimation of crop fractional vegetation cover: The use of noise equivalent as an indicator of performance of vegetation indices, *International Journal of Remote Sensing*, 34, 1–13, 2013, doi: 10.1080/01431161.2013.793868.

13. Gitelson, A.A., Peng, Y., and Huemmrich, K.F., Relationship between fraction of radiation absorbed by photosynthesizing maize and soybean canopies and NDVI from remotely sensed data taken at close range and from MODIS 250 m resolution data, *Remote Sensing of Environment*, 147, 108–120, 2014.

14. Gitelson, A.A., Peng, Y., Vina, A., Arkebauer, T., and Schepers, J.S., Efficiency of chlorophyll in gross primary productivity: A proof of concept and application in crops, *Journal of Plant Physiology*, 201, 101–110, 2016, http://dx.doi.org/10.1016/j.jplph.2016.05.019

15. Liu, X., Guanter, L., Liu, L., Damm, A., Malenovský, Z., Rascher, U., Peng, D., Du, S., and Gastellu-Etchegorry, J.-P., Downscaling of solar-induced chlorophyll fluorescence from canopy level to photosystem level using a random forest model, *Remote Sensing of Environment*, 2018, https://doi.org/10.1016/j.rse.2018.05.035

16. Peng, Y., Nguy-Robertson, A., Arkebauer, T., and Gitelson, A.A., Assessment of canopy chlorophyll content retrieval in maize and soybean: Implications of hysteresis on the development of generic algorithms, *Remote Sensing*, 9, 226, 2017, doi: 10.3390/rs9030226.

17. Inoue, Y., Guerif, M., Baret, F., Skidmore, A., Gitelson, A., Schlerf, M., and Olioso, A., Simple and robust methods for remote sensing of canopy chlorophyll content: A comparative analysis of hyperspectral data for different types of vegetation, *Plant, Cell and Environment*, 39, 2609–2623, 2016, doi: 10.1111/pce.12815.

18. Ciganda, V.S., Gitelson, A.A., and Schepers, J., How deep does a remote sensor sense? Expression of chlorophyll content in a maize canopy, *Remote Sensing of Environment*, 126, 240–247, 2012.

19. Baret, F., Houlès, V., and Guérif, M., Quantification of plant stress using remote sensing observations and crop models: The case of nitrogen management, *Journal of Experimental Botany*, 58, 869–880, 2007.

20. Yoder, B.J., and Pettigrew-Crosby, R.E., Predicting nitrogen and chlorophyll content and concentrations from reflectance spectra (400–2500 nm) at leaf and canopy scales, *Remote Sensing of Environment*, 53, 199–211, 1995.

21. Oppelt, N., and Mauser, W., Hyperspectral monitoring of physiological parameters of wheat during a vegetation period using AVIS data, *International Journal of Remote Sensing*, 25, 145–159, 2004.

22. Clevers, J.G.P.W., and Kooistra, L., Using hyperspectral remote sensing data for retrieving canopy chlorophyll and nitrogen content, *IEEE Journal of Selected Topics in Applied Earth Observations and Remote Sensing*, 5, 574–583, 2012.

23. Clevers, J.G.P.W., and Gitelson, A.A., Remote estimation of crop and grass chlorophyll and nitrogen content using red-edge bands on Sentinel-2 and -3, *International Journal of Applied Earth Observation and Geoinformation*, 23, 344–351, 2013.

24. Schlemmer, M., Gitelson, A.A., Schepers, J., Ferguson, R., Peng, Y., Shanahan, J., and Rundquist, D.C., Remote estimation of nitrogen and chlorophyll contents in maize at leaf and canopy levels, *International Journal of Applied Earth Observation and Geoinformation*, 25, 47–54, 2013.

25. Inoue, Y., Sakaiya, E., Zhu, Y., and Takahashi, W., Diagnostic mapping of canopy nitrogen content in rice based on hyperspectral measurements, *Remote Sensing of Environment*, 126, 210–221, 2012.

26. Viña, A., Gitelson, A.A., Nguy-Robertson, A.L., and Peng, Y., Comparison of different vegetation indices for the remote assessment of green leaf area index of crops, *Remote Sensing of Environment*, 115, 3468–3478, 2011, doi: 10.1016/j.rse.2011.08.010.

27. Nguy-Robertson, A.L., Peng, Y., Gitelson, A.A., Arkebauer, T.J., Pimstein, A., Herrmann, I., Karnieli, A., Rundquist, D.C., and Bonfil, D.J., Estimating green LAI in four crops: Potential of determining optimal spectral bands for a universal algorithm, *Agricultural and Forest Meteorology*, 192–193, 140–148, 2014.

28. Nguy-Robertson, A.L., Gitelson, A.A., Peng, Y., Viña, A., Arkebauer, T.J., and Rundquist, D.C., Green leaf area index estimation in maize and soybean: Combining vegetation indices to achieve maximal sensitivity, *Agronomy Journal*, 104, 1336–347, 2012.

29. Kira, O., Nguy-Robertson, A.L., Arkebauer, T.J., Linker, R., and Gitelson, A.A., Informative spectral bands for remote green LAI estimation in C3 and C4 crops, *Agricultural and Forest Meteorology*, 218–219, 243–249, 2016, http://dx.doi.org/10.1016/j.agrformet.2015.12.064

30. Cai, W., Yankun, Y., and Shao, X., A variable selection method based on uninformative variable elimination for multivariate calibration ofnear-infrared spectra, *Chemometr. Intell. Lab.*, 90, 188–194, 2008.

31. Centner, V., Massart, D.-L., de Noord, O.E., de Jong, S., Vandeginste, B.M., and Sterna, C., Elimination of uninformative variables for multivariate calibration, *Anal. Chem.*, 68, 3851–3858, 1996, http://dx.doi.org/10.1021/ac960321m

32. Kira, O., Nguy-Robertson, A.L., Arkebauer, T.J., Linker, R., and Gitelson, A.A., Toward generic models for green LAI estimation in maize and soybean: Satellite observations, *Remote Sensing*, 9, 318, 2017, doi: 10.3390/rs9040318.

33. Nguy-Robertson, A.L., and Gitelson, A.A., Algorithms for estimating green leaf area index in C3 and C4 crops for MODIS, Landsat TM/ETM+, MERIS, Sentinel MSI/OLCI, and Venµs sensors, *Remote Sensing Letters*, 6, 5, 360–369, 2015, http://dx.doi.org/10.1080/2150704X.2015.1034888

34. Guindin-Garcia, N., Gitelson, A.A., Arkebauer, T.J., Shanahan, J., and Weiss, A., An evaluation of MODIS 8 and 16 day composite products for monitoring maize green leaf area index, *Agricultural and Forest Meteorology*, 161, 15–25, 2012.

35. Verrelst, J., Rivera, J.P., Gitelson, A., Delegido, J., Moreno, J., and Camps-Valls, G., Spectral band selection for vegetation properties retrieval using Gaussian processes regression, *International Journal of Applied Earth Observation and Geoinformation*, 52, 554–567, 2016, http://dx.doi.org/10.1016/j.jag.2016.07.016

36. Gitelson, A.A., Peng, Y., Arkebauer, T.J., and Schepers, J., Relationships between gross primary production, green LAI, and canopy chlorophyll content in maize: Implications for remote sensing of primary production, *Remote Sensing of Environment*, 144, 65–72, 2014.

37. Peng, Y., and Gitelson, A.A., Application of chlorophyll-related vegetation indices for remote estimation of maize productivity, *Agricultural and Forest Meteorology*, 151, 1267–1276, 2011, doi: 10.1016/j.agrformet.2011.05.005.

38. Peng, Y., and Gitelson, A.A., Remote estimation of gross primary productivity in soybean and maize based on total crop chlorophyll content, *Remote Sensing of Environment*, 117, 440–448, 2012, doi: 10.1016/j.rse.2012.10.021.

39. Rossini, M., Migliavacca, M., Galvagno, M., Meroni, M., Cogliati, S., Cremonese, E., Fava, F. et al., Remote estimation of grassland gross primary production during extreme meteorological seasons, *International Journal of Applied Earth Observation and Geoinformation*, 29, 1–10, 2014.

40. Sakamoto, T., Gitelson, A.A., Wardlow, B.D., Verma, S.B., and Suyker, A.E., Estimating daily gross primary production of maize based only on MODIS WDRVI and shortwave radiation data, *Remote Sensing of Environment*, 115, 3091–3101, 2011.

41. Gitelson, A.A., Peng, Y., Masek, J.G., Rundquist, D.C., Verma, S., Suyker, A., Baker, J.M., Hatfield, J.L., and Meyers, T., Remote estimation of crop gross primary production with Landsat data, *Remote Sensing of Environment*, 121, 404–414, 2012.

42. Peng, Y., Gitelson, A.A., and Sakamoto, T., Remote estimation of gross primary productivity in crops using MODIS 250 m data, *Remote Sensing of Environment*, 128, 186–196, 2013.

43. Gitelson, A.A., Peng, Y., Rundquist, D.C., Suyker, A., and Verma, S.B. 2015. Remote estimation of gross primary productivity in maize and soybean: From close range to satellites. In: J.V. Stafford (ed.), *Precision Agriculture'15* (pp. 183–190), Wageningen, Netherlands: Wageningen Academic Publishers, 2015, http://www.wageningenacademic.com/doi/10.3920/978-90-8686-814-8_22

44. Gitelson, A.A., Verma, S.B., Vina, A., Rundquist, D.C., Keydan, G., Leavitt, B., Arkebauer, T.J., Burba, G.G., and Suyker, A.E., Novel technique for remote estimation of CO2 flux in maize, *Geophysical Research Letters*, 30, 9, 1486, 2003, doi: 10.1029/2002GL016543.

45. Gitelson, A.A., Viña, A., Verma, S.B., Rundquist, D.C., Arkebauer, T.J., Keydan, G., Leavitt, B., Ciganda, V., Burba, G.G., and Suyker, A.E., Relationship between gross primary production and chlorophyll content in crops: Implications for the synoptic monitoring of vegetation productivity, *Journal of Geophysical Research*, 111, D08S11, 2006, doi: 10.1029/2005JD006017.

46. Peng, Y., Gitelson, A.A., Keydan, G.P., Rundquist, D.C., and Moses, W.J., Remote estimation of gross primary production in maize and support for a new paradigm based on total crop chlorophyll content, *Remote Sensing of Environment*, 115, 978–989, 2011, doi: 10.1016/j.rse.2010.12.001.

47. Wu, C., Niu, Z., Tang, Q., Huang, W., Rivard, B., and Feng, J., Remote estimation of gross primary production in wheat using chlorophyll-related vegetation indices, *Agricultural and Forest Meteorology*, 149, 1015–1021, 2009.

48. Sakowska, K., Juszczak, R., and Gianelle, D., Remote sensing of grassland biophysical parameters in the context of the sentinel-2 satellite mission, *Journal of Sensors*, 2016, Article ID 4612809, http://dx.doi.org/10.1155/2016/4612809

49. Harris, A., and Dash, J., The potential of the MERIS terrestrial chlorophyll index for carbon flux estimation, *Remote Sensing of Environment*, 114, 1856–1862, 2010.

50. Harris, A., and Dash, J., A new approach for estimating northern peatland gross primary productivity using a satellite-sensor-derived chlorophyll index, *Journal of Geophysical Research*, 116, G04002, 2011, doi: 10.1029/2011JG001662.

51. Rossini, M., Cogliati, S., Meroni, M., Migliavacca, M., Galvagno, M., Busetto, L., Cremonese, E. et al., Remote sensing-based estimation of gross primary production in a subalpine grassland, *Biogeosciences*, 9, 2565–2584, 2012, doi: 10.5194/bg-9-2565-2012.

52. Rossini, M., Migliavacca, M., Galvagno, M., Meroni, M., Cogliati, S., Cremonese, E., Fava, F. et al., Remote estimation of grassland gross primary production during extreme meteorological seasons, *International Journal of Applied Earth Observation and Geoinformation*, 29, 1–10, 2014.

53. Sakowska, K., Vescovo, L., Marcolla, B., Juszczak, R., Olejnik, J., and Gianelle, D., Monitoring of carbon dioxide fluxes in a subalpine grassland ecosystem of the Italian Alps using a multispectral sensor, *Biogeosciences*, 11, 4695–4712, 2014.

54. Suyker, A.E., and Verma, S.B., Gross primary production and ecosystemrespiration of irrigated and rainfed maize–soybean cropping systems over 8 years, *Agricultural and Forest Meteorology*, 165, 12–24, 2012, http://dx.doi.org/10.1016/j.agrformet.2012.05.021

55. Nguy-Robertson, A., Suyker, A., and Xiao, X., Modeling gross primary production of maize and soybean croplands using light quality, temperature, water stress, and phenology, *Agricultural and Forest Meteorology*, 213, 160–172, 2015, http://dx.doi.org/10.1016/j.agrformet.2015.04.008

56. Gitelson, A.A., Arkebauer, T.J., and Suyker, A.E., Convergence of daily light use efficiency in irrigated and rainfed C3 and C4 crops, *Remote Sensing of Environment*, 217, 30–37, 2018, https://doi.org/10.1016/j.rse.2018.08.007

57. Zhang, Y., Xiao, X., Wolf, S., Wu, J., Wu, X., Gioli, B., Wohlfahrt, G. et al., Spatio-temporal convergence of maximum daily light use efficiency based on radiation absorption by canopy chlorophyll, *Geophysical Research Letters*, 45, 3508–3519, 2018, https://doi.org/10.1029/2017GL076354.

58. Field, C.B., Ecological scaling of carbon gain to stress and resources availability. In: H.A. Mooney, W.E. Winner, and E.J. Pell (eds), *Response of Plants to Multiple Stresses* (pp. 35–65). London, UK: Academic Press, 1991.

59. Goetz, S.J., and Prince, S.D., Modelling terrestrial carbon exchange and storage: Evidence and implications of functional convergence in light-use efficiency. In: A.H. Fitter, and D. Raffaelli (eds), *Advances in Ecological Research* (vol. 28, pp. 57–92) San Diego, CA, USA: Academic Press, 1999.

2 Hyperspectral Assessment of Ecophysiological Functioning for Diagnostics of Crops and Vegetation

Yoshio Inoue, Roshanak Darvishzadeh, and Andrew Skidmore

CONTENTS

2.1 INTRODUCTION

Ecophysiological functioning of vegetation is an essential basis for plant production, food security, and carbon exchange between land surface and the atmosphere. Plant productivity varies to a large extent based on the ecophysiological functioning under changing environment and management. Carbon and water balances in ecosystems are strongly controlled by ecophysiological functioning such as photosynthesis and transpiration (Nobel, 2005).

Accordingly, systematic monitoring, assessment, and prediction of the ecophysiological functioning in various scales are critical for food security and environmental policy making. Geospatial information on crop growth and ecophysiological functioning is useful for site-specific management and smart agriculture that seek to achieve higher crop yield and quality with less labor and agrochemical inputs. Regional or global scale information on plant growth, carbon exchange, and water balance would support policy making for better management of ecosystems and environment (de Leeuw et al., 2010; Vadrevu et al., 2018).

Assessment of photosynthetic functioning is one of the most important bases for the diagnosis and prediction of plant growth, as well as carbon exchange between ecosystems and the atmosphere. A great deal of research effort has been directed toward the assessment of ecophysiological variables and plant productivity using remote sensing in optical, thermal, and microwave wavelength domains together with modeling approaches (e.g., reviews by Moran et al., 1997; Inoue, 2003; Olioso et al., 2005). Remotely sensed optical signatures have proven useful for estimating ecological variables such as leaf area index (LAI) and the absorptivity of photosynthetically active radiation (fAPAR) that affect photosynthetic capacity (Daughtry et al., 1983; Asrar et al., 1989; Huete, 1988). Thermal remote sensing (e.g., brightness temperature) provides critical information on transpiration and/or water stress via energy balance of vegetation (Jackson et al., 1981; Inoue et al., 1990; Moran et al., 1994). Microwave signatures (e.g., backscattering coefficient) can provide structural information of plant canopies such as leaf area index and biomass depending on their frequency and incidence angle (Le Toan et al., 1997; Inoue et al., 2002; Inoue et al., 2014).

Nevertheless, the hyperspectral spectral reflectance in the optical domain is the richest information source for ecophysiological functioning owing to the close relationships of biophysical and biochemical properties of vegetation with the solar radiation (300–3300 nm). The typical seasonal change of hyperspectral reflectance in rice paddy (Figure 2.1) clearly depicts the spectral response to vegetation, water, and soil associated with plant growth. Such relationships are inherent because the biological properties, structures, and functions of vegetation would have been optimized in response to the high-resolution ($\sim$1 nm) spectral change of solar radiation for the survival strategies. Accordingly, hyperspectral reflectance data would provide various ecophysiological and biochemical information although the data richness is not always favorable for predictive purposes.

In this chapter, methodologies for and insights about the use of hyperspectral reflectance data for assessment of ecophysiological functioning are discussed based on a range of case studies with ground-based and airborne hyperspectral measurements.

2.2 LINKAGE OF ECOPHYSIOLOGICAL FUNCTIONING AND BIOPHYSICAL PROPERTIES WITH HYPERSPECTRAL REFLECTANCE

One of the most important functions of vegetation is photosynthesis. The net photosynthetic rate per leaf area (Pn) or net primary production (NPP) is the basis for plant production and crop yield. Photosynthesis is composed of several steps that work together in harmonized manner with high overall efficiency (Nobel, 2005). Pn is determined by (1) the photosynthetically active radiation absorbed by leaf chlorophyll (Chl-a and Chl-b); (2) water and carbon dioxide supplies; and (3) the photochemical and biochemical efficiency for producing carbohydrates. Accordingly, the amount of Chl-a and b provides essential information on the photons available for photosynthesis.

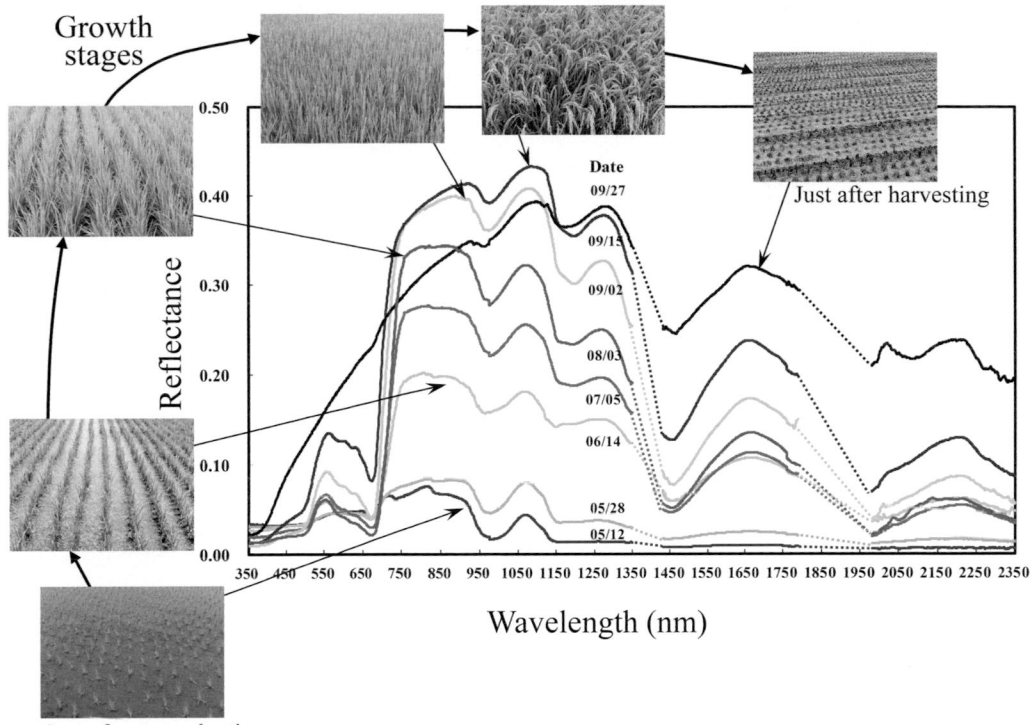

FIGURE 2.1 Typical seasonal change of hyperspectral reflectance in a rice paddy. Date (MM/DD) indicates month and date. Note: Spectral lines indicate the average spectrum of 10–30 measurements over individual canopy surfaces. Number of repetitions depends on the heterogeneity of each surface. This applies to all spectral lines for plant canopies throughout this chapter.

The de-epoxidation state of the xanthophyll cycle also provides physiological information on the photochemical activity related to heat dissipation (Demmig-Adams and Adams, 1996).

The absorptivity of pure Chl a and b has unique spectrum in 400–700 nm wavelength regions; Chl-a and Chl-b have relatively narrow peaks in blue (430 nm and 455 nm) and red (662 nm and 644 nm), respectively (Nobel, 2005). In plant leaves, these peaks shift toward longer wavelengths by 10–30 nm due to the interactions between chlorophyll molecules and the surrounding molecules such as proteins and lipids in the chloroplast membranes as well as the adjacent water molecules. Accordingly, the actual absorption peak of plant leaves is usually 670–680 nm (Figure 2.2).

On the other hand, the photochemical functionality may be related to the spectral change around 531 nm because a small spectral shift is associated with the de-epoxidation state of the xanthophyll cycle (Gamon et al., 1992; Peñuelas et al., 1995). This is the biological basis for photochemical reflectance index (PRI), where the shift of absorption spectra between zeaxanthin and violaxanthin is assumed to be detected by the difference of reflectance at around 531 nm (Figure 2.3).

Water is another essential basis for ecophysiological functioning of vegetation, and has significant influences on the photosynthetic efficiency through stomatal conductance and photochemical processes (Nobel, 2005). Since water has broad light absorption bands around 1450 and 1950 nm, reflectance spectra of leaves is strongly affected by leaf water content (Figure 2.4, Inoue et al., 1993; Ceccato et al., 2002). Accordingly, the water conditions can be detected by broad-band multispectral data by such as NDWI (normalized difference water index) using shortwave infrared bands (Hunt and Rock, 1989). Additionally, several hyperspectral signatures such as R970 nm, first derivative at 1121 nm, and the spectral shift at approximately 2010 nm were also closely related to leaf water status (Inoue et al., 1993). Peñuelas and Inoue (1999) showed that the ratio of reflectance at 970 nm

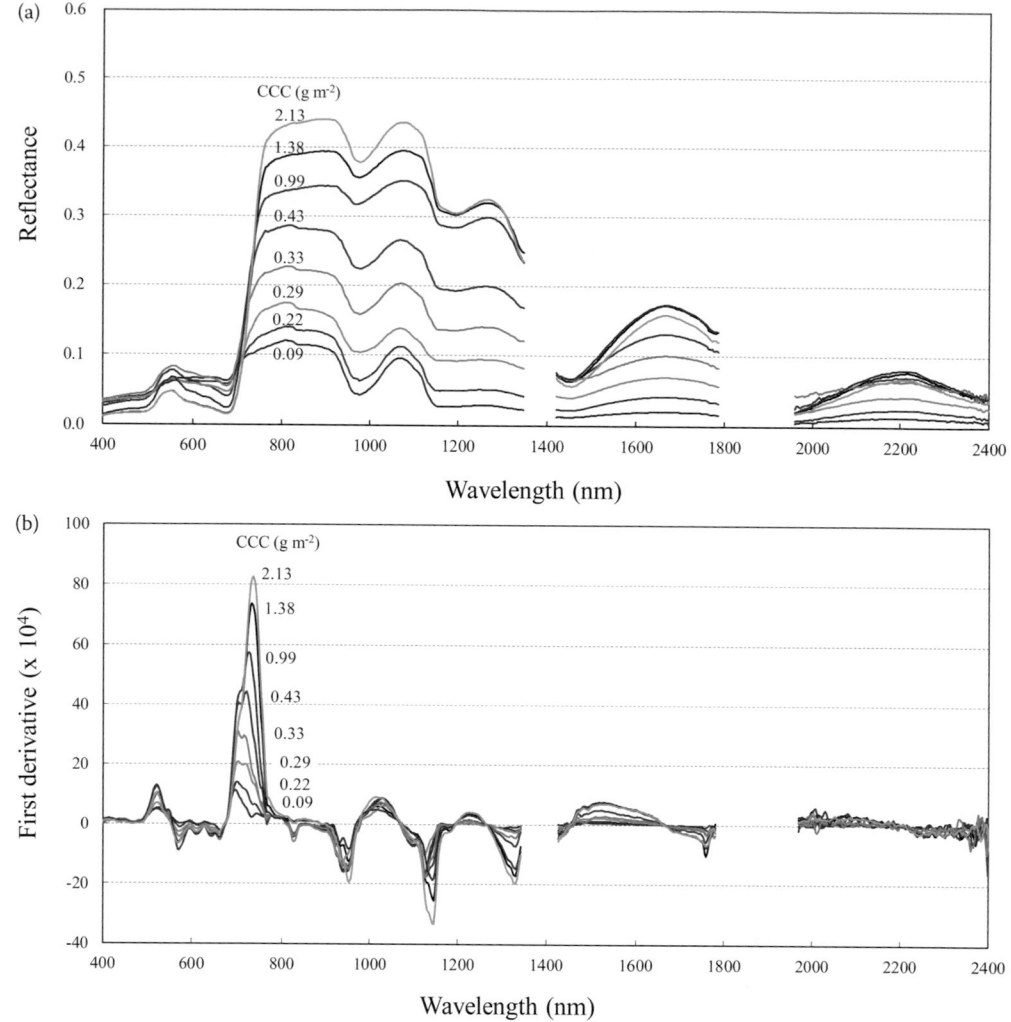

FIGURE 2.2   Typical reflectance spectra (a) and derivative spectra (b) for rice canopies with different canopy chlorophyll content (CCC). Data in two wavelength regions around 1400 and 1900 nm are eliminated because of the low incoming solar energy due to atmospheric water vapor. (After Inoue, Y. et al., 2016. *Plant, Cell and Environment*, 39, 2609–2623.)

(R_{970}) to R_{900} was closely correlated with leaf water condition. These findings would be useful for assessment of water stress of vegetation (Ceccato et al., 2002; Roberto et al., 2012).

Some additional spectral features related to specific pigments and biochemical functioning such as carotenoids, anthocyanin, and fluorescence would also be the basis for hyperspectral remote sensing of ecophysiological functioning (Blackburn, 1998; Gitelson et al., 2001; Grace et al., 2007). A number of researchers suggested that these signatures would be promising, although many more studies would be needed for accurate and robust monitoring of functioning of vegetation (Sims and Gamon, 2002; Ustin et al., 2009).

These unique spectral responses are the promising key for remote sensing of ecophysiological functioning of vegetation based on their inherent relations with the functioning. Hyperspectral measurement is necessary for detection and quantification of these spectral signatures since these narrow peaks can easily be masked or obscured in broad-band data. Especially in canopy or larger scales, subtle spectral changes associated with ecophysiological changes would also be obscured by

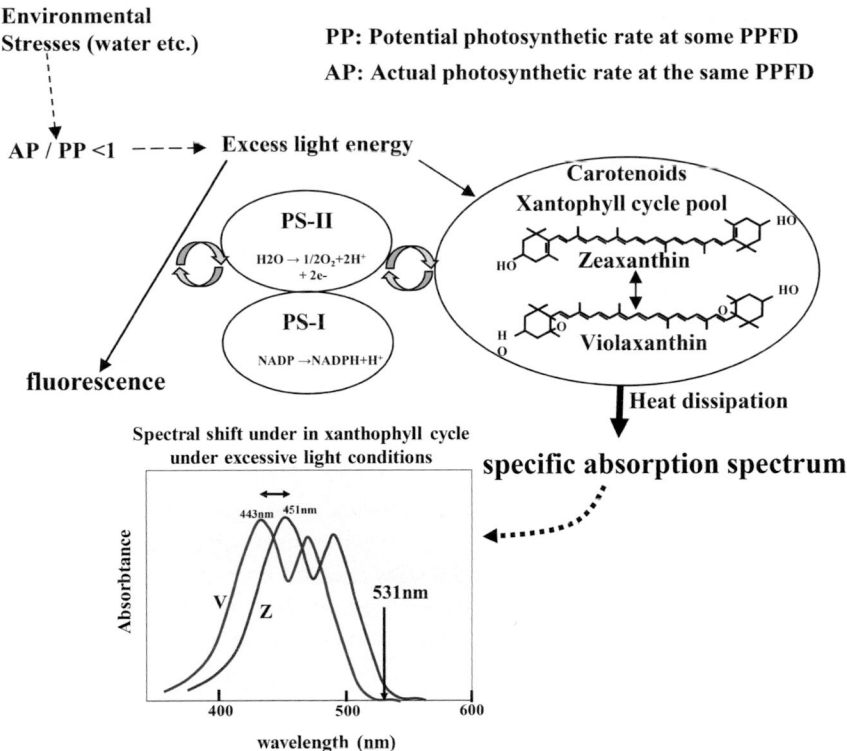

FIGURE 2.3 Schematic representation of spectral change associated with the xanthophyll cycle in response to environmental conditions.

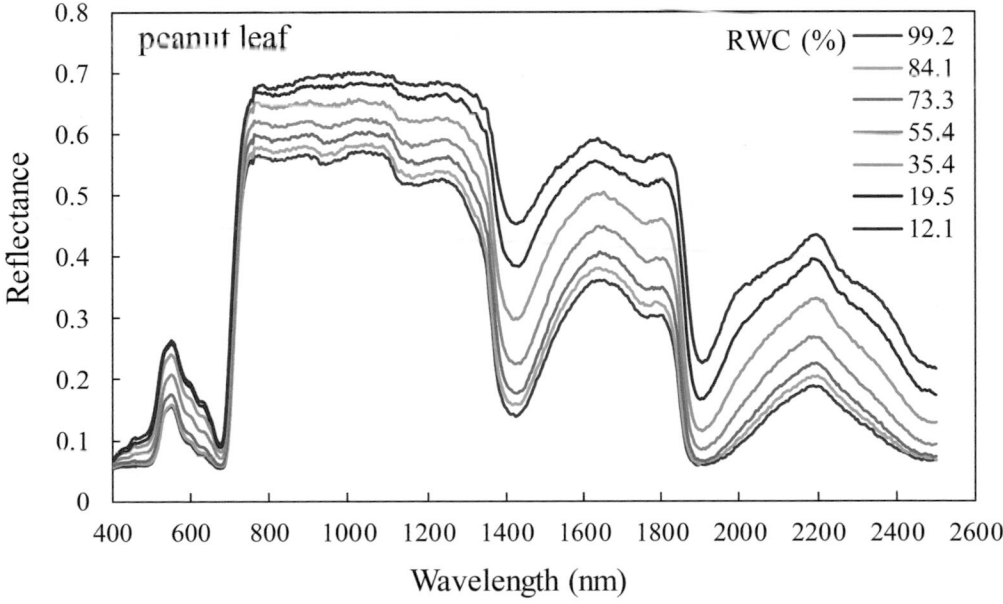

FIGURE 2.4 Typical change of reflectance spectra in response to relative water content (RWC) of a plant leaf. RWC is the ratio of actual water content to full-turgor water content. Each line indicates the single spectrum taken for a leaf with the corresponding RWC during consecutive desiccation.

various confounding factors such as canopy structure and background soils (Ceccato et al., 2002; Dash and Curran, 2004: Garbulsky et al., 2010).

2.3 METHODOLOGIES TO GENERATE PREDICTIVE MODELS FROM HYPERSPECTRAL DATA

It is essential to extract ecophysiological information from remotely sensed data for diagnostics and prediction of vegetation dynamics. A great deal of efforts have been spent to derive useful information from remotely sensed signatures. It is true that multispectral or broadband data can provide useful information on vegetation to some extent. Many vegetation indices have been developed for assessment of individual biophysical variables using multispectral data (Castro-Esau et al., 2006; Cammarano et al., 2014; Karnieli et al., 2013). However, hyperspectral datasets would provide much more opportunities for detection and modeling of ecophysiological, biochemical, and photochemical status due to the inherent linkages of narrow band signatures with ecophysiological functioning (see Section 2.2).

Additionally, hyperspectral data allow a range of advanced analytical methods such as derivative analysis, multivariable analysis, simulation for virtual sensor specification, and inversion of physically based models. Here, major methodologies enabled by hyperspectral measurement are discussed. Although they are semiempirical, data-driven, or physically based approaches, all types of methodologies are based on statistical or empirical modeling or parameterizations to a certain extent. Therefore, appropriate methodologies would have to be chosen for individual scientific or operational purposes.

2.3.1 GENERALIZED SPECTRAL INDEX APPROACH

In earlier stage of remote sensing studies, a great deal of effort was spent to devise various spectral indices for applications of satellite and airborne multispectral sensors to classification or quantification of vegetation and other land surfaces. Accordingly, a number of spectral indices have been proposed with specific names such as NDVI, EVI, SAVI, OSAVI, TSAVI, MCARI, and so forth (see Section 2.5). Generally, these indices are calculated using a few waveband data since the normalization of reflectance values by using multiple wavebands is effective in reducing the influence of errors or uncertainty due to atmospheric and background differences, as well as in enhancing and/or linearizing the spectral response to target vegetation (e.g., Huete, 1988; Gitelson et al., 2005; Qi et al., 2012).

However, the wavelengths and bandwidths for those indices are often specific to sensors used in individual studies, and their naming is also arbitrary. This implies that predictive models are also sensor specific and new calibration datasets are always required for each sensor, which would hamper standardization and require extra labor and cost. In addition, such target-oriented naming is insufficient to represent the multiple capability of hyperspectral indices. Alternatively, the generalized spectral index approach using hyperspectral data would allow more precise and flexible application of knowledge on the relationships between hyperspectral data and ecophysiological variables (see Section 2.2). The most typical formulations and expressions of generalized spectral indices are the normalized difference spectral index (NDSI) and ratio spectral index (RSI). Their definitions of the are given by the following equations:

$$NDSI(x, y) = (y - x)/(x + y) \qquad (2.1)$$

where x and y are reflectance (R_i and R_j) or first derivative (D_i and D_j) values at i and j nm over the whole hyperspectral range. For example, the most frequently used vegetation index, Landsat-NDVI, is equivalent to $NDSI(R_{660}, R_{830})$. Note that R or D should be denoted using the value of the central wavelength in units of nanometers. The derivative processing is effective for enhancing weak spectral features and extracting critical wavelengths by reducing the influence of trends or low-frequency noise (Demetriades-Shah et al., 1990). Similarly, another transformation, RSI, is defined as

$$RSI(x, y) = x/y \qquad (2.2)$$

where both R and D values were also used for x and y.

The contour maps of predictive ability (r^2, RMSE, etc.) for all NDSI or RSI indices using the thorough combinations of two wavebands can provide overview information on the optimal wavebands and bandwidths. The NDSI and RSI formulations are linked to each other but the their response (such as linearity) to a target variable is different. Similar formulations have been employed in many studies to explore useful indices (e.g., Liu et al., 2003; Gitelson and Merzlyak, 1997; Mutanga and Skidmore, 2004; Inoue et al., 2008a,b, 2012b, 2016), but their expressions were not necessarily suitable for standardization, comprehensive comparisons, or wider operational applications.

Additionally, similar approaches can be extended to the various formulations using additional parameters and/or multiple wavebands such as OSAVI (Rondeaux et al. 1996) and TCARI (Daughtry et al. 2000).

$$TCARI = 3\left[(R_{700} - R_{670}) - 0.2(R_{700} - R_{550})\left(\frac{R_{700}}{R_{670}}\right)\right] \qquad (2.3)$$

$$OSAVI = \frac{(1 + 0.16)(R_{800} - R_{670})}{(R_{800} + R_{670} + 0.16)} \qquad (2.4)$$

The use of nonlinear models (e.g., logarithmic, polynomial) would allow better fitting, and so might result in improvement of predictive ability. Nevertheless, such nonlinear fitting or use of increased number of wavebands usually implies the inclusion of a larger number of model parameters and/or less sensitivity in marginal range of the target variable. Accuracy, robustness, and simplicity have to be balanced depending on sensor systems and applications.

2.3.2 MULTIVARIABLE REGRESSION APPROACH

The major advantage of hyperspectra is the large number of wavebands at fine spectral resolution, which are suitable for multivariable statistical analysis as in the laboratory chemometric method (Spiegelman et al., 1998). However, despite the data richness of hyperspectral data, multicollinearity due to the high correlation between wavebands is the inherent problem of multivariable statistical analysis. Multiple linear regression (MLR) is a simple and widely used method in early study stages, but is not suitable for predictive purposes because it is prone to be unstable due to multicollinearity and strongly affected by the number of sample data as well as the number of independent variables (Grossman et al., 1996; Inoue et al., 2008a).

Both principal component regression (PCR) and partial least-squares regression (PLSR) are able to reduce the effects of multicollinearity. The PCR and PLSR methods have similar structures and would be suitable for the analysis of redundant data such as hyperspectra that include a large number of independent variables correlating with each other. Nevertheless, PLSR may be more suitable than PCR for predictive purposes because the principal components in PCR are determined based on the variance of independent variables only, whereas latent variables in PLSR are determined by taking the covariance between the independent and target variables into consideration (Norgaard et al., 2000). PLSR have been used to estimate biochemical variables such as nitrogen content (e.g., Martin et al. 2008). However, according to theoretical and experimental considerations (Spiegelman et al., 1998; Inoue et al., 2008a, 2012b, 2016), the PLSR method may not always provide the best solutions because some of the wavebands are not informative of the target variable or may even disturb useful signatures. Thus, the interval PLSR (iPLSR) may be the most feasible methods for predictive modeling using hyperspectra (Norgaard et al., 2000; Leardi and Norgaard, 2004). In iPLSR, inclusion or exclusion of an individual waveband is determined by its statistical contribution to the root mean squares error (RMSE) in cross validation; that is, an optimal subset of wavebands is explored through the iterative PLSR trials using the thorough subsets of all wavebands so that the RMSE is minimized.

2.3.3 Machine Learning Approach

Machine learning methods such as artificial neural network (ANN) and support vector machine (SVM) would be applied for estimating ecophysiological functioning from hyperspectral data. In general, the ANN consists of interconnected multiple layers for transformation from input data to output data based on optimization of nonlinear models. The SVM is one of the supervised classification algorithms, which constructs a set of kernel feature space of higher dimensionality (hyperplane) for classification of input data by maximizing the distance (margin) between samples. By the introduction of the kernel function, SVM can also be applied to nonlinear problems (Camps-Valls et al., 2006).

These methods require no assumptions on the statistical data distribution and linearity of the relationship between variables (Hansen and Schjoerring 2003; Ali et al. 2015). These approaches have been applied to classification problems using hyperspectral data, but would be applicable to assessment of remote sensing of ecophysiological variables (Del Frate et al. 2003; Durbha et al., 2007; Notarnicola et al., 2008; Ali et al., 2015). They would be useful especially for analysis and/or prediction of biophysical or ecophysiological variables that are influenced by various factors where influences are not formulated explicitly.

However, accuracy and applicability of these predictive models by such data-driven methods are highly dependent on the size and quality of the training datasets (Doktor et al. 2014; Ali et al. 2015). Especially in ecophysiological applications for agricultural and ecosystem problems, acquisition of *in situ* measurements often requires time-consuming, careful, or expensive field work. This would affect the quality and quantity of good training dataset for the application of machine learning methods. In addition, predictive models based on data-driven methods are basically site and sensor specific. Accordingly, we have to be careful about the applicability of such empirical models as well as the accuracy and computational speed (Meroni et al., 2004).

Recent trends in science and technology suggest that these data-driven methods (so-called artificial intelligence, AI) would be powerful tools to solve various problems provided that a large datasets (so-called "big data") are available (Ali et al. 2015). Since the ecophysiological functioning of vegetation is related to many biochemical and photochemical processes, and their relationships are often nonlinear, machine learning methods might provide better predictive ability than statistical and physically based models. It might allow nonexperts to derive moderate predictive models without expertise in a specific field of science. Nevertheless, the necessity of scientific knowledge and expertise should be emphasized in three aspects: (1) optimization of the quantity and quality of training data; (2) appropriate selection of input variables; and (3) evaluation of the soundness and importance of causal relationships or parameterization.

2.3.4 Use of Physically Based Reflectance Models

If a target ecophysiological variable is incorporated explicitly in canopy reflectance models, inversion of such models using hyperspectral measurements may be useful. For example, a physically based reflectance model PROSAIL can simulate the hyperspectral reflectance (1 nm resolution) of a canopy as a function of several ecophysiological variables such as leaf chlorophyll $a + b$ concentration, equivalent water thickness, leaf structural parameter, LAI, and mean leaf inclination angle (LAD: leaf angle distribution) in addition to sun zenith angle and soil reflectance (Jacquemoud et al., 2009;

Bacour et al., 2002). Several approaches such as numerical optimization methods (e.g., Meroni et al., 2004), lookup table approaches (e.g., Combal et al., 2003), and artificial neural networks (e.g., Schlerf and Atzberger, 2006) have been applied for inversion of the model.

The inversion method is attractive because the models describe the underlying mechanisms of canopy reflectance and are basically free from requirements of calibration. However, some inherent or technical constraints such as uncertainty in finding an optimum solution (ill-posed problem), strong need for accurate reflectance, and availability of data for model training or parameterization (e.g., Combal et al., 2003; Darvishzadeh et al., 2008b) may need attention. Nevertheless, physically based models are quite useful for investigating the possible effects of changes in ecophysiological and biophysical status as well as in the measurement configurations to the hyperspectral reflectance of a canopy (Féret et al., 2011). At present, important ecophysiological variables incorporated in physically based models are still limited. It would be useful to incorporate some additional variables such as leaf nitrogen content, light use efficiency and grain protein content as explicit state variables.

2.4 SPECTRAL ASSESSMENT OF PHOTOSYNTHETIC CAPACITY AND EFFICIENCY IN CROPS

The potential photosynthetic capacity in terms of energy is limited by the photosynthetically active radiation (PAR) absorbed by chlorophyll. The close relationship between dry matter production (as NPP) and seasonally integrated APAR was first reported by Shibles and Weber (1966). Accordingly, the use of fAPAR in a simple ecological model (e.g., Monteith, 1977) has been employed in a range of applications for rough assessment of net primary productivity (NPP). In such growth models, leaf area index (LAI) has been used to estimate fAPAR to estimate daily APAR since a close exponential relationship was found in ecological studies (Monsi and Sakei, 1953). Subsequently, due to the inherent close relationship between spectral reflectance (e.g., NDVI) and fAPAR, the APAR in the growth models has been estimated by fAPAR × PAR directly by using remote sensing data (e.g., Potter et al., 1993; Ruimy et al., 1994; Goetz et al., 1999; Turner et al., 2002). However, the actual productivity is affected by the efficiency in use of APAR in ecophysiological processes (light use efficiency: LUE). LUE is controlled by environmental conditions such as temperature and water and nutrient availability via physiological and photochemical processes (e.g., Peñuelas and Filella, 1998).

$$NPP = LUE \times fAPAR \times PAR \qquad (2.5)$$

This simple model has been used in remote sensing of plant productivity based on semiempirical relationships of spectral indices with fAPAR (capacity) and LUE (efficiency). Accordingly, accurate and robust assessment of these variables is the basis for better understanding and modeling of photosynthetic productivity. Nevertheless, the structure and determinants of both fAPAR and LUE would have to be investigated in more detail in the aspects of ecophysiological and photochemical processes.

2.4.1 PHOTOSYNTHETIC CAPACITY

The fAPAR has been estimated from spectral indices such as NDVI (e.g., Kumar and Monteith, 1982; Asrar et al., 1989; Baret and Guyot, 1991) and applied to production models (e.g., Ruimy et al., 1994; Nouvellon et al., 2000). The relationship is little affected by pixel heterogeneity, LAI, or variation in leaf orientation and optical properties (e.g., Pinter et al., 1985). In fact, the NDVI-fAPAR relationship is rather robust for different crops during the vegetative growth stages (Figure 2.5). However, the relationship is affected by background and bidirectional effects (Myneni and Williams, 1994) and by phenological stages and senescence. The NDVI–fAPAR relationship is largely different between

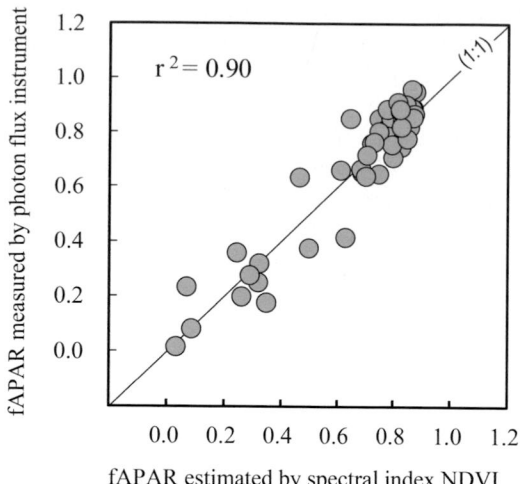

FIGURE 2.5 Comparison of fAPAR (fraction of absorbed photosynthetically active radiation) values estimated by spectral index NDVI and *in situ* measurements during vegetative growth stages in three crops (rice, soybean and corn). (After Inoue, Y. and Olioso, A. 2006. *Journal of Geophysical Research*, 111(D24), D24S91, doi:10.1029/2006JD007469.)

before and after heading (anthesis) in most crops (e.g., Daughtry et al., 1983; Asrar et al., 1989; Inoue and Iwasaki, 1991; Inoue et al., 1998; Choudhury, 2000).

An analysis by Inoue et al. (2008b) clearly showed the limitation of NDVI based on generalized spectral index approach using hyperspectral data and *in situ* fAPAR measurements during the entire growth season (Figure 2.6). According to the NDSI-map, that is, a contour map of predictive ability, the NDVI with Landsat bands was not the best predictor of fAPAR due to the distinct difference before and after the heading stage. Alternatively, the NDSI(R_{720}, R_{420}) had the high and consistent predictive ability throughout entire growth stages. This analysis was based on a dataset for rice, but the results would be common to the other crops in Figure 2.5.

Note that the total photosynthetically active radiation absorbed by a plant canopy (APARt) is the maximum energy source for photosynthesis used in the simple growth models (e.g., Monteith, 1977). However, the APARt is not always used fully for photosynthesis because of the absorption by nonphotosynthetic or senescent plant elements. Accordingly, the assessment of fAPARc, that is, the part of APARt absorbed by chlorophyll and used for photosynthesis (APARc) should be another important target of remote sensing. In other words, the traditional LUE represents an overall efficiency that integrates the biophysical efficiency in photon absorption by a canopy and the photochemical efficiency (LUEc) in use of absorbed photons. Separate estimation of fAPARc and LUEc would be more suitable to assess the dynamic change of photosynthetic activity of a canopy.

Another variable related to the photosynthetic capacity is P_{max} which represents the maximum photosynthetic rate at the saturating APAR. P_{max} was found to be moderately correlated with RVI using red and near-infrared bands, GRI (green ratio index) using green and red ands, and MSAVI (modified soil adjusted vegetation index; Qi et al., 1994), and four new indices, NDSI(R_{518}, R_{676}), NDSI(R_{620}, R_{623}), NDSI(R_{620}, R_{637}), and NDSI(R_{750}, R_{761}) ($r^2 = 0.65$–0.66). In principle, P_{max} may not be highly correlated with actual P_n since P_{max} is the potential photosynthesis at the highest incident radiation without any stresses. Nevertheless, it is interesting that these wavelengths are all within the wavelength region of chlorophyll a and b absorption. These results suggest that conventional indices using red and near-infrared wavelength regions (e.g., RVI, GRI, and MSAVI) could be useful for a rough assessment of P_{max}.

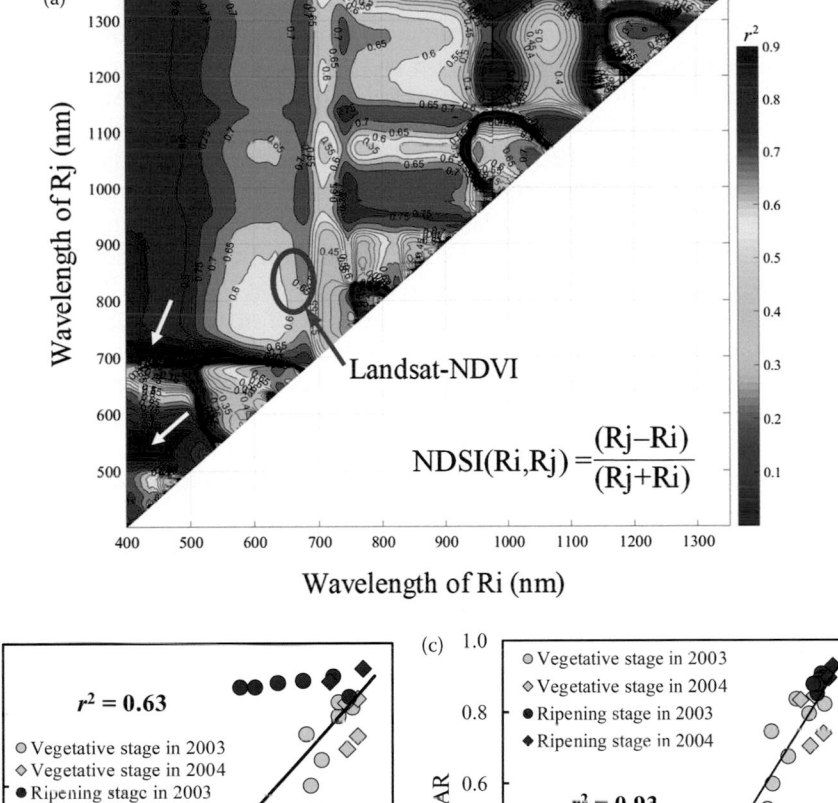

FIGURE 2.6 NDSI map of predictive ability for fAPAR in rice canopies (a) and relationships of spectral indices with the *in situ* measurements of fAPAR with (b) NDVI and (c) NDSI(R_{720}, R_{420}). The white arrows indicate the most significant spots. (After Inoue, Y. et al., 2008b. *Remote Sensing of Environment*, 112, 156–172.)

2.4.2 PHOTOSYNTHETIC EFFICIENCY

The light use efficiency LUE may be assumed to be constant under nonstressed conditions, but it is affected by phenological stages and by abiotic or biotic stresses via ecophysiological processes (e.g., Kiniry et al., 1989; Sinclair, 1994; Peñuelas and Filella, 1998; Choudhury, 2000). Hence, it would be inappropriate to assume constant LUE, especially for assessment of canopy carbon exchange at short temporal resolution (e.g., daily or shorter).

The photochemical reflectance index PRI has been found to be closely related to photosynthetic radiation use efficiency of plant leaves (Gamon et al., 1992; Peñuelas et al., 1995).

$$PRI = (R_{531} - R_{570})/(R_{531} + R_{570}) \qquad (2.6)$$

The relationship can be explained based on the photochemical reactions (see Section 2.2), and has been investigated in a wide range of vegetation, sensors, and scales (Garbulsky et al., 2010). Investigated vegetations ranged from uniform crops to heterogenous forest, and sensors also ranged from ground-based or airborne hyperspectral sensors (e.g., ASD, CASI, AVIRIS) to coarse satellite sensors (e.g., MODIS). The LUE or related ecophysiological variables were measured by photosynthetic instruments (IRGA, PAM) for leaves, estimated directly by eddy covariance methods (flux tower) for ecosystems, or estimated indirectly from the other measurements for larger scales. The available evidence indicates that the PRI may be a useful indicator of ecophysiological variables closely related to the photosynthetic efficiency at the leaf and canopy levels over a wide range of species (Garbulsky et al., 2010). However, a consistent and robust PRI–LUE relationship is not established for several reasons, such as uncertainty of *in situ* measurements of LUE and/ or various confounding factors (Garbulsky et al., 2010). Apparent close relationships obtained for broadband and coarse spatial resolution sensors might be indirect and attributable to the linkage of LUE with other biophysical variables. Accordingly, the generalized application of PRI, especially to satellite sensors, may be questionable. At least, hyperspectral or narrow-band reflectance data as well as the reliable *in situ* measurement of LUE would be needed for precise analysis of PRI–LUE relationships.

Generally, actual LUE values have been estimated by photosynthetic instruments for leaves or by EC (eddy covariance) method for ecosystems. Nevertheless, the spatial representativeness of LUE for the vegetated area is a difficult issue especially in heterogeneous ecosystems. Moreover, the matching of image pixels and footprint of EC is critical for precise analysis, especially in the analysis of satellite images. Thus, precise datasets of hyperspectral data and concurrent LUE data for a uniform crop vegetation with fewer confounding factors would be useful to create definitive evidence for remote sensing of photosynthetic efficiency. Here, the applicability of PRI and other potential spectral indicators for remote sensing of LUE are discussed based on the studies on crops.

2.4.2.1   Leaf Scale Assessment Based on Hyperspectral and Photosynthetic Measurements

Light use efficiency or photosynthetic parameters may be accurately determined by the light (PPFD: photosynthetically active photon flux density) vs. net photosynthesis (P_n μmolCO$_2$ m^{-2} s^{-1}) curve, which is expressed by the following equation:

$$P_n = P_s(1-\exp[-\varphi \cdot \text{PPFD}]) - R \tag{2.7}$$

where P_n is the net photosynthetic rate; P_s is the saturated photosynthetic rate; φ is the quantum use efficiency; and R is the dark respiration. LUE (μmolCO$_2$ μmolPhotons^{-1}) is calculated as the ratio of P_n to PPFD. Figure 2.7 shows some examples of PPFD–P_n curves for soybean leaves (*Glycine max* L. Merr.) and concurrent reflectance spectra of leaves under a wide range of PPFD, leaf chlorophyll index (CI), and soil water content (θ) (Inoue and Peñuelas, 2006). This dataset allowed analysis of the relationships of photosynthetic parameters (P_n, P_s, φ, and LUE) with spectral indices (e.g., NDVI, WI, GRI, SIPI, NPQI, SWWI, and WI/NDVI), where the water index was WI = R_{900}/R_{970}, green ratio index GRI = R_{830}/R_{550}, structural independent pigment index SIPI = $(R_{800} - R_{445})/(R_{800} + R_{680})$, normalized phaeophytinization index NPQI = $(R_{415} - R_{435})/(R_{415} + R_{435})$, shortwave infrared water index SWWI = R_{800}/R_{1650}, and WI/NDVI (Inoue et al., 1993; Peñuelas and Filella, 1998; Peñuelas and Inoue, 1999). These spectral indices were assumed to have close or moderate correlations with photosynthetic parameters, but detailed correlation analysis among all variables revealed that PRI–LUE was the only highly significant relationship for inferring the photochemical status directly from the reflectance spectra (Figure 2.8). The PRI–LUE relationship was significantly close under each level of soil water contents, although its slope was strongly affected by soil moisture (water stress). In this case, a consistent PRI–LUE relationship under changing soil water conditions was derived successfully by normalizing the influence of soil water content on the PRI–LUE relationship. Contrarily, chlorophyll concentration had no significant effect on LUE as well as the PRI–LUE

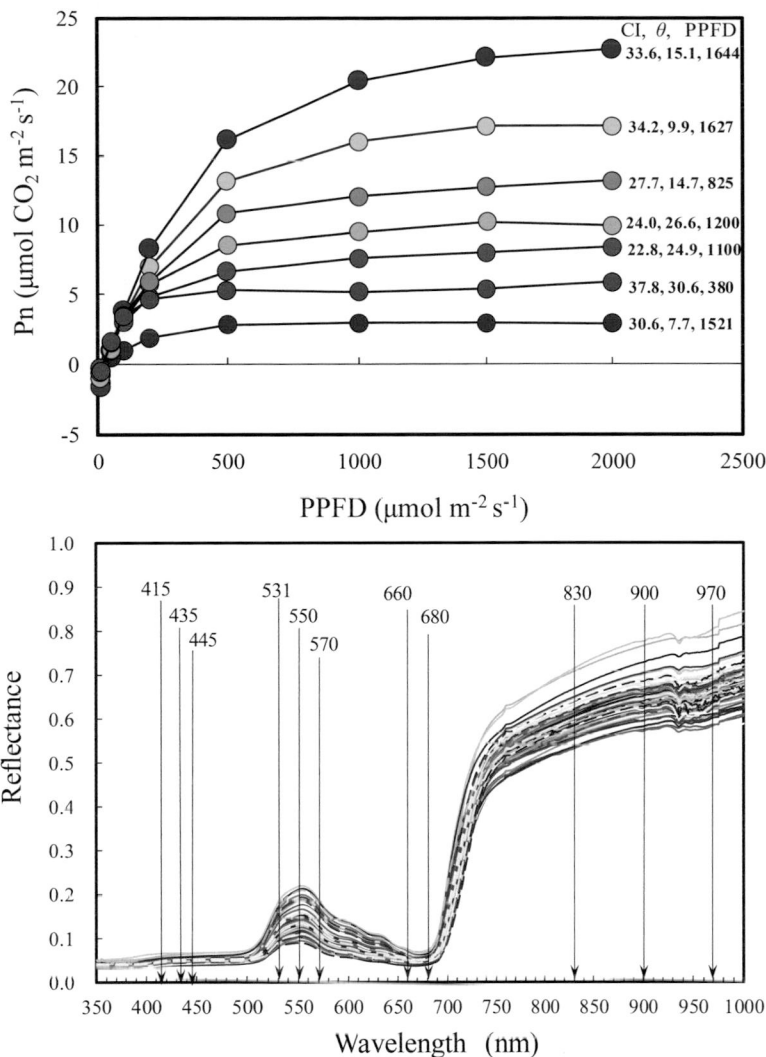

FIGURE 2.7 Light–photosynthesis curves of soybean leaves with different chlorophyll content (CI), soil water content (θ), and photosynthetically active radiation (PPFD) (upper), and spectral reflectance of the leaves (lower). The arrows labeled with wavelength are main spectral wavebands selected based on previous studies on spectral characteristics of plant leaves. Each curve indicates the average of 10 repeated measurements (After Inoue, Y. and Peñuelas, J. 2006. *International Journal of Remote Sensing*, 27, 5249–5254.)

relationship, while the range of CI was large. This may be because both LUE and PRI indicate the photosynthetic functioning of chloroplasts, although the CI simply indicates the density of chloroplast per unit leaf area. These results suggest that PRI would be able to detect relatively rapid photochemical changes caused by PPFD changes, whereas the effects of slower changes such as those in soil water content and chlorophyll content would have to be incorporated differently in predictive modeling (Inoue and Peñuelas, 2006).

The radiation use efficiency (RUE gDM MJ^{-1}) is usually defined as the ratio of dry matter increment to the integrated APAR for a day or longer term (Shibles and Weber, 1966). Accordingly, RUE is more robust and representative of plant productivity at a canopy scale than is LUE (μmolCO$_2$ μmolPhotons^{-1}) which is a dynamic measure of photosynthetic functioning at leaf scale. It would be

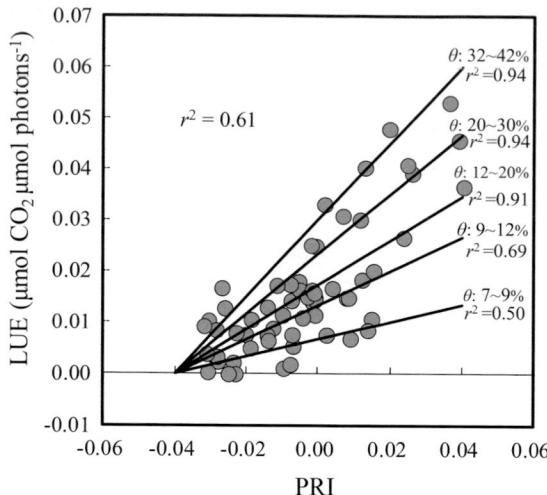

FIGURE 2.8 Relationship between light use efficiency (LUE) and photochemical reflectance index (PRI) as affected by soil water content θ. Regression lines were determined for data points grouped by soil water content; 7%–9%, 9%–12%, 12%–20%, 20%–30%, 32%–42%. (After Inoue, Y. and Peñuelas, J. 2006. *International Journal of Remote Sensing*, 27, 5249–5254.)

reasonable to utilize the LUE as an indicator for leaf-scale diagnosis of photosynthetic functioning, whereas RUE would be suitable for assessment of integrated efficiency at a day or longer temporal resolutions. Scaling from leaf to canopy is an important but difficult issue especially for this type of narrow-band indices due to various confounding factors (Barton and North, 2001). Therefore, both temporal and spatial resolutions have to be carefully considered for spectral assessment of LUE and RUE.

2.4.2.2 Canopy Scale Assessment Based on Hyperspectral and Eddy Covariance Measurements

During the past decade, LUE values estimated from the CO_2 flux and environmental data at flux towers have been used for the analysis of the PRI–LUE relationship (Garbulsky et al., 2010). However, most flux towers were installed in natural or seminatural ecosystems such as grassland, rangeland, and forest, where the spatial and species heterogeneity is large. In addition, it is difficult to acquire precise *in situ* hyperspectral reflectance data in such ecosystems. Therefore, seasonal datasets of the precise measurements of hyperspectral reflectance data and flux data acquired in uniform crop ecosystems such as rice fields would be useful for detailed ecophysiological and remote sensing analysis. The experimental study by Inoue et al. (2008b) may be the first such analysis.

In their analysis, the CO_2 flux data at intervals of 30 min and micrometeorological data as well as the hyperspectral reflectance data for the entire growth stages were used (Figures 2.9 and 2.10). The net ecosystem exchange of CO_2 (NEE_{CO2}) was low for several weeks after transplantation, but CO_2 uptake at midday became obvious even at the low LAI. The midday peak value of NEE_{CO2} increased with increasing LAI and decreased rapidly during the late ripening period. The photosynthetic rate was much lower during the ripening stage than during the early vegetative stage, with similar or larger values of green LAI. Generally, the combination of a footprint model (Kljun et al., 2004) with micrometeorological data and remotely sensed images is useful to confirm the spatial representativeness of the flux data for the ecosystem (Inoue and Olioso, 2006).

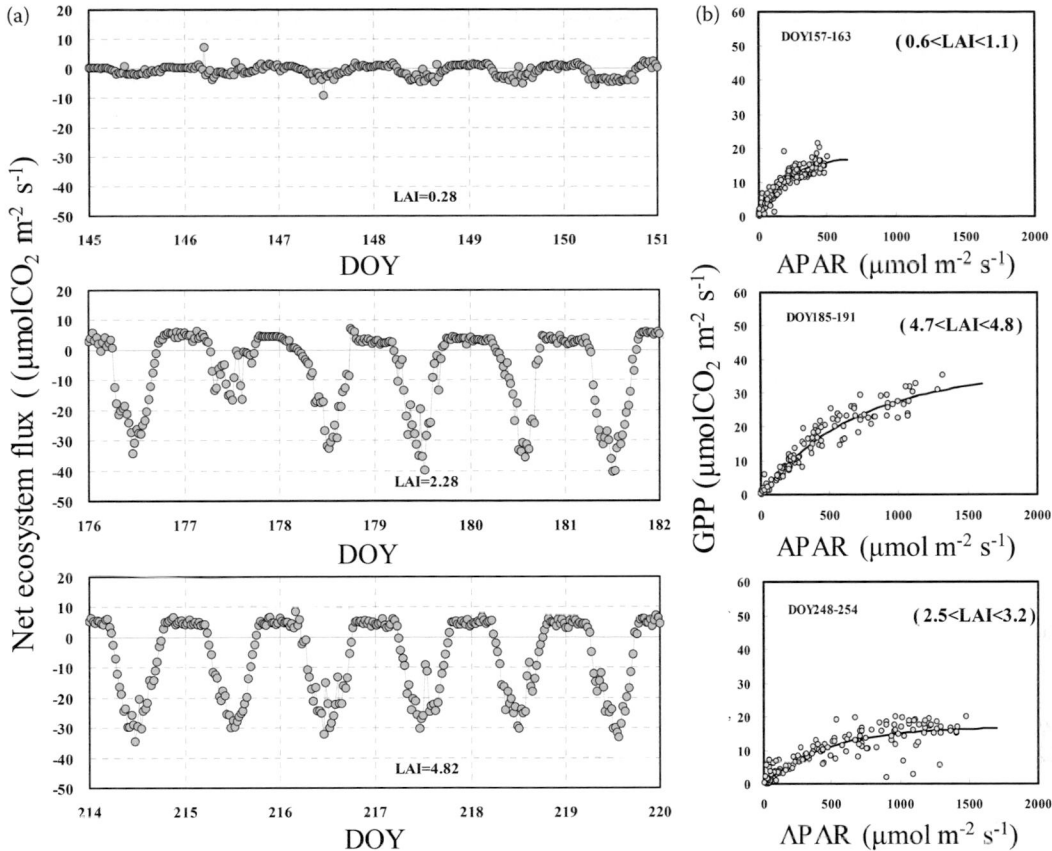

FIGURE 2.9 Seasonal change of net ecosystem CO_2 exchange (NEE_{CO2}) in a rice field measured by eddy covariance method (a), and APAR–GPP relationship (b) during early (upper), middle (middle), and late (lower) growth stages. (After Inoue, Y. et al., 2008b. *Remote Sensing of Environment*, 112, 156–172.)

Photosynthetic efficiency parameters can be determined from these datasets by using the asymptotic exponential equation as follows:

$$GPP = P_{max} (1 - exp[-\varphi \cdot APAR/P_{max}]) \qquad (2.8)$$

where GPP is the gross primary productivity, φ is the initial slope of the curve (quantum efficiency), and P_{max} is the maximum photosynthetic rate at the saturating APAR (Figure 2.9b). Two indicators of canopy light use efficiency at the time of remote sensing measurements were estimated from half-hourly data of CO_2 and photon flux data.

$$LUE_N = NEE_{CO2}/APAR \qquad (2.9)$$

$$LUE_G = GPP/APAR \qquad (2.10)$$

The generalized spectral index approach was applied to explore new spectral indices and to evaluate the predictive ability of previous indices. Figure 2.11 shows the NDSI map of predictive ability of thorough combination of two wavebands for assessment of LUE_G. The map can provide an overview of the statistical significance of all NDSIs for selecting the optimal central wavelength and bandwidth. There are several narrow peaks (reddish) and deep troughs (bluish). Some steep walls

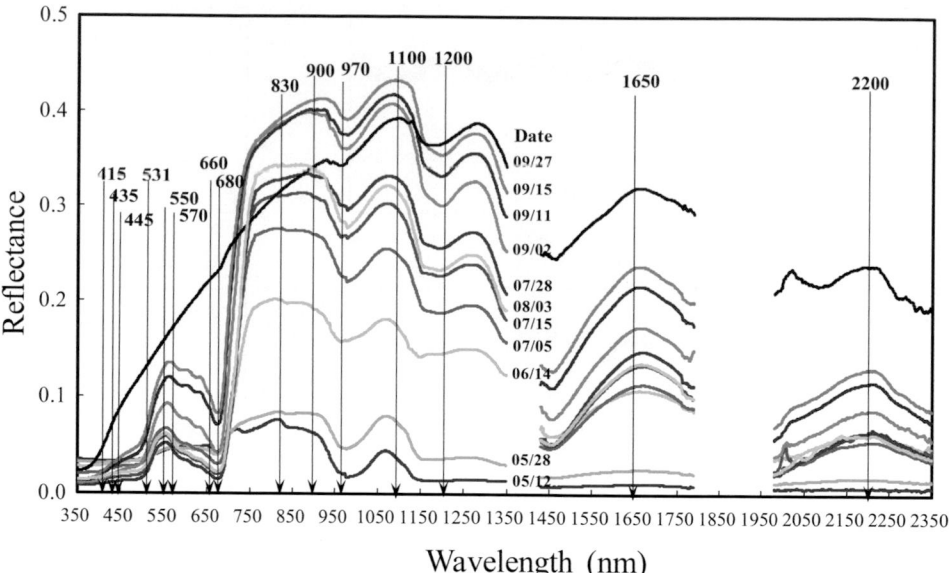

FIGURE 2.10 Seasonal change of reflectance spectra in a rice canopy associated with the growth shown in Figure 2.8. Date numbers indicate month/day. Wavelengths indicated with arrows were used to derive conventional spectral indices. (After Inoue, Y. et al., 2008b. *Remote Sensing of Environment*, 112, 156–172.)

are also found (e.g., around the 530 vs. 560 nm region). Therefore, selection of effective wavelengths should be made carefully so as not to span such steep-wall regions, since even a small bias could largely reduce the potential significance.

The LUE_G was best correlated with $NDSI(R_{710}, R_{410})$, $NDSI(R_{710}, R_{520})$, and $NDSI(R_{530}, R_{550})$ derived from nadir measurements, while most conventional indices were less well correlated. R_{550}, R_{710}, and R_{1122} were selected for multiple indices presumably because they have some specific role in ecophysiological processes that directly or indirectly affect photosynthetic efficiency at a canopy scale. R_{710} is positioned around the red-edge wavelengths, which represent the amount of chlorophyll. Therefore, R_{710} may have a role in normalizing the amount of vegetation when it is used with other visible wavelengths such as 410 or 530 nm. The performance of PRI was surprisingly poor in this analysis, but R_{530} may have significant role in detection of photosynthetic rate because $NDSI(R_{530}, R_{550})$ was highly correlated with LUE_G. The use of R_{550} nm instead of R_{570} nm in PRI as a reference wavelength may significantly improve the estimation of photosynthetic efficiency. R_{1122} may be related to the amount of canopy nitrogen because 1120 nm is the wavelength of lignin absorption (Curran, 1989), and lignin/cellulose content is inversely related to nitrogen content during the growing season.

The quantum efficiency φ was best correlated with $NDSI(R_{450}, R_{1330})$. It is interesting that $NDSI(R_{403}, R_{830})$ and $NDSI(R_{420}, R_{970})$ using very far wavelengths were also highly correlated with φ, whereas some other NDSIs use nearby wavelengths such as $NDSI(R_{933}, R_{940})$, $NDSI(R_{933}, R_{948})$, and $NDSI(R_{1053}, R_{1058})$. The bandwidth is relatively broad for the former NDSIs (5–80 nm) and narrow for the latter NDSIs (3 nm). The WI using R_{970}, a weak absorption peak of water, proved useful for detecting leaf water status (Peñuelas and Inoue, 1999). This may explain the high correlations of WI with φ and LUE_G, since the leaf water status strongly affects photosynthetic efficiency. Although no previous literature suggested the role of the 930–950 nm region, it may have potential for estimating φ.

The results described were obtained using nadir spectral data, but off-nadir measurements would have additional potential since the photosynthetic efficiency is related the photochemical processes or photosynthetic rate per unit leaf area and oblique reflectance measurement would allow acquisition

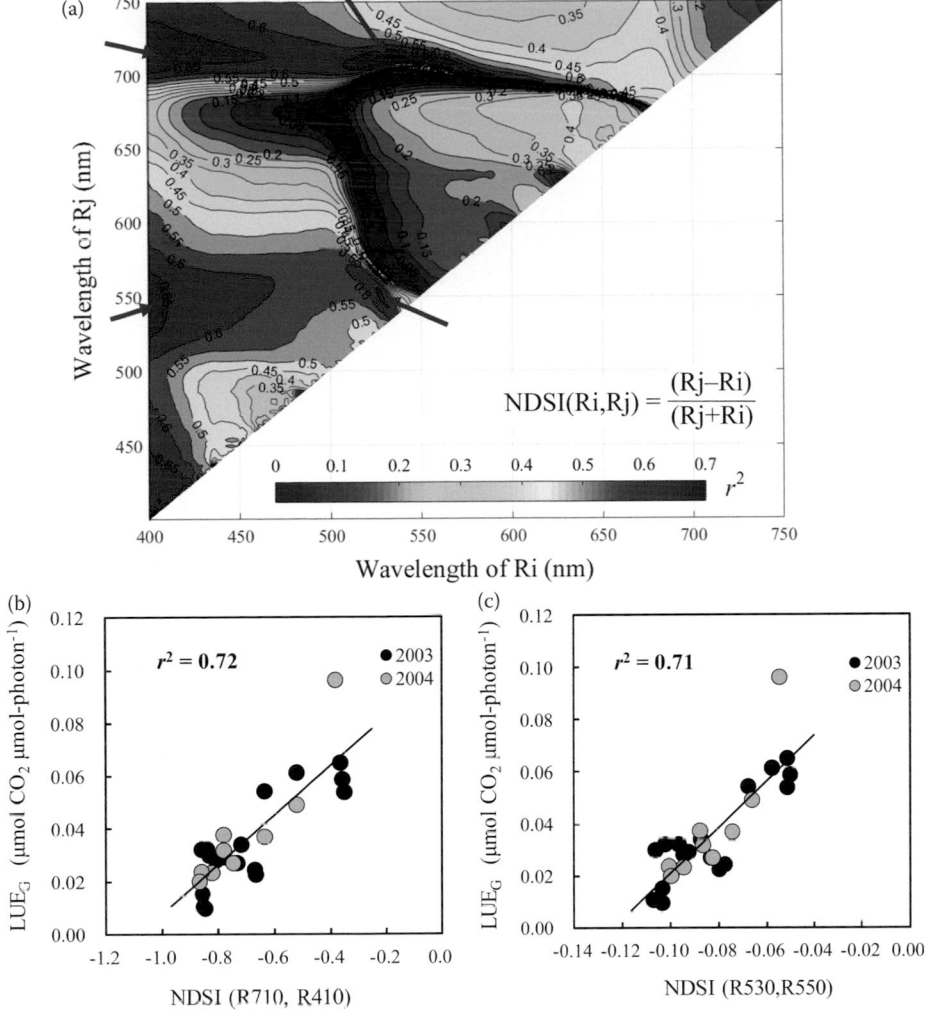

FIGURE 2.11 NDSI map of predictive ability for LUE_G in rice canopies (a) and relationships of promising spectral indices with the LUE_G (b,c). The arrows in the NDSI map indicate most significant positions. (After Inoue, Y. et al., 2008b. *Remote Sensing of Environment*, 112, 156–172.)

of spectral data of leaves only. Recent advances in drone (unmanned aerial vehicle: UAV) platforms enable flexible measuring configurations including off-nadir observation (Verhoeven, 2011; Roosjen et al., 2017). The analysis using off-nadir measurements suggested promising capability of new spectral indices for assessment of LUE_G or LUE_N (Figure 2.12). Overall, off-nadir measurements were more closely related to the efficiency parameters than nadir measurements. Figure 2.12 suggested the region consisting of 400–440 nm vs. 400–500 nm had rather high predictive ability, while there is a deep trough around the 410–530 nm vs. 500–580 nm region. It is interesting that the 400–420 nm regions showed a high correlation ridge along R_{422}. Figure 2.13 shows a close-up view of two significant parts in the contour map of r^2 between φ and NDSIs derived from off-nadir measurements. A ridge of high r^2 was found along 413 nm, while the other regions showed poor correlations. It is noteworthy that R_{435} included in the NQPI proved to be significant in assessment of LUE. In contrast, relatively large regions had high r^2 values around the 1000–1120 nm vs. 1100–1130 nm and the 1130–1150 nm vs. 1160–1200 nm regions (Figure 2.13b).

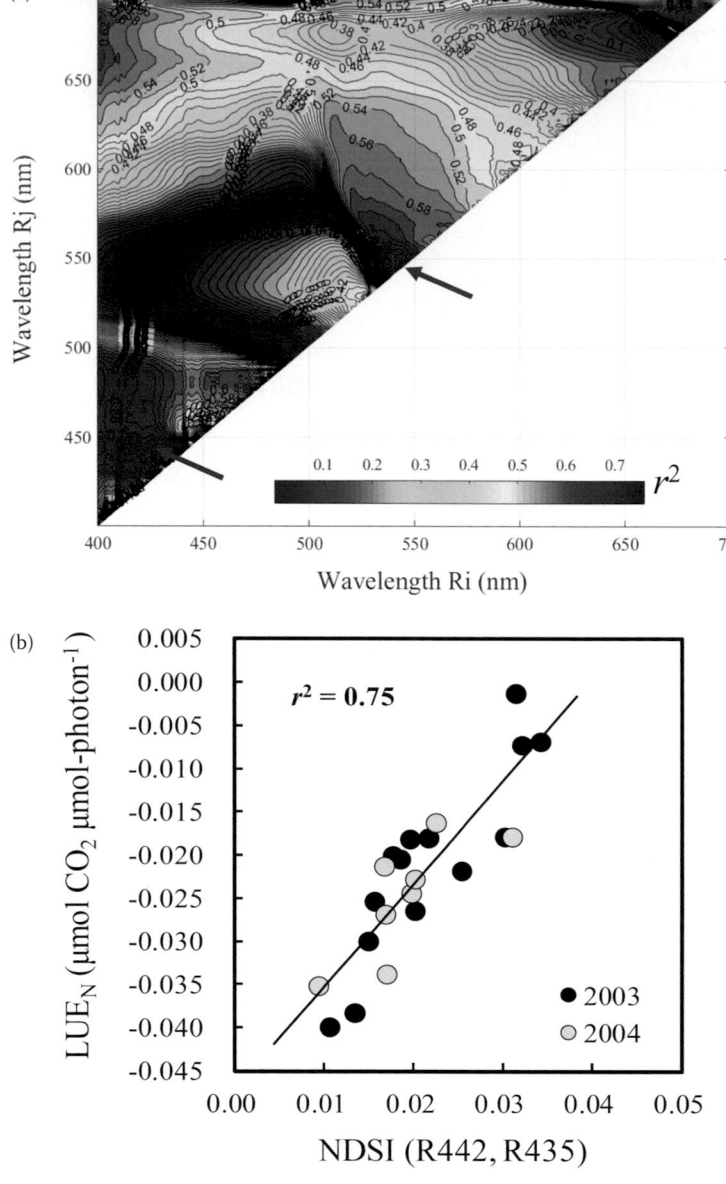

FIGURE 2.12 NDSI map of predictive ability for LUE_N in rice canopies (a) and relationships of promising spectral indices with the LUE_N (b). The arrows in the NDSI map indicate most significant positions. This result was derived from off-nadir observations. (After Inoue, Y. et al., 2008b. *Remote Sensing of Environment*, 112, 156–172.)

These experimental results would provide useful insights for the assessment of radiation use efficiency and photosynthetic capacity in other ecosystems. Nevertheless, the underlying ecophysiological mechanisms for the predictive ability of spectral indices are not well identified at the moment, except for the well-known absorption features of major pigments and materials such as chlorophyll, xanthophyll, carotenoids, and water (Gamon et al., 1992; Inoue et al., 1993; Peñuelas and Filella, 1998, Daughtry et al., 2000). Further studies are needed to examine their applicability to wider range of ecosystems and to investigate the ecophysiological and photochemical mechanisms to systematically account for the effects of biotic and abiotic stresses.

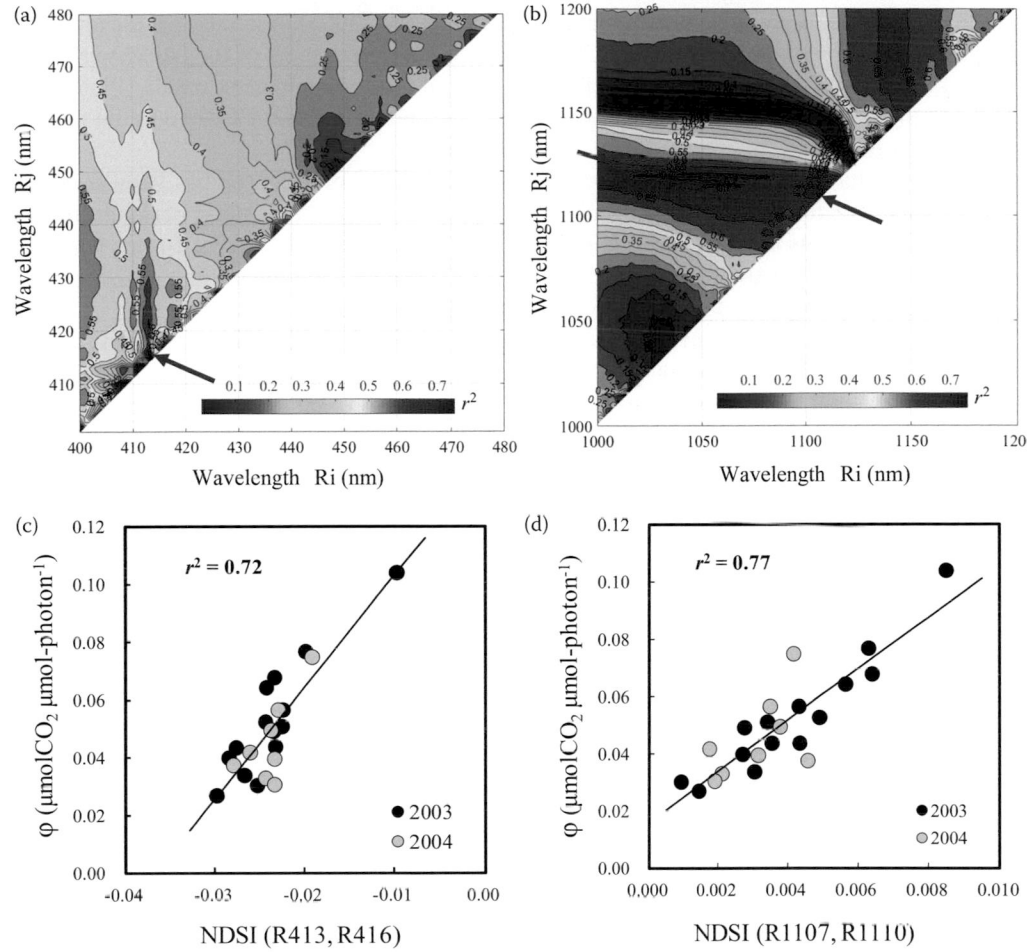

FIGURE 2.13 NDSI maps of predictive ability for quantum efficiency (φ) in rice canopies (a,b) and relationships of promising spectral indices with the φ (c,d). The arrows in NDSI-maps indicate most significant positions. These results were derived from off-nadir observations. (After Inoue, Y. et al., 2008b. *Remote Sensing of Environment*, 112, 156–172.)

2.5 SPECTRAL ASSESSMENT OF DIAGNOSTIC INFORMATION FOR CROP MANAGEMENT

2.5.1 CHLOROPHYLL CONTENT FOR GROWTH MONITORING

The canopy-scale photosynthetic capacity is determined to a large extent by the absorptivity of PAR by chlorophyll (fAPARc). The fAPARc is determined by the density and spatial distribution of chlorophyll pigments within the canopy. The 3D distribution of chlorophyll within a canopy is often expressed by using the area of green leaves (LAI), leaf chlorophyll content (LCC) and leaf angle distribution (LAD). In a wide range of canopy photosynthetic models, the potential photosynthetic rate per leaf area (P_n) is used with LAI and LAD to express the 3D distribution of photosynthesis (De Pury and Farquhar, 1997; Roy et al., 2001). In simple models, the LAI has been used to represent the photosynthetic capacity of a canopy with a constant value of LUE by assuming that the effects of variability in LCC, LAD, and clumping are negligible (Monteith, 1977; De Pury and Farquhar, 1997; Choudhury, 2000). However, determination of green LAI is not simple or is rather arbitrary because the green and nongreen threshold is unclear. In addition, the potential P_n is strongly controlled by the chlorophyll ($a + b$) concentration per unit leaf area LCC (Nobel, 2005). Accordingly, the amount of chlorophyll content per unit land surface (CCC) is the major determinant of the fAPARc although the spatial distribution of chlorophyll would affect the light absorption to some extent (Pinter et al., 1985; Jacquemoud et al., 2009). Thus, accurate determination of the total CCC would be the most essential and robust basis for ecophysiological modeling.

On the other hand, the chlorophyll content or greenness of crop leaves has been used as a diagnostic indicator for fertilizer management owing to the close relationship between nitrogen and chlorophyll contents in green leaves (Houlès et al., 2007; Inoue et al., 2012b). In rice leaves, for example, 75%–85% of the total nitrogen is included in chloroplast throughout the growing period although a part of nitrogen is allocated into nongreen parts such as grain (Morita, 1978).

Therefore, the accurate and timely assessment of ecosystem CCC by remote sensing is vital for a wide range of agricultural and ecological applications such as diagnostics for crop management, assessment of degradation/exuberance of terrestrial vegetation, and monitoring of ecosystem carbon balance. This is why chlorophyll has been one of the major targets of optical remote sensing (e.g., Vogelmann et al., 1993; Blackburn, 1998; Daughtry et al., 2000; Dash and Curran, 2004; Gitelson et al., 2005). A number of spectral indices have been proposed specifically for assessment of chlorophyll content in leaf or canopy scales (e.g., Inada, 1985; Shibayama and Akiyama, 1986; Daughtry et al., 2000; Broge and Leblanc, 2000; Richardson et al., 2002; Dash and Curran, 2004; Gitelson et al., 2005; Delegido et al., 2008). The pioneering study by Inada (1985) was used as a basis for the handheld chlorophyll meter SPAD-502 (Minolta). The critical role of the red edge wavelength region for leaf and canopy chlorophyll status has been recognized by many studies (e.g., Vogelmann et al., 1993; Gitelson et al., 2005). However, comparative studies on accuracy and robustness of various methods using diverse hyperspectral data and in situ measurements are still very few (Martin et al. 2008, Inoue et al., 2016).

The study by Inoue et al. (2016) made a comprehensive comparison of generalized spectral index approach (NDSIs and RSIs) and multivariable regression approach (PLSR and iPLSR) for assessment of CCC using diverse hyperspectral datasets taken in different locations and species. Their comprehensive study revealed the relative advantages and disadvantages of the majority of spectral models in the aspects of accuracy, linearity, robustness, simplicity, and versatility (Figure 2.14).

Overall, previous indices proposed for CCC using red edge and green wavebands, such as VOG-3, GMI-2, $CI_{red\text{-}edge}$, CI_{green}, and MTCI, proved to have moderate to excellent predictive ability. Nevertheless, an SI-based model using RSI(R_{815}, R_{704}), that is, the ratio of reflectance values at 815 and 704 nm, was found to be superior to all other models in overall predictive ability of CCC (Figures 2.14 and 2.15). The λ_{rep} model using the red edge position proved to have a limitation because actual wavelength shift might be beyond the range assumed by the model. Generally, spectral models proposed for leaf-scale variables (e.g., MCARI, PRI, SIPI) did not show good predictive ability for assessment of CCC as suggested in preceding studies (le Maire et al. 2008).

It is interesting that PLSR and iPLSR models using much larger number of wavebands proved to be inferior to the index-based models, especially in versatility. This may be attributable to the over-fitting to the calibration dataset. Their performance was poor, especially in different plant types (Figure 2.15). Multivariable regression methods (e.g., PLSR) and machine learning methods (e.g., support vector machine and artificial neural network) can be applied to hyperspectral data (Hansen and Schjoerring, 2003; Ali et al., 2015). However, the applicability of multivariable regression models to different sensors and/or different types of vegetation proved unstable. Accuracy and applicability of data-driven models by machine learning methods are highly dependent on the size and quality of the training datasets (Doktor et al., 2014; Ali et al., 2015).

The soundness of the SI-based models was inferred by simulation results using a physically based reflectance model (PROSAIL) under various canopy conditions including plant types (canopy geometry), LCC, LAI, and soil background (Figure 2.16). The generalized SI approach has unique advantages in simplicity, interpretability, robustness, and applicability compared to these methods. Additionally, the SI contour map approach using hyperspectral data can provide clear overview for

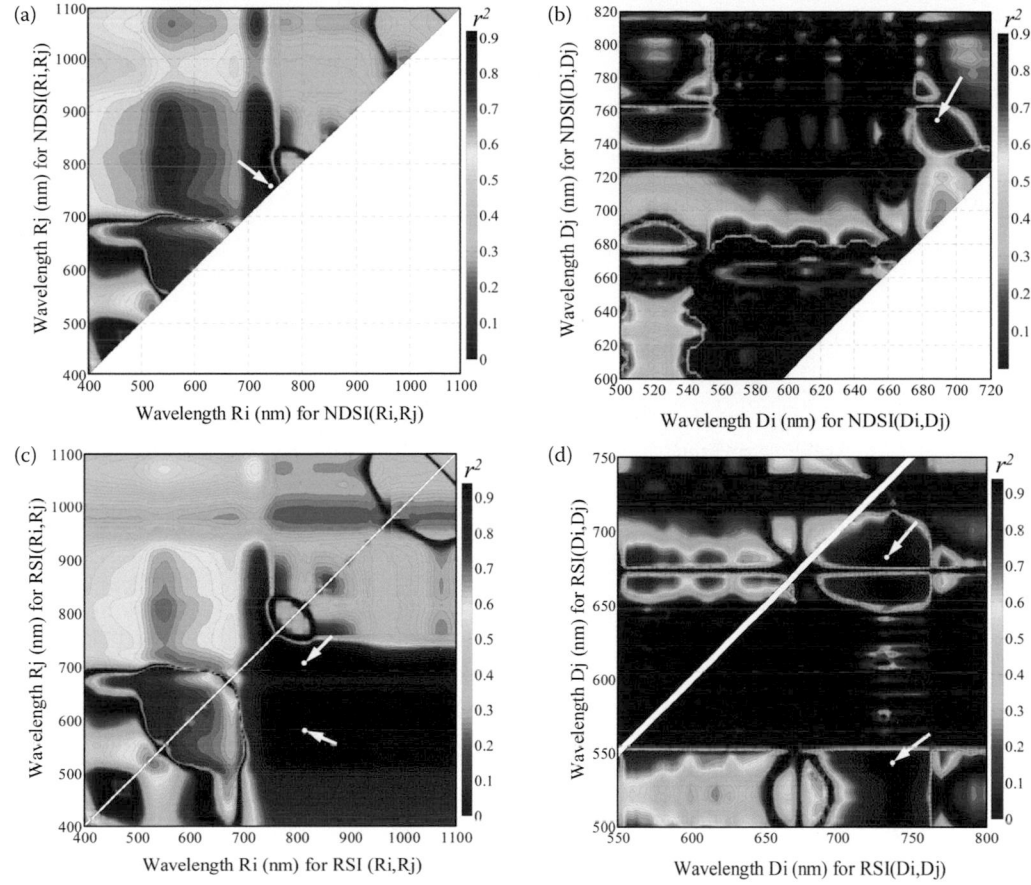

FIGURE 2.14 NDSI and RSI maps of predictive ability for canopy chlorophyll content (CCC) in rice canopies using reflectance spectra (a,c) and first derivative spectra (b,d). The white arrows indicate the most significant spots. (After Inoue, Y. et al., 2016. *Plant, Cell and Environment*, 39, 2609–2623.)

selecting optimal wavebands and bandwidths for various sensors. The best RSI model would be used as a simple and robust algorithm for the canopy-scale chlorophyll meter and/or remote sensing of CCC in ecosystem and regional scales. The model is now applied to the diagnostics for fertilizer management in rice and wheat using both satellite and drone-based image data (Inoue, 2017; Inoue and Yokoyama, 2017).

2.5.2 NITROGEN CONTENT FOR FERTILIZER MANAGEMENT

Nitrogen is essential for higher photosynthetic functioning and productivity in plants. The significant increase in crop yield over the past century is nearly proportional to the increase in nitrogen fertilizer applied to farmland (e.g., Dobermann and Cassman, 2004; Nishio, 2005). However, nitrogen fertilizer is also a significant nonpoint source of water and atmospheric pollution. Groundwater contamination with nitrate nitrogen (NO_3-N) is a serious problem in many countries (e.g., MAFF-UK, 2000; Nishio, 2005; Inoue et al., 2012a). Furthermore, Ishijima et al. (2007) demonstrated that nitrogen fertilizer is an important source of nitrous oxide (N_2O) in the atmosphere.

Therefore, agricultural management practices should aim for efficient use of nitrogen to achieve high-yielding and high-quality crop production with minimal environmental impacts. Site-specific farm management, that is, precision agriculture based on remote sensing, is a promising approach to attaining such intelligent crop management practices (e.g., Moran et al., 1997; Inoue, 2003;

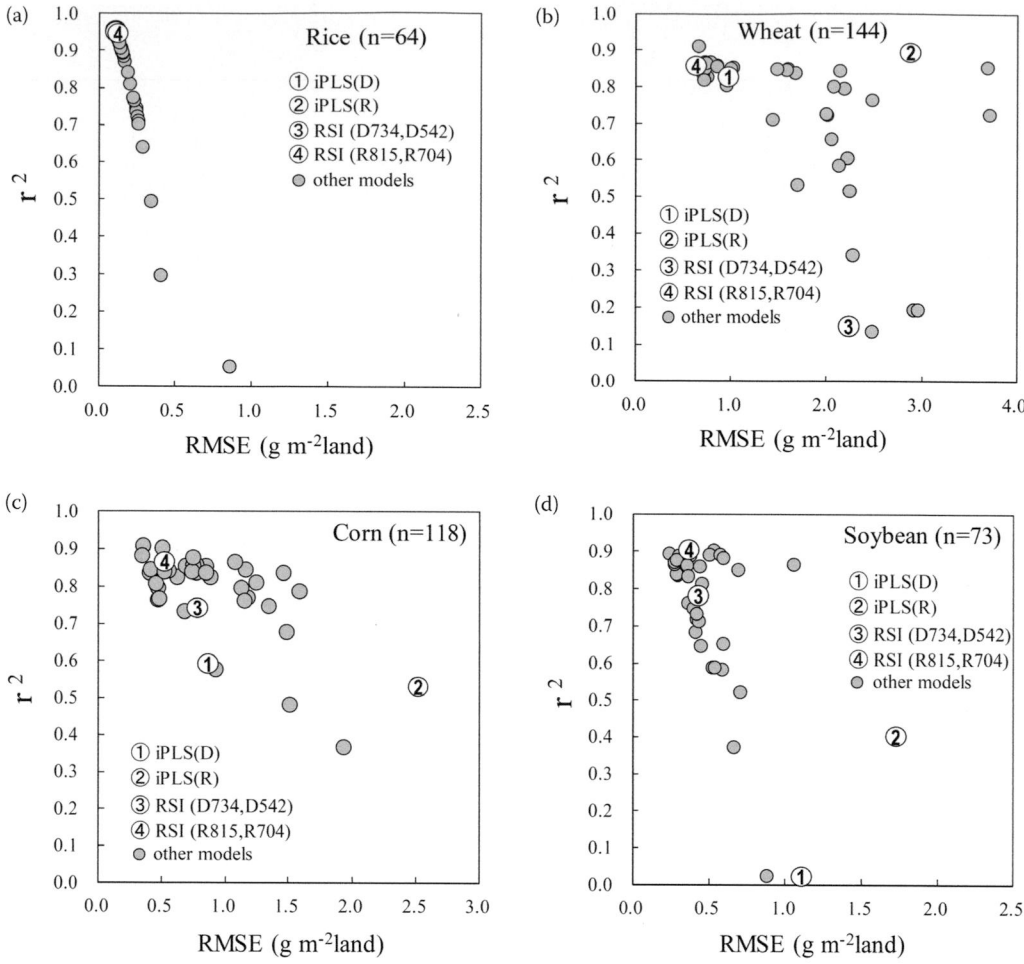

FIGURE 2.15 Applicability of 40 spectral models determined for rice a dataset to the other datasets (rice, wheat, corn and soybean) as indicated by coefficient of determination (r^2) and root mean-square error (RMSE). The best four spectral models obtained for the rice dataset are indicated by symbols with number. Definition of the previous models can be found in the references Vogelmann et al. (1993), Gitelson and Merzlyak (1997), Gitelson et al. (1997), Zarco-Tejada et al. (2001), Peñuelas et al. (1995), Rouse et al. (1974), Huete (1988), Gitelson et al. (2005), Gitelson et al. (2003), Sims and Gamon (2002), Daughtry et al. (2000), Dash and Curran (2004), Huete et al. (2002), Jongschaap and Booij (2004), Lee et al. (2008), Inoue et al. (2012b), and Wang et al. (2003). (After Inoue, Y. et al., 2016. *Plant, Cell and Environment*, 39, 2609–2623.)

Bongiovanni and Lowenberg-Deboer, 2004). Rice (*Oryza sativa* L.) is an important crop for global food security, and the assessment of canopy nitrogen content (CNC) in rice is critical for growth diagnosis and precision management to generate higher yield and better grain quality while also minimizing adverse environmental impacts. Generally, agricultural applications such as precision farming are relatively more demanding than the other applications in terms of accuracy, timeliness, and spatial resolution (Moran et al., 1997; Inoue, 2003; Baret et al., 2007; Asner et al., 2011).

A number of remote sensing approaches have been proposed for the assessment of CNC (e.g., Takahashi et al., 2000; Lee et al., 2008; Sripada et al., 2008; Zhu et al., 2008), but their accuracy and robustness may be inadequate for practical use at regional scales. Previous spectral indices may not be optimized using the merits of data richness and continuity of hyperspectra (Inoue et al., 2008a). Moreover, chemometric approaches such as a partial least-squares regression (PLSR; e.g., Takahashi et al., 2000) may not always be useful.

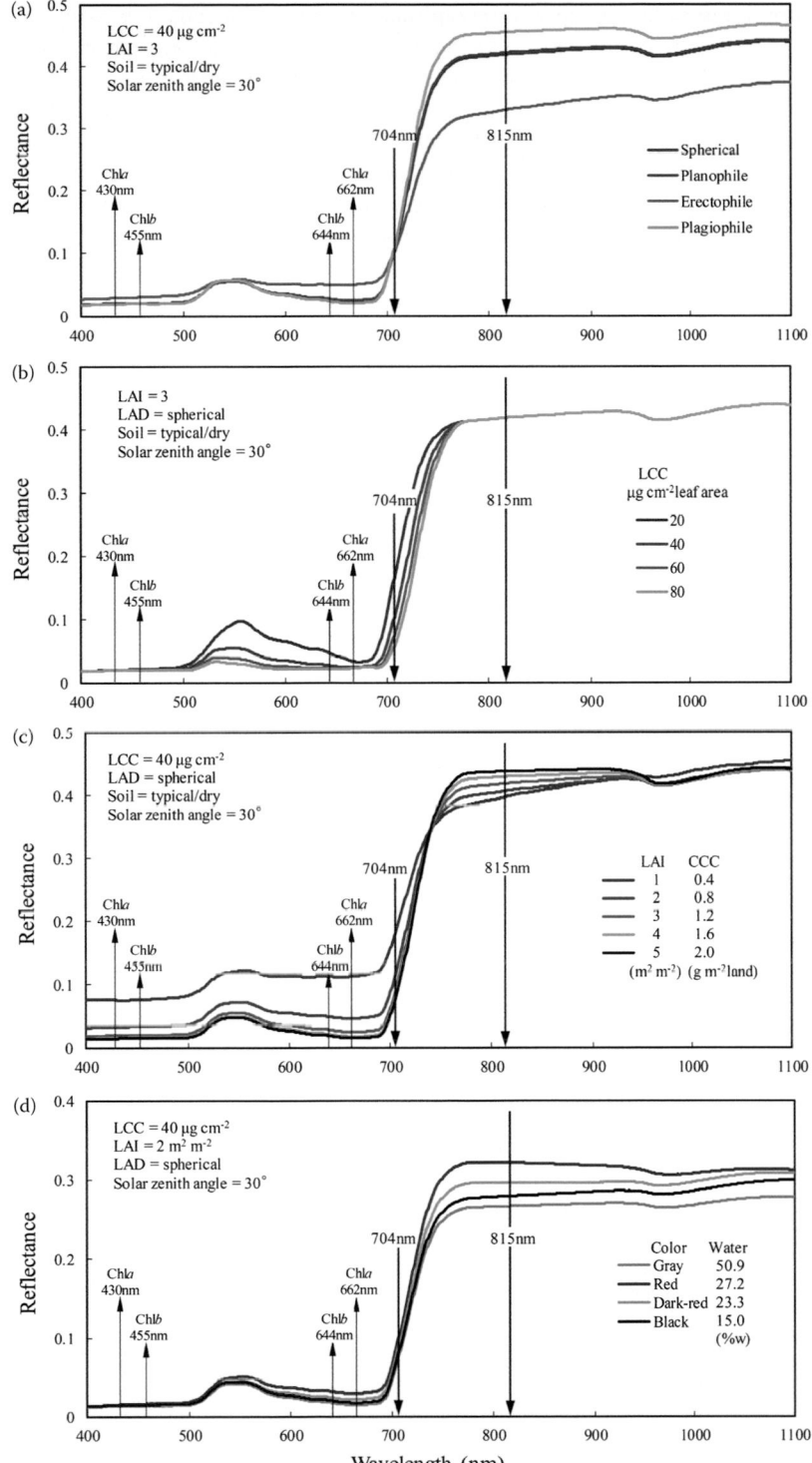

FIGURE 2.16 Reflectance spectra simulated by a physically based canopy reflectance mode PROSAIL under a wide range of plant and soil conditions, namely, different leaf angle distribution (a), leaf chlorophyll concentration (LCC) (b), leaf area index (LAI) (c), and soil color and water content (d). (After Inoue, Y. et al., 2016. *Plant, Cell and Environment*, 39, 2609–2623.)

The study by Inoue et al. (2012b) on the optimal algorithms for remote sensing of CNC in rice (*Oryza sativa* L.) based on diverse *in situ* plant measurements and ground-based and airborne hyperspectral data represents an interesting case study for hyperspectral remote sensing of CNC. Their comprehensive analysis based on the generalized spectral index approach (NDSIs and RSIs) and multivariable regression approach (PLSR and iPLSR) revealed a predictive ability for spectral methods and found promising new spectral indices (Figure 2.17). Among the 20 indices that use the reflectance values at two wavelengths, three indices explored in this study, that is, NDSI(R_{825}, R_{735}), RSI(R_{825}, R_{735}), and SAVI(R_{825}, R_{735}), showed significantly high predictive ability (small RMSE). Although most indices from the literature used the red edge bands, only one index, the CI red edge (Gitelson et al., 2005), showed comparable predictive ability to the three indices. Out of the 12 indices using three to five wavebands, mSR705, VOG-2, MTCI, and VOG-3 showed relatively high predictive ability. Most of these indices also used the red edge wavebands in part, but they were not superior to the three indices using R_{825} and R_{735}. Nevertheless, when derivatives were available, a

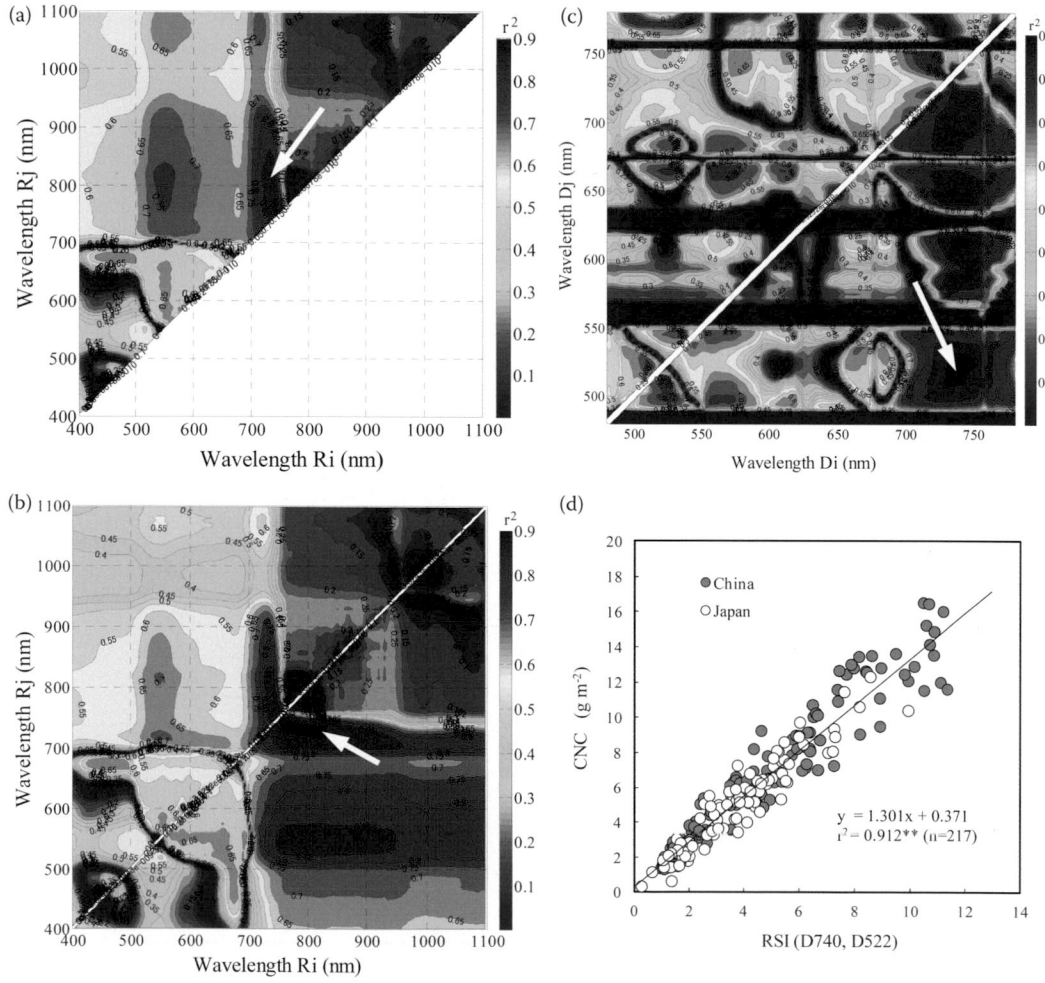

FIGURE 2.17 NDSI and RSI maps of predictive ability for canopy nitrogen content (CNC). (a) NDSI using reflectance values; (b) RSI using reflectance values, (c) RSI using derivative values, and d) relationship between RSI(D_{740}, D_{522}) and CNC for the datasets in China and Japan. ** Indicates the statistical significance at 0.001. The white arrows indicate the most significant spots. (After Inoue, Y. et al. 2012b. *Remote Sensing of Environment*, 126, 210–221.)

simple ratio index RSI(D_{740}, D_{522}) using the first derivative (D) values at 740 and 522 nm, was found to be most accurate and robust for the assessment of CNC. It is reasonable that those spectral indices designed for canopy chlorophyll content would have great potential in assessment of CNC because of the close relationship between nitrogen and chlorophyll contents in green leaves (Morita, 1978). However, optimal spectral models for nitrogen and chlorophyll contents would be different because their relative relationship changes depending on growth stage and nutritive conditions and because the reflectance spectra change differently in response to the ecophysiological changes in nitrogen and chlorophyll contents.

The multivariable models showed a high predictive ability; however, iPLSR using selected wavebands (24% of all wavebands) was superior to PLSR using all wavebands. The predictive ability of iPLSR was comparable to the simple index RSI(R_{825}, R_{735}). The independent validation using the airborne dataset also demonstrated the excellent performance of the simple index RSI(D_{740}, D_{522}) in diagnostic mapping of CNC at a regional scale.

Since nitrogen is a key element for protein and chlorophyll in green plants and thus important for animal nutrition, precise monitoring of CNC would be useful in various applications such as crop and grass management and habitat assessment. These results suggest the strong potential of the new SIs for the assessment of nitrogen content in various plant canopies. The model is now applied to the diagnostics for fertilizer management in rice at regional and farm scales using satellite or drone-based multispectral images (Inoue, 2017; Inoue and Yokoyama, 2017; see Section 2.7.2).

2.5.3 PROTEIN CONTENT OF RICE GRAIN FOR PRODUCTION OF HIGH-QUALITY RICE

The protein content of grain (GCP) in rice and wheat is an important quality parameter that affects the price of products (Sakaiya and Inoue, 2013). Higher GCP is valuable in wheat, but lower GCP is preferable in rice because protein content is inversely related to the taste of cooked rice. For example, in rice, GCP is a critical indicator in quality control for regional branding and marketing strategies in several countries such as Japan, Korea, and Taiwan.

However, GCP varies considerably between regions and individual paddies because it is affected by ecophysiological processes such as translocation of nitrogen as well as environmental and management practices during the growing season. Therefore, acquisition of spatial information on GPC in each rice paddy before harvesting is quite useful for regional branding strategies. Such information is used for optimizing the harvest scheduling for better quality, and as a basis for optimizing fertilizer management in the following season. This may be a unique application of remote sensing in agriculture, but would provide some insights for extending the application opportunities of remotely sensed data.

Since the GCP is not incorporated explicitly in any biophysical or physiological models, the generalized spectral index approach using hyperspectral data would be the most feasible to create the optimal predictive algorithms (Inoue et al., 2008a). Figure 2.18 shows the NDSI map of predictive ability for GCP created from hyperspectral data by an airborne sensor (CASI) during the ripening period. The combinations of wavebands in the 530–600 and 710–1050 nm regions have high predictive ability and two hot spots are found on the NDSI map. The best capability was found around the peak of NDSI(R_{970}, R_{570}) with a spectral width of ± 20 nm. The significant contribution of R_{970} nm was due to the apparent relation of water content with GCP since water has a specific absorption peak at 970 nm (see Section 2.2). The wavelength region for Landsat NDVI with Landsat bands is between the two peaks and showed relatively low predictive power. The problem of multicollinearity (over-fitting) in MLR is clearly shown in Figure 2.19, where RMSE is smallest in calibration but largest in validation results. It is interesting that the best NDSI(R_{970}, R_{570}) has comparable or better predictive ability than PLSR using the whole spectra or MLR using four selected bands. Another important result is that the iPLSR with band selection had higher ability than PLSR; meaning that the use of whole hyperspectra

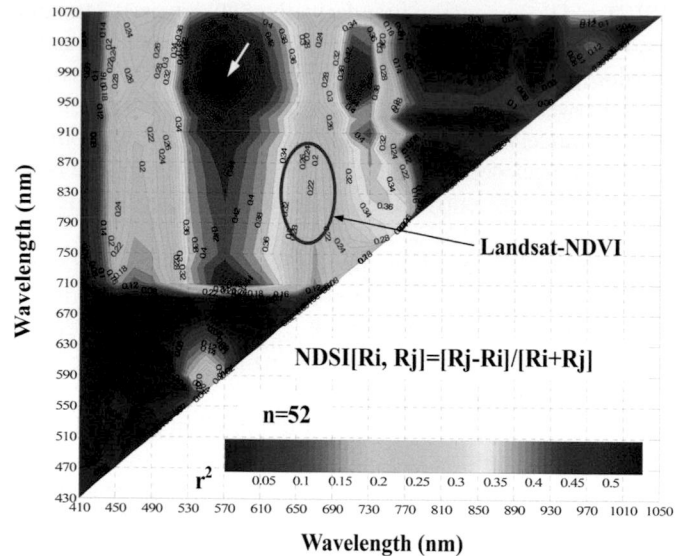

FIGURE 2.18 A contour map of coefficient of determination (r^2) between protein content of rice grain (GPC) and NDSIs. The ellipse indicates the spectral region equivalent to the NDVI using Landsat bands (660 ± 30 and 830 ± 70 nm). (After Inoue, Y. et al., 2008a. *Journal of the Remote Sensing Society of Japan*, 28, 1–14.)

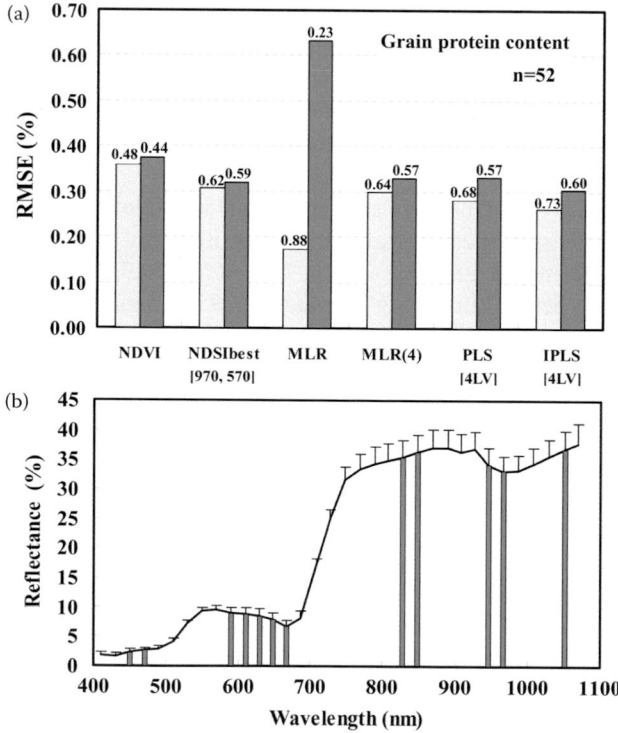

FIGURE 2.19 Comparison of predictive ability in several methods for hyperspectral assessment of grain protein content in rice. (a) RMSEs in calibration (left bars) and in validation (right bars) are shown with coefficient of determination (R^2). The NDSI(R_{970}, R_{570}) was the best NDSI extracted from the NDSI map in Figure 2.18. The values with "LV" in the PLS methods indicate the number of latent variables. (b) Selected 12 wavelengths for the iPLSR model in the average spectra with standard deviation bars for sampled paddy fields. (After Inoue, Y. et al., 2008a. *Journal of the Remote Sensing Society of Japan*, 28, 1–14.)

does not always provide the best result or may be even worse for prediction, presumably because of some erroneous or disturbing spectral bands. The iPLSR method may be useful, but simple indices selected by systematic methods such as NDSI maps would also be useful when only a small number of wavebands are available. These SI models are now applied successfully to the regional-scale management of high-quality rice production using satellite images (Inoue, 2017).

2.6 SPECTRAL ASSESSMENT OF BIOCHEMICAL AND BIOPHYSICAL PROPERTIES IN GRASSES AND TREES

Biochemical and biophysical traits provide understanding about the functional strategies of plants (Colgan et al., 2015). Due to their role in the photosynthetic process, they have close interactions and co-vary across plant functional types (Wright et al., 2004). Many of these traits have been recognized as fundamental functional traits for biodiversity monitoring (Pereira et al., 2013; Skidmore et al., 2015). Biodiversity research from species diversity is moving toward functional diversity (Tilman, 2001), therefore, accurate assessment of these traits is of prime importance.

Detailed hyperspectral studies have revealed that in natural ecosystems such as forests and grasslands, the spectral signature of plants, neglecting the peripheral factors, are mainly altered by change in their biochemical and biophysical properties (Ferwerda et al., 2005; Mutanga and Skidmore, 2007; Cho et al., 2008; Darvishzadeh et al., 2009; Wang et al., 2015b; Ali et al., 2016b; Neinavaz et al., 2016b). These changes are mainly caused by plant growth, health status and stress caused by infestation, pest, and diseases (Hinzman et al., 1986; Zarco-Tejada et al., 2002; Abdullah et al., 2018). Confounding effects such as canopy architecture and background soil are among the most prominent factors affecting the reflectance of vegetation canopies, and are particularly pronounced in sparse canopies. For instance, the effect of soil brightness on the canopy reflectance in a sparse canopy is presented in Figure 2.20. As can be observed from this figure, the variation of background soil in a vegetative canopy has led to differences in canopy reflectance and has caused distinct reflectance offsets in the first derivatives.

A large number of studies have investigated the relationships among the leaf traits. The strengths of these relationships vary across ecosystems and functional types and depend on whether they are expressed on mass or area basis (Wright et al., 2004; Asner and Martin, 2009; Homolová et al., 2013).

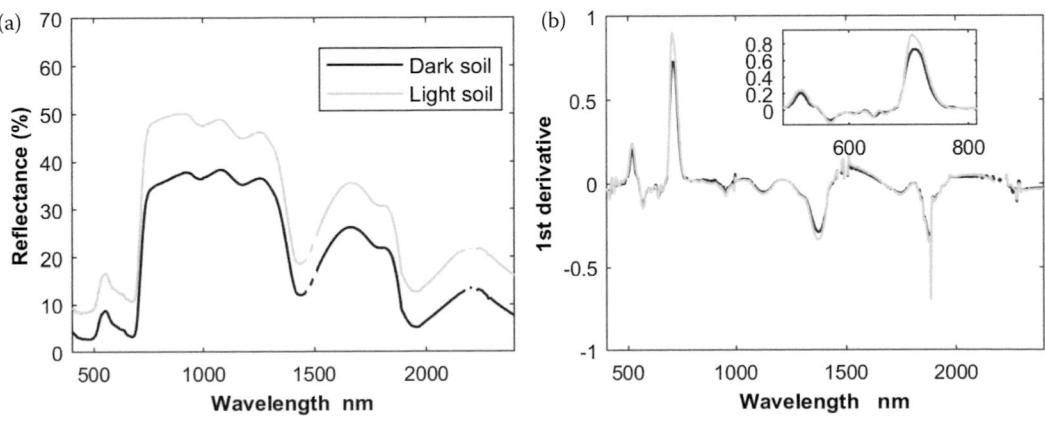

FIGURE 2.20 (a) Spectral reflectance for *Asplenium nidus* canopy with an LAI of 1.5 in dark and light soils; (b) first derivative of the reflectance. (From Darvishzadeh, R. et al. 2008a. *International Journal of Applied Earth Observation and Geoinformation*, 10, 358–373.)

However, it has been observed that area-based leaf traits are generally retrieved with higher accuracy from optical remote sensing data (Wang et al., 2015a), which is due to their role in photosynthesis and processes such as light interception and transpiration that expressed as flux per unit leaf surface area (Hikosaka, 2004; Lloyd et al., 2013).

Among the leaf traits, chlorophyll is the most prominent leaf biochemical that an important role in verifying vegetation physiological status and health, and has been recognized as an indicator for photosynthetic capacity, productivity and detection of vegetation stress (e.g., Carter, 1994; Boegh et al., 2002; Abdullah et al., 2018) (see Section 2.6.1 and Figure 2.21). A wide range of studies have observed moderately strong relationships between chlorophyll and foliar nitrogen, for example (Field and Mooney, 1986; Mutanga and Skidmore, 2007; Kokaly et al., 2009). Therefore, leaf chlorophyll has been suggested as an operational proxy for nitrogen content and vice versa (Muñoz-Huerta et al., 2013). Foliar nitrogen is also related to leaf mass per area (LMA) across different ecosystems, for example (Rosati et al., 2000; Wright et al., 2004; Wang et al., 2015a). LMA is the inverse of specific leaf area (SLA) and provides information on the spatial variation of photosynthetic capacity in addition to leaf nitrogen (Pierce et al., 1994). Since LMA is highly related to the amount of leaf water content (Ali et al., 2016c) (see Section 2.6.2), it can be concluded that leaf water content is linked to other leaf traits such as nitrogen and chlorophyll. As such its relation with foliar nitrogen is demonstrated by Wang et al. (2015a). However, this relationship is vigorous due to the strong temporal dynamic of leaf water content in leaves (Ustin, 2013).

Among canopy biophysical parameters, leaf area index (LAI), in addition to leaf angle distribution and canopy gaps, plays a key role in the magnitude of canopy reflectance (Darvishzadeh et al., 2008a) (see Section 2.6.3). The interrelationship among canopy biophysical traits in forest stands has been documented in several studies. For instance, strong relationships between LAI and biomass, stem density, diameter at breast height (DBH), above-ground net primary productivity (NPP) and canopy cover fraction have been observed (e.g., Hall et al., 1995; Lefsky et al., 1999; Naesset et al., 2005; Schlerf and Atzberger, 2006; Huesca et al., 2016). Moreover, LAI has shown significant correlations with a number of leaf traits such as SLA and leaf nitrogen (Pierce et al., 1994), and leaf dry matter (Ali et al., 2016a). Although the strengths of these relationships are ecosystem and species dependent,

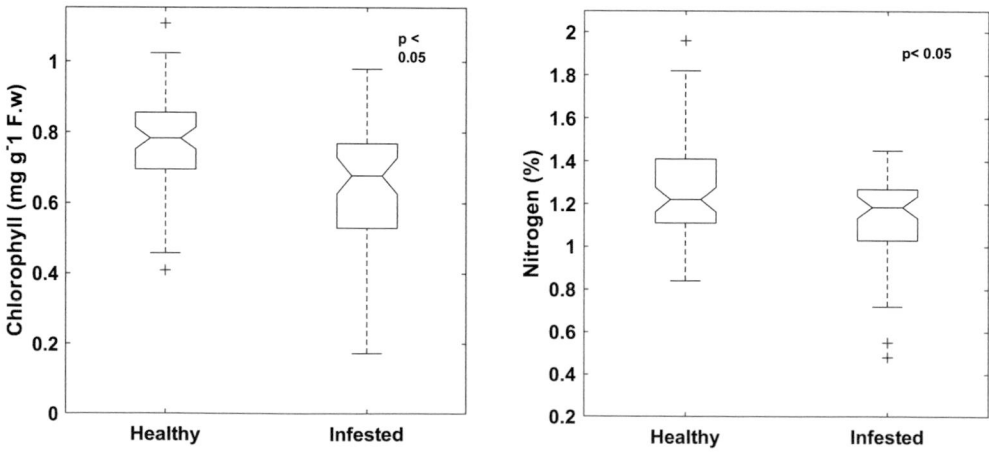

FIGURE 2.21 Distribution of measured chlorophyll and nitrogen concentration for healthy needles and needles infested by bark beetle at the green attack stage. There is a significant difference ($p < 0.05$) in chlorophyll and foliar nitrogen concentration between healthy and infested samples. Infestation decreased mean chlorophyll and nitrogen. (From Abdullah, H. et al., 2018. *International Journal of Applied Earth Observation and Geoinformation*, 64, 199–209.)

these co-variation relations have a key role in evaluation and understanding of their spectral responses (Ollinger et al., 2008).

Studies performed in grasslands and forest (see Section 2.6.2) on leaf traits such as specific leaf area (SLA), leaf mass per area (LMA), and water content have shown sensitivity to wavelengths in the shortwave infrared (SWIR) region of the electromagnetic spectrum (1200–2500 nm) (Ali et al., 2017a,b; Mirzaie et al., 2014; Romero et al., 2012). While the spectral bands sensitive to foliar chlorophyll and carotenoids, exist in the visible and red edge region (400–750 nm) (Curran, 1989; Dawson et al., 1999), leaf nitrogen has been related to different spectral regions (Curran, 1989; Fourty et al., 1996; Kokaly et al., 2009; Ollinger et al., 2008; Knyazikhin et al., 2013; Wang et al., 2017). This may be explained by the interrelation and interactions of leaf nitrogen with other leaf and structural traits. Thus Wang et al. (2017) demonstrated that in spruce stands the sensitive bands for estimating foliar nitrogen in the visible region are mainly related to chlorophyll absorption, while in the near-infrared (NIR) and SWIR regions, the bands sensitive to foliar nitrogen are explained by the existence of protein and dry matter content as well as biological associations between nitrogen and structural traits.

Knowledge of biophysical and biochemical traits and their variations in natural ecosystems is critical for many management practices such as biodiversity monitoring, climate modeling, and carbon cycle assessment. The following subsections present a few scenarios in which, utilizing hyperspectral measurements, accurate estimations of these parameters in the forest and grasslands have been made.

2.6.1 Leaf Biochemical Properties for Detection of Insect Outbreaks in Forest

European spruce bark beetle (*Ips typographus* (L.)) is the most aggressive forest insect pest in the European forests, and they can destroy many more forested areas than all other natural disturbances. The extent of bark beetle damage has been huge, and trees have been killed over tens of millions of hectares, causing a great economic loss in timber production. In addition, a large amount of public money has been expended in clearing the fallen trees. Further, insect outbreaks have resulted in significant ecological changes in terms of the forests' structure and composition, wildlife habitats, and degradation of large areas within these forests. To meet the information requirements to minimize economic losses and to preclude further mass outbreaks, early detection of insect outbreak is vital (in a period in which trees are yet to show visual signs of infestation stress). This information plays a vital role in forest management and is critical when developing sustainable forest management policies and conservation strategies. Visual inspections during field surveys and pheromone traps traditionally have been used for detection of bark beetle infestation in Norway spruce forests. However, these methods are subjective, time-consuming, and ineffective for low-level infestation (i.e., not an outbreak). Consequently, acquisition of remote sensing data is a proper alternative for developing new techniques for monitoring and detecting this forest disturbance.

Physiological studies have indicated that biochemical variables such as leaf water, chlorophyll, and nitrogen contents are the main properties of the plants that are sensitive to both environmental conditions and insect infestation and have a direct impact on leaf and canopy optical properties (Gitelson et al., 2003; Munoz-Huerta et al., 2013). Accordingly, biochemical variables play a key role in plant growth and photosynthesis and are considered as indicators of plant health and stress. In recent decades, many studies have shown the role of remote sensing, in particular hyperspectral imagery, in measuring vegetation biochemical variables such as chlorophyll and nitrogen content in different ecosystems (Asner and Martin, 2009; Asner et al., 1998; Darvishzadeh et al., 2008b; Wang et al., 2015a). Therefore, hyperspectral measurements integrated with biochemical properties have been considered to study the forest insect outbreaks. Abdullah et al. (2018), have used foliar reflectance and biochemical propitiates (chlorophyll and nitrogen content) to detect early stage of bark beetle infestation in the Norway spruce forests in Central Europe. In their study, partial

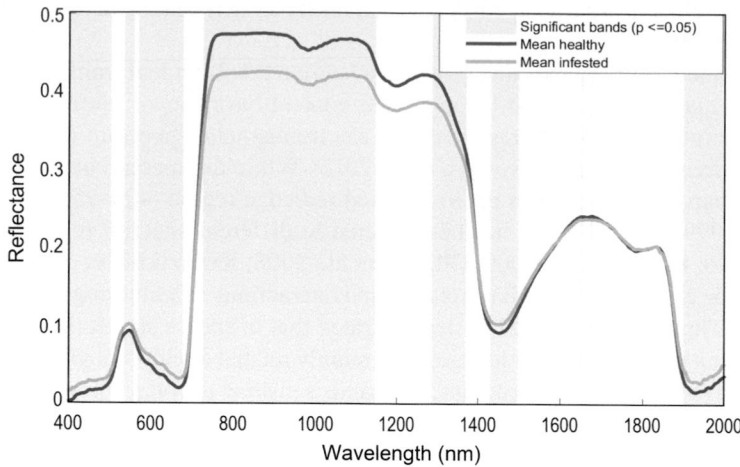

FIGURE 2.22 Mean reflectance spectra of healthy and infested leaves of European spruce at the green attack stage (see Fig. 2.22). Tinted areas depict the location of wavebands displaying is a significant difference between healthy and infested spectra. (From Abdullah, H. et al., 2018. *International Journal of Applied Earth Observation and Geoinformation*, 64, 199–209.)

least-square regression (PLSR) was utilized to assess the impact of infestation stress on the retrieval accuracy of foliar biochemical properties. They found that bark beetle infestation at the early, so-called green attack, stage (when the leaves are still green and not visibly stressed) affects the leaf spectral response as well as leaf biochemical properties and their retrievals from hyperspectral measurements (Figure 2.22). They concluded that the assessment of retrieval accuracy of these two biochemical components could be used as an indicator for the efficient landscape-wide detection of the early stage of bark beetle infestation.

Not only chlorophyll and nitrogen content but also leaf water content can be used as a proxy to detect bark beetle infestation. Cheng et al. (2010) used leaf water content to detect pre-visible mountain pine beetle damage in lodgepole pine needles using hyperspectral measurements. They concluded that the measurable water deficit of infested tree samples was detectable from a narrow spectral region (1318–1322 nm) and could be used to distinguish between healthy and infested trees by mountain pine beetle.

Although detecting the early phase of insect outbreak in the Norway spruce forests is challenging using remote sensing data, due to the lack of apparent visual symptoms on the affected trees, quantifying biochemical variables from hyperspectral measurements is promising and presents a powerful tool to determine the damage caused by bark beetle green attack at the leaf level. In other words, the information obtained about biochemical variables using hyperspectral measurements is sufficient to detect insect outbreaks in the forest at an early phase.

2.6.2 Assessment of Leaf Functional Traits in Forest (SLA and LDMC)

Quantifying functional traits in natural communities is pivotal to understanding the spatial and temporal distribution of biodiversity, ecosystem services, and plant community productivity (Cadotte et al., 2011; Lavorel et al., 2011). Since human survival relies on economic benefits and services provided by ecosystems, it is evident that ecosystem functions are a top conservation priority. Scientists believe that better conservation and restoration decisions can be made by measuring and understanding functional traits (Cadotte et al., 2011). Remotely sensed data can play a critical role in acquiring such functional traits data over broad spatial scales nondestructively and repeatedly. Ali et al. (2016a,b, 2017a,b) elucidated the potential of remotely sensed data to quantify two fundamental leaf functional traits (i.e., leaf dry matter content [LDMC] and specific leaf area [SLA]) in a mixed forest at leaf, canopy, and landscape scales by using statistically and physically based predictive models.

The leaf-level study by Ali et al. (2016b) through inversion of the most commonly used leaf radiative transfer model PROSPECT (Jacquemoud and Baret, 1990) revealed accurate estimation of the two leaf traits in a wide range of samples collected from broadleaf and conifer forest stands. Validation of the two leaf trait estimations showed the high coefficient of determination (R^2) between the predicted and ground measured values ($R^2 = 0.83$ for LDMC and $R^2 = 0.89$ for SLA). The results showed that the PROSPECT_4 leaf model accurately simulates spectral information of samples from mixed mountain forest and can be used to retrieve the biochemical content of leaves/needles directly and indirectly through inversion over a range of vegetation types. It highlighted the fact that LDMC and SLA are quantitatively represented by leaf spectra. This leaf-level result sheds light on extending or upscaling the application of remotely sensed data in order to accurately estimate the two functional traits at canopy and landscape scales.

There are no well-developed methods for fast and accurate retrieval of LDMC and SLA at canopy or larger scales. Hence, both the statistical and physical approaches were investigated in order to map the two leaf traits in a mixed mountain forest, using the hyperspectral HySpex airborne images and the multispectral Landsat-8 Operational Land Imager (OLI) data (Ali et al., 2017b). The spectral features that have high correlations with LDMC and SLA were identified by applying continuous wavelet analysis to INFORM-simulated canopy spectra (Atzberger, 2000). The application of continuous wavelet analysis resulted in six sensitive wavelet features for LDMC and four for SLA being in the top 1% strongly correlated wavelet features. The results indicated the capability of the wavelet transformation to identify the most sensitive spectral features from a large hyperspectral dataset and the robustness of wavelet transformation in creating a higher correlation of wavelet features to vegetation variables compared to use of nontransformed original spectra (Figure 2.23). This may be attributed to the effectiveness of wavelet analysis in decomposing the trait's absorption features into various scales of narrow- and broad-band absorption features and identifying those that correlate best with the variation in the traits' concentration.

The suitability of indices for SLA retrieval at leaf and canopy levels for heterogeneous mountain forest has been evaluated by Ali et al. (2017b). The correlation tended to be lower at the canopy level, and many of band combinations that showed a high correlation to SLA at leaf level became

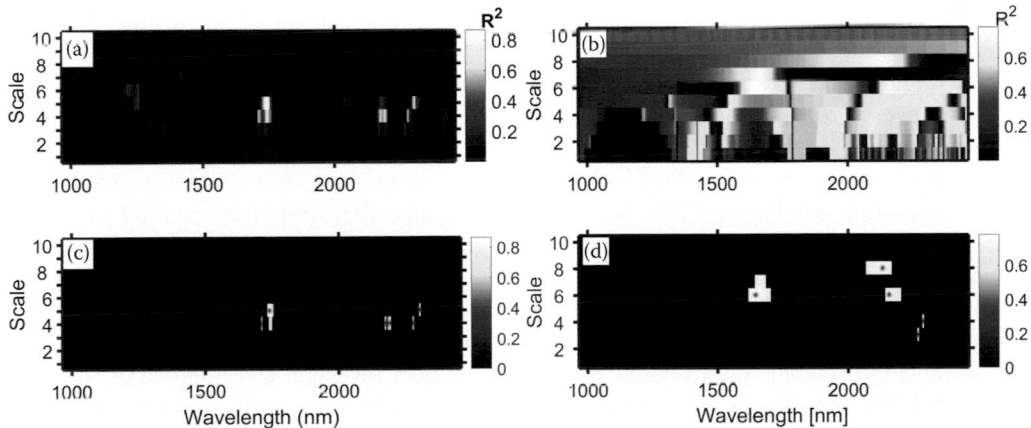

FIGURE 2.23 Correlation scalograms for the identification of wavelet features that significantly correlate with (a) leaf dry matter content (LDMC) and (b) specific leaf area (SLA). Scalograms are derived from continuous wavelet analysis of simulated spectra. Brightness represents the coefficient of determination (R^2) relating wavelet power to LDMC and SLA. Colored regions in scalograms (c) and (d) depict the wavelet features with the top 1% greatest R^2 values for LDMC and SLA. (From Ali, A.M. et al., 2016c. *ISPRS Journal of Photogrammetry and Remote Sensing*, 122, 68–80.)

insensitive at canopy scale. Although only narrow-band indices were examined, our findings suggest that bands with a wider wavelength range (broad bands) can be utilized. This implies that remote sensing data from coarse spectral resolution sensors may suffice for SLA estimation.

In the study by Ali et al. (2017b), SLA was retrieved from the recently launched Landsat-8 imagery by empirical methods (individual bands and vegetation indices) and the inversion of the INFORM radiative transfer model through wavelet transformation and lookup table approaches in order to explore the potential of medium-resolution multispectral satellite images to predict SLA for a mixed mountain forest (Ali et al., 2017b). It was evaluated whether the methods earlier studied at leaf and canopy scale using hyperspectral field and airborne data can be utilized at a larger scale using medium-resolution satellite data. Both the statistical and INFORM inversion predicted SLA with satisfactory accuracy.

The findings and methods of these studies have the potential to produce useful information from hyperspectral and multispectral remote sensing data about leaf functional traits at local, regional, and global scales. The results further confirmed the applicability of remote sensing to elucidate variation in general ecological leaf functional traits across relevant spatial and temporal scales. This will facilitate regular monitoring of biodiversity, particularly with respect to natural or anthropogenic changes in ecosystem functioning.

2.6.3 Analysis of Biophysical and Biochemical Traits in Grasslands

In natural heterogeneous canopies such as grasslands, where the species diversity is high, the reflectance is often a mixture of different species; hence, remote sensing applications are challenging. A detailed investigation is required to assess the aptitude of remote sensing models when it comes to the combination of different plant species in varying proportions. The utility of hyperspectral remote sensing in predicting leaf and canopy characteristics such as LAI and canopy and leaf chlorophyll content in a heterogeneous Mediterranean grassland by different modeling approaches (statistical and radiative transfer models) at laboratory, field, and airborne levels was studied by Darvishzadeh et al. (2008a,b).

A number of key observations about the effects of canopy heterogeneity on retrieval of traits were made for analysis using statistical models. It was observed that unlike most common vegetation indices where the near-infrared (NIR) region is the keystone, wavebands from the short-wave infrared (SWIR) region contained most relevant information about canopy LAI, and the "hot spots" (regions where the sensitive band combinations exist) mostly occurred in this spectral region at laboratory, field, and airborne levels. This emphasized that the vegetation indices that do not include the SWIR spectral region may be less satisfactory for LAI estimation (Darvishzadeh et al., 2008b). Moreover, the results suggested that not only the choice of vegetation index but also prior knowledge of plant structural components and the background soil are important when using vegetation indices for LAI estimation (see Figure 2.20). Accordingly, landscape stratification is required when using hyperspectral imagery for large-scale mapping of vegetation biophysical variables.

Radiative transfer models have rarely been used for studying heterogeneous grassland canopies. The widely used PROSAIL (Verhoef, 1984, 1985; Jacquemoud and Baret, 1990; Kuusk, 1991) model by means of a lookup table (LUT) was utilized by Darvishzadeh et al. (2008b) with a number of stratifications to overcome the ill-posed nature of the model inversion and improve the retrievals of LAI, and leaf and canopy chlorophyll content. Their results showed that canopy chlorophyll content was predicted with the highest accuracy ($R^2 = 0.70$, normalized RMSE = 0.18), while LAI was estimated with moderate accuracy ($R^2 = 0.59$, normalized RMSE = 0.18). Heterogeneity had a pronounced effect on inversion algorithm and the retrieval of these parameters. Hence, the stratification of data based on the number of species increased the estimation accuracies of these parameters. The accuracy systematically fell each time sample plots with more (up to four) species were included in the inversion process. This demonstrated the limitations of the PROSAIL radiative transfer model when the spectral reflectance stems from a rather heterogeneous vegetation canopy condition. When a limited number of wavelengths corresponding to various vegetation parameters and bands showing low average absolute errors were used for model inversion, the relationships between estimated and measured grass variables were comparable to those obtained from all wavebands. This indicated that a carefully chosen spectral subset contains sufficient information for a successful model inversion and can improve the ill-posed nature of the model inversion.

Utilizing canopy reflectance, leaf chlorophyll content could not be estimated with an acceptable accuracy exploiting either of the modelling approaches. Leaf biochemical variables suffer from poor signal propagation from leaf to canopy scale (Yoder and Pettigrew-Crosby, 1995; Jacquemoud et al., 1996; Asner et al., 1998), with only few exceptions (Jacquemoud et al., 2009; Xiao et al., 2014). Therefore, their retrieval from canopy reflectance is challenging (Darvishzadeh et al., 2008b). At the canopy level, biophysical variables such as leaf area index (LAI) and canopy architecture (e.g., leaf angle distribution) presented major contributions to the canopy reflectance, mainly in near-infrared (NIR) and SWIR regions (Darvishzadeh et al., 2009; Ali et al., 2016a). Among these parameters, perhaps LAI has the most prominent role in defining the magnitude of the canopy reflectance, such that increase in LAI values increased the canopy reflectance in NIR and SWIR regions (Figure 2.24a).

The above findings illustrate some possibilities for estimating and mapping LAI and chlorophyll in a rather heterogeneous grassland. However, the application of the methods developed to other heterogeneous biomass and parameters needs to be evaluated using different hyperspectral data sets to explore the scale and sensor effects as well as phenological influences. It was observed that the heterogeneity (species diversity) almost disappeared at the airborne level, as using airborne data the retrieval accuracy of the studied variables increased using either the statistical or the physical model. However, heterogeneity is strongly scale dependent and relative.

2.6.4 Response of LAI to Thermal Hyperspectral Data

The thermal infrared region (TIR 8–14 µm) has traditionally been treated differently from the other parts of the electromagnetic spectrum due to the differences in acquisition, calibration, and processing of the data over this region. Despite the recent advances in the field of thermal remote sensing, this domain has hardly been used for vegetation studies. Consequently, there still is limited and insufficient information regarding the responses of plant traits over TIR domain. The existing

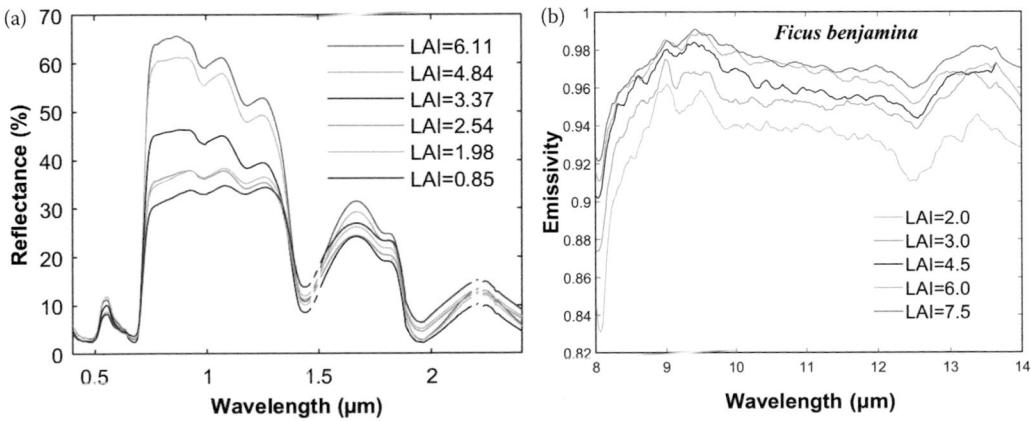

FIGURE 2.24 (a) Canopy spectral reflectance of *Asplenium nidus* in visible, near infrared, and short-wave infrared regions corresponding to LAI values between 0.87 and 6.11. (b) The canopy emissivity spectra in *Ficus benjamina*. Changes in canopy emissivity spectra become smaller when LAI values exceed 4.0 (m² m⁻²). ([a] From Darvishzadeh, R. et al., 2009. *International Journal of Remote Sensing*, 30, 6199–6218; [b] Neinavaz, E. et al., 2016a. *ISPRS Journal of Photogrammetry and Remote Sensing*, 119, 390–401.)

misconceptions over the TIR data are related to relatively inaccessible instruments with high signal-to-noise ratio, and subtle emissivity spectral variations in plants (Ribeiro da Luz and Crowley, 2010). Further, little is known by the ecological and remote sensing communities about plant physiology in association with plant spectral features in the TIR domain (Quattrochi and Luvall, 1999). However, none of these difficulties has reduced the importance of the TIR region, which benefits from a wide atmospheric window with large transparency offering high potential for remote sensing vegetation studies (Clerbaux et al., 2011).

As a result, Neinavaz et al. (2016a,c) explored the potential of hyperspectral TIR canopy measurements to measure and quantify the canopy LAI as the key canopy biophysical property. LAI is strongly associated with numerous ecosystem processes (e.g., water balance and evapotranspiration) and plays a vital role in climate and terrestrial ecosystem models. Additionally, LAI has recently been proposed as one of the essential biodiversity variables (EBVs) that are potentially suitable for satellite monitoring of biodiversity and progress toward the Aichi Biodiversity Targets (Pereira et al., 2013; Skidmore et al., 2015). Moreover, the demand for LAI monitoring over large areas has increased in recent decades due to rising concern over habitat degradation and climate change.

The effect of changing LAI values on canopy reflectance has been investigated and confirmed in a number of studies (Asner et al., 1998; Darvishzadeh et al., 2009) in which it was demonstrated that with increasing LAI the canopy reflectance spectra increases noticeably in the NIR and SWIR regions (Figure 2.24a). The effect of LAI on plant emissivity spectra has only recently been investigated using TIR hyperspectral data by Neinavaz et al. (2016a). In their study the response of vegetation canopies with various LAI values to canopy emissivity spectra was investigated. It was observed that there is a positive relationship between canopy emissivity spectra and LAI and that the emissivity rises with increasing LAI values. This increment was more pronounced beyond the wavelength region of 9 μm (Figure 2.24b). Further, their results revealed that the species with similar LAI values could have distinct canopy emissivity spectra, suggesting that LAI is not the sole parameter that affects the canopy emissivity spectra and that other traits also contribute to these spectra.

The potential of TIR hyperspectral data for estimation of LAI was further explored for four structurally different species in Neinavaz et al. (2016c). The results of their study revealed that LAI and emissivity relations are only statistically significant in some wavebands. Vegetation indices, partial least-square regression (PLSR) and artificial neural networks (ANN) were applied for LAI retrieval in different species; their retrieval accuracies were comparable and the predicted accuracies were highly species dependent and varied among species. The most sensitive wavebands across species were in the 10–12 μm range in combination with the bands in the 8–11 μm range. The predicted LAI from the pooled dataset had lower accuracy than the ones obtained for individual species resulting from variation of biochemical and structural parameters across the species studied.

It is expected that other parameters, particularly those modulated from biochemical and biophysical variables, are likely to have more contribution to canopy emissivity spectra. Recent studies performed in mixed forest stands using TIR remotely sensed data demonstrated that canopy emissivity is a meaningful measure for retrieving vegetation modulated properties, particularly at the canopy level (Neinavaz, 2016a,b,c).

2.7 GENERIC ROLE OF HYPERSPECTRAL DATA FOR ESTIMATION OF ECOPHYSIOLOGICAL FUNCTIONING

2.7.1 GENERIC USE OF HYPERSPECTRAL DATA FOR PRESENT, FUTURE, AND PAST SENSORS

In general, calibration and validation of predictive models are essential for higher accuracy and applicability in assessment of biophysical and ecophysiological variables from remotely sensed data. In past decades, a large number of papers have reported predictive algorithms and models using various sensors having different numbers of spectral bands, central wavelength, and bandwidth (Thenkabail et al., 2012). However, their predictive accuracy and applicability are often inconsistent between datasets because of the differences in sensor specifications. A predictive model based on a dataset using a sensor cannot be applied directly to the other sensors. For example, the widely used spectral index NDVI (normalized difference vegetation index) derived from different sensors would not be equivalent to each other because of the differences in position and width of the red and near-infrared wavebands. Some sensors have the red edge band but the others do not, while the band is most useful for assessment of chlorophyll content. Accordingly, calibration and validation procedures are required for individual sensors based on sufficient number of spectral data and reliable *in situ* measurements of biophysical or ecophysiological variables. Such datasets usually have to be acquired by laborious and careful procedures, which would hamper the wide and generalized application of remote sensing. Furthermore, acquisition of such datasets is impossible from the various sensors used in the past even if a large volume of images are available in open archives.

However, precise hyperspectral data of sufficient spectral resolution (1–3 nm) acquired together with the diverse range of target variables would play a generic role in deriving the optimal algorithms and predictive models for various types of sensors of the past, present, and future. Hyperspectral datasets obtained in close range can especially be an accurate basis for various applications because uncertainties due to atmospheric conditions and spatial heterogeneity within a pixel are negligible. Optimal algorithms and predictive models can be derived as default from such datasets for various sensor specifications of satellite-, aircraft-, and drone-based multispectral sensors of the past, present, and future. This hyperspectral approach would be realized through three methods: (1) generalized spectral index method, (2) advanced use of physically based reflectance models, and (3) data-driven machine learning methods. The second and third methods have several constraints as discussed in Section 2.3. Therefore, the first method would be the simplest and most flexible approach.

Figure 2.25 shows a schematic of the generic use of hyperspectral datasets for assessment of ecophysiological functioning of crops and vegetation based on the generalized spectral index method. The generalized spectral index approach, such as $NDSI(R_i, R_j)$, $RSI(R_i, R_j)$ and/or their derivatives using the whole spectral wavebands (see Section 2.3.1), allows one to choose optimal sensors and wavebands from available satellite or drone sensors. In addition, most suitable algorithms and predictive models (formulas and parameters) can be determined for any sensors, even those used in the past. Note that spectral data for this approach have to be precise reflectance data at the target surfaces, so that appropriate radiometric and atmospheric collections are needed.

2.7.2 STRATEGIC APPLICATION OF HYPERSPECTRAL DATA TO DRONE-, AIRBORNE-, AND SATELLITE-BASED MULTISPECTRAL IMAGE SENSORS

The number of earth observation satellites with high spatial resolution (under 10 m) has been increasing consistently during the past decade, and is further increasing at a rapid pace. A constellation of hundreds of small satellites is already realized (e.g., Dove satellites by Planet) and many similar programs are planned (Lal et al., 2017). Accordingly, global land surfaces can be observed once a day at high spatial resolution. On the other hand, the availability of small unmanned aerial vehicles (termed drones, UAVs, or UASs) is expanding rapidly worldwide (Zhang and Kovacs, 2012; Stöcker et al., 2017). This platform environment favors the advanced operational applications of remote sensing to assessment of ecophysiological functioning of crops and vegetation. The high spatial resolution and timely data acquisition are the unique advantages of drone-based remote sensing.

Under such a platform environment, the hyperspectral datasets would play a critical role for generating predictive models for any sensors with different specifications, regardless of the availability of satellite hyperspectral sensors (see Section 2.7.1). The generalized spectral index approach would greatly reduce the laborious and careful tasks for creation and validation of predictive models required for advanced use of various sensors with diverse specifications (number of spectral bands, central wavelength, and bandwidth).

Figure 2.26 shows application examples of the generic hyperspectral approach to diagnostic mapping of canopy nitrogen contents by satellite- and drone-based multispectral sensors. The optimal

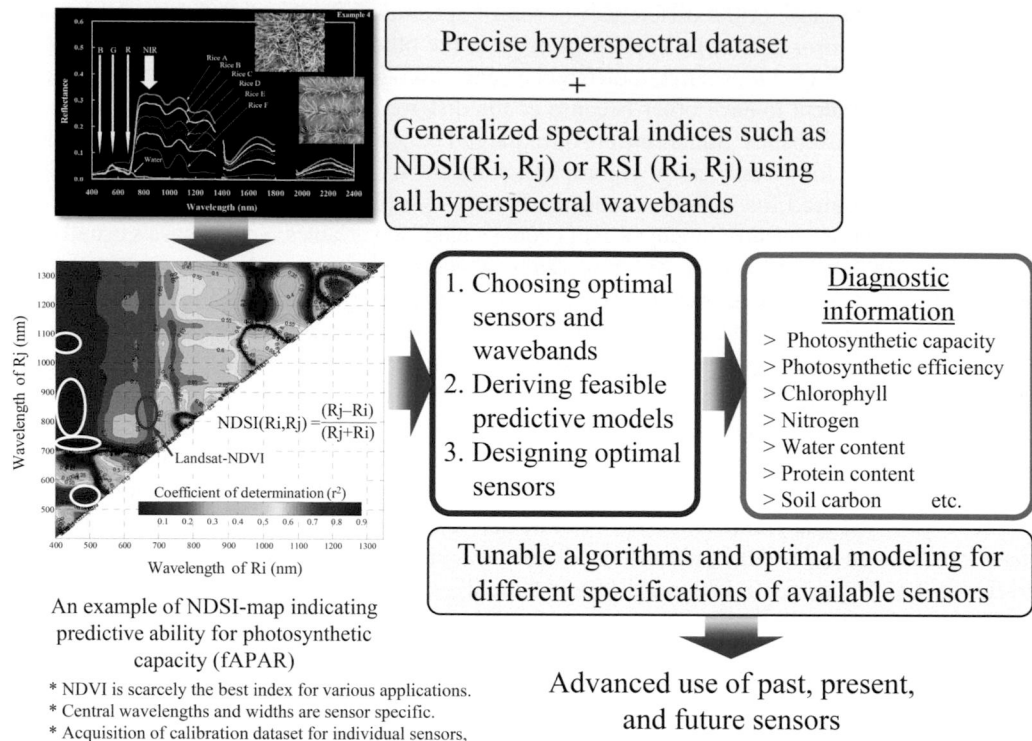

An example of NDSI-map indicating
predictive ability for photosynthetic
capacity (fAPAR)

* NDVI is scarcely the best index for various applications.
* Central wavelengths and widths are sensor specific.
* Acquisition of calibration dataset for individual sensors,
  locations, and targets is laborious and in efficient.

FIGURE 2.25 Schematic representation of the generic role of a hyperspectral dataset for assessment of biophysical and ecophysiological functioning of vegetation. (After Inoue, Y. 2017. *Journal of the Remote Sensing Society of Japan*, 37, 213–223.)

algorithms and predictive models for each sensor were created and validated based on hyperspectral datasets and applied to each sensor for mapping. The scatterplot with each map indicates the results of simple validation with independent *in situ* datasets. These results confirm the robust applicability of the generic approach based on hyperspectral datasets for creation of optimal algorithms and predictive models for individual band specifications. For example, in prediction of chlorophyll content, different optimal (best, second-best, etc.) models can be created in accordance with sensor specifications even with or without the red edge bands. Nevertheless, we should be careful about the quality of multispectral images since all algorithms and predictive models are affected by the accuracy of spectral reflectance. Appropriate radiometric and atmospheric corrections are required for satellite images, and precise radiometric correction and ortho-mosaic processing are needed for drone-based images (Inoue, 2017; Inoue and Yokoyama, 2017).

2.8   CONCLUSIONS AND PERSPECTIVES: DIAGNOSTIC INFORMATION FOR SMART AGRICULTURE AND ENVIRONMENTAL MANAGEMENT

So-called "precision agriculture" (PA) has emerged as one of the advanced technologies in recent decades (Stafford, 2000). GPS and agricultural machines equipped with sensors and variable-rate apparatus are essential for PA. In addition, georeferenced information from remote sensing can play a significant role in PA (Moran et al., 1997; Zhang and Kovacs, 2012) since site-specific management based on the spatial heterogeneity within each field is the central issue for PA. "Smart agriculture" (SA) is a new trend following PA, which is more advanced style of agriculture based on sensing and robotics technologies as well as information and communication technologies (ICT) including big

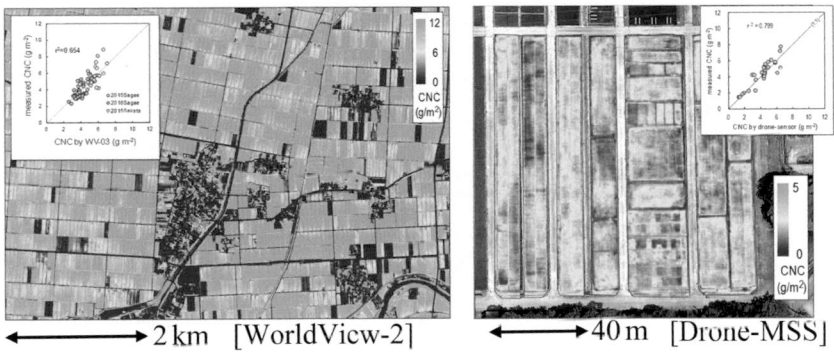

Model generator

Precise hyperspectral datasets with diverse *in-situ* ecophysiological data

Generalized spectral indices such as NDSI(Ri, Rj) or RSI (Ri, Rj) using all hyperspectral wavebands

Optimized sensor-specific predictive models

IKONOS, QuickBird, WorldView-2&3, SPOT-5/6, RapidEye, GeoEye-1, Sentinel-2A, Dove, etc. various Drone-based sensors.

←————→ 2 km [WorldView-2]   ←————→ 40 m [Drone-MSS]

FIGURE 2.26 Application of a generic model generator based on hyperspectral datasets. Examples shows the diagnostic maps of canopy nitrogen content at regional scale by WorldView-2 (left) and drone-based multispectral sensors (right). Optimal predictive models for each sensor were derived from the hyperspectral dataset and calibrated and validated before application to each multispectral sensor. Scatterplots show the independent validation of multispectral estimates using *in situ* data. (After Inoue, Y. 2017. *Journal of the Remote Sensing Society of Japan*, 37, 213–223; Inoue, Y. and Yokoyama, M. 2017. *Journal of the Remote Sensing Society of Japan*, 37, 224–235.)

data and artificial intelligence (Inoue, 2017; Figure 2.27). By making the most of these technologies, SA seeks higher yield and quality with less labor and inputs. Geoinformation from remote sensing will be even more critical in SA as a unique source of spatial data and diagnostic information for optimization of planning and management practices. Accordingly, in SA, ecophysiological information on crop growth such as chlorophyll and nitrogen contents, water stress index, maturity stage, and photosynthetic productivity and/or on soil fertility is needed for optimization of fertilizer application, irrigation management, and harvest scheduling as well as for yield prediction.

On the other hand, remotely sensed information plays an important role in various aspects of environmental policy making (de Leeuw et al., 2010; IPCC, 2014). Especially under conditions of climate change such as increasing atmospheric CO_2 and global warming, mitigation and adaptation strategies are of great concern worldwide. The dynamics of terrestrial vegetation is strongly related to exchange of CO_2, energy, and water between atmosphere and land surface, in which ecophysiological functioning of vegetation plays a critical role. Accordingly, the diagnostic geoinformation on

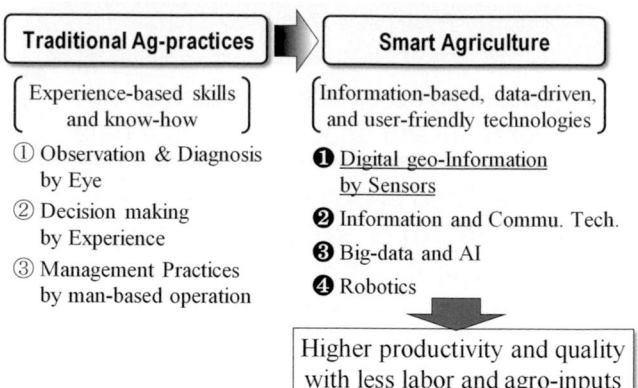

FIGURE 2.27 Innovative technologies needed for transformation from traditional agriculture to smart agriculture, and the role of diagnostic information from remote sensing in smart agriculture. (After Inoue, Y. 2017. *Journal of the Remote Sensing Society of Japan*, 37, 213–223.)

ecophysiological functioning of vegetation at regional or global scales is an essential basis for environmental sciences and policy making. Remotely sensed data are prerequisite for spatiotemporal assessment of carbon cycle in terrestrial ecosystems (Inoue et al., 2010; Thapa et al., 2015; Vadrevu et al., 2018; Figure 2.28). Assessment of biodiversity is also one of the important application areas for hyperspectral remote sensing (e.g., Nagendra et al., 2013; Feilhauer et al., 2017).

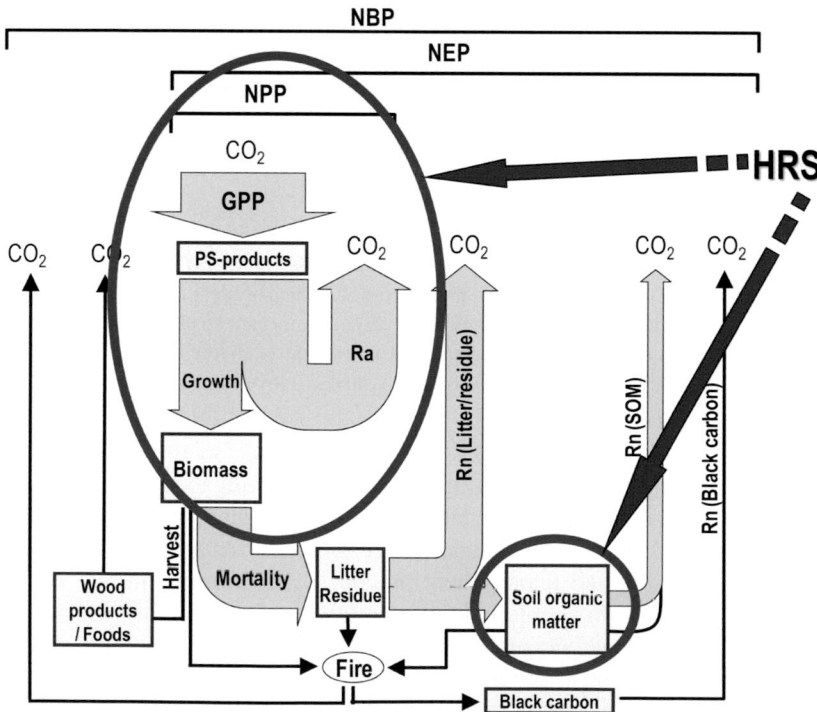

FIGURE 2.28 Hyperspectral assessment of ecophysiological functioning at ecosystem scales for vegetation and carbon cycle sciences as a basis for policy making for land use and environmental conservation. HRS: Hyperspectral remote sensing; GPP, gross primary production; NPP, net primary production; NBP, net biome production; NEP, net ecosystem production. (After Inoue, Y. et al., 2010. *International Journal of Applied Earth Observation and Geoinformation*, 12, 287–297.)

Hyperspectral data enables plant functional information to be acquired that is impossible to acquire with other types of sensors. Further studies on hyperspectral remote sensing of vegetation functioning are needed not only for hyperspectral satellite sensors such as EnMAP and HyspIRI (Staenz and Held, 2012), but also for advanced use of a wide range of multispectral sensors.

Hyperspectral remote sensing using spaceborne, airborne, drone-based, and ground-based sensors will play significant roles in these agricultural, ecological, and biogeological sciences and applications. Interdisciplinary research studies between the disciplines of physics, ecophysiology, and computing technologies can be productive.

ACKNOWLEDGMENTS

This work was supported in part by MEXT, JSPS and CSTI-SIP (NARO), Japan, respectively.

Author contribution: This chapter was conceived and supervised by Inoue. Section 2.6 was written by Darvishzadeh and Skidmore, and all the other sections were written by Inoue.

REFERENCES

Abdullah, H., Darvishzadeh, R., Skidmore, A.K., Groen, T.A., and Heurich, M. 2018. European spruce bark beetle (*Ips typographus*, L.) green attack affects foliar reflectance and biochemical properties. *International Journal of Applied Earth Observation and Geoinformation*, 64, 199–209.

Ali, I., Greifeneder, F., Stamenkovic, J., Neumann, M., and Notarnicola, C. 2015. Review of machine learning approaches for biomass and soil moisture retrievals from remote sensing data. *Remote Sensing*, 7, 16398–16421.

Ali, A.M., Darvishzadeh, R., Skidmore, A.K., and van Duren, I. 2016a. Effects of canopy structural variables on retrieval of leaf dry matter content and specific leaf area from remotely sensed data. *IEEE Journal of Selected Topics in Applied Earth Observations and Remote Sensing*, 9, 898–909.

Ali, A.M., Darvishzadeh, R., Skidmore, A.K., Duren, I. van, Heiden, U., and Heurich, M. 2016b. Estimating leaf functional traits by inversion of PROSPECT: Assessing leaf dry matter content and specific leaf area in mixed mountainous forest. *International Journal of Applied Earth Observation and Geoinformation*, 45, A, 66–76.

Ali, A.M., Skidmore, A.K., Darvishzadeh, R., van Duren, I., Holzwarth, S., and Mueller, J. 2016c. Retrieval of forest leaf functional traits from HySpex imagery using radiative transfer models and continuous wavelet analysis. *ISPRS Journal of Photogrammetry and Remote Sensing*, 122, 68–80.

Ali, A.M., Darvishzadeh, R., and Skidmore, A.K. 2017a. Retrieval of specific leaf area from Landsat-8 surface reflectance data using statistical and physical models. *IEEE Journal of Selected Topics in Applied Earth Observations and Remote Sensing*, 10, 3529–3536.

Ali, A.M., Darvishzadeh, R., Skidmore, A.K., and van Duren, I. 2017b. Specific leaf area estimation from leaf and canopy reflectance through optimization and validation of vegetation indices. *Agricultural and Forest Meteorology*, 236, 162–174.

Asner, G.P., Wessman, C.A., Schimel, D.S., and Archer, S. 1998. Variability in Leaf and Litter Optical Properties: Implications for BRDF Model Inversions Using AVHRR, MODIS, and MISR. *Remote Sensing of Environment*, 63, 243–257.

Asner, G.P. and Martin, R.E. 2009. Airborne spectranomics: Mapping canopy chemical and taxonomic diversity in tropical forests. *Frontiers in Ecology and the Environment*, 7, 269–276.

Asner, G.P., Martin, R.E., Knapp, D.E., Tupayachi, R., Anderson, C., Carranza, L., Martinez, P., Houcheime, M., Sinca, F., and Weiss, P. 2011. Spectroscopy of canopy chemicals in humid tropical forests. *Remote Sensing of Environment*, 115, 3587–3598.

Asrar, G., Myneni, R.B., and Kanemasu, E.T. 1989. Estimation of plant-canopy attributes from spectral reflectance measurements. In: Asrar, G. (Ed.), *Theory and Application of Optical Remote Sensing*. Wiley Interscience, New York, USA, pp. 252–296.

Atzberger, C. 2000. Development of an invertible forest reflectance model: The INFOR-Model, In: Buchroithner, M.F. (Ed.), *A Decade of Trans-European Remote Sensing Cooperation: Proceedings of the 20th EARSeL Symposium*. Dresden, Germany, 14–16 June, 2000, pp. 2039–2044.

Bacour, C., Jacquemoud, S., Tourbier, Y., Dechambre, M., and Frangi, J.P. 2002. Design and analysis of numerical experiments to compare four canopy reflectance models. *Remote Sensing of Environment*, 79, 72–83.

Baret, F. and Guyot, G. 1991. Potentials and limits of vegetation indices for LAI and APAR assessment. *Remote Sensing of Environment*, 35, 161–173.

Baret, F., Houlès, V., and Guérif, M. 2007. Quantification of plant stress using remote sensing observations and crop models: The case of nitrogen management. *Journal of Experimental Botany*, 58, 869–880.

Barton, C.V.M. and North, P.R.J. 2001. Remote sensing of canopy light use efficiency using the photochemical reflectance index—Model and sensitivity analysis. *Remote Sensing of Environment*, 78, 264–273.

Blackburn, G.A. 1998. Quantifying chlorophylls and carotenoids at leaf and canopy scales: An evaluation of some hyperspectral approaches. *Remote Sensing of Environment*, 66, 273–285.

Boegh, E., Soegaard, H., Broge, N., Hasager, C.B., Jensen, N.O., Schelde, K., and Thomsen, A. 2002. Airborne multispectral data for quantifying leaf area index, nitrogen concentration, and photosynthetic efficiency in agriculture. *Remote Sensing of Environment*, 81, 179–193.

Bongiovanni, R. and Lowenberg-Deboer, J. 2004. Precision agriculture and sustainability. *Precision Agriculture*, 5, 359–387.

Broge, N.H. and Leblanc, E. 2000. Comparing prediction power and stability of broadband and hyperspectral vegetation indices for estimation of green leaf area index and canopy chlorophyll density. *Remote Sensing of Environment*, 76, 156–172.

Cadotte, M.W., Carscadden, K., and Mirotchnick, N. 2011. Beyond species: Functional diversity and the maintenance of ecological processes and services. *Journal of Applied Ecology*, 48, 1079–1087.

Cammarano, D., Fitzgerald, G.J., Casa, R., and Basso, B. 2014. Assessing the robustness of vegetation indices to estimate wheat N in Mediterranean environments. *Remote Sensing*, 6, 2827–2844.

Camps-Valls, G., Bruzzone, L., Rojo-Alvarez, J.L., and Melgani, F. 2006. Robust support vector regression for biophysical variable estimation from remotely sensed images. *IEEE Geoscience and Remote Sensing Letters*, 3, 339–343.

Carter, G.A. 1994. Ratios of leaf reflectances in narrow wavebands as indicators of plant stress. *International Journal of Remote sensing*, 15, 697–703.

Castro-Esau, K. L., Sánchez-Azofeifa, G. A., and Rivard, B. 2006. Comparison of spectral indices obtained using multiple spectroradiometers. *Remote Sensing of Environment*, 103, 276–288.

Ceccato, P., Gobron, N., Flasse, S., Pinty, B., and Tarantola, S. 2002. Designing a spectral index to estimate vegetation water content from remote sensing data: Part 2. Validation and applications. *Remote Sensing of Environment*, 82, 198–207.

Cheng, T., Rivard, B., Sánchez-Azofeifa, G.A., Feng, J., and Calvo-Polanco, M. 2010. Continuous wavelet analysis for the detection of green attack damage due to mountain pine beetle infestation. *Remote Sensing of Environment*, 114, 899–910.

Cho, M.A., Skidmore, A.K., and Atzberger, C.G. 2008. Towards red-edge positions less sensitive to canopy biophysical parameters for leaf chlorophyll estimation using properties optique spectrales des feuilles PROSPECT and scattering by arbitrarily inclined leaves SAILH simulated data. *International Journal of Remote Sensing*, 29, 2241–2255.

Choudhury, B.J. 2000. A sensitivity analysis of the radiation use efficiency for gross photosynthesis and net carbon accumulation by wheat. *Agricultural and Forest Meteorology*, 101, 217–234.

Clerbaux, C., Drummond, J.R., Flaud, J.-M., and Orphal, J. 2011. Using thermal infrared absorption and emission to determine trace gases. *The Remote Sensing of Tropospheric Composition from Space*, Springer, pp. 123–151.

Colgan, M.S., Martin, R.E., Baldeck, C.A., and Asner, G.P. 2015. Tree foliar chemistry in an African savanna and its relation to life history strategies and environmental filters. *PLoS One*, 10(5), e0124078.

Combal, B., Baret, F., Weiss, M., Trubuil, A., Macé, D., Pragnère, A., Myneni, R., Knyazikhin, Y., and Wang, L. 2003. Retrieval of canopy biophysical variables from bidirectional reflectance using prior information to solve the ill-posed inverse problem. *Remote Sensing of Environment*, 84, 1–15.

Curran, P.J. 1989. Remote sensing of foliar chemistry. *Remote Sensing of Environment*, 30, 271–278.

Darvishzadeh, R., Skidmore, A., Atzberger, C., and van Wieren, S. 2008a. Estimation of vegetation LAI from hyperspectral reflectance data: Effects of soil type and plant architecture. *International Journal of Applied Earth Observation and Geoinformation*, 10, 358–373.

Darvishzadeh, R., Skidmore, A., Schlerf, M., and Atzberger, C. 2008b. Inversion of a radiative transfer model for estimating vegetation LAI and chlorophyll in a heterogeneous grassland. *Remote Sensing of Environment*, 112, 2592–2604.

Darvishzadeh, R., Atzberger, C., Skidmore, A.K., and Abkar, A.A. 2009. Leaf area index derivation from hyperspectral vegetation indices and the red edge position. *International Journal of Remote Sensing*, 30, 6199–6218.

Dash, J. and Curran, P.J. 2004. The MERIS terrestrial chlorophyll index. *International Journal of Remote Sensing*, 25, 5403–5413.

Daughtry, C.S.T., Gallo, K.P., and Bauer, M.E. 1983. Spectral estimates of Solar radiation intercepted by corn canopies. *Agronomy Journal*, 75, 527–531.

Daughtry, C.S.T., Walthall, C.L., Kim, M.S., de Colstoun, E.B., and McMurtrey, J.E. III. 2000. Estimating corn leaf chlorophyll concentration from leaf and canopy reflectance. *Remote Sensing of Environment*, 74, 229–239.

Dawson, T.P., Curran, P.J., North, P.R.J., and Plummer, S.E. 1999. The propagation of foliar biochemical absorption features in forest canopy reflectance: a theoretical analysis. *Remote Sensing of Environment*, 67, 147–159.

de Leeuw, J., Georgiadou, Y., Kerle, N., de Gier, A., Inoue, Y., Ferwerda, J., Smies, M., and Narantuya, D. 2010. The function of remote sensing in support of environmental policy. *Remote Sensing*, 2, 1731–1750.

De Pury, D.G.G. and Farquhar, G.D. 1997. Simple scaling of photosynthesis from leaves to canopies without the errors of big-leaf models. *Plant, Cell and Environment*, 20, 537–557.

Del Frate, F., Ferrazzoli, P., and Schiavon, G. 2003. Retrieving soil moisture and agricultural variables by microwave radiometry using neural networks. *Remote Sensing of Environment*, 84, 174–183.

Delegido, J., Fernndez, G., Gand, S., and Moreno, J. 2008. Retrieval of chlorophyll content and LAI of crops using hyperspectral techniques: Application to PROBA/CHRIS data. *International Journal of Remote Sensing*, 29, 7107–7127.

Demetriades-Shah, T.H., Steven, M.D., and Clark, J.A. 1990. High resolution derivative spectra in remote sensing. *Remote Sensing of Environment*, 33, 55–64.

Demmig-Adams, B.B. and Adams, W.W. 1996. The role of xanthophyll cycle carotenoids in the protection of photosynthesis. *Trends in Plant Science*, 1, 21–26.

Dobermann, A. and Cassman, K. G. 2004. Environmental dimensions of fertilizer nitrogen: What can be done to increase nitrogen use efficiency and ensure food security? In: Mosier, A.R., Syers, J.K., and Freney, J.R. (Eds.), *Agriculture and the Nitrogen Cycle*. Island Press, Washington, USA, pp. 261–278.

Doktor, D., Lausch, A., Spengler, D., and Thurner, M. 2014. Extraction of plant physiological status from hyperspectral signatures using machine learning methods. *Remote Sensing*, 6, 12247–12274.

Durbha, S.S., King, R.L., and Younan, N.H. 2007. Support vector machines regression for retrieval of leaf area index from multiangle imaging spectroradiometer. *Remote Sensing of Environment*, 107, 348–361.

Feilhauer, H., Somers, B., and van der Linden, S. 2017. Optical trait indicators for remote sensing of plant species composition: Predictive power and seasonal variability. *Ecological Indicators*, 73, 825–833.

Féret, J.-B., François, C., Gitelson, A., Asner, G.P., Barry, K.M., Panigada, C., Richardson, A.D., and Jacquemoud, S. 2011. Optimizing spectral indices and chemometric analysis of leaf chemical properties using radiative transfer modeling. *Remote Sensing of Environment*, 115, 2742–2750.

Ferwerda, J.G., Skidmore, A.K., and Mutanga, O. 2005. Nitrogen detection with hyperspectral normalized ratio indices across multiple plant species. *International Journal of Remote sensing*, 26, 4083–4095.

Field, C. and Mooney, H.A. 1986. The photosynthesis–nitrogen relationship in wild plants, In: Givnish, T.J. (Ed.), *On the Economy of Plant Form and Function*. Cambridge University Press., London, UK, pp. 25–55.

Fourty, T., Baret, F., Jacquemoud, S., Schmuck, G., and Verdebout, J. 1996. Leaf optical properties with explicit description of its biochemical composition: Direct and inverse problems. *Remote Sensing of Environment*, 56, 104–117.

Gamon, J.A., Peñuelas, J., and Field, C.B. 1992. A narrow-waveband spectral index that tracks diurnal changes in photosynthetic efficiency. *Remote Sensing of Environment*, 41, 35–44.

Garbulsky, M.F., Peñuelas, J., Gamon, J., Inoue, Y., and Filella, I. 2010. The photochemical reflectance index (PRI) and the remote sensing of leaf, canopy and ecosystem radiation use efficiencies. *Remote Sensing of Environment*, 115, 281–297.

Gitelson, A.A. and Merzlyak, M.N. 1997. Remote estimation of chlorophyll content in higher plant leaves. *International Journal of Remote Sensing*, 18, 2691–2697.

Gitelson, A.A., Merzlyak, M.N., and Chivkunova, O.B. 2001. Optical properties and nondestructive estimation of anthocyanin content in plant leaves. *Photochemistry and Photobiology*, 74, 38–45.

Gitelson, A.A., Gritz, U., and Merzlyak, M.N. 2003. Relationships between leaf chlorophyll content and spectral reflectance and algorithms for non-destructive chlorophyll assessment in higher plant leaves. *Journal of Plant Physiology*, 160, 271–282.

Gitelson, A.A., Viña, A., Rundquist, D.C., Ciganda, V., and Arkebauer, T.J. 2005. Remote estimation of canopy chlorophyll content in crops. *Geophysical Research Letters*, 32, L08403, doi:10.1029/2005GL022688.

Goetz, S.J., Prince, S.D., Goward, S.N., Thawley, M.M., and Small, J. 1999. Satellite remote sensing of primary production: An improved production efficiency modeling approach. *Ecological Modeling*, 122, 239–255.

Grace, J., Nichol, C., Disney, M., Lewis, P., Quaife, T., and Bowyer, P. 2007. Can we measure terrestrial photosynthesis from space directly, using spectral reflectance and fluorescence? *Global Change Biology*, 13, 1484–1497.

Grossman, Y.L., Ustin, S.L., Jacquemoud, S., Sanderson, E.W., Schmuck, G., and Verdebout, J. 1996. Critique of stepwise multiple linear regression for the extraction of leaf biochemistry information from leaf reflectance data. *Remote Sensing of Environment*, 56, 182–193.

Hall, F.G., Shimabukuro, Y.E., and Huemmrich, K.F. 1995. Remote sensing of forest biophysical structure using mixture decomposition and geometric reflectance models. *Ecological Applications*, 5, 993–1013.

Hansen, P.M. and Schjoerring, J.K. 2003. Reflectance measurement of canopy biomass and nitrogen status in wheat crops using normalized difference vegetation indices and partial least squares regression. *Remote Sensing of Environment*, 86, 542–553.

Hikosaka, K. 2004. Interspecific difference in the photosynthesis-nitrogen relationship: Patterns, physiological causes, and ecological importance. *Journal of Plant Research*, 117(6), 481–494.

Hinzman, L.D., Bauer, M.E., and Daughtry, C.S.T. 1986. Effects of nitrogen fertilization on growth and reflectance characteristics of winter wheat. *Remote Sensing of Environment*, 19, 47–61.

Homolová, L., Malenovský, Z., Clevers, J.G.P.W., García-Santos, G., and Schaepman, M.E. 2013. Review of optical-based remote sensing for plant trait mapping. *Ecological Complex*, 15, 1–16.

Houlès, V., Guérif, M., and Mary, B. 2007. Elaboration of a nitrogen nutrition indicator for winter wheat based on leaf area index and chlorophyll content for making nitrogen recommendations. *European Journal of Agronomy*, 27, 1–11.

Huesca, M., García, M., Roth, K.L., Casas, A., and Ustin, S.L. 2016. Canopy structural attributes derived from AVIRIS imaging spectroscopy data in a mixed broadleaf/conifer forest. *Remote Sensing of Environment*, 182, 208–226.

Huete, A.R. 1988. A soil vegetation adjusted index (SAVI). *Remote Sensing of Environment*, 25, 295–309.

Huete, A.R., Didan, K., Miura, T., Rodriguez, E.P., Gao, X., and Ferreira, L.G. 2002. Overview of the radiometric and biophysical performance of the MODIS vegetation indices. *Remote Sensing of Environment*, 83, 195–213.

Hunt, E.R., Jr. and Rock, B.N. 1989. Detection of changes in leaf water content using near- and middle-infrared reflectances. *Remote Sensing of Environment*, 30, 43–54.

Inada, K. 1985. Spectral ratio of reflectance for estimating chlorophyll content of leaf. *Japanese Journal of Crop Science*, 154, 261–265.

Inoue, Y., Kimball, B.A., Jackson, R.D., Pinter, P.J. Jr., and Reginato, R.J. 1990. Remote estimation of leaf transpiration rate and stomatal resistance based on infrared thermometry. *Agricultural and Forest Meteorology*, 51, 21–33.

Inoue, Y. and Iwasaki, K. 1991. Spectral estimation of radiation absorptance and leaf area index in corn canopies as affected by canopy architecture and growth stage. *Japanese Journal of Crop Science*, 60, 578–580.

Inoue, Y., Morinaga, S., and Shibayama, M. 1993. Non-destructive estimation of water status of intact crop leaves based on spectral reflectance measurements. *Japanese Journal of Crop Science*, 62, 462–469.

Inoue, Y., Moran, M.S., and Horie, T. 1998. Analysis of spectral measurements in Rice paddies for predicting rice growth and yield based on a simple crop simulation model. *Plant Production Science*, 1, 269–279.

Inoue, Y., Kurosu, T., Maeno, H., Uratsuka, S., Kozu, T., Dabrowska-Zielinska, K., and Qi, J. 2002. Season-long daily measurements of multifrequency (Ka, Ku, X, C, and L) and full-polarization backscatter signatures over paddy rice field and their relationship with biological variables. *Remote Sensing of Environment*, 81, 194–204.

Inoue, Y. 2003. Synergy of remote sensing and modeling for estimating ecophysiological processes in plant production. *Plant Production Science*, 6, 3–16.

Inoue, Y. and Olioso, A. 2006. Estimating dynamics of ecosystem CO_2 flux and biomass production in agricultural field by synergy of process model and remotely sensed signature. *Journal of Geophysical Research*, 111(D24), D24S91, doi:10.1029/2006JD007469.

Inoue, Y. and Peñuelas, J. 2006. Relationship between light use efficiency and photochemical reflectance index in soybean leaves as affected by soil water content. *International Journal of Remote Sensing*, 27, 5249–5254.

Inoue, Y., Miah, G., Sakaiya, E., Nakano, K., and Kawamura, K. 2008a. NDSI map and IPLS using hyperspectral data for assessment of plant and ecosystem variables. *Journal of the Remote Sensing Society of Japan*, 28, 1–14.

Inoue, Y., Peñuelas, J., Miyata, A., and Mano, M. 2008b. Normalized difference spectral indices for estimating photosynthetic efficiency and capacity at a canopy scale derived from hyperspectral and CO_2 flux measurements in rice. *Remote Sensing of Environment*, 112, 156–172.

Inoue, Y., Kiyono, Y., Asai, H., Ochiai, Y., Qi, J., Olioso, A., Shiraiwa, T., Horie, T., Saito, K., and Dounagsavanh, L. 2010. Assessing land use and carbon stock in slash-and-burn ecosystems in tropical mountain of Laos based on time-series satellite images. *International Journal of Applied Earth Observation and Geoinformation*, 12, 287–297.

Inoue, Y., Dabrowska-Zierinska, K., and Qi, J. 2012a. Synoptic assessment of environmental impact of agricultural management: A case study on nitrogen fertilizer impact on groundwater quality, using a fine-scale geoinformation system. *International Journal of Environmental Studies*, 69, 443–460.

Inoue, Y., Sakaiya, E., Zhu, Y., and Takahashi, W. 2012b. Diagnostic mapping of canopy nitrogen content in rice based on hyperspectral measurements. *Remote Sensing of Environment*, 126, 210–221.

Inoue, Y., Sakaiya, E., and Wang, C. 2014. Capability of C-band backscattering coefficients from high-resolution satellite SAR sensors to assess biophysical variables in paddy rice. *Remote Sensing of Environment*, 140, 257–266.

Inoue, Y., Guérif, M., Baret, F., Skidmore, A., Gitelson, A., Schlerf, M., Darvishzadeh, R., and Olioso, A. 2016. Simple and robust methods for remote sensing of canopy chlorophyll content: A comparative analysis of hyperspectral data for different types of vegetation. *Plant, Cell and Environment*, 39, 2609–2623.

Inoue, Y. 2017. Remote sensing of plant and soil information by high-resolution optical satellite sensors and its applications to smart agriculture. *Journal of the Remote Sensing Society of Japan*, 37, 213–223.

Inoue, Y. and Yokoyama, M. 2017. Drone-based remote sensing of crops and soils and its application to smart agriculture. *Journal of the Remote Sensing Society of Japan*, 37, 224–235.

IPCC. 2014. Summary for policymakers. In: Field, C.B. and Barros, V.R. et al. (Eds), *Climate Change 2014: Impacts, Adaptation, and Vulnerability. Part A: Global and Sectoral Aspects. Contribution of Working Group II to the Fifth Assessment Report of the Intergovernmental Panel on Climate Change.* Cambridge University Press, Cambridge, UK, pp. 1–32.

Ishijima, K., Sugawara, S., Kawamura, K., Hashida, G., Morimoto, S., Murayama, S., Aoki, S., and Nakazawa, T. 2007. Temporal variations of the atmospheric nitrous oxide concentration and its $\delta^{15}N$ and $\delta^{18}O$ for the latter half of the 20th century reconstructed from firn air analyses. *Journal of Geophysical Research*, 112, D03305, doi:10.1029/2006JD007208.

Jackson, R.D., Idso, S.B., Reginato, R.J., and Pinter, P.J., Jr. 1981. Canopy temperature as a crop water stress indicator. *Water Resources Research*, 17, 1133–1138.

Jacquemoud, S. and Baret, F. 1990. PROSPECT: A model of leaf optical properties spectra. *Remote Sensing of Environment*, 34, 75–91.

Jacquemoud, S., Ustin, S.L., Verdebout, J., Schmuck, G., Andreoli, G., and Hosgood, B. 1996. Estimating leaf biochemistry using the PROSPECT leaf optical properties model. *Remote Sensing of Environment*, 56, 194–202.

Jacquemoud, S., Verhoef, W., Baret, F., Bacour, C., Zarco-Tejada, P.J., Asner, G.P., François, C., and Ustin, S.L. 2009. PROSPECT+SAIL: A review of use for vegetation characterization. *Remote Sensing of Environment*, 113, S56–S66.

Jongschaap, R.E. and Booij, R. 2004. Spectral measurements at different spatial scales in potato: Relating leaf, plant and canopy nitrogen status. *International Journal of Applied Earth Observation and Geoinformation*, 5, 204–218.

Karnieli, A., Bayarjargal, Y., Bayasgalan, M., Mandakh, B., Dugarjav, Ch., Burgheimer, J., Khudulmur, S., Bazha, S.N., and Gunin, P.D. 2013. Do vegetation indices provide a reliable indication of vegetation degradation? A case study in the Mongolian pastures. *International Journal of Remote Sensing*, 34, 6243–6262.

Kiniry, J.R., Jones, C.A., O'Toole, J.C., Blanchet, R., Cabelguenne, M., and Spanel, D.A. 1989. Radiation-use efficiency in biomass accumulation prior to grain-filling for five grain-crop species. *Field Crops Research*, 20, 51–64.

Kljun, N., Kastner-Klein, P., Fedorovich, E., and Rotach, M.W. 2004. Evaluation of Lagrangian footprint model using data from wind tunnel convective boundary layer. *Agricultural and Forest Meteorology*, 127, 189–201.

Knyazikhin, Y., Lewis, P., Disney, M.I., Stenberg, P., Mottus, M., Rautiainen, M., Kaufmann, R.K. et al. 2013. Reply to Townsend et al.: Decoupling contributions from canopy structure and leaf optics is critical for remote sensing leaf biochemistry. *Proceedings of the National Academy of Sciences*, 110, E1075–E1075.

Kokaly, R.F., Asner, G.P., Ollinger, S.V., Martin, M.E., and Wessman, C.A. 2009. Characterizing canopy biochemistry from imaging spectroscopy and its application to ecosystem studies. *Remote Sensing of Environment*, 113, S78–S91.

Kumar, M. and Monteith, J.L. 1982. Remote sensing of crop growth. In: Smith, H. (Ed.), *Plants and the Daylight Spectrum*. Academic Press, London, UK, pp. 133–144.

Kuusk, A. 1991. The hot-spot effect in plant canopy reflectance. In: R.B. Myneni and J. Ross (Eds.), *Photon–Vegetation Interactions*. Springer-Verlag, New York, USA, pp. 139–159.

Lal, B., Balakrishnan, A., Picard, A., Corbin, B., Behrens, J., Green, E., and Myers, R. 2017. *Trends in small satellite technology and the role of the NASA small spacecraft technology program*. Report of Science and Technology Policy Institute. https://www.nasa.gov/sites/default/files/atoms/files/nac_march2017_blal_ida_sstp_tagged.pdf. (Accessed December 15, 2017)

Lavorel, S., Grigulis, K., Lamarque, P., Colace, M., Garden, D., Girel, J., Pellet, G., and Douzet, R. 2011. Using plant functional traits to understand the landscape distribution of multiple ecosystem services. *Journal of Ecology*, 99, 135–147.

Leardi, R. and Norgaard, L. 2004. Sequential application of backward interval partial least squares and genetic algorithms for the selection of relevant spectral regions. *Journal of Chemometrics*, 18, 486–497.

Lee, Y., Yang, C., Chang, K., and Shen, Y. 2008. A simple spectral index using reflectance of 735 nm to assess nitrogen status of rice canopy. *Agronomy Journal*, 100, 205–212.

Lefsky, M.A., Cohen, W.B., Acker, S.A., Parker, G.G., Spies, T.A., and Harding, D. 1999. Lidar remote sensing of the canopy structure and biophysical properties of douglas-fir Western Hemlock forests. *Remote Sensing of Environment*, 70, 339–361.

le Maire, G., François, C., Soudani, K., Berveiller, D., Pontailler, J.Y., Bréda, N., Genet, H., Davi, H., and Dufrêne, E. 2008. Calibration and validation of hyperspectral indices for the estimation of broadleaved forest leaf chlorophyll content, leaf mass per area, leaf area index and leaf canopy biomass. *Remote Sensing of Environment*, 112, 3846–3864.

Le Toan, T., Ribbes, F., Wang, L.F., Nicolas, F., Ding, K.H., Kong, J.A., Fujita, M., and Kurosu, T. 1997. Rice crop mapping and monitoring using ERS-1 data based on experiment and modeling results. *IEEE Transactions on Geoscience and Remote Sensing*, 35, 41–56.

Liu, W., Baret, F., Gu, X., Zhang, B., Tong, Q., and Zheng, L. 2003. Evaluation of methods for soil surface moisture estimation from reflectance data. *International Journal of Remote Sensing*, 24, 2069–2083.

Lloyd, J., Bloomfield, K., Domingues, T.F., and Farquhar, G.D. 2013. Photosynthetically relevant foliar traits correlating better on a mass vs an area basis: Of ecophysiological relevance or just a case of mathematical imperatives and statistical quicksand? *New Phytologist*, 199, 311–321.

Martin, M.E., Plourde, L.C., Ollinger, S.V., Smith, M.-L., and McNeil, B.E. 2008. A generalizable method for remote sensing of canopy nitrogen across a wide range of forest ecosystems. *Remote Sensing of Environment*, 112, 3511–3519.

Meroni, M., Colombo, R., and Panigada, C. 2004. Inversion of a radiative transfer model with hyperspectral observations for LAI mapping in poplar plantations. *Remote Sensing of Environment*, 92, 195–206.

Ministry of Agriculture, Fisheries and Food of UK (MAFF-UK) 2000. *Fertilizer Recommendations for Agricultural and Horticultural crops* (RB209). 7th Edn., 1–175pp.

Mirzaie, M., Darvishzadeh, R., Shakiba, A., Matkan, A.A., Atzberger, C., and Skidmore, A. 2014. Comparative analysis of different uni- and multi-variate methods for estimation of vegetation water content using hyperspectral measurements. *International Journal of Applied Earth Observation and Geoinformation*, 26, 1–11.

Monsi, M. and Saeki, T. 1953. Uber den Lichtfactor in den Pflanzengesellshaften und seine Bedeutung fur dir Stoffproduktion. *Japanese Journal of Botany*, 14, 22–52.

Monteith, J.L. 1977. Climate and the efficiency of crop production in Britain. *Philosophical Transactions of the Royal Society of London B*, 281, 277–294.

Moran, M.S., Clarke, T.R., Inoue, Y., and Vidal, A. 1994. Estimating crop water deficit using the relation between surface-air temperature and spectral vegetation index. *Remote Sensing of Environment*, 49, 246–263.

Moran, M.S., Inoue, Y., and Barnes, E.M. 1997. Opportunities and limitations for image-based remote sensing in precision crop management. *Remote Sensing of Environment*, 61, 319–346.

Morita, K. 1978. A physiological study on the dynamic status of leaf nitrogen in rice plants. *Bulletin of Hokuriku Agricultural Experimental Station*, 21, 1–61.

Muñoz-Huerta, R.F., Guevara-Gonzalez, R.G., Contreras-Medina, L.M., Torres-Pacheco, I., Prado-Olivarez, J., and Ocampo-Velazquez, R. V. 2013. A review of methods for sensing the nitrogen status in plants: Advantages, disadvantages and recent advances. *Sensors*, 13, 10823–10843.

Mutanga, O. and Skidmore, A.K. 2004. Narrow band vegetation indices overcome the saturation problem in biomass estimation. *International Journal of Remote Sensing*, 25, 3999–4014.

Mutanga, O. and Skidmore, A.K. 2007. Red edge shift and biochemical content in grass canopies. *ISPRS Journal of Photogrammetry and Remote Sensing*, 62(1), 34–42.

Myneni, R.B. and Williams, D.L. 1994. On the relationship between FAPAR and NDVI. *Remote Sensing of Environment*, 49, 200–211.

Naesset, E., Bollandsas, O.M., and Gobakken, T. 2005. Comparing regression methods in estimation of biophysical properties of forest stands from two different inventories using laser scanner data. *Remote Sensing of Environment*, 94, 541–553.

Nagendra, H., Lucas, R., Honrado, J.P., Jongman, R.H.G., Tarantino, C., Adamo, M., and Mairota, P. 2013. Remote sensing for conservation monitoring: Assessing protected areas, habitat extent, habitat condition, species diversity, and threats. *Ecological Indicators*, 33, 45–59.

Neinavaz, E., Skidmore, A.K., Darvishzadeh, R., and Groen, T.A. 2016a. Retrieval of leaf area index in different plant species using thermal hyperspectral data. *ISPRS Journal of Photogrammetry and Remote Sensing*, 119, 390–401.

Neinavaz, E., Skidmore, A.K., Darvishzadeh, R., and Groen, T.A. 2016b. Leaf area index retrieved from thermal hyperspectral data. ISPRS Commission VII, WG VII/3. *International Archives of the Photogrammetry, Remote Sensing and Spatial Information Sciences—ISPRS Archives*, 99–105.

Neinavaz, E., Darvishzadeh, R., Skidmore, A. K., and Groen, T. A. 2016c. Measuring the response of canopy emissivity spectra to leaf area index variation using thermal hyperspectral data. *International Journal of Applied Earth Observation and Geoinformation*, 53, 40–47.

Nishio, M. 2005. *Agriculture and Environmental Pollution—Technology and Policy for Soil Environment -*. Rural Culture Association, Tokyo, Japan, 439pp.

Nobel, P.S. 2005. *Physicochemical and Environmental Plant Physiology* (3rd Ed.). Elsevier Academic Press, Amsterdam, Netherlands, 567pp.

Norgaard, L., Saudland, A., Wagner, J., Nielsen, J.P., Munck, L., and Engelsen, S.B. 2000. Interval partial least-squares regression (iPLS): A comparative chemometric study with an example from near-infrared spectroscopy. *Applied Spectroscopy*, 54, 413–419.

Notarnicola, C., Angiulli, M., and Posa, F. 2008. Soil moisture retrieval from remotely sensed data: Neural network approach versus Bayesian method. *IEEE Transactions on Geoscience and Remote Sensing*, 46, 547–557.

Nouvellon, Y., Seen, D.L., Rambal, S., Begue A., Moran, M.S., Kerr, Y., and Qi, J. 2000. Time course of radiation use efficiency in a shortgrass ecosystem: Consequences for remotely sensed estimation of primary production. *Remote Sensing of Environment*, 71, 43–55.

Ollinger, S. V., Richardson, A.D., Martin, M.E., Hollinger, D.Y., Frolking, S.E., Reich, P.B., Plourde, L.C. et al. 2008. Canopy nitrogen, carbon assimilation, and albedo in temperate and boreal forests: Functional relations and potential climate feedbacks. *Proceedings of the National Academy of Sciences*, 105, 19336–19341.

Olioso, A., Inoue, Y., Ortega-Farias, S., Demarty, J., Wigneron, J-P. Braud, T., Jacob, F. et al. 2005. Future directions for advanced evapotranspiration modeling: Assimilation of remote sensing data into crop simulation models and SVAT models. *Irrigation and Drainage Systems*, 19, 377–412.

Peñuelas, J., Filella, I., and Gamon, J.A. 1995. Assessment of plant photosynthetic radiation-use efficiency with spectral reflectance. *New Phytologist*, 131, 291–296.

Peñuelas J., Filella I. 1998. Visible and near-infrared reflectance techniques for diagnosing plant physiological status. *Trends in Plant Science*, 3, 151–156.

Peñuelas, J. and Inoue, Y. 1999. Reflectance indices indicative of changes in water and pigment content of peanut and wheat leaves. *Photosynthetica*, 36, 355–360.

Pereira, H.M., Ferrier, S., Walters, M., Geller, G.N., Jongman, R.H.G., Scholes, R.J., Bruford, M.W. et al. 2013. Essential biodiversity variables. *Science*, 80–339, 277–278.

Pierce, L., Running, S.W., and Walker, J. 1994. Regional-scale relationships of leaf area index to specific leaf area and leaf nitrogen content. *Ecological Applications*, 313–321.

Pinter, P.J., Jackson, R.D., Ezra, C.E., and Gausman, H.W. 1985. Sun angle and canopy architecture effects on the reflectance of six wheat cultivars. *International Journal of Remote Sensing*, 6, 1813–1825.

Potter, C.S., Randerson, J.T., Field, C.B., Matson, P.A., Vitousek, P.M., Mooney, H.A., and Klooster, S.A. 1993. Terrestrial ecosystem production: A process model based on global satellite and surface data. *Global Biogeochemical Cycles*, 7, 811–841.

Qi, J., Chehbouni, A., Huete, A.R., Kerr, Y.H., and Sorooshian, S. 1994. A modified soil adjusted vegetation index. *Remote Sensing of Environment*, 48, 119–126.

Qi, J., Inoue, Y., and Wiangwang, N. 2012. Hyperspectral remote sensing in global change studies. In: Thenkabail, P.S., Lyon, J.G., and Huete, A. (Eds.), *Hyperspectral Remote Sensing of Vegetation*. CRC Press, New York, USA, pp. 70–89.

Quattrochi, D.A. and Luvall, J.C. 1999. Thermal infrared remote sensing for analysis of landscape ecological processes: Methods and applications. *Landscape Ecology*, 14, 577–598.

Ribeiro da Luz, B. and Crowley, J.K. 2010. Spectral reflectance and emissivity features of broad leaf plants: Prospects for remote sensing in the thermal infrared (8.0–14.0 μm). *Remote Sensing of Environment*, 109, 393–405.

Richardson, A.D., Duigan, S.P., and Berlyn, G.P. 2002. An evaluation of noninvasive methods to estimate foliar chlorophyll content. *New Phytologist*, 153, 185–194.

Roberto, C., Lorenzo, B., Michele, M., Micol, R., and Cinzia, P. 2012. Optical remote sensing of vegetation water content. In: Thenkabail, P.S., Lyon, J.G., and Huete, A. (Eds.), *Hyperspectral Remote Sensing of Vegetation*. CRC Press, New York, USA, pp. 227–244.

Romero, A., Aguado, I., and Yebra, M. 2012. Estimation of dry matter content in leaves using normalized indexes and PROSPECT model inversion. *International Journal of Remote Sensing*, 33, 396–414.

Rondeaux, G., Steven, M., and Baret, F. 1996. Optimization of soil-adjusted vegetation indices. *Remote Sensing of Environment*, 55, 95–107.

Roosjen, P.P.J., Suomalainen, J.M., Bartholomeus, H.M., Kooistra, L., and Clevers, J.G.P.W. 2017. Mapping reflectance anisotropy of a potato canopy using aerial images acquired with an unmanned aerial vehicle. *Remote Sensing*, 9, 417.

Rosati, A., Day, K.R., and DeJong, T.M. 2000. Distribution of leaf mass per unit area and leaf nitrogen concentration determine partitioning of leaf nitrogen within tree canopies. *Tree Physiology*, 20, 271–276.

Rouse, J.W., Haas, R.H. Jr, Schell, J.A., and Deering, D.W. 1974. Monitoring vegetation systems in the Great Plains with ERTS. *Third ERTS-1 Symposium*, Washington, DC: NASA, 09–317.

Roy, J., Saugier, B., and Mooney, A.H. (Eds.) 2001. *Terrestrial Global Productivity*. Academic Press, London, UK, 573pp.

Ruimy, A., Saugier, B., and Dedieu, G. 1994. Methodology for the estimation of terrestrial net primary production from remotely sensed data. *Journal of Geophysical Research*, 99, 5263–5283.

Sakaiya, E. and Inoue, Y. 2013. Operational use of remote sensing for harvest management of rice. *Journal of The Remote Sensing Society of Japan*, 33(3), 185–199.

Schlerf, M. and Atzberger, C. 2006. Inversion of a forest reflectance model to estimate structural canopy variables from hyperspectral remote sensing data. *Remote Sensing of Environment*, 100, 281–294.

Shibayama, M. and Akiyama, T. 1986. A spectroradiometer for field use. VI. Radiometric estimation for chlorophyll index of rice canopy. *Japanese Journal of Crop Science*, 55, 433–438.

Shibles, R.M. and Weber, R.C. 1966. Interception of solar radiation and dry matter production by various soybean planting patterns. *Crop Science*, 6, 55–59.

Sims, D.A. and Gamon, J.A. 2002. Relationships between leaf pigment content and spectral reflectance across a wide range of species, leaf structures and developmental stages. *Remote Sensing of Environment*, 81, 337–354.

Sinclair, T.R. 1994. Limits to crop yield? In: Boote, K.J., Bennet, J.M., Sinclair, T.R., and Paulsen, G.M. (Eds.), *Physiology and Determination of Crop Yield*. American Society of Agronomy, Madison, WI, USA, pp. 509–532.

Skidmore, A.K., Pettorelli, N., Coops, N.C., Geller, G.N., Hansen, M., Lucas, R., Mücher, C.A. et al. 2015. Environmental science: Agree on biodiversity metrics to track from space. *Nature*, 523, 403–405.

Spiegelman, C.H., McShane, M.J., Goetz, M.J., Motamedi, M., Yue, Q.L., and Coté, G.L. 1998. Theoretical justification of wavelength selection in PLS calibration: Development of a new algorithm. *Analytical Chemistry*, 70, 35–44.

Sripada, R.P., Schmidt, J.P., Dellinger, A.E., and Beegle, D.B. 2008. Evaluating multiple indices from a canopy reflectance sensor to estimate corn N requirements. *Agronomy Journal*, 100, 1553–1561.

Staenz, K. and Held, A. 2012. Summary of current and future terrestrial civilian hyperspectral spaceborne systems. *Proceeding of IEEE International Geoscience and Remote Sensing Symposium (IGARSS) 2012*, July 22–27, 2012, Munich, Germany, 123–125.

Stafford, J.V. 2000. Implementing precision agriculture in the 21st century. *Journal of Agricultural Engineering Research*, 76, 267–275.

Stöcker, C., Bennett, R., Nex, F., Gerke, M., and Zevenbergen, J. 2017. Review of the current state of UAV regulations. *Remote Sensing*, 9, 459.

Takahashi, W., Nguyen-Cong, V., Kawaguchi, S., and Minamiyama, M. 2000. Statistical models for prediction of dry weight and nitrogen accumulation based on visible and near-infrared hyper-spectral reflectance of rice canopies. *Plant Production Science*, 3, 377–386.

Thapa, R.B., Watanabe, M., Motohka, T., and Shimada, M. 2015. Potential of high-resolution ALOS–PALSAR mosaic texture for aboveground forest carbon tracking in tropical region. *Remote Sensing of Environment*, 160, 122–133.

Thenkabail, P.S., Lyon, J.G. and Huete, A. (Eds.) 2012. *Hyperspectral Remote Sensing of Vegetation*. CRC Press, New York, USA, 705pp.

Tilman, D. 2001. Functional Diversity. In: Levin, S.A. (Ed.), *Encyclopedia of Biodiversity*. Academic Press, Oxford, UK, pp. 109–120.

Turner, D.P., Gower, S.T., Stith, T., Cohen, W.B., Gregory, M., and Maiersperger, T.K. 2002. Effects of spatial variability in light use efficiency on satellite-based NPP monitoring. *Remote Sensing of Environment*, 80, 379–406.

Ustin, S.L., Gitelson, A.A., Jacquemoud, S., Schaepman, M., Asner, G.P., Gamon, J.A., and Zarco-Tejada, P. 2009. Retrieval of foliar information about plant pigment systems from high resolution spectroscopy. *Remote Sensing of Environment*, 113, 67–77.

Ustin, S.L. 2013. Remote sensing of canopy chemistry. *Proceedings of the National Academy of Sciences*, 110, 804–805.

Vadrevu, K., Ohara, T., and Justice, C. (Eds.) 2018. *Land-Atmospheric Interactions in Asia*. Springer & Cham, Cham, Switzerland, 500pp.

Verhoef, W. 1984. Light scattering by leaf layers with application to canopy reflectance modeling: The SAIL model. *Remote Sensing of Environment*, 16(2), 125–141.

Verhoef, W. 1985. Earth observation modeling based on layer scattering matrices. *Remote Sensing of Environment*, 17(2), 165–178.

Verhoeven, G. 2011. Taking computer vision aloft – Archaeological three-dimensional reconstructions from aerial photographs with photoscan. *Archaeological Prospection*, 18, 67–73.

Vogelmann, J.E., Rock, B.N., and Moss, D.M. 1993. Red edge spectral measurements from sugar maple leaves. *International Journal of Remote Sensing*, 14, 1563–1575.

Wang, S.H., Ji, Z.J., Liu, S.H., Ding, Y.F., and Cao, W.X. 2003. Relationships between balance of nitrogen supply-demand and nitrogen translocation and senescence of different position leaves on rice. *Agricultural Sciences in China*, 2, 747–751.

Wang, Z., Skidmore, A.K., Darvishzadeh, R., Heiden, U., Heurich, M., and Wang, T. 2015a. Leaf nitrogen content indirectly estimated by leaf traits derived from the PROSPECT model. *IEEE Journal of Selected Topics in Applied Earth Observations and Remote Sensing*, 8, 3172–3182.

Wang, Z., Skidmore, A.K., Wang, T., Darvishzadeh, R., and Hearne, J. 2015b. Applicability of the PROSPECT model for estimating protein and cellulose + lignin in fresh leaves. *Remote Sensing of Environment*, 168, 205–218.

Wang, Z., Skidmore, A.K., Wang, T., Darvishzadeh, R., Heiden, U., Heurich, M., Latifi, H., and Hearne, J. 2017. Canopy foliar nitrogen retrieved from airborne hyperspectral imagery by correcting for canopy structure effects. *International Journal of Applied Earth Observation and Geoinformation*, 54, 84–94.

Wright, I.J., Reich, P.B., Westoby, M., Ackerly, D.D., Baruch, Z., Bongers, F., Cavender-Bares, J. et al. 2004. The worldwide leaf economics spectrum. *Nature*, 428, 821–827.

Xiao, Y., Zhao, W., Zhou, D., and Gong, H. 2014. Sensitivity analysis of vegetation reflectance to biochemical and biophysical variables at leaf, canopy, and regional scales. *IEEE Transactions on Geoscience and Remote Sensing*, 52, 4014–4024.

Yoder, B.J. and Pettigrew-Crosby, R.E. 1995. Predicting nitrogen and chlorophyll content and concentration from reflectance spectra (400–2500 nm) at leaf and canopy scales. *Remote Sensing of Environment*, 53, 199–211.

Zarco-Tejada, P.J., Miller, J.R., Noland, T.L., Mohammed, G.H., and Sampson, P.H. 2001. Scaling-up and model inversion methods with narrowband optical indices for chlorophyll content estimation in closed forest canopies with hyperspectral data. *IEEE Transactions on Geoscience and Remote Sensing*, 39, 1491–1507.

Zarco-Tejada, P.J., Miller, J.R., Mohammed, G.H., Noland, T.L., and Sampson, P.H. 2002. Vegetation stress detection through chlorophyll a + b estimation and fluorescence effects on hyperspectral imagery. *Journal of Environmental Quality*, 31, 1433–1441.

Zhang, C. and Kovacs, J.M. 2012. The application of small unmanned aerial systems for precision agriculture: A review. *Precision Agriculture*, 13, 693–712.

Zhu, Y., Yao, X., Tian, Y., Liu, X., and Cao, W. 2008. Analysis of common canopy vegetation indices for indicating leaf nitrogen accumulations in wheat and rice. *International Journal of Applied Earth Observation and Geoinformation*, 10, 1–10.

3 Spectral and Spatial Methods for Hyperspectral and Thermal Image-Analysis to Estimate Biophysical and Biochemical Properties of Agricultural Crops

Yafit Cohen and Victor Alchanatis

CONTENTS

3.1 INTRODUCTION

The ultimate goals of hyperspectral and thermal sensing in precision agriculture are to estimate biophysical and biochemical properties of agricultural crops (BB-PACs: phonetically pronounced as bee-bee-pax) and to delineate and characterize homogeneous management zones for optimal agricultural management such fertilization, irrigation, or other agrotechnical operations. On the face of it, the use of hyperspectral and thermal remote sensing for precision agriculture seems to be similar to their use for natural vegetation. But for natural vegetation, remote sensing is widely used for classification of natural vegetation types, while in precision agriculture it is aimed at quantification of BB-PACs. The different goals are pursued by adapted analysis approaches. This chapter concentrates on three characteristics of hyperspectral (HS) images: First, their unique *spectral* properties, namely the narrow bandwidths and the plethora of the bands, as opposed to wider and limited number of bands in other broad-band spectral sensing systems such most multi-spectral satellite images; Second, the *spatial* attribute of hyperspectral images, as opposed to point spectral measurements of other spectral systems; and third, the state-of-the-art algorithms for hyperspectral image processing that show the added value of spatial information when combined with spectral information for mapping plant BB-PACs. In addition we present thermal panchromatic imaging as an image type complementary to the hyperspectral images.

Modern agricultural crop production relies on close monitoring of the crop status. This enables efficient management of available resources for profitable and environmentally friendly agricultural practice. Widely used monitoring tools are mainly based on point sampling of biophysical and biochemical properties of the crop. Numerous crop properties have been studied over the years and act as indicators of the crop condition. Local and global growth protocols have been developed based on these measured biophysical and biochemical properties. For example, irrigation management of cotton is widely based on the height measurement of the plants at selected points; this is a biophysical property that can be easily measured by simple means, but it is labor intensive and is based on selected sampled spots. Another example is fertilization management in potatoes, where nitrate content in the petiole is used as an indication for the fertilizer requirement. Table 3.1 lists some important biophysical and biochemical properties that are used in agricultural crop growing protocols.

These examples illustrate the great importance of monitoring biophysical and biochemical properties of agricultural crops. The desire to upgrade from point measurements to maps with high density of data has brought remote sensing to the front of the technologies that can fulfill such a mission.

A number of other crop health conditions related to crop protection, such as pest damage, plant diseases, and weed infestation, are also expressed through changes in the biophysical and biochemical properties of the crop. Several reports in the literature show the contribution of remote sensing techniques in the detection of plant diseases [1,2], pest damage [3,4], and weed infestation [5–7]. All studies report that hyperspectral remote sensing can detect the phenomena assuming that they are the factor that causes the anomalies in the field. This chapter will focus on sensing plant properties related to manageable agricultural resources such as irrigation and fertilization and will not discuss the issues of sensing plant properties related to plant diseases, pest damage, and weed infestation.

This chapter is divided into four main parts. The first (Section 3.2) describes the general characteristics of spectral and thermal sensing of agricultural crops. The second (Sections 3.3 and 3.4) reviews the most prominent methods of hyperspectral and thermal data processing to model and enhance quantification of BB-PACs. The third part (Section 3.5) describes how these methods

TABLE 3.1

Biophysical and Biochemical Properties of Crops That Serve as Indicators for Agricultural Crop Management

Property (BB-PAC)	Example Crops	Agrotechnical Management Parameter
	Biophysical	
Biomass [kg m^{-1}]	Wheat, rice, corn	Fertilization
Leaf area index/crop cover [no units/%]	Wheat, soybean, corn, cotton	Fertilization
Crop height [m]	Cotton, wheat	Irrigation, application of growth regulators
Canopy volume [m^3]	Orchards, wheat	Irrigation, fertilization
Yield [kg m^{-1}]	Wheat, corn, cotton	–
Stomata conductance [mmol s^{-1}]	Vineyards	Irrigation
Leaf/stem water potential [MPa]	Cotton, orchards, vineyards	Irrigation
Flowering intensity [relative units]	Orchards	Growth regulators, mechanical thinning
	Biochemical	
Nitrogen content [%N]	Corn, wheat, potatoes	Fertilization
Chlorophyll content [µg cm^{-2}]	Corn, wheat, cotton	Fertilization
Salinity [mmol]	Cotton	Water quality management; not used in practice
Leaf water content [%]	Wheat, potato	Irrigation
Leaf macro-elements such as phosphorus (P) and potassium (K) [mg kg^{-1}]	Olives	Fertilization, not used in practice

are applied to predict specific biophysical and selected biochemical properties of agricultural crops. The last part of the chapter, presents approaches to integrate the hyperspectral data with hyperspatial attributes of the hyperspectral images to enhance their potential to delineate management zones. Additionally, in a few places in this chapter, the complementary characteristics of the VIS–NIR–SWIR and the thermal ranges are described and discussed.

3.2 SPECTRAL AND THERMAL SENSING OF AGRICULTURAL CROP PROPERTIES: A GENERAL CHARACTERIZATION

3.2.1 VISIBLE AND NEAR-INFRARED (VIS–NIR–SWIR) RANGE

Spectral characteristics of green vegetation have very prominent features: two valleys in the visible portion of the spectrum are determined by the pigments contained in the plant. Chlorophyll absorbs strongly in the blue (450 nm) and red (680 nm) regions, also known as the chlorophyll absorption bands. This is the reason for the human eye perceiving healthy vegetation as green. When the plant is subjected to stress that hinders normal growth and chlorophyll production, there is less absorption in the red and blue regions and the amount of reflection in the red waveband increases. In some cases where stress is severe, the stress can be sensed by human eyes.

The spectral reflectance signature has a dramatic increase in the reflection for healthy vegetation at around 700 nm. In the near-infrared (NIR) between 700 and 1300 nm, a plant leaf typically reflects between 40% and 60%, of the incident radiation; the rest is transmitted, with only about 5% being adsorbed. For comparison, the reflectance in the green range reaches to 15%–20% of the incident radiation.

This high reflectance in the NIR is due to scattering of the light in the intercellular volume of the leaves' mesophyll. Structural variability in leaves in this range allows one to differentiate between species, even though they might look the same in the visible region. Beyond 1300 nm the energy incident upon the vegetation is largely absorbed or reflected with very little transmission of energy. Water absorption bands are mostly noted at around 760, 970, 1200, 1470, 1900, and 2870 nm can be used for plant water content estimation.

3.2.2 FAR OR THERMAL INFRARED (TIR)

The water pathway from the stem to leaf evaporation sites is essential for maintaining leaf water balance, allowing stomata to stay open, and resulting in carbon capture. Evapotranspiration is the process in which water stored in the soil or vegetation is converted from the liquid into the vapor phase and is transferred to the atmosphere. Evapotranspiration decreases plant temperature. Stomatal regulation plays a key role in plant response to water stress. As plant stomata close, evapotranspiration rate decreases; the energy heat balance between the vegetation and its environment is changing and leaf temperature rises. Thus, leaf temperature may be used as an indicator of plant water status and plant health. Leaf temperature can be sensed by measuring the far-infrared or thermal infrared (8–14 μm) radiation they emit. First attempts to apply canopy temperature for assessing plant water status and plant health status were made in the 1960s [8]. The availability of thermal cameras led to a significant evolution of the thermal remote sensing in the 2000s [9].

3.3   SPECTRAL ANALYSIS METHODS

Spectral and thermal remote sensing provide important information on agricultural crops. The link between the biophysical and biochemical properties of the crops and the sensing data is based on intensive data processing of the remotely sensed data using a variety of methods. A large number of processing methods have been developed over the last decades that differ in their underlying physical assumptions, the mathematical models, and how direct or indirect is the link between the data and the property. It is of great importance to understand well the basic methodology of data processing in order to ensure that the limitations and the advantages of each of method are used properly in interpretation and application of real situations. In this section, we review the most prominent methods of hyperspectral and thermal data processing to model and enhance quantification of biophysical and biochemical properties of agricultural crops. This section does not address the use of specific bands for specific properties but describes the underlying methodology for building specific relationships.

A number of methods are commonly used for analyzing spectral data to extract BB-PACs. The source of the data may be a point spectral sensor as well as a hyperspectral imaging camera. In the latter case, each pixel is regarded as a single point measurement. In both cases, there are hundreds of narrow spectral bands, with bandwidth around 1–10 nm. There are three main methods for spectral analysis: (a) bands selection, (b) spectral indices, (c) linear and nonlinear multivariate statistics and models.

Selection of individual/set of bands and the use of spectral indices were mainly developed in the arena of remote sensing, whereas multivariate statistical methods are mainly developed in the chemometrics arena.

3.3.1 SPECTRAL BANDS SELECTION

Spectral bands selection comprises a methodology for choosing hyperspectral bands that provide sufficient, but not redundant, information to classification or prediction algorithms, using practical amount of computational resources. There are two conceptually different approaches to band selection: unsupervised and supervised. Unsupervised methods order the spectral bands without training, based on generic information evaluation approaches. They are usually very fast and computationally efficient, and can provide information for clustering an image to classes of common spectral signatures. Supervised methods require training data in order to build an internal predictive model. They are usually more computationally intensive than unsupervised methods and can provide quantitative models for predicting BB-PACs [10].

Unsupervised methods for spectral band selection include the use of such methods as principal components analysis [11] and band–band correlation [12]. Supervised methods include the use of such methods as correlation of the spectral bands with the BB-PAC studied [13], and stepwise discriminant analysis [11] to extract the number of independent wavelengths that can explain the variability of the measured BB-PAC. Both methods result in an optimum number of spectral bands that contain unique information.

3.3.2 SPECTRAL INDICES

Spectral indices assume that the combined interaction between a small number of wavelengths is enough to describe the biochemical or biophysical interaction between light and matter. The simplest form of index is a simple ratio (SR), where the ratio between two wavelengths is indicative for a BB-PAC under investigation. The typical form of a SR index is

$$I = \frac{R_{\lambda_i}}{R_{\lambda_j}}$$

where I is the index value, R_{λ_i} and R_{λ_j} are the reflectance values in wavelength λ_i and λ_j respectively.

Enhanced SRs are the normalized difference spectral indices (NDI or NDSI), which also exploit the difference between two distinct wavelengths, but normalize it using the following equation:

$$I = \frac{R_{\lambda_i} - R_{\lambda_j}}{R_{\lambda_i} + R_{\lambda_j}}$$

Another category of spectral indices comprises integrated indices (or derivative indices), where more than two wavelengths are combined to produce a value that is correlated with BB-PACs. Integrated indices are usually specific to a certain BB-PAC and sometimes to the crop that they were developed for. An extensive compilation of all three index categories can be found in Li et al. [14].

3.3.3 MULTIVARIATE METHODS

Spectral indices that are based on a small number of bands are indicators of irregular conditions and provide evidence that an anomaly is present. Despite their widespread use it has not been possible to design an index that is sensitive only to a desired variable and totally insensitive to all other vegetation parameters [15]. Thus, if the factor or the cause of the anomaly in the field is known, then some of the spectral indices may be able to quantify the level or the severity of the anomaly. The advantage of the whole spectral signature of the crop is that it contains information that can be used to identify the cause for the spectral changes in the light reflected from the canopy as well as to quantify it.

Multivariate statistics assumes that there is an underlying relation between the spectral signature of the crop and its biochemical or biophysical properties. Statistical tools extract this underlying relationship as a model that is often a linear model. The large number of independent variables (wavelengths) together with the high colinearity between the variables (spectral bands) do not permit the use of common multivariate methods such as multivariate linear regression (MLR) based on least squares, before prior selection of the most indicative independent wavelengths. Therefore, methods that overcome these constraints were developed over the years. Increasing numbers of multivariate methods were adopted for processing spectral data and hyperspectral images for agriculture. Here we list a few that have been used recently for BB-PAC estimation and describe in more details the most common used for that purpose. The spectral angle mapper (SAM) algorithm determines the spectral similarity between two spectra by calculating the angle between them, treating them as vectors. The Artificial Neural Network (ANN) is a nonparametric nonlinear model that uses neural networks spreading between layers and simulates human brain receptors and information processing. ANN is a learning classification method based on large labeled (tagged) samples and is affected by the complexities of the network structure and the sample making it prone to over-learning and reducing the ability for generalization. Spectral vector machine (SVM), is a pattern recognition method which is also based on statistical learning theory [16]. Another machine learning algorithm is the random forest (RF) which is designated for classification or regression tasks [17]. For classification tasks, it is operated by constructing a multiple decision trees based on iterative selection of training samples. As the random selection is sensitive to selected feature dimensions (or insensitive to some feature dimensions), the trees can gain accuracy as they grow without suffering from over-fitting.

Additionally, there are two more common methods: principal components regression (PCR), which has a core of unsupervised data extraction, and partial least-squares regression (PLSR), which is a supervised method. Both methods produce a linear model.

Partial least-squares regression is related to both PCR and MLR, and can be thought of as occupying a middle ground between them [18]. PCR finds factors that capture the greatest amount of variance in the predictor variables (spectra). MLR seeks to find a single factor that best correlates predictor variables with predicted variables (BB-PACs). PLS attempts to find factors that both capture variance in the predictor variables (spectra) and achieve correlation while avoiding the colinearity between spectral bands. In other words, PLS attempts to find factors (called latent variables) that maximize the amount of variation explained in the spectra that is relevant for predicting the BB-PAC. This is in contrast to PCR, where the factors (called principal components) are selected solely based on the amount of variation that they explain in spectra. In mathematical terms, the difference between them is the objective function that is used to optimize the calculation of the regression coefficients. Unsupervised methods tend to minimize only the inter-class (between classes) variance based on the spectral curves of the samples. Supervised methods either minimize the variance of the intra-class (within the class) variance or a combination of the inter-class and intra-class variance.

Wavelets are a group of functions that vary in complexity and mathematical properties and that are used to dissect data into different frequency components and then characterize each component with a resolution appropriate to its scale. Wavelet analysis of a reflectance spectrum is performed by scaling and shifting the wavelet function to produce wavelet coefficients that are assigned to different frequency components. By selecting appropriate wavelet coefficients, a spectral model can be established between the coefficients and biochemical concentrations. Hence, wavelet analysis has the potential to capture the information contained within high-resolution spectra and offers the prospect of developing robust, generic methods for pigment determinations [19,20].

It should be noted that most of these methods are confined to classification and detection problems and are not often used for quantitative estimation of crop characteristics from hyperspectral data.

3.4   THERMAL ANALYSIS METHODS

Currently, thermal cameras provide either panchromatic images or multi/super-spectral images in the range 3–14 μm. This section concentrates on the analysis of panchromatic thermal images in the range of 8–14 μm that are used in most agricultural studies. Since they are panchromatic images there is no dimensionality complexity. For BB-PACs estimation, the core of the thermal image analysis is to convert the surface temperature to meaningful water status indices. Maes and Steppe [9] provide a comprehensive review on ground-based thermal imaging for estimating evapotranspiration and water shortage stress, while this section focuses on thermal imaging analysis for water status estimation and mapping.

3.4.1 Computation of Crop Water Stress Index

The use of canopy temperature as an indicator of plant water status is not new and was popularized by Idso and colleagues [e.g., 21,22]. Since canopy temperature is affected by both plant water status and environmental conditions, water stress indices that calibrate the environmental conditions were developed. The crop water stress index (CWSI) based on canopy temperature [22] has become an acceptable index to map in-field variability of crop water status using thermal images. CWSI is defined as a fraction of the canopy temperature between dry (upper) and wet (lower) baselines under ambient conditions. It can be calculated by:

$$\text{CWSI} = \frac{T_{\text{canopy}} - T_{\text{wet}}}{T_{\text{dry}} - T_{\text{wet}}}$$

where T_{canopy} is the canopy temperature, T_{wet} is the temperature of a fully transpiring leaf, and T_{dry} is the temperature of a nontranspiring leaf. For irrigation scheduling, CWSI mapping based on aerial thermal images should be simple to compute in order to be used in the routine of irrigation management. Accordingly, there are two main challenges in CWSI computations that researchers have addressed in the last two decades: (i) development of a methodology for accurate extraction of canopy temperature; and (ii) the setup and formulation of baselines (T_{wet} and T_{dry}) that can be accurately measured, extracted, or computed.

3.4.1.1 Canopy Temperature Extraction

An object-oriented methodology for pure canopy temperature extraction using merely thermal images has been suggested for trees [23–25]. The methodology suits some orchard structures that have soil between crop rows, since the canopy temperature is well differentiated from the exposed soil. Nevertheless, in orchards that have grass in between the crop rows, thermal imaging cannot easily be used to differentiate between grass and tree canopy pixels. Fusing a digital surface model or information on the rows' location with the thermal images is suggested to address this challenge.

For field crops like cotton and wheat, the main challenge is the extraction of mixed pixels of canopy and soil. Methodologies that combine multispectral (MS) images in the VIS-NIR range with thermal images have been developed to extract canopy pixels [26,27]. An empirical methodology for canopy temperature extraction using only thermal images and air temperature was developed for field crops by Meron et al. [28]. This methodology requires only thermal images and air temperature and may be suited for orchards as well.

3.4.1.2 Forms of Wet and Dry Baselines

Empirical and theoretical (analytical) forms of wet and dry baselines have been proposed and used for CWSI calculation and mapping, as summarized in [9,29]. For large scale CWSI mapping, dry baseline temperature was used solely in its empirical form, that is, air temperature $+ X°C$. The canopy–air temperature difference is unique for each crop in each region, but it is relatively stable and indifferent to changes in vapor pressure deficit (VPD). In comparison, wet baseline temperature determination is a greater challenge for researchers as it is highly dependent on vapor pressure; thus it can be found mostly in its empirical (as a function of air temperature and VPD) and theoretical (based on the energy balance [30]) forms. Berliner et al. [31] and Taghvaeian et al. [32] have shown the potential in using a well-watered reference plot to measure the wet baseline temperature, but they used it for canopy and air temperature difference index and not for CWSI. This approach uses the crop as a bioindicator but, instead of using a single leaf as in ground thermal imaging [25], it uses a set of pixels from a field. Another bioindicator wet reference, named statistical or virtual reference, has been suggested more recently, which uses the average temperature of the coolest 5%–10% of the canopy pixels [23,29,33,34]. The statistical reference assumes that at the time of thermal imaging, there are areas in the field that are over-irrigated.

3.4.2 SATELLITE AND AERIAL THERMAL IMAGES

A trade-off exists between satellite and aerial thermal imaging in terms of cover area and spatiotemporal resolution. Satellites provide images covering large areas at a low cost per area unit and have thus became a common tool used by farmers. Currently, satellite-based images in the VIS–NIR range have a relatively high spatiotemporal resolution, but in the thermal range their finest resolution (60 m in Landsat) and their long revisit time are often not appropriate for irrigation management (Table 3.2). Aerial-based thermal images, which, theoretically, can be acquired on a daily basis, have high spatial resolution but are limited by their cover area (limited capacity), and are thus expensive per area unit. A revival of thermal imaging for water status mapping has been sensed lately with the increasing availability of compact, low-cost uncooled microbolometer-based thermal focal plane arrays. These cameras can be easily mounted on unmanned airborne vehicles (UAVs) (or even integrated into smartphones). However, being noncooled, they suffer from temporally and spatially dependent changes that require constant calibration of both the gain and offset. With the absence of a means of internal calibration, they cannot be used for trustworthy assessment of canopy temperature and thus reliable estimation of the crop water status. Most of the compact thermal cameras are very sensitive (thermal resolution of 0.1°C degrees or more). Yet, only a few have sufficient accuracy (±2°C and better) while most of them lack it (±5°C and worse). Even the more accurate compact cameras suffer from a significant drift of the temperature. To our knowledge, currently there is no compact low-cost camera that has appropriate calibration hardware. New publications may have paved the way for retroactive calibration of such cameras [35,36] but their applicability in real conditions have yet to be proved. Enhancement of the spatial resolution of the aerial thermal images can also be achieved by employing super-resolution algorithms that exploit the vast overlap between sequential images [37]. If applicable, aerial thermal images may be acquired from higher altitudes and cover much larger areas. For larger fields, it was proposed that sharpening methods would be adjusted and applied for thermal images of Sentinel-3, which has a revisit time of a few days [37]. Finally, a novel approach was introduced that fused aerial thermal images with satellite MS images in the VIS–NIR–SWIR range [37]. A similar method was employed to upscale aerial hyperspectral images for natural vegetation monitoring [38].

TABLE 3.2
Satellite Thermal Bands, Their Spatial Resolution and Revisit Time

Satellite	Spatial Resolution (meters)	Revisit Time (days)
Landsat 7 and 8	60–100 (30)[a]	16 (8)[b]
ASTER	90	16
MODIS	1000	1
Sentinel-3 (SLSTR)[c]	1000	1

[a] Bands are acquired at 60 or 100 m resolution (in Landsat 7 and 8, respectively), but are resampled to 30 m in delivered data product.
[b] Landsat 8 satellite images the entire Earth every 16 days in an 8-day offset from Landsat 7.
[c] Sea and Land Surface Temperature Radiometer.

3.5 PREDICTION OF BB-PACS

3.5.1 PREDICTION OF BIOPHYSICAL PROPERTIES

3.5.1.1 Leaf Area Index, LAI

Green leaf area index (LAI) is a key variable used by crop physiologists and modelers for estimating foliage cover, as well as forecasting crop growth and yield. The exposed area of living leaves plays a key role in various biophysical processes such as plant transpiration and CO_2 exchange. Because LAI is functionally linked to the canopy spectral reflectance, its retrieval from remote sensing data has prompted many investigations and studies over the years [19,39–44]. Most of these studies have relied on empirical relationships between the ground-measured LAI and observed spectral responses.

The most common index to estimate LAI and its counterparts, the crop cover and biomass, is the normalized differential vegetation index (NDVI) [15], which expresses the normalized ratio between the reflected energy in the red chlorophyll absorption region and the reflected energy in the NIR mesophyll scattering region. Yet, it is well documented that the NDVI approaches saturation asymptotically under conditions of moderate-to-high above-ground biomass [43], it therefore may be a good predictor only for low to medium LAIs (0–4).

Linear regression analysis of single bands and two-band combinations of pseudo NDVIs (NDSIs) have shown the importance of the red-edge spectral region (700–740 nm) and the short-wave infrared (SWIR) spectral region, and the advantage of narrow bands over traditional broad bands in LAI prediction [13,41,42,45,46]. A major problem in the use of indices to estimate LAI arises from the fact that canopy reflectance, in the visible and NIR, is strongly dependent also on chlorophyll content of the canopy [e.g., 47]. Moreover, both variables have similar effects on canopy reflectance, particularly in the spectral region from the green (550 nm) to the red edge (740 nm). To uncouple the LAI effect, Haboudane et al. [40] developed two indices: the modified triangular vegetation index (MTVI2) and the modified chlorophyll absorption ratio (MCARI1). Prediction algorithms based on these two indices were applied for CASI hyperspectral image over fields of soybean, corn, and wheat and showed excellent agreements between modeled and measured LAI.

Other studies exploit wider ranges of the spectra or even the whole spectrum to improve LAI prediction. Delegido et al. [48] have shown that the spectra between 500 and 750 nm can be fitted with good precision to third-degree polynomials and that there was strong correlation between one of the coefficients and LAI values that ranged from 0 to 7. This is a significant improvement over other methods since it covers the whole range of LAI (0–7) and is not limited to low (0–2) and medium (2–4) LAI. Multivariate and PLS regression models based on selected narrow bands or the whole spectrum, respectively, have been shown to be comparable or better LAI predictors than narrow-band NDIs [41,42]. While narrow-band NDVI had strong correlation in LAI range of 0–3 and explained 80% of the LAI variability, the multivariate regression of wider range had a very high correlation in LAI range of 0–6 and explained 90% of the variability [41,49]. In another study, several multivariate methods were used to predict LAI in soybean, namely, RF, ANN, SVM, and PLS [50], and all methods demonstrated similar performance measures, explaining around 70% of the LAI variability.

3.5.1.2 Biomass

Forecasting and estimating of crop production using remote sensing has great consequence on food provision management and is fundamental to applications of precision agriculture. In-season biomass estimation from remote sensing for yield forecasting and variable rate applications has been a challenge for various studies.

Biomass and LAI have similar effects on spectral characteristics and studies have shown similarities as well as some differences in the estimation of both crop properties using spectral measurements and hyperspectral images. Correlation coefficients between spectral reflectance in discrete narrow bands and LAI and biomass in various crops presented similar shapes [13,45]. Spectral bands that are best suited for characterizing LAI and biomass were determined by Thenkabail et al. [13,45] and no significant differences on their prediction accuracy was found. Yet, while no improvement was achieved by using PLS regression models for LAI estimation, PLS models significantly improved the prediction of biomass by lowering the RMSE by 22%, compared to the best narrow-band indices [42]. Correspondingly, PLS models using the spectral range of 350–2500 nm were found to better predict wheat dry biomass compared to common vegetation indices: R^2 values of 0.80 and 0.50, respectively [51]. A recent study, that used snapshot hyperspectral images from a hyperspectral camera mounted on a UAV, showed no advantage of PLS models over the best narrow-band indices ($R^2 = 0.5$) to predict biomass of winter wheat canopies, but a significant improvement was achieved where crop heights and the full spectra were combined in a PLS model ($R^2 = 0.78$) [52].

3.5.1.3 Water Status

Crop water status is a key biophysical property that is used to manage irrigation, as well as to evaluate crop health. In most cases, it is directly associated with water availability in the soil, and when this is not the case (i.e., water availability is not the limiting factor), water status becomes an indicator of crop health. For example, when salinity is a limiting factor of water uptake, crop water status becomes an indicator of salinity stress. Similarly, plant diseases that damage water flow in the plant affect the crop water status, which becomes an indicator of the disease's presence or its severity.

Crop water status is a function of soil water availability, hydraulic resistance along the flow path, plant water capacitance, and meteorological conditions that determine atmospheric evaporative demand [53]. Crop water status can be quantified by measuring either leaf water content or leaf and stem water potential. The spectral characteristics of water can be used to quantify the water content in the leaves. For wavelengths sensitive to water absorption (760, 970, 1450, 1940, and 2950 nm), leaf reflectance decreases as water content increases. Numerous studies have shown the ability of spectral indices to determine leaf relative water content (RWC), for example, the early study of Hunt and Rock [54], the study of Ceccato et al. [55], and more recent studies such as [56–58]. In a few studies, attempts to use indices as algebraic expressions of reflectance values for specific wavelengths did not yield significant relationships at the canopy level [51,59]. Nevertheless, when methods that use the whole spectrum were analyzed, canopy water content could be predicted from remotely sensed data. Namely, PLS models based on the first derivative of the spectrum in the range 350–2500 nm predicted water content with R^2 of 0.87, while spectral indices with exponential model achieved $R^2 = 0.2$ [51]. PLS models of spectral curves were found best predictors of RWC in comparison to various spectral indices and of other multivariate spectral models like MLR [60]. In addition, when the water absorbance band at 970 nm was considered, leaf water content was successfully predicted based on the slope (first derivative) of the spectral curve at 1015–1050 nm ($R^2 = 0.97$ for simulated data and 0.68 for field data) [61]. Other methods that consider the entire wavelength spectrum between 700 and 1300 nm showed that nonlinear models based on radial basis functions produce considerably better results than linear regression models (relative error of 4% and 17%, respectively) [62]. This outcome might indicate the existence of a complex dependency relationship between reflectance and leaf water content. It might also explain the poor results obtained by some methods based on indices in other studies.

Leaf water potential (LWP) in crops and stem water potential (SWP) in orchards are important biophysical parameters that indicate the ability of the crop to transfer water from soil to leaf [63]. The reports in the literature show limited ability to remotely estimate them using hyperspectral sensing in the VIS/NIR region since they express the physical status of water potential in the plant tissue [59,64,65]. Nevertheless, they affect the status of the leaves' stomata, which control the evapotranspiration process and affect leaf temperature. An important consequence of the stomatal closure that occurs when plants are subject to water stress is that energy dissipation is decreased, so leaf temperature tends to rise [30]. As mentioned above (in Section 3.4), the most common and widely utilized thermal index is the CWSI [22]. In the last decade, thermal crop sensing technologies have been widely used as tools for monitoring and mapping crop water status in various orchards [34], grapevines [23,66], olives [67,68], almonds [69], and other various crops [26,29,70,71]. Furthermore, thermal sensors and imaging have been employed for uniform and variable-rate irrigation management [72–76]. There are only a few studies that combined hyperspectral spectral images in the VIS–NIR range with thermal images for the prediction of various BB-PACs [e.g., 77,78]. A recent study integrated thermal imaging with hyperspectral sensors to assess their relationship with water status and grain yield of wheat cultivars [68]. The results show that the normalized relative canopy temperature (NRCT) alone was closely and significantly associated with RWC, with canopy water content and with grain yield ($R^2 = 0.81$ and $R^2 = 0.87$). The data fusion model of PLSR based on selected spectral indices improved the yield prediction under three irrigation regimes ($R^2 = 0.97$). A scientific report* on the fusion of hyperspectral images in the range of 400–980 nm and panchromatic thermal images in the range of 8–14 μm for estimating and mapping nitrogen level and water status has shown that the two ranges have complementary characteristics. In this study, a two-factor experiment of different nitrogen and irrigation treatments was conducted in potato fields, and it was found that (1) the spectral index NDI [79] extracted from the hyperspectral images acquired on two different dates was significantly affected by nitrogen treatments; (2) the water index 900/970 [80] was not affected by irrigation treatments; and (3) canopy temperature was sensitive to irrigation treatments while insensitive to nitrogen treatment. From the plentiful studies that have assessed either thermal images or spectral reflectance sensing and imaging or both to estimate water status, it can be concluded that thermal imaging has the ability to detect minor and mild water stress while spectral sensing technologies in the VIS–NIR range are more capable of detecting water stress in more advanced stages.

3.5.2   PREDICTION OF BIOCHEMICAL PROPERTIES

3.5.2.1   Chlorophyll Content

The most commonly used biochemical property of crops is chlorophyll content. It reflects the general condition of the crop, since chlorophyll is the producing "factory" of the crop. Changes in chlorophyll may indicate limited availability of important elements, among a wide possibility of options or other biotic or abiotic stresses. Chlorophyll deficiency can be detected by remote sensing, using specific spectral indices. Nevertheless, detection of chlorophyll deficiency is not an indicator of the cause that induced the deficiency.

Chlorophyll-specific spectral indices can be divided into two categories: (a) indices based on chlorophyll absorption in the blue (around 450 nm) and red (around 680 nm) spectral region and (b) indices that are based on the displacement of the red edge inflection point (700–740 nm). Several reports in the literature describe the use of simple and combined spectral indices for leaf chlorophyll estimation [20,81]. Among the set of indices tested, index combinations such as modified chlorophyll absorption ratio index/optimized soil-adjusted vegetation index (MCARI/OSAVI), triangular chlorophyll index/OSAVI, Moderate Resolution Imaging Spectrometer terrestrial chlorophyll index/ improved OSAVI (MSAVI), and red-edge model/MSAVI seemed to be relatively consistent and more stable as estimators of crop chlorophyll content [20].

Chlorophyll content was also estimated using wavelet decomposition on hyperspectral data. In the context of remote sensing of foliar chlorophyll, wavelet analysis has the potential to capture much more of the information contained with reflectance spectra than previous analytical approaches that use a small number of optimal wavebands. This approach was found to be more reliable than simple linear regression analysis when linking chlorophyll to the reflectance measured. This was observed both for leaf-level measurements as well as top of canopy measurements (peach trees) [82]. The wavelet-based approach outperformed models based on untransformed spectra (such as stepwise derivative) and a range of existing spectral indices. While wavelet-based models yielded 1:1 relationships between measured and predicted chlorophyll content in the range of 0–60 μg cm^{-2} (with R^2 of 0.88), other methods (including indices and first derivative) saturated above 30 μg cm^{-2} [83]. These findings indicate that wavelet analysis warrants further investigation as a method for extracting meaningful quantitative information from hyperspectral data.

---

* https://drive.google.com/open?id=1krIEGIc1BOYKr77AUXwr6NZjHDpiVCDp

Refinements in the technique for quantifying chlorophyll could explore the use of new wavelet functions or combinations of functions, multiple scales of wavelet coefficients, alternative methods for calculating derivatives prior to wavelet decomposition and different approaches to the selection of wavelet coefficients during model calibration. The value of wavelet analysis of spectra for quantifying leaf chlorophyll in principle has been demonstrated; it is now important that this is tested in practice and that the generality of the technique for hyperspectral remote sensing of vegetation is explored, particularly at the canopy and landscape scales [83].

The approach providing the highest predictive accuracy was that using multiple regression models based on wavelet coefficient energy feature vectors. This was closely followed by multiple regression models derived from the energy feature vectors of the nth-largest wavelet coefficients, which in turn was closely followed by stepwise regression models based on wavelet coefficients. The predictive accuracy of the stepwise regression models derived from narrow-band reflectance was substantially lower than that of the wavelet-based approaches, and the simple ratio and normalized difference ratio spectral indices had the poorest performance by some margin [84]. A number of techniques have been developed for red edge position extraction in the past. A more recent one suggested by Dong et al. [85] is a wavelet-based technique.

Leaves that suffer from chlorophyll shortage may have discolored spots, resulting in a range of spectral properties from a single leaf. Liu et al. [86] have demonstrated that the entropy, standard deviation of the spectral properties used, and spatial features were very good indicators of the leaf chlorophyll content. They concluded that spatial information can be used to retrieve chlorophyll content, with an accuracy equivalent to that of spectral information, and can provide information that spectral reflectivity cannot provide.

3.5.2.2 Nitrogen Content

Nitrogen deficiency is one of the most important conditions to be detected, since it directly affects the productivity of the crop. An additional reason that makes the detection of nitrogen deficiency very important is the fact that nitrogen leaches under the root zone when irrigation or water management is not appropriate, creating conditions that are suboptimal for crop growth.

Nitrogen (N) indices can be divided into indices that are based on wavelengths in the visible and the NIR region, and indices that include specific nitrogen absorption wavelengths in the SWIR. The additional value of using SWIR-based indices has been shown in studies on wheat in which a firm advantage was revealed for the proposed SWIR-based indices in their ability and sensitivity to predict N content in potato leaves [87].

Many hyperspectral vegetation indices (VIs) have been developed to estimate crop nitrogen status at leaf and canopy levels. They have been evaluated for different growth stages and years using data from both nitrogen experiments and farmers' fields. Furthermore, to identify alternative promising hyperspectral VIs, evaluation of all possible two-band combinations of SRs and NDIs has been performed. The results indicated that best-performing published and newly identified VIs included simple ratios in the red edge region and in the blue region [11,14]. Red edge and NIR bands were more effective for nitrogen estimation at early growing stage, but visible bands, especially ultraviolet, violet, and blue bands, were more sensitive at later growing stage.

Across sites, years, cultivars, and growth stages, the combination of wavelengths in the blue range (370 and 400 nm), as either simple ratio or an NDI, performed most consistently in both experimental and field data for wheat [14]. Together with green, red, NIR, and red edge, the blue range was found sensitive to nitrogen content in rice in another extensive study that integrated ground based spectral data from different sites and years [88]. Yet, in the same study, it was found to be insensitive to nitrogen content when Hyperion images were utilized. For Hyperion images, only the red edge and NIR were found sensitive.

In their study for detecting nitrogen stress in two potato cultivars, Tyler et al. [89] reported that canopy-scale spectral data can distinguish between N treatments better than tissue samples and that among several spectral indices Medium Resolution Imaging Spectrometer (MERIS) terrestrial chlorophyll index (MTCI) [90] was the best spectral index to be used for variable rate nitrogen prescriptions in potatoes. MTCI extracted from Hyperion satellite images exhibited the best relation (logarithmic) to N content also in rice in comparison to more than 50 published two- and three-bands indices [88]. In terms of prediction ability, MTCI performed slightly better than other published indices but significantly worse than two and three-band indices proposed by Tian et al. [88]. Moreover, the two new indices performed well using ground spectra, modeled airborne visible/ infrared imaging spectrometer (AVIRIS) spectra, Hyperion spectra and acquired Hyperion images. Despite that, a newer study that analyzed the spatial variability of chlorophyll and N content of rice from Hyperion imagery in India found different relationships between N content and the indices suggested by Tian et al. [91]. Moreover, their modified index has shown significantly wider range of predicted N content than the index suggested by Tian et al. and was thus better for mapping the spatial variability of N content.

The majority of the indices predicting N content are based on indirect indicators, mostly chlorophyll content, which is proven to be physiologically linked to N content. Herrmann et al., explored the performance of new N spectral indices dependent upon the SWIR (1200–2500 nm), and particularly the 1510 nm band because it is related to N content [92]. The results revealed a firm advantage for the SWIR-based indices in their ability to predict and in their sensitivity to N content. The best index is one that combines information from the 1510 and 660 nm bands, but no significant differences were found among the new SWIR-based indices.

Beyond the differences between crops, sites, and years, growth stage also had a significant influence on the performance of different vegetation indices and on the selection of sensitive wavelengths for leaf nitrogen estimation. The observed interchangeability of wavelengths and indices along growth stages and cultivars may be addressed by multivariate methods, which make use of the whole spectrum and not only selected wavelengths. For instance, multivariate methods were used to estimate leaf nitrogen content based on narrow-band spectral data in potatoes. PLSR analysis has resulted in a stronger correlation between predicted and measured leaf nitrogen content ($R^2 = 0.95$) than the nitrogen-specific transformed chlorophyll absorption reflectance index (TCARI) ($R^2 = 0.82$), even though in both models data from narrow bands was used [93]. Moreover, with PLS the improved correlation was achieved with a single model for both the vegetative and the tuber-bulking periods, while the TCARI yielded a different model for each period [93]. In the same study, when the number of wavelengths was reduced from 400 to 11, and the bands' bandwidth was broadened from 1.3 to 20–40 nm, in order to simulate the Venµs satellite data, the accuracy of the spectral model was decreased ($R^2 = 0.78$), yet still included both vegetative and tuber-bulking periods. Similar results were obtained for nitrogen prediction in winter wheat. Models based on NDVI had an exponential characteristic, which implies saturation for high nitrogen values, and low coefficient of determination ($R^2 = 0.15$). When the derivative of the spectrum between 350 and 2500 nm was used in conjunction with PLSR models, the coefficient of determination was significantly better ($R^2 = 0.82$).

Another approach was suggested to address the observed interchangeability of wavelengths and indices along crops, cultivars, growth stages, years, and sites. A nitrogen sufficiency index (NSI) was applied to leaf N concentration and spectral indices/models to normalize them for comparative purposes between spectral indices and PLS prediction models [89]. Applying the NSI formula to spectral data made it insensitive to external factors such as cultivar and growth stage. In practice, it means that for a proper use of hyperspectral data the farmer should keep N-rich areas within the field [94] to be used for NSI estimation and N prescription maps for variable N application rate. Adapting the N-rich areas concept to commercial production practices might seem to be straight forward but the N-rich plants are likely to develop differently from the remaining field and do not represent the normal canopy. The virtual reference concept uses a histogram to characterize and display the spectral data from which the vegetation index of adequately fertilized plants can be identified [95]. As described in Section 3.5.3, this approach was also suggested and successfully tested for wet-temperature reference for water status estimation.

3.5.3 Suggested Approaches in Predicting BB-PACs

Key BB-PACs like LAI, chlorophyll level, and water status have major effect on transition zones of the spectra reflectance curve such as the red edge and water absorption bands. Thus, the narrow band widths of hyperspectral data allows for better estimation of crop properties compared with the relatively coarse bandwidths acquired with multispectral scanners. While hyperspectral images in the VIS-NIR range provide a tool for estimating and mapping various BB-PACs in the fields, they are limited in assessing and mapping crop water status parameters that are essential for irrigation management. To that end, integration of images from the thermal range is required.

With the advances in hyperspectral technologies, practical issues related to data volumes and data-processing emerged. The processing complexity and the statistical concerns of colinearity and over-fitting entailed in spectral analysis have led to the widespread adoption of the dimensionality reduction approach. Various narrow-band indices were developed and were shown to improve the broad-band indices. Step-wise discriminant analysis was used in many studies to select a few optimal bands for characterizing agricultural crop variables. In general this type of analysis had demonstrated the importance of the red, the red edge, and the SWIR regions and, to lesser extent, the blue, green, and NIR regions. These findings together with the high cost of hyperspectral systems and the analytical complexity promoted the development of super-spectral platforms such as the Rapid-Eye, the World-View2, and the Venμs (launched on July 2017).

Despite the similarities found in the literature, the selected bands were not identical for the same crop property in different sites nor were they identical to different crop properties in the same site. Moreover, beyond the differences between crops, sites, and years, growth stage also has been shown to have a significant influence on the performance of different vegetation indices. Finally, beyond issues related to calibration, accuracies, and operational characteristics of the sensors, the leaf or canopy reflectance, in the visible–NIR–SWIR ranges, is highly dependent on both biophysical and biochemical properties. Moreover, several properties have similar effects on canopy reflectance. It means, for example, that while N content is the desired property, information provided by analyzed spectral measurements would be biased by factors other than N, such as water status, stand density, and pests. In view of this, we doubt the utility of pursuing the approach of the best set of bands or the best index. In other words, in our opinion, there is no global set of bands or global index for predicting any BB-PACs. For this reason, we do not provide a table that summarizes specific bands, spectral indices, and spectral ranges for the various key BB-PACs. Instead we present Figure 3.1, which generally and very coarsely shows single bands and band ranges that have been used in the cited studies for estimating LAI and biomass; water content; and nitrogen and chlorophyll levels. Rather than using a set of single bands or indices, this overview strongly demonstrates the advantage of hyperspectral systems that provide contiguous spectra using multivariate analysis techniques. With recent developments in compact hyperspectral sensors and compact uncooled thermal cameras, combined with available UAV that can carry them [36,96,97], new horizons for hyperspectral and thermal data are opened. Snapshot hyperspectral cameras were used to create radiometrically calibrated hyperspectral data [98,99] and even provide 3D hyperspectral information for vegetation monitoring [99]. To make these systems affordable for agricultural stakeholders, methodologies that fuse hyperspectral or thermal aerial imaging with MS satellite imaging should be developed [37,38].

Whether analyzed contiguous spectra or spectral indices are used, normalization procedures such as the well-fertilized reference plots for N level estimation or statistical wet reference for water status estimation, seem to be inevitable. In the view of the cited studies, it is deduced that absolute estimation of any BB-PACs is unachievable unless a reference area or reference data are available. Based on such references, the approach of a sufficiency index that was introduced for nitrogen variable rate application (nitrogen sufficiency index, NSI) should be utilized for other BB-PACs.

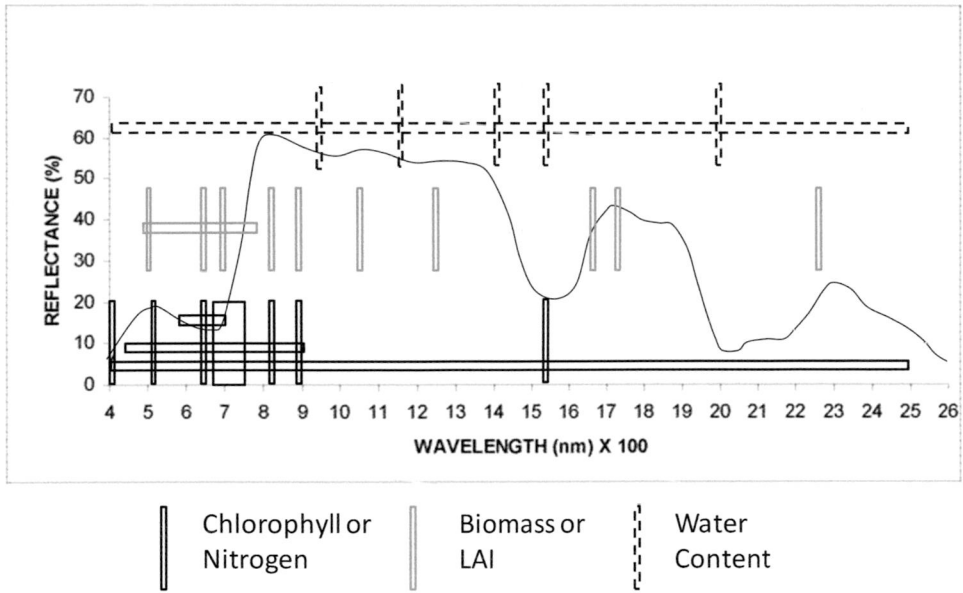

FIGURE 3.1   Spectral bands and spectral ranges that were used in various studies to estimate key BB-PACs. Vertical bars refer to single bands and horizontal lines refer to band range.

3.6   SPATIAL METHODS

In the early days of hyperspectral imaging, hyperspectral data processing techniques focused on analyzing the spectral data without incorporating information on the spatially adjacent data. In other words, hyperspectral data were usually treated not as images but as unordered listings of spectral measurements with no particular spatial arrangement [100]. The importance of analyzing both spectral and spatial patterns has been identified as a desired goal by many scientists devoted to multidimensional data analysis. This type of processing has been approached from various points of view representing different levels of combination between spectral and spatial information.

Nearly all of the methods combining spectral and spatial information were developed for land cover classification. General reviews and illustrations of spectral–spatial classification methods of hyperspectral imagery can be found in Fauvel et al. [101] and Plaza et al. [100]. In general, the spectral–spatial methods seek to reduce the salt-and-pepper appearance of the classification; to use spatial characteristics (such as entropy and standard deviation (STD)) and spatial features (such as size and shape) to enhance the separation ability between classes; and to perform image segmentation prior to the classification to define a spatial neighborhood for the pixels. There is a basic difference between land use/cover classification and estimation of BB-PACS. Land use/cover types are discrete elements with relatively well-defined borders. Moreover, most of them have relatively distinct spectral signature. In comparison, biophysical and biochemical crop properties are continuous variables, with smooth differences in spectral signature and with amorphous shapes.

The spatial and temporal variability of soil and crop factors within a field is the factual base of precision agriculture [102]. Opportunities exist to use airborne hyperspectral and thermal imaging for mapping the spatial variability of crop properties in agricultural fields. Maps of nitrogen and water status can then be used to delineate management zones for fertilization and irrigation variable-rate application. Delineating management zones involves spatial filtering to reduce effects of noise in measurements of individual factors and removal of excessive details in within-field variability to simplify the shapes and size of the zones. Methods combining spectral and spatial information in studies designated to estimate levels of crop biophysical and biochemical properties and to divide them into homogenous/

TABLE 3.3
Nitrogen Treatments Applied in the Potato Field in Spring 2007

Nitrogen Treatment	N Rate (kg ha^{-1})	Percentage N Rate Relative to Commercial Rate	Yield
T100%	400	100	52 c
T75%	300	75	52 c
T50%	200	50	52 c
T25%	100	25	45 b
T0%	0	0	36 a

management zones are scarce. Indeed, spectral models manipulated over hyperspectral images were used to create maps of biophysical and biochemical crop properties [40,103,104] or further to partition fields into management zones based on spectral properties [105]. In the creation of the maps, smoothing operations were applied for reducing the speckle effect. Yet, all of the maps were created merely based on a pixel-by-pixel spectral data without incorporating information on the neighboring pixels. Here we describe potential approaches and illustrate methods for combining spectral and spatial information for segmentation of hyperspectral images based on spectral-based crop properties.

3.6.1 HYPERSPECTRAL DATA SET

To illustrate the potential that lies in some of the described methods we used two aerial hyperspectral images taken over an experimental potato plot under different nitrogen treatments and over a commercial potato field. For the commercial field, an aerial thermal image was also available. The experimental plot and the commercial field were planted with *cv. Desiree* in Kibbutz Ruhama, Israel (31.388N, 34.598E). To assess a range of N levels, five treatments were applied (Table 3.3) with four replicates. Each replicate was 18 m wide (six rows) by 50–100 m long. More details on the overall study can be found in [93]. The experimental plot represents a controlled area with a relatively wide range of N levels with known borders. In contrast, the commercial field represents spatial variability that its range and spatial pattern are unknown in advance.

Hyperspectral images above the experimental plot and the commercial field were acquired on May 25, 2007 and April 24, 2012, respectively. AISA Eagle hyperspectral imaging push broom sensor (Spectral Imaging Ltd., Oulu, Finland) in the range of 400–970 nm, with 420 bands with spectral resolution of 1.3 nm. The image was acquired from 500 m height and had 1 m spatial resolution. Preprocessing of the image included selection of every second band of the original 420 bands and smoothing of the 210-band spectra of the new cube, with a 15-points window.

Figures 3.2 and 3.3 are an RGB and false color images of the experimental plot derived from the narrow-band hyperspectral image. The false color is overlaid by the borders of the N treatments.

3.6.2 SPATIAL INFORMATION AS A PREPROCESSING TOOL

Individual spectra of the same object or property taken from neighboring pixels in the hyperspectral image present relatively high variability. Figure 3.4 shows individual spectra for regions of interest (ROIs) taken from sub-plots 75% and 0% along with the mean spectra (thick black line). These hyperspectral data are rather noisy in comparison to spectra collected using a spectrometer [106]. This is primarily a result of how hyperspectral data are collected. In most spectrometers, a single measurement is actually the mean of several independent spectra that were collected over a small area, which greatly reduces the noise in the spectra.

Reduction of the noise is essential for calibrating a spectrally based model. However, in an aerial hyperspectral image, each pixel of a hypercube is a single spectrum of a relatively wide area. To

FIGURE 3.2 RGB (670 nm, 550 nm and 420 nm) image of the experimental plot.

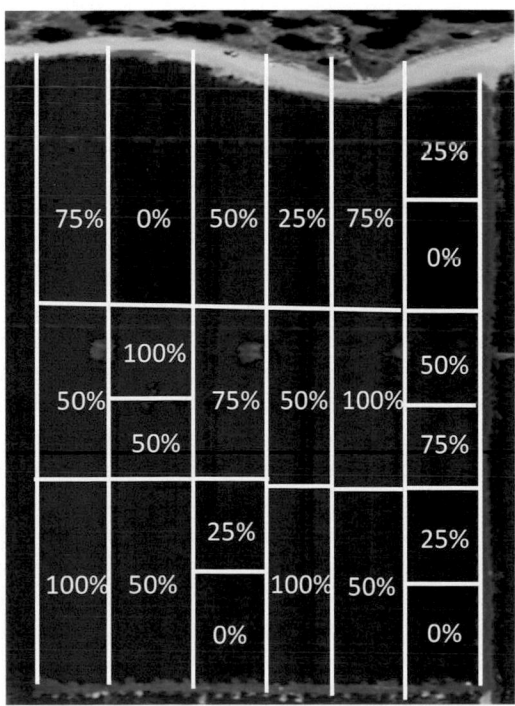

FIGURE 3.3 False color image of IR (750 nm), red (670 nm) and green (550 nm) bands of the experimental plot overlaid by the borders of the N treatments.

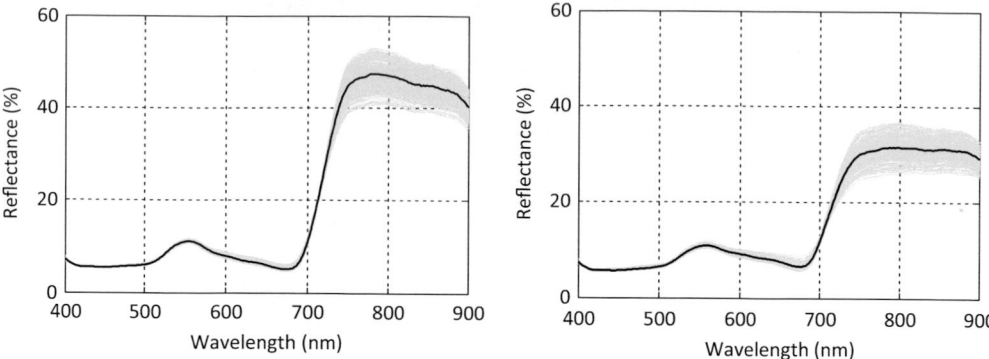

FIGURE 3.4   Individual spectra for ROIs taken from sub-plots of T75% (left) and T0% (right) along with the mean spectra (thick black line).

reduce the spectral noise in calibrating models, Lawrence et al. [106] suggested using a spatially averaged ROI spectrum and then applying the model on a pixel-by-pixel basis. Lawrence [106] used manual selection of ROI spectra from a close range hyperspectral image to calibrate a PLS spectral model for contaminant detection on poultry carcasses. Manual selection of homogenous ROIs from an aerial hyperspectral image of a field is problematic and might suffer from subjectivity.

3.6.3   Spatial Information to Improve Spectral Classification

Spatial context was suggested as a second step for the refinement of results obtained by spectrallybased techniques. This approach consists of three parts: (1) a pixel-by-pixel spectral classification; (2) definition of a pixel neighborhood (surrounding each pixel); and (3) performance of a local operation so that if there is strong evidence that individual spectra of pixels in a neighborhood are spectrally homogenous they are included in the same cluster. This approach was developed for MS images and extended by Jimenes et al. [107] to airborne hyperspectral sensors. The developed classifier is an unsupervised modification of the supervised extraction and classification of homogenous objects (ECHO). Based on the dataset, the developed algorithm, called UnECHO, automatically estimates the required threshold of homogeneity level of the entire neighborhood without input from the human analyst [107]. When applied to urban and rural areas, the UnECHO successfully uncovered spatial structures and significantly improved spectral classifications (C-means or maximum likelihood [ML]).

Another example of this approach is the Markov random field (MRF) in which spatial characterization is performed by modeling the spatial neighborhood of a pixel as a spatially distributed random process. The MRF attempts to make regularization via the minimization of an energy function using known land covers and their prior probabilities. Similarly to Jimenes et al. [107], Plaza et al. [100] developed an unsupervised version of this methodology. They used a neuro-fuzzy classifier to perform classification in the spectral domain and to compute a first approximation of the posterior probabilities of classes. The output of this step is then fed to the MRF spatial analysis stage, which was performed using a maximum likelihood probabilistic reclassification. The performance of the MRF in classifying urban land cover types was compared with the results of the first stage, that is, a neuro-fuzzy classifier. Similar classification accuracies were achieved mainly because the spatial analysis stage reassigned only border pixels to different classes.

In both classification methods—the UnECHO and the modified MRF—the neighborhoods are determined in advance and do not account for the real size and shape of the objects in the image. This kind of division might not be suitable for the gradual change of biophysical and biochemical properties of crops over the field. Several segmentation approaches were suggested to be performed prior to the hyperspectral image classification. The segmentation techniques such as partitional clustering and hierarchical segmentation [108,109] partition an image into homogeneous regions with different sizes based on a homogeneity criterion [101]. These nondeterministic approaches may be more suitable for mapping homogeneous zones of biophysical and biochemical properties of crops in a field (an illustration of such a technique is provided in the next section). Another alternative that may be adopted for segmentation of homogeneous zones in a field is the geostatistical approach [110,111]. Lark [110] suggested a spatially weighted averaging of the class memberships within a local neighborhood based on the variogram. Although the generation of spatially coherent regions (SCR), developed by Lark [110], was initially applied for limited dimensionality of nonspectral data (multitemporal yield data), it can be adapted to hyperspectral images.

3.6.3.1 Analysis of Hyperspectra Images of an Experimental Plot

To initially investigate the ability to generate SCRs in the spectral domain, we realized and applied the SCR for the HS images of the experimental plot [112]. A fuzzy C-means classification into seven classes was applied (Figure 3.5). Assuming that in the experimental plot the main effect on the reflectance is the nitrogen level, the classes were labeled with a nitrogen level based on a visual inspection and prior knowledge of the N treatments. Despite the noisy result, the fuzzy classification captured differences in N levels: similar N levels in individual pixels were assigned with similar classes. Yet, the classes do not fully match with N treatments. Major difference exists between the two lowest N treatments (0% and 25%) and the other three treatments (50%–100%) while minor or no difference exists between the 50%, 75% and 100% treatments [93]. They also did not differ in their yield (Table 3.2). Additionally, it seems that the western part of the experimental plot is populated with more vital plants than the eastern part.

Following the spectral classification, the variogram half-range was calculated and used as the neighborhood radius for refining the spectral classification (Figure 3.6). The resulted SCR significantly reduced the speckle effect by uncovering most of the spatial structures of the sub-plots with different N levels. If the affecting factor responsible for the variability in the field is known in advance, the resulting SCRs can be used as management zones for variable-rate application. If not, they can be used both for selecting spectra free of noise for calibrating the spectral model [106] or for implementing a validated model for crop properties estimation.

Since the neighborhood is not determined by geometrical shapes, the borders between sub-plots are not crisp but rather fuzzy. This result implies that this type of flexible neighborhood definition is more suitable for the real situation in the field where changes in N levels are gradual and not sharp.

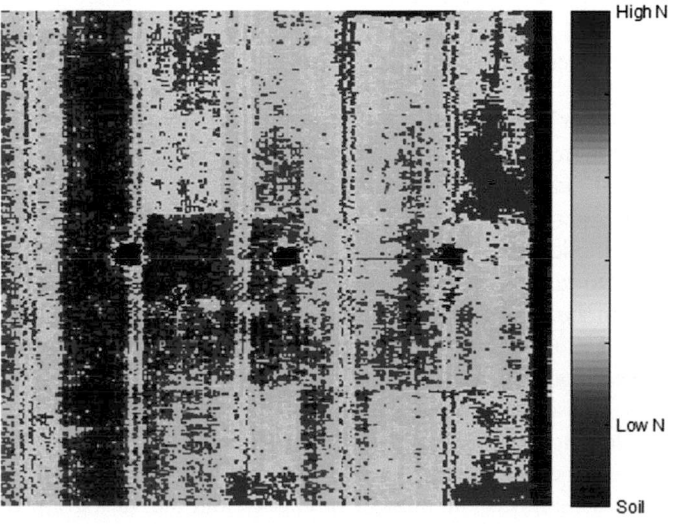

FIGURE 3.5   A fuzzy C-means classification of the 210-band HS image of the experimental plot, 7 classes.

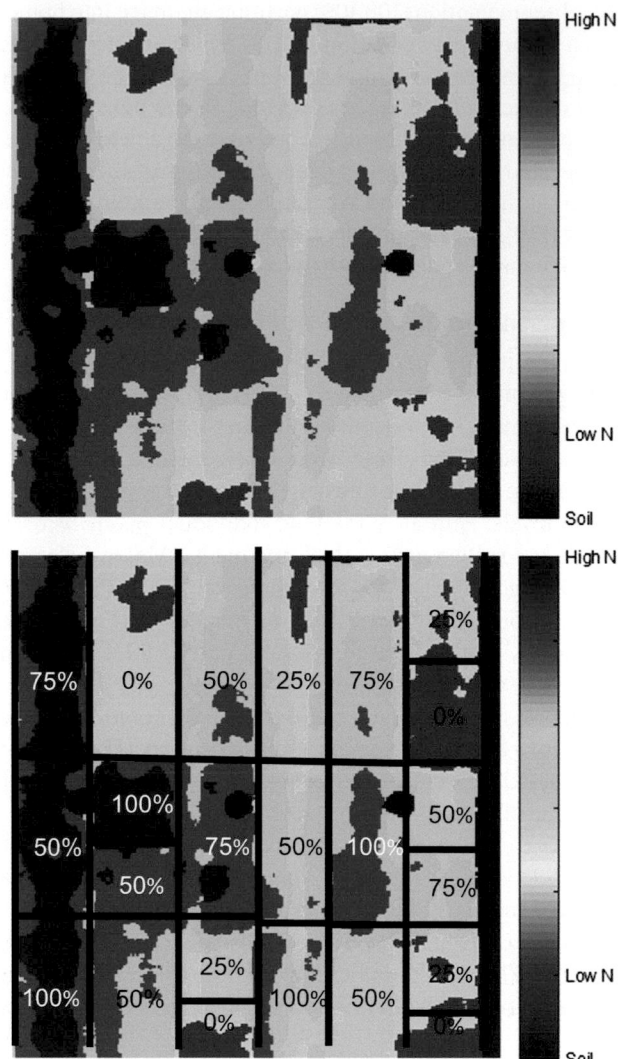

FIGURE 3.6   Spatially coherent regions of the fuzzy C-means.

3.6.3.2   Analysis of Hyperspectral and Thermal Images of a Commercial Field

The spectral vegetation indices—NDVI and NDI [79]—were calculated from the aerial HS images (Figure 3.7). The spectral indices and the thermal images (Figure 3.7) exhibit some variability in the commercial field. The eastern side has higher NDVI and NDI values with lower temperature. The SCR methodology was implemented on NDVI and NDI maps and on the raw thermal image of the commercial potato field to examine its ability to delineate homogeneous zones of N levels and water status levels. Similarly to the analysis of the experimental plot hyperspectral image, the image of the commercial field was classified by the fuzzy C-means classification. Since, unlike the manipulated experimental plot, the variability in a commercial field is unknown in advance, the optimized number of classes was calculated based on the change in goodness variance of fit (GVF) with the increasing number of classes. Following the spectral classification, the variogram was calculated to determine the neighborhood radius for refining the spectral classification.

Despite the visual resemblance, the number of classes and the variogram of each image were different, leading to different homogeneous zones. The NDVI, NDI and canopy temperature were classified into 2, 3, and 4 classes, respectively.

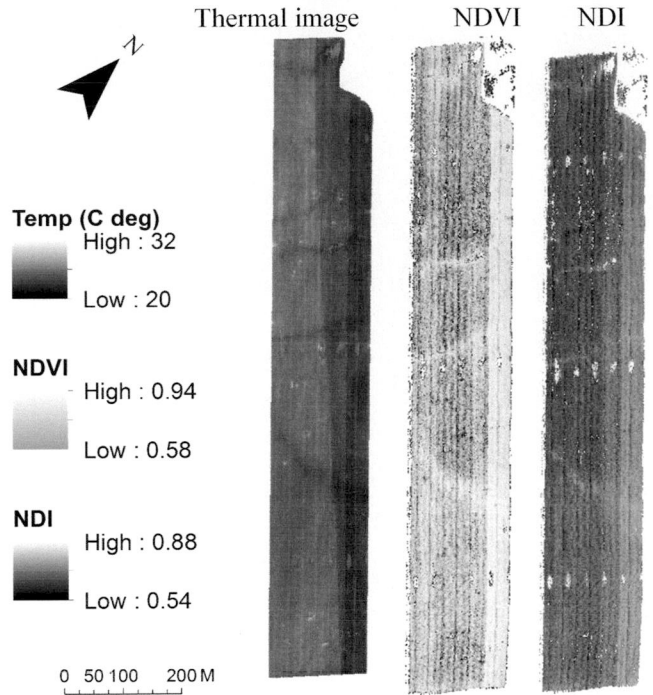

FIGURE 3.7 Aerial thermal image and NDVI and NDI maps of the commercial potato field. The two latter maps were extracted from an aerial HS image.

Figure 3.8 presents the major segments following the omission of very small segments. The SCR algorithm seems to capture the variability in the field but did not necessarily follow the borders between zones perceived by visual inspection of the images. The NDVIs of the two classes were very high, with minor differences. In this range, NDVI saturates and can hardly be associated with potato crop parameters [113]. The NDI image showed relatively high values [89] yet the differences between the lowest and the highest classes may be of consequence for nitrogen level [89]. The thermal image was divided into four classes and, according to the CWSI values, two of them were over-irrigated and two of them had the optimal CWSI value before irrigation. The results show that while there were differences in nitrogen levels the main source of the variability in this field was water status. The main conclusions are as follow: (1) hyperspectral image in the VIS–NIR range and thermal images are complementary; (2) integrating the spatial attribute to the analysis contributes to reveal the variability in BB-PACs and is required for delineating homogeneous zones for variable-rate fertilization and irrigation. However, some of the variability that may be revealed by the image analysis might be of no consequence to the variable-rate application and further expert inspection is required prior to the creation of prescription maps.

3.6.4 FUSION OF SPECTRAL AND SPATIAL INFORMATION

The previous approaches separate spatial from spectral information, and thus the two types of information are not treated simultaneously. Plaza et al. [100] suggested incorporating spatial context into the SVM spectral classifier. In this method, a pixel entity is redefined simultaneously both in the spectral and spatial domains by applying some feature extraction to its surrounding area, which yields spatial (contextual) features such as the mean or standard deviation per spectral band. These separated entities lead to two different kernel matrices that can be summed and introduce cross-information features in the formulation. When applied for land cover classification in an agricultural area, the contextual SVM showed classification accuracy of 95%. It outperformed a spectral classifier

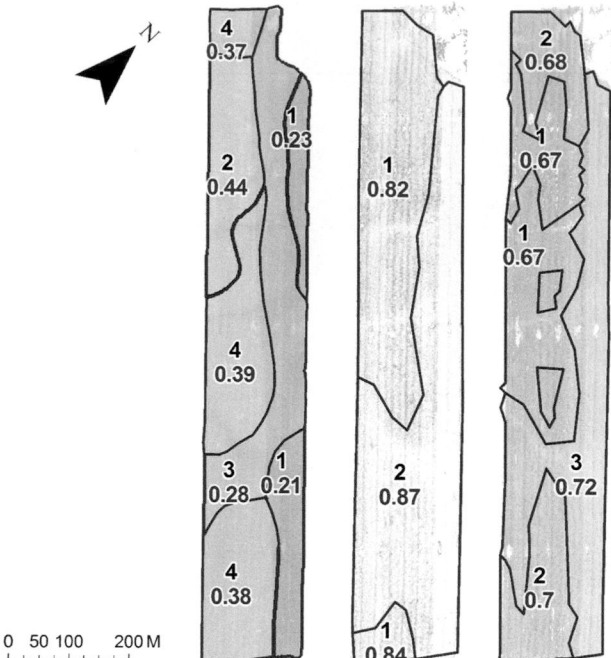

FIGURE 3.8   Homogeneous zones based on the NDVI, NDI and canopy temperature images. Red lines present borders of the homogeneous zones.

based on Euclidean distance and performed much better than other methods like ECHO that use spectral and spatial information to classify homogeneous objects. Similarly to the UnECHO and the modified MRF, the contextual SVM is based on a predefined neighborhood of $N \times N$ windows.

Another approach for fusing spectral and spatial information is a multiscale or hierarchical segmentation. Hierarchical segmentation is based on sequential optimization to produce a hierarchical data-driven decomposition of the picture with no restriction on segment shapes [114]. Beamlet analysis is a framework for multiscale image analysis in which line segments play a role analogous to the role played by points in wavelet analysis [115]. The beamlet-decorated recursive dyadic partitioning (BD-RDP) is one realization of the beamlet analysis. While partitioning with basic RDP is limited to square elements, the BD-RDP allows that some of its squares (optionally) are decorated by a beamlet, that is, can be partitioned not only by squares but also by other geometrical shapes. In comparison to the basic RDP, this additional flexibility allows the BD-RDP to approximate an image more accurately with much fewer segments. The BD-RDP was originally designated to one-dimensional images by implementing two main steps: a spreading phase, where the image is partitioned into its smallest parts according to a quad-tree structure and a folding phase, where the tree is folded up according to a target function. The target function has to serve the idea of minimum variation between the original and reassigned values with a penalty for the number of different segments. The BD-RDP does not require a priori knowledge of the number of segments.

Levi et al. [116] introduced two enhancements of the algorithm, which were illustrated using the hyperspectral of the experimental plot by Cohen et al. [112]. First, it was modified to suit multidimensional images based on the Euclidian distance between vectors, and second, a merging neighbor's phase was added that checks the possibility of merging segments that belong to different dyadic squares using the target function. The multidimensional three-step BD-RDP was applied to the 210-band hyperspectral image of the experimental plot and the segmentation result is shown in Figure 3.9. As it is not a classification, the color of each segment was determined by its average reflectance in an IR band (750 nm) to partially demonstrate the differences in N levels.

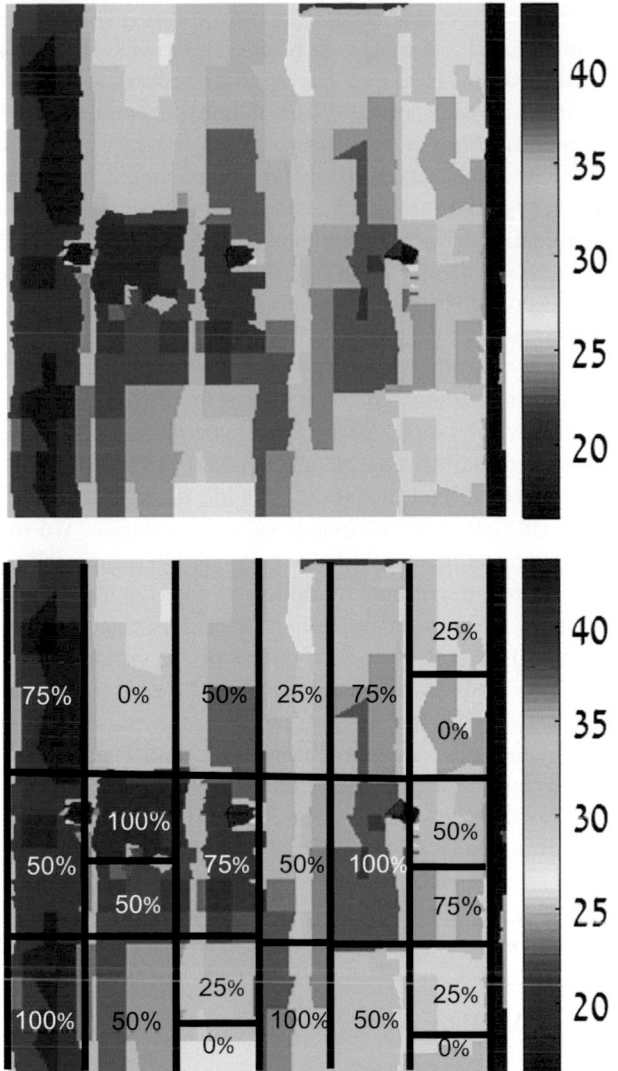

FIGURE 3.9 Multidimensional beamlet-decorated recursive dyadic partitioning of the 210-band hyperspectral image of the experimental plot. The value of each segment is the average of the reflectance in an IR band (750 nm).

The multiscale segmentation successfully uncovered the spatial structures in the image according to differences in N levels. Unlike the SCR classification, the beamlet analysis results in homogeneous segments and needs further analysis to classify the segments according to their N level. For that a fuzzy C-means or a calibrated PLS model can be applied.

3.6.5 INTEGRATING SPECTRAL AND SPATIAL ATTRIBUTES: A SUMMARY

Hyperspectral images are distinguished from point spectral measurements by their added spatial aspect. Nevertheless, hyperspectral images are usually not treated as images but as lists of spectral measurements with no particular spatial arrangement. Estimation of crop properties using hyperspectral images was based only on spectral information. Precision agriculture necessitates the partition of the field into homogeneous zones. Similar to land use segmentation and classification, delineation of homogeneous zones would benefit from incorporating spectral and spatial information. In this chapter we have introduced different levels of integration of spatial information in estimating crop properties. The ability to integrate and fuse spatial analysis with spectral information was initially demonstrated using an hyperspectral image of an experimental potato plot.

In general, the suggested methods were effective in classifying N levels and uncovered spatial structures that coincided with the blocks of the different treatments, but no quantitative evaluation was done. In other studies, spatial information derived from hyperspectral images was found to be valuable in land use classification. The contribution of spatial analysis for BB-PACs estimation is yet to be studied. For that, research is needed to investigate existing methods and adjust them or develop new methodologies to incorporate spatial and spectral information.

While integrated spectral/spatial algorithms hold great promise for in-season management zone delineation using hyperspectral images, they also introduce computational challenges. Since agrotechnical decisions are made routinely by the farmer once or twice a week, a temporal aspect should also be taken into consideration. With rapid developments, satellites constellations like PlanetLab are already providing daily high–spatial resolution MS images with NIR and RGB bands. Other satellites like the Sentinel2 and the Venµs provide more than 10 bands in the VIS–NIR range in a 2- to 5-day revisit time with reasonable spatial resolution. We may foresee that in the near future frequent hyperspectral images will also be available from satellites. In the meantime, a multiscale approach may address the limited resolutions (temporal and spatial) of the hyperspectral and thermal satellite images. UAV hyperspectral and thermal imaging systems could be used as sampling systems, directed to spots that set wide ranges of the targeted BB-PACs. Thereafter, the associations between the UAV and MS satellite images can be used to extrapolate to wider areas.

In order to fully exploit hyperspectral images, processing methods that can take advantage of their enhanced spectral, spatial and temporal features are required. Parallel processing hardware has necessarily become a requirement to speed up processing performance and to satisfy high computational requirements. As a result, the future potential of hyperspectral image processing methods will also be largely defined by their suitability for being implemented in parallel [100]. The hierarchical segmentation approach that haas been presented in this chapter is suitable for parallel processing hardware and thus the routine of in-season management zone delineation utilizing this approach may be speeded up to meet the timeline requirements of the agrotechnical applications.

3.7 CONCLUDING REMARKS

Hyperspectral remote sensing systems enable the collection of several hundred spectral bands in a single acquisition, thus producing detailed spectral and spatial data. The narrow band widths of hyperspectral data allow for better estimation of crop properties compared with the relatively coarse bandwidths acquired with multispectral scanners. While hyperspectral images in the VIS–NIR range provide a tool for estimating and mapping various BB-PACs in the field, they are limited in assessing and mapping crop water status parameters that are essential for irrigation management. For that end, integration of images from the thermal range is required. The literature on the use of hyperspectral and thermal imaging for predicting BB-PACs is massive and cannot be fully summarized in any framework. In this chapter we have reviewed a small portion of these studies and tried to draw guiding principles on how to extend the spectral and spatial properties of hyperspectral and thermal data for management zone delineation toward variable-rate applications.

1. Rather than using sets of single bands or indices, hyperspectral systems that provide contiguous spectra are highly advantageous when analyzed with multivariate techniques.
2. Thermal range and the VIS–NIR–SWIR spectrum should be integrated to decouple the effect of water status and chlorophyll or nitrogen level.
3. Compact hyperspectral sensors and compact uncooled thermal cameras suited for UAVs are already available, opening new possibilities. But to make these systems affordable for

agricultural stakeholders, methodologies that fuse hyperspectral or thermal aerial imaging with MS satellite imaging should be developed.

4. Analysis of hyperspectral and thermal images should integrate both their spectral and spatial attributes for the delineation of management zones.

REFERENCES

1. Mahlein, A.K. et al., Spectral signatures of sugar beet leaves for the detection and differentiation of diseases. *Precision Agriculture*, 11(4) p. 413–431.
2. Liu, Z.Y., H.F. Wu, and J.F. Huang, Application of neural networks to discriminate fungal infection levels in rice panicles using hyperspectral reflectance and principal components analysis. *Computers and Electronics in Agriculture*, 72(2) p. 99–106.
3. Yang, Z. et al., Differentiating stress induced by greenbugs and Russian wheat aphids in wheat using remote sensing. *Computers and Electronics in Agriculture*, 2009. 67(1–2) p. 64–70.
4. Reisig, D.D. and L.D. Godfrey, Remotely sensing arthropod and nutrient stressed plants: a case study with nitrogen and cotton aphid (Hemiptera: Aphididae). *Environmental Entomology*, 2010. 39(4) p. 1255–1263.
5. López-Granados, F., Weed detection for site-specific weed management: mapping and real-time approaches. *Weed Research*: 51, p. 1–11.
6. Shapira, U. et al., Weeds detection by ground-level hyperspectral imaging. in *10th International Conference on Precision Agriculture*, 2010.
7. Thorp, K.R. and L.F. Tian, A review on remote sensing of weeds in agriculture. *Precision Agriculture*, 2004. 5(5) p. 477–508.
8. Fuchs, M. and C.B. Tanner, Infrared thermometry of vegetation1. *Agronomy Journal*, 1966. 58(6) p. 597–601.
9. Maes, W.H. and K. Steppe, Estimating evapotranspiration and drought stress with ground-based thermal remote sensing in agriculture: A review. *Journal of Experimental Botany*, 2012. 63(13) p. 4671–4712.
10. Bajcsy, P. and P. Groves, Methodology for hyperspectral band selection. *Photogrammetric Engineering and Remote Sensing*, 2004. 70(7) p. 793–802.
11. Jain, N. et al., Use of hyperspectral data to assess the effects of different nitrogen applications on a potato crop. *Precision Agriculture*, 2007. 8(4–5) p. 225–239.
12. Thenkabail, P.S. et al., Accuracy assessments of hyperspectral waveband performance for vegetation analysis applications. *Remote Sensing of Environment*, 2004. 91(3–4) p. 354–376.
13. Thenkabail, P.S., R.B. Smith, and E. De Pauw, Evaluation of narrowband and broadband vegetation indices for determining optimal hyperspectral wavebands for agricultural crop characterization. *Photogrammetric Engineering and Remote Sensing*, 2002. 68(6) p. 607–621.
14. Li, F. et al., Evaluating hyperspectral vegetation indices for estimating nitrogen concentration of winter wheat at different growth stages. *Precision Agriculture*, 2010. 11(4) p. 335–357.
15. Tucker, C.J., Red and photographic infrared linear combinations for monitoring vegetation. *Remote Sensing of Environment*, 1979. 8 p. 127–150.
16. Cortes, C. and V. Vapnik, Support-vector networks. *Machine Learning*, 1995. 20(3) p. 273–297.
17. Breiman, L., Random forests. *Machine Learning*, 2001. 45(1) p. 5–32.
18. Wold, S. et al., The collinearity problem in linear-regression—the partial least-squares (PLS) approach to generalized inverses. *Siam Journal on Scientific and Statistical Computing*, 1984. 5(3) p. 735–743.
19. Baret, F. and G. Guyot, Potentials and limits of vegetation indexes for LAI and APAR assessment. *Remote Sensing of Environment*, 1991. 35(2–3) p. 161–173.
20. Haboudane, D. et al., Remote estimation of crop chlorophyll content using spectral indices derived from hyperspectral data. *IEEE Transactions on Geoscience and Remote Sensing*, 2008. 46(2) p. 423–437.
21. Idso, S.B. et al., Normalizing the stress-degree-day parameter for environmental variability. *Agricultural Meteorology*, 1981. 24(C) p. 45–55.
22. Jackson, R.D. et al., Canopy temperature as a crop water stress indicator. *Water Resources Research*, 1981. 171 p. 133–138.
23. Baluja, J. et al., Assessment of vineyard water status variability by thermal and multispectral imagery using an unmanned aerial vehicle (UAV). *Irrigation Science*, 2012. 30(6) p. 511–522.
24. Cohen, Y. et al., Use of aerial thermal imaging to estimate water status of palm trees. *Precision Agriculture*, 2012. 13(1) p. 123–140.
25. Zarco-Tejada, P.J., V. González-Dugo, and J.A.J. Berni, Fluorescence, temperature and narrow-band indices acquired from a UAV platform for water stress detection using a micro-hyperspectral imager and a thermal camera. *Remote Sensing of Environment*, 2012. 117(0) p. 322–337.

26. Rud, R. et al., Crop water stress index derived from multi-year ground and aerial thermal images as an indicator of potato water status. *Precision Agriculture*, 2014. 15(3) p. 273–289.

27. Jones, H.G. et al., Use of infrared thermography for monitoring stomatal closure in the field: Application to grapevine. *Journal of Experimental Botany*, 2002. 53(378) p. 2249–2260.

28. Meron, M. et al., Crop water stress mapping for site-specific irrigation by thermal imagery and artificial reference surfaces. *Precision Agriculture*, 2010. 11(2) p. 148–162.

29. Cohen, Y. et al., Mapping water status based on aerial thermal imagery: Comparison of methodologies for upscaling from a single leaf to commercial fields. *Precision Agriculture*, 2017. 18(5) p. 801–822.

30. Jones, H.G., Use of infrared thermometry for estimation of stomatal conductance as a possible aid to irrigation scheduling. *Agricultural and Forest Meteorology*, 1999. 95(3) p. 139–149.

31. Berliner, P., D.M. Oosterhuis, and G.C. Green, Evaluation of the infrared thermometer as a crop stress detector. *Agricultural and Forest Meteorology*, 1984. 31(3–4) p. 219–230.

32. Taghvaeian, S. et al., Conventional and simplified canopy temperature indices predict water stress in sunflower. *Agricultural Water Management*, 2014. 144 p. 69–80.

33. Alchanatis, V. et al., Evaluation of different approaches for estimating and mapping crop water status in cotton with thermal imaging. *Precision Agriculture*, 2010. 11(1) p. 27–41.

34. Gonzalez-Dugo, V. et al., Using high resolution UAV thermal imagery to assess the variability in the water status of five fruit tree species within a commercial orchard. *Precision Agriculture*, 2013. 14(6) p. 660–678.

35. Klapp, I., S. Papini, and N. Sochen, Radiometric imaging by double exposure and gain calibration. *Applied Optics*, 2017. 56(20) p. 5639–5647.

36. Ribeiro-Gomes, K. et al., Uncooled thermal camera calibration and optimization of the photogrammetry process for UAV applications in agriculture. *Sensors (Switzerland)*, 2017. 17(10).

37. Cohen, Y. et al., Future approaches to facilitate large-scale adoption of thermal based images as key input in the production of dynamic irrigation management zones. *Advances in Animal Biosciences*, 2017. 8(2) p. 546–550.

38. Asner, G.P. et al., Progressive forest canopy water loss during the 2012–2015 California drought. *Proceedings of the National Academy of Sciences*, 2016. 113(2) p. E249–E255.

39. Aparicio, N. et al., Spectral vegetation indices as nondestructive tools for determining durum wheat yield. *Agronomy Journal*, 2000. 92(1) p. 83–91.

40. Haboudane, D. et al., Hyperspectral vegetation indices and novel algorithms for predicting green LAI of crop canopies: Modeling and validation in the context of precision agriculture. *Remote Sensing of Environment*, 2004. 90(3) p. 337–352.

41. Lee, K.S. et al., Hyperspectral versus multispectral data for estimating leaf area index in four different biomes. *Remote Sensing of Environment*, 2004. 91(3–4) p. 508–520.

42. Hansen, P.M. and J.K. Schjoerring, Reflectance measurement of canopy biomass and nitrogen status in wheat crops using normalized difference vegetation indices and partial least squares regression. *Remote Sensing of Environment*, 2003. 86(4) p. 542–553.

43. Gitelson, A.A., Wide dynamic range vegetation index for remote quantification of biophysical characteristics of vegetation. *Journal of Plant Physiology*, 2004. 161(2) p. 165–173.

44. Danner, M. et al., Retrieval of biophysical crop variables from multi-angular canopy spectroscopy. *Remote Sensing*, 2017. 9(7).

45. Thenkabail, P.S., R.B. Smith, and E. De Pauw, Hyperspectral vegetation indices and their relationships with agricultural crop characteristics. *Remote Sensing of Environment*, 2000. 71(2) p. 158–182.

46. Darvishzadeh, R. et al., Leaf Area Index derivation from hyperspectral vegetation indices and the red edge position. *International Journal of Remote Sensing*, 2009. 30(23) p. 6199–6218.

47. Zarco-Tejada, P.J. et al., Assessing vineyard condition with hyperspectral indices: Leaf and canopy reflectance simulation in a row structured discontinuous canopy. *Remote Sensing of Environment*, 2005. 99(3) p. 271–287.

48. Delegido, J. et al., Retrieval of chlorophyll content and LAI of crops using hyperspectral techniques: Application to PROBA/CHRIS data. *International Journal of Remote Sensing*, 2008. 29(24) p. 7107–7127.

49. Wang, F.-m., J.-f. Huang, and Z.-h. Lou, A comparison of three methods for estimating leaf area index of paddy rice from optimal hyperspectral bands. *Precision Agriculture*, 2011. 12(3) p. 439–447.

50. Yuan, H. et al., Retrieving soybean leaf area index from unmanned aerial vehicle hyperspectral remote sensing: Analysis of RF, ANN, and SVM regression models. *Remote Sensing*, 2017. 9(4).

51. Pimstein, A., A. Karnieli, and D.J. Bonfil, Wheat and maize monitoring based on ground spectral measurements and multivariate data analysis. *Journal of Applied Remote Sensing*, 2007. 1 p. 16.

52. Yue, J. et al., Estimation of winter wheat above-ground biomass using unmanned aerial vehicle-based snapshot hyperspectral sensor and crop height improved models. *Remote Sensing*, 2017. 9(7).

53. Naor, A., I. Klein, and I. Doron, Stem water potential and apple size. *Journal of the American Society for Horticultural Science*, 1995. 120(4) p. 577–582.

54. Hunt, E.R. and B.N. Rock, Detection of changes in leaf water-content using near-infrared and middle-infrared reflectances. *Remote Sensing of Environment*, 1989. 30(1) p. 43–54.

55. Ceccato, P. et al., Designing a spectral index to estimate vegetation water content from remote sensing data: Part 1—Theoretical approach. *Remote Sensing of Environment*, 2002. 82(2–3) p. 188–197.

56. El-Hendawy, S.E. et al., Spectral assessment of drought tolerance indices and grain yield in advanced spring wheat lines grown under full and limited water irrigation. *Agricultural Water Management*, 2017. 182 p. 1–12.

57. Kim, D.M. et al., Highly sensitive image-derived indices of water-stressed plants using hyperspectral imaging in SWIR and histogram analysis. *Scientific Reports*, 2015. 5.

58. Zhang, F. and G.S. Zhou, Estimation of canopy water content by means of hyperspectral indices based on drought stress gradient experiments of maize in the North Plain China. *Remote Sensing*, 2015. 7(11) p. 15203–15223.

59. Rodriguez-Perez, J.R. et al., Evaluation of hyperspectral reflectance indexes to detect grapevine water status in vineyards. *American Journal of Enology and Viticulture*, 2007. 58(3) p. 302–317.

60. Das, B. et al., Comparison of different uni- and multi-variate techniques for monitoring leaf water status as an indicator of water-deficit stress in wheat through spectroscopy. *Biosystems Engineering*, 2017. 160 p. 69–83.

61. Clevers, J., L. Kooistra, and M.E. Schaepman, Estimating canopy water content using hyperspectral remote sensing data. *International Journal of Applied Earth Observation and Geoinformation*, 2010. 12(2) p. 119–125.

62. Ordonez, C. et al., Functional statistical techniques applied to vine leaf water content determination. *Mathematical and Computer Modelling*, 2010. 52(7–8) p. 1116–1122.

63. Jarvis, P.G., The interpretation of the variations in leaf water potential and stomatal conductance found in canopies in the field. *Philosophical Transactions of the Royal Society of London. Series B, Biological Sciences*, 1976. 273(927) p. 593–610.

64. Rapaport, T. et al., Combining leaf physiology, hyperspectral imaging and partial least squares-regression (PLS-R) for grapevine water status assessment. *ISPRS Journal of Photogrammetry and Remote Sensing*, 2015. 109 p. 88–97.

65. Pôças, I. et al., Predicting grapevine water status based on hyperspectral reflectance vegetation indices. *Remote Sensing*, 2015. 7(12).

66. Müller, M. et al., Use of thermal and visible imagery for estimating crop water status of irrigated grapevine. *Journal of Experimental Botany*, 2007. 58(4) p. 827–838.

67. Agam, N. et al., Spatial distribution of water status in irrigated olive orchards by thermal imaging. *Precision Agriculture*, 2013. 15(3) p. 346–359.

68. Berni, J.A.J. et al., Mapping canopy conductance and CWSI in olive orchards using high resolution thermal remote sensing imagery. *Remote Sensing of Environment*, 2009. 113(11) p. 2380–2388.

69. Gonzalez-Dugo, V. et al., Almond tree canopy temperature reveals intra-crown variability that is water stress-dependent. *Agricultural and Forest Meteorology*, 2012. 154-155(0) p. 156–165.

70. O'Shaughnessy, S.A. et al., Using radiation thermography and thermometry to evaluate crop water stress in soybean and cotton. *Agricultural Water Management*, 2011. 98(10) p. 1523–1535.

71. Tilling, A.K. et al., Remote sensing of nitrogen and water stress in wheat. *Field Crops Research*, 2007. 104(1–3) p. 77–85.

72. O'Shaughnessy, S.A., S.R. Evett, and P.D. Colaizzi, Dynamic prescription maps for site-specific variable rate irrigation of cotton. *Agricultural Water Management*, 2015. 159 p. 123–138.

73. Rosenberg, O. et al., Are thermal images adequate for irrigation management?, in *12th International Conference on Precision Agriculture*. 2014 Sacramento, California, USA.

74. Osroosh, Y. et al., Automatic irrigation scheduling of apple trees using theoretical crop water stress index with an innovative dynamic threshold. *Computers and Electronics in Agriculture*, 2015. 118 p. 193–203.

75. Steele, D.D., B.L. Gregor, and J.B. Shae, Irrigation scheduling methods for popcorn in the northern Great Plains. *Transactions of the ASAE*, 1997. 40(1) p. 149–155.

76. Prenger, J.J. et al., Plant response-based irrigation control system in a greenhouse: System evaluation. *Transactions of the ASAE*, 2005. 48(3) p. 1175–1183.

77. Berni, J.A.J. et al., Thermal and narrowband multispectral remote sensing for vegetation monitoring from an unmanned aerial vehicle. *Ieee Transactions on Geoscience and Remote Sensing*, 2009. 47(3) p. 722–738.

78. Elsayed, S. et al., Thermal imaging and passive reflectance sensing to estimate the water status and grain yield of wheat under different irrigation regimes. *Agricultural Water Management*, 2017. 189 p. 98–110.

79. Datt, B., Visible/near infrared reflectance and chlorophyll content in Eucalyptus leaves. *International Journal of Remote Sensing*, 1999. 20(14) p. 2741–2759.

80. Penuelas, J. et al., Estimation of plant water concentration by the reflectance Water Index WI (R900/R970). *International Journal of Remote Sensing*, 1997. 18(13) p. 2869–2875.

81. Bannari, A. et al., Potential of hyperion EO-1 hyperspectral data for wheat crop chlorophyll content estimation. *Canadian Journal of Remote Sensing*, 2008. 34 p. S139–S157.

82. Kempeneers, P. et al., Generic wavelet-based hyperspectral classification applied to vegetation stress detection. *Ieee Transactions on Geoscience and Remote Sensing*, 2005. 43(3) p. 610–614.

83. Blackburn, G.A. and J.G. Ferwerda, Retrieval of chlorophyll concentration from leaf reflectance spectra using wavelet analysis. *Remote Sensing of Environment*, 2008. 112(4) p. 1614–1632.

84. Blackburn, G.A., Wavelet decomposition of hyperspectral data: A novel approach to quantifying pigment concentrations in vegetation. *International Journal of Remote Sensing*, 2007. 28(12) p. 2831–2855.

85. Li, D. et al., WREP: A wavelet-based technique for extracting the red edge position from reflectance spectra for estimating leaf and canopy chlorophyll contents of cereal crops. *Isprs Journal of Photogrammetry and Remote Sensing*, 2017. 129 p. 103–117.

86. Liu, B. et al., Combining spatial and spectral information to estimate chlorophyll contents of crop leaves with a field imaging spectroscopy system. *Precision Agriculture*, 2017. 18(4) p. 491–506.

87. Herrmann, I. et al., SWIR-based spectral indices for assessing nitrogen content in potato fields. *International Journal of Remote Sensing*, 31(19) p. 5127–5143.

88. Tian, Y.C. et al., Assessing newly developed and published vegetation indices for estimating rice leaf nitrogen concentration with ground- and space-based hyperspectral reflectance. *Field Crops Research*, 2011. 120(2) p. 299–310.

89. Nigon, T.J. et al., Hyperspectral aerial imagery for detecting nitrogen stress in two potato cultivars. *Computers and Electronics in Agriculture*, 2015. 112(Supplement C) p. 36–46.

90. Dash, J. and P.J. Curran, The MERIS terrestrial chlorophyll index. *International Journal of Remote Sensing*, 2004. 25(23) p. 5403–5413.

91. Moharana, S. and S. Dutta, Spatial variability of chlorophyll and nitrogen content of rice from hyperspectral imagery. *Isprs Journal of Photogrammetry and Remote Sensing*, 2016. 122 p. 17–29.

92. Herrmann, I. et al., SWIR-based spectral indices for assessing nitrogen content in potato fields. *International Journal of Remote Sensing*, 2010. 31(19) p. 5127–5143.

93. Cohen, Y. et al., Leaf nitrogen estimation in potato based on spectral data and on simulated bands of the VEN mu S satellite. *Precision Agriculture*, 2010. 11(5) p. 520–537.

94. Samborski, S.M., N. Tremblay, and E. Fallon, Strategies to make use of plant sensors-based diagnostic information for nitrogen recommendations. *Agronomy Journal*, 2009. 101(4) p. 800–816.

95. Holland, K.H. and J.S. Schepers, Use of a virtual-reference concept to interpret active crop canopy sensor data. *Precision Agriculture*, 2013. 14(1) p. 71–85.

96. Faiçal, B.S. et al., An adaptive approach for UAV-based pesticide spraying in dynamic environments. *Computers and Electronics in Agriculture*, 2017. 138 p. 210–223.

97. Pantazi, X.E. et al., Evaluation of hierarchical self-organising maps for weed mapping using UAS multispectral imagery. *Computers and Electronics in Agriculture*, 2017. 139 p. 224–230.

98. Yang, G. et al., The DOM generation and precise radiometric calibration of a UAV-mounted miniature snapshot hyperspectral imager. *Remote Sensing*, 2017. 9(7).

99. Aasen, H. et al., Generating 3D hyperspectral information with lightweight UAV snapshot cameras for vegetation monitoring: From camera calibration to quality assurance. *ISPRS Journal of Photogrammetry and Remote Sensing*, 2015. 108 p. 245–259.

100. Plaza, A. et al., Recent advances in techniques for hyperspectral image processing. *Remote Sensing of Environment*, 2009. 113(Supplement 1) p. S110–S122.

101. Fauvel, M. et al., Advances in spectral-spatial classification of hyperspectral images. *Proceedings of the IEEE*, 2013. 101(3) p. 652–675.

102. Zhang, N., M. Wang, and N. Wang, Precision agriculture: A worldwide overview. *Computers and Electronics in Agriculture*, 2002. 36(2) p. 113–132.

103. Haboudane, D. et al., Integrated narrow-band vegetation indices for prediction of crop chlorophyll content for application to precision agriculture. *Remote Sensing of Environment*, 2002. 81(2–3) p. 416–426.

104. Miao, Y.X. et al., Combining chlorophyll meter readings and high spatial resolution remote sensing images for in-season site-specific nitrogen management of corn. *Precision Agriculture*, 2009. 10(1) p. 45–62.

105. Liu, J.G. et al., Variability of seasonal CASI image data products and potential application for management zone delineation for precision agriculture. *Canadian Journal of Remote Sensing*, 2005. 31(5) p. 400–411.

106. Lawrence, K.C. et al., Partial least squares regression of hyperspectral images for contaminant detection on poultry carcasses. *Journal of Near Infrared Spectroscopy*, 2006. 14(4) p. 223–230.

107. Jimenez, L.O. et al., Integration of spatial and spectral information by means of unsupervised extraction and classification for homogenous objects applied to multispectral and hyperspectral data. *IEEE Transactions on Geoscience and Remote Sensing*, 2005. 43(4) p. 844–851.

108. Tilton, J.C. et al., Best merge region-growing segmentation with integrated nonadjacent region object aggregation. *IEEE Transactions on Geoscience and Remote Sensing*, 2012. 50(11) p. 4454–4467.

109. Tarabalka, Y., J.A. Benediktsson, and J. Chanussot, Spectral and spatial classification of hyperspectral imagery based on partitional clustering techniques. *IEEE Transactions on Geoscience and Remote Sensing*, 2009. 47(8) p. 2973–2987.

110. Lark, R.M., Forming spatially coherent regions by classification of multi-variate data: An example from the analysis of maps of crop yield. *International Journal of Geographical Information Science*, 1998. 12(1) p. 83–98.

111. Nansen, C., A.J. Sidumo, and S. Capareda, Variogram analysis of hyperspectral data to characterize the impact of biotic and abiotic stress of maize plants and to estimate biofuel potential. *Applied Spectroscopy*, 2010. 64(6) p. 627–636.

112. Cohen, S. et al., Combining spectral and spatial information from aerial hyperspectral images for delineating homogenous management zones. *Biosystems Engineering*, 2013. 114(4) p. 435–443.

113. Herrmann, I. et al., LAI assessment of wheat and potato crops by VENµS and Sentinel-2 bands. *Remote Sensing of Environment*, 2011. 115(8) p. 2141–2151.

114. Beaulieu, J.M. and M. Goldberg, Hierarchy in picture segmentation: A stepwise optimization appraoch. *IEEE Transactions on Pattern Analysis and Machine Intelligence*, 1989. 11(2) p. 150–163.

115. Donoho, D. and X. Huo, Beamlets and multiscale image analysis, in *Multiscale and Multiresolution Methods*, T.J. Barth, T. Chan, and R. Haimes, Editors. 2002, Springer Lecture Notes in Computational Science and Engineering. p. 149–196.

116. Levi, O., S. Cohen, and Z. Mharaby, Effective hyper-spectral image segmentation using multi-scale geometric analysis, in *IADIS Multi Conference on Computer Science and Information Systems 2010*. 2010 Freiburg, Germany.

4 Spectral and 3D Nonspectral Approaches to Crop Trait Estimation Using Ground and UAV Sensing

Helge Aasen and Georg Bareth

CONTENTS

4.1 INTRODUCTION

4.1.1 MOTIVATION

A number of plant characteristics or traits are valuable for understanding vigor, production, and reproduction. The green leaf area index (LAI) is a key variable for monitoring crop physiology and foliage cover (Alchanatis and Cohan, 2012). Authors indicate the strong relation between LAI and crop biomass, which results in similar vegetation indices (VIs) used for characterizing both being valuable. Because of the strong relationship between LAI and biomass and the difficulty of deriving LAI measurements, LAI can also be estimated as a function of biomass. Additionally, (i) biomass is a key crop parameter for optimized nitrogen (N) management in precision farming (Lemaire and Gastal, 1997; Lemaire and Meynart, 1997; Mistele and Schmidhalter, 2008); (ii) biomass is necessary for computing nitrogen or nutrient uptake (Li et al. 2010); and (iii) biomass is a predictor

103

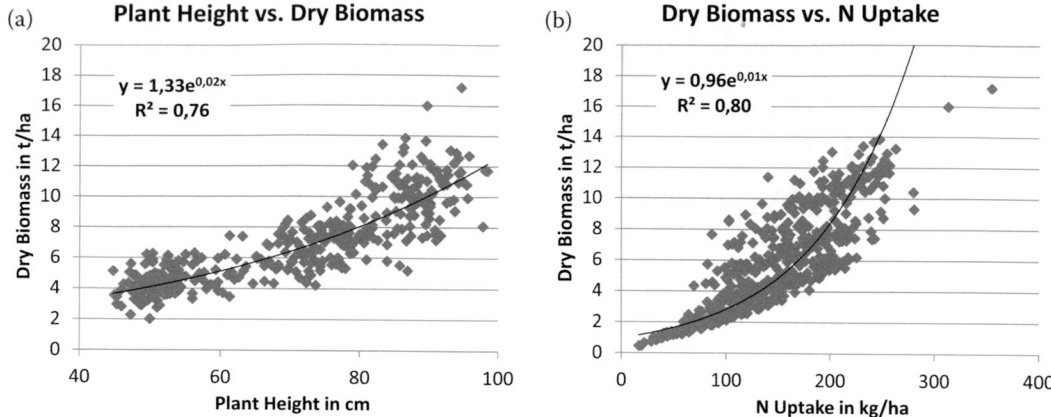

FIGURE 4.1 Relationship of plant height (PH) to dry biomass (a) and of dry biomass of winter wheat to N uptake (b) from destructively obtained field data (unpublished data by Li et al., 2008, and Gnyp et al., 2014).

for estimating total carbon assimilation in terms of net primary production (Gower et al., 1999; Reed et al., 2008). Measures of biomass and LAI can support management of plant production. But it is the combination of such measures of plant traits that can supply information for biological, ecological, agronomic, and management studies and understanding.

Many of these plant traits can be measured remotely and facilitation of their collection over large areas, through either direct measures or related measures or indicators, is very advantageous. It is well known that most VIs using the spectral range between 350 and 1000 nm are of great utility but tend to saturate in later growing stages from canopy closure (Haboudane et al., 2004; Li et al., 2008; Alchanatis and Cohan, 2012) and therefore are of limited use for vegetation monitoring. Vegetation indices that are not as strongly affected by saturation affects are, for example, the normalized ratio index (NRI) using all possible VIS/NIR/SWIR narrow-band combinations, or the hyperspectral VI named *GnyLi,* which is also used in the NIR/SWIR domain (Mutanga and Skidmore, 2004; Koppe et al., 2010; Gnyp et al., 2014). As well as the progress in improving VIs for monitoring biophysical parameters, the combination or fusion of different sensor data—for example, microwave and multi- or hyperspectral sensor data—has further improved vegetation monitoring and especially biomass monitoring (Pohl and Van Genderen, 1998; Lu, 2006; Koppe et al., 2012).

In recent years, 3D data has also gained attention for crop trait extraction (Hoffmeister, 2016). While tree height has long been acknowledged to help characterize forest productivity (Skovsgaard and Vanclay, 2008), this has not been the case for crop monitoring. In our case, the motivation to combine 3D sensing approaches with multi- or hyperspectral data acquisition for crop monitoring emerged (i) from the promising multivariate analysis presented by Koppe et al. (2012) and (ii) from the field data set acquired in 2006 and 2007 by Li et al. (2008) and Gnyp et al. (2014). These datasets showed that manually measured plant height (PH) is a good nonspectral predictor for biomass and nitrogen uptake (c.f. Figure 4.1a). More recently, similar findings were reported by Marshall and Thenkabail (2015) and Pittman et al. (2015).

4.1.2 Scope

In this chapter, we focus on uses of 3D information in combination with spectral data for evaluation of crop traits. First, the concept of crop trait estimation from 3D data based on multitemporal digital surface modeling is introduced, along with the tools for terrestrial laser scanning and unmanned aerial vehicle (UAV) imaging. Then, spectral 3D sensing with UAVs and spectral 2D imagers is addressed as these combine synergistically to supply spectral and 3D data. We close by reviewing and discussing crop trait estimates and valuable characteristics that can be derived by fusion of 3D and spectral data.

4.2 CROP TRAIT ESTIMATING WITH 3D DATA

The findings described in Figure 4.1 motivated the search for methods that could capture 3D data of crop canopies nondestructively to complement spectral sensing techniques for crop trait estimation. At the same time the concept of using multitemporal digital surface models (DSMs) for crop growth estimation was born (Hoffmeister et al., 2010). Here the concept is introduced and the current, most common techniques for estimating accurate 3D data in ultrahigh spatial resolution (~1–2 cm) of crop canopies, namely terrestrial laser scanning (TLS) and structure from motion (SfM), and their application are examined.

4.2.1 DIGITAL SURFACE MODELS FROM TERRESTRIAL LASER SCANNING

From our work and the findings others, nonspectral crop parameters such as crop height can serve as robust predictors for certain crop traits such as biomass (Figure 4.1). The demand for accurate and nondestructive PH sensing in ultrahigh spatial resolutions has given added emphasis to the search. At the same time in the early 2000s, laser scanning was already being successfully applied for capturing precise 3D textures and surfaces of objects with approximately 1–2 cm accuracy (Wiedemann, 2001; Girardeau-Montaut et al., 2005; Alshawabkeh et al., 2007). LiDAR (light detection and ranging) is an active proximal and remote sensing method using the travel time of laser beams to measure the distance to an object. LiDAR became a key technology in capturing 3D point clouds and consequently objects in 3D with millimeter or centimeter accuracy. In remote sensing, LiDAR is applied from airborne platforms by airborne laser scanning (ALS) or from elevated ground positions by terrestrial laser scanning (TLS) (van Leeuwen et al., 2011; Tilly et al., 2014). From 2007 on, the first investigations of using TLS for crop trait monitoring were published and were well received (Omasa et al., 2007; Ehlert et al., 2008; Hoffmeister et al., 2010; Eitel et al., 2011). In Figure 4.2, an example of a field campaign with a TLS is shown.

(a)   (b)   (c)

FIGURE 4.2 (a) TLS, the Riegl LMS-Z420i, with a RGB camera and a RTK receiver is shown; (b) TLS systems are used on a tripod or as shown on other elevated platforms; (c) RTK-GPS are used to georeference reflective cylinders which are used as reference objects for merging 3D point clouds from different locations. (From Tilly, N. et al. 2015a. *Remote Sens.* 7, 11449–11480. https://doi.org/10.3390/rs70911449.)

The terrain surface, which is commonly covered by vegetation, buildings, or other objects, can be represented as a DSM at a given spatial resolution. In this case the terrain was represented without such vegetation, buildings, or other objects and the term digital terrain model (DTM) or digital elevation model (DEM) is used. From the TLS-acquired 3D point cloud data, DTMs and DSMs can be computed. In 2007, Hoffmeister et al. (2010) started to work on 3D monitoring of crop height and crop growth on field scale at ultrahigh resolution (<2 cm) with TLS. The authors used a multistationary TLS approach to generate the 3D data of a sugarbeet field. From the 3D point cloud, multitemporal DSMs at a resolution of <2 cm were generated and the term crop surface model (CSM) was introduced to describe these ultrahigh-resolution DSMs of crop canopies. By calculating the difference between a DTM and a DSM, for crops also called a CSM, the crop height can be directly computed (Hoffmeister et al., 2010). This procedure can be repeated for different phenological stages. In Figure 4.3, the CSM approach is shown. After sowing and before crop emergence, a DTM was acquired from TLS-derived 3D point cloud data. The DTM represents the spatially continuous zero crop height. This can also be done by selecting bare ground points between the plants and interpolating a DTM (Geipel et al., 2014; Aasen et al., 2015). In characteristic phenological stages, CSMs are computed from the multitemporal 3D point cloud data. By using spatial analysis, the DTM is subtracted from the CSMs, resulting in pixel-wise crop height data (Figure 4.3a). Furthermore, pixel-wise crop growth rates (Figure 4.3b) from one stage to another stage can be computed by subtracting an earlier CSM (t_1) from a later CSM (t_2).

This concept of acquiring multitemporal CSMs for crop height determination at field scales was successfully demonstrated in the studies of Tilly et al. (2014), and extended to field phenotyping applications (Tilly et al., 2015a). Hoffmeister et al. (2016) compared the results for TLS-derived crop heights to in-field variations derived from orthophotos and soil data and found significant correlations. Today, TLS is increasingly used to track plant growth. In Friedli et al. (2016), PH was extracted directly from point clouds instead of first going through the process of generating DSMs. The authors investigated different height percentiles to estimate the height of maize and soybean in a field phenotyping experiment during the diurnal cycle and at different dates. They found that in most cases the 99% percentile showed the best correlation with manual measurements. Similar approaches have also been published for other phenotyping experiments for soybean (Kirchgessner et al., 2017)

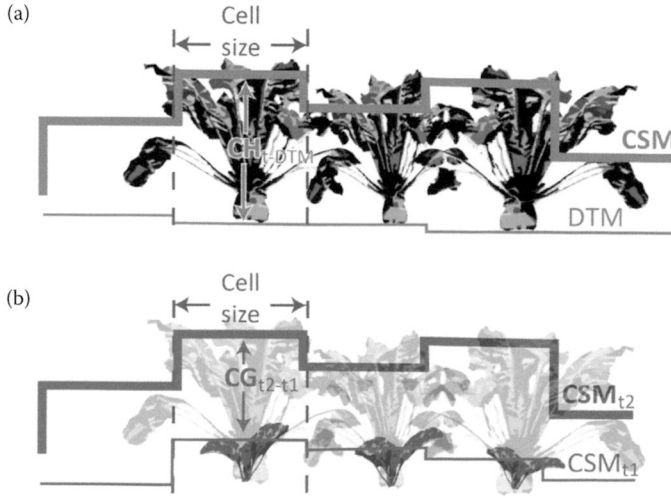

FIGURE 4.3 Concept of TLS-derived digital terrain model (DTM) and crop surface models (CSM) to determine crop height (CH) and crop growth (CG). (From Hoffmeister, D., 2014. Feasibility studies of terrestrial laser scanning in Coastal Geomorphology, Agronomy, and Geoarchaeology (PhD thesis). Universität zu Köln, Cologne, Germany. With permission.)

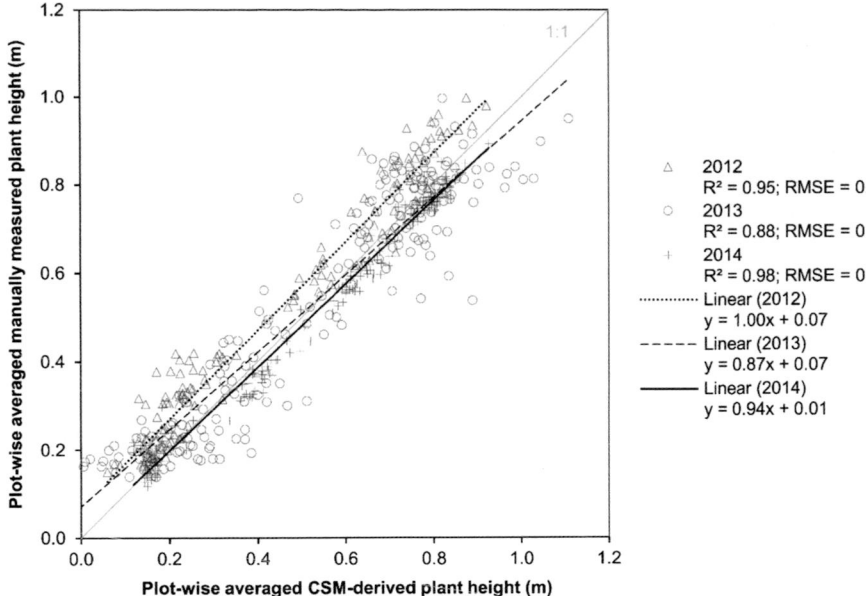

FIGURE 4.4 Regression of the mean CSM-derived and manual measured PHs for barley (2012: $n = 131$; 2013: $n = 196$; 2014: $n = 180$). (From Tilly, N. et al. 2015a. *Remote Sens.* 7, 11449–11480. https://doi.org/10.3390/rs70911449).

and wheat (Kronenberg et al., 2017; Madec et al., 2017). In Jimenez-Berni et al. (2018) the approach was extended to estimate ground cover and above-ground biomass from TLS-derived point clouds.

However, the robustness of TLS-derived CSMs for nondestructive and direct determination of crop height was investigated and demonstrated in a number of studies (Hämmerle and Höfle, 2014; Tilly et al., 2014, 2015a; Friedli et al., 2016; Hoffmeister et al., 2016). In Figure 4.4, a case study from a three-year duration barley experiment is shown in which CSMs were derived by TLS. R^2 between manually measured and CSM-derived crop height ranged around 0.9.

4.2.2 3D INFORMATION FROM UAVs WITH STRUCTURE FROM MOTION

Unmanned aerial vehicles (UAVs) have also been referred to as drones, unmanned aerial/aircraft systems (UAS), or remotely piloted aircraft systems (RPAS). Their advent and operational implementation have transformed the way we collect geospatial data. By being very flexible and relatively cheap, they complement traditional ground, air, and satellite remote sensing when several hectares to square kilometers need to be surveyed. Currently, a wide range of UAV platforms and sensors are available (Colomina and Molina, 2014; Salamí et al., 2014; Pajares, 2015; Toth and Józków, 2016). They have been applied in precision agriculture (Zhang and Kovacs, 2012; Adão et al., 2017), forestry (Sanchez-Azofeifa et al., 2017), spatial ecology (Anderson and Gaston, 2013), ecohydrology (Vivoni et al., 2014), and other environmental monitoring applications (Warner and Cracknell, 2017). This has led to an ongoing revolution in spectral remote sensing, since more and more researchers and others can collect their own spectral geodata (Aasen et al., 2018).

UAVs are able to carry a large variety of sensing systems and allow collection of RGB, spectral, thermal, as well as LiDAR data. An overview over different sensors and applications can be found in several sources (Colomina and Molina, 2014; Gago et al., 2015; Pajares, 2015). Depending on the flying altitude and sensor configuration, the data may have ultrahigh spatial resolutions of a few centimeters (e.g., Lucieer et al., 2014a; Turner et al., 2014; Madec et al. 2017) to a few millimeters (Malenovský et al., 2017; Roth et al., 2018).

This high-resolution data can also be used for crop monitoring (Herwitz et al. 2004, Zhang and Kovacs 2012). Major advances in sensor development, computer vision, and photogrammetric solutions have profited from increasing computing capabilities. This has led to the implementation of algorithms such as the scale invariant feature transform operator (Lowe, 2004), bundle adjustment and point density matching, and procedures such as SfM (structure from motion) (Ullman, 1979) into consumer-grade software packages that allow derivation of the relative position and orientation of images based on the features visible within these images. Additionally, the 3D topography of a scene can be reconstructed and point clouds can be derived with a density similar to the ground sampling distance of the images (Triggs et al., 2000; Snavely et al., 2008; Haala, 2013; Luhmann et al., 2014; Remondino et al., 2014; Snavely, 2016). This not only allows compilation of individual images into a scene but also generates 3D point cloud data for a digital surface model.

Although SfM was originally developed for oblique terrestrial images (Snavely et al., 2008), it can also be applied to overlapping 2D images captured by UAVs. Usually an RGB camera is mounted on a gimbal that compensates for the movements of the UAV. Several types of cameras have been mounted on UAVs. They range from small and lightweight compact cameras (Dandois and Ellis, 2013; Bendig et al., 2014, 2015; Geipel et al., 2014; Rasmussen et al., 2016; Alexandridis et al., 2017; Berra et al., 2017; Roth et al., 2018), but larger digital single-lens reflex (Lucieer et al., 2014b; Turner et al., 2014; Jagt et al., 2015; Roth and Streit, 2017), and system cameras (Zarco-Tejada et al., 2014) have also been used since they carry larger chips than the compact cameras, which increases the light efficiency. Integrated systems have also appeared on the market. For example, in 2013, the Phantom 2 vision UAV with a 14-megapixel camera was introduced by DJI. Their current model Phantom 4 pro features a 20-megapixel camera with a 25 mm (1 inch) sensor (DJI, 2017). With a UAV-based 2D imaging systems and the SfM approach (referred to here as UAV/SfM), 3D data and DSMs in ultrahigh spatial resolutions of <2 cm can be acquired. Thus, UAVs are well suited to generate multitemporal DSMs of crop fields and field phenotyping experiments.

The potential of multitemporal, ultrahigh-resolution DSMs, also termed crop surface models (CSMs), for monitoring crop traits was presented in the previous section on TLS. The TLS-based CSM approach described in Figure 4.3 was transferred to UAV-based image acquisition and SfM-generated DSMs by Bendig et al. (2013). Figure 4.5 symbolizes the resulting UAV-derived DTM and CSMs data. Three different phenological stages (t_1, t_2, and t_3) and bare soil before crop emergence (t_0) of the same crop-covered spatial area are shown with the three corresponding CSMs, CSM_1,

Multi-temporal Crop Surface Models

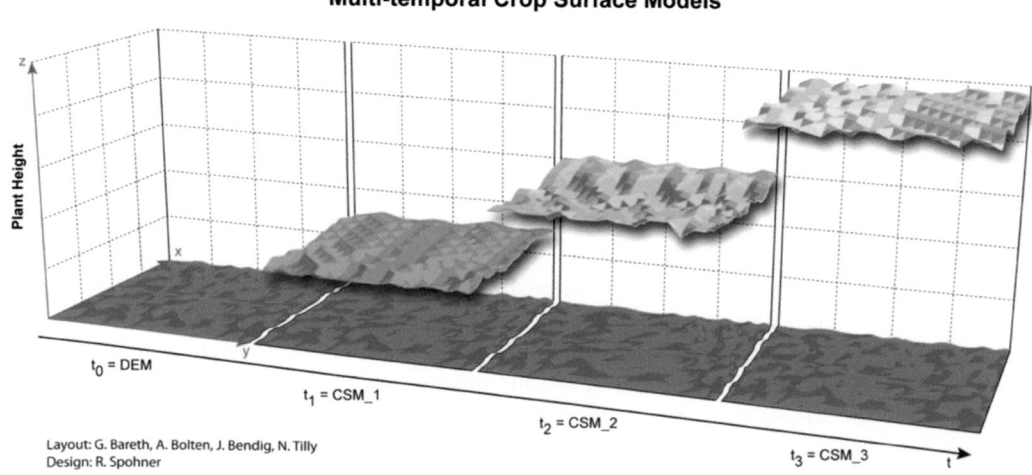

Layout: G. Bareth, A. Bolten, J. Bendig, N. Tilly
Design: R. Spohner

FIGURE 4.5 Concept of UAV-derived crop surface models (CSMs) for crop height and crop growth monitoring. (From Bendig et al., 2014. *Remote Sens.* 6, 10395–10412. https://doi.org/10.3390/rs61110395.)

CSM_2, and CSM_3, respectively. The three CSMs represent in ultrahigh resolution the digital surface of the crop canopy. In the example of Figure 4.5, crops are growing in general over the whole temporal period but showing spatial heterogeneity in crop growth. This spatial crop growth pattern is clearly visible in the DSMs.

To calculate the crop heights (t_1, t_2, and t_3), the digital elevation model (DEM) representing the DTM is subtracted from the CSMs (t_1, t_2, and t_3). To calculate the growth rate, the crop height difference between different phenological stages was computed. Using Geographic Information System (GIS) software raster analysis tools, the following operations can be applied to the data shown in Figure 4.5:

$CH_{t1} = CSM\_1 - DEM$	DTM/DEM: digital terrain model
$CH_{t2} = CSM\_2 - DEM$	(bare soil or plant surface before plant growth starts)
$CH_{t3} = CSM\_3 - DEM$	CSM: crop surface model
$CG_{t1\ to\ t2} = CSM\_2 - CSM\_1$	CH: crop height
$CG_{t2\ to\ t3} = CSM\_3 - CSM\_2$	CG: crop growth
$CG_{t1\ to\ t3} = CSM\_3 - CSM\_1$	

Bareth et al. (2015b) transferred the UAV/SfM-based CSM approach to monitor managed grasslands. From this grassland study in the Eifel region of Western Germany, two DSMs are visualized in Figure 4.6. The DSM-1 represents the grassland canopy surface of late March 2014 before grass started to grow. Hence, almost no spatial height variability was visible within and between the investigated plots. Differences resulted from topographic conditions of the experiment field. In contrast, the DSM-2 of Figure 4.6 represents the grassland canopy surface of early June 2014. The various growing rates in the differently managed plots as well as the spatial heterogeneity within the plots are now clearly visible in the DSM-2 and by visually comparing DSM-1 and DSM-2. Consequently, multitemporal and ultrahigh-resolution DSMs enable the investigation of spatiotemporal crop growth patterns.

In Figure 4.7 the DSM-derived PH data from the DSMs shown in Figure 4.6 at a spatial resolution of 2 cm for the managed grassland experiment in Germany is presented (Bareth et al., 2015b). The different fertilization strategies are clearly affecting plant growth. The spatial variability in the plots is a result of biotic disturbances (e.g., molehills), floristic characteristics, or even lodging. The GIS-processing steps to compute pixel-wise absolute PH from the DSMs were rather basic as described above. The DSM-1 of Figure 4.6 was subtracted from the DSM-2, which represents the later growth stage.

Besides plant growth and PH per pixel, plot-wise statistics can be computed. In Figure 4.8, the histograms of a zonal statistic analysis are shown for plot 6 and plot 9 of Figure 4.7. Additionally, descriptive statistics such as MIN (minimum), MAX (maximum), MEAN, and STD (standard deviation) are included. From the spatial pattern of the two plots, different characteristic such as spatial PH heterogeneity are obvious. While both plots (plots 6 and 9) have very similar mean PH values of approx. 11 cm, the PH distribution within the plots showed a very different pattern. This is visualized in Figure 4.8 by plotting the count of PH classes. Compared with plot 6, plot 9 was characterized by a more homogeneous PH distribution. Consequently, the ultrahigh resolution of PH also provides a kind of roughness and density information if it is analyzed for zones or plots (Bareth et al. 2016).

The potential monitoring of the aforementioned canopy surface roughness or density, which might represent the fractional vegetation cover (FVC), was discussed by Bareth et al. (2016) and is summarized in Figure 4.9. Nadir-looking imaging sensors, Figure 4.9a which are generally used in UAV-based monitoring, produce different zonal statistics compared with oblique viewing sensors, Figure 4.9b, such as TLS or tractor-mounted ultrasonic sensors. The mean PH or histogram pattern computed from UAV-derived, ultrahigh-resolution DSMs for a given grid or plot size by zonal

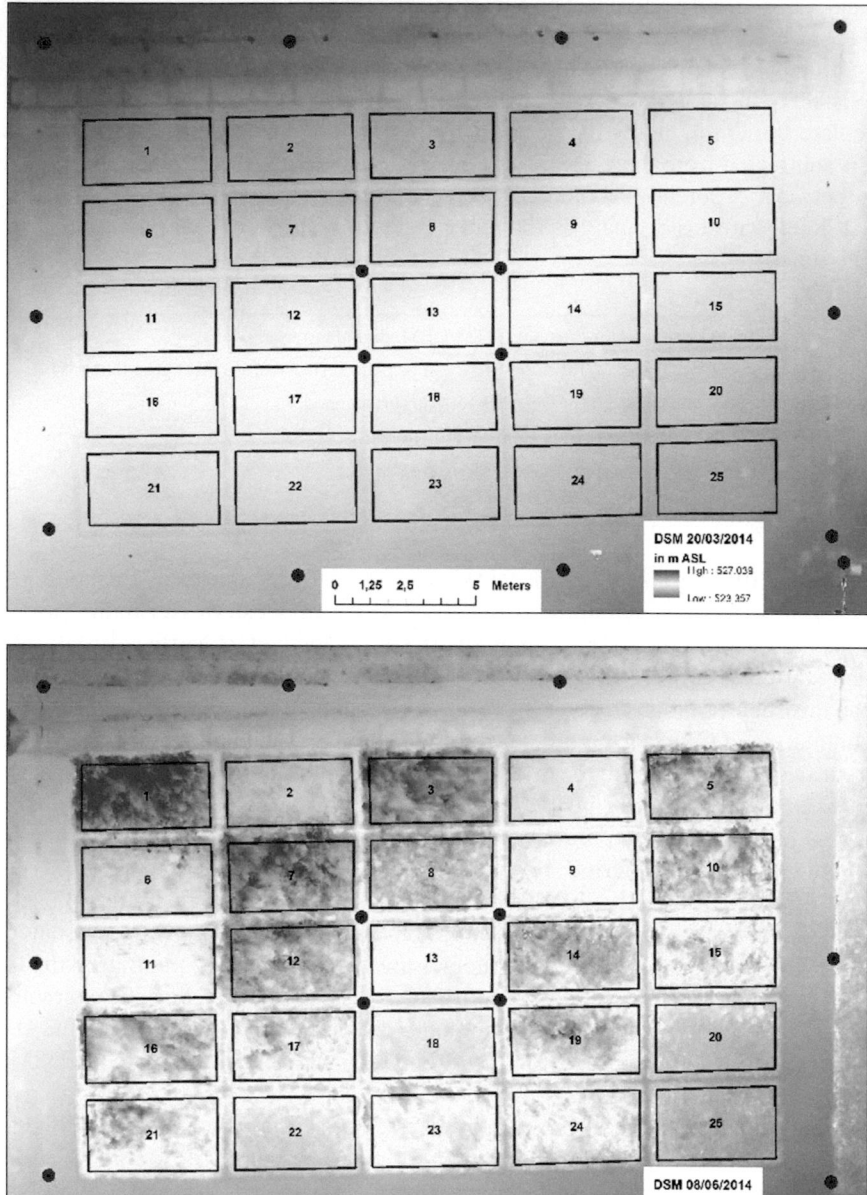

FIGURE 4.6 Digital surface models (DSMs) of a grassland experiment at a spatial resolution of 2 cm: DSM-1 (above) in meters above sea level (ASL) represents the grassland canopy surface before plant growth starts after winter; DSM-2 (below) in meters ASL represents the grassland canopy surface in early summer. Red polygons represent the individual experiment plots and the red points are ground control points.

statistics also depends on the viewing angle of the sensor. As can be seen in Figure 4.9, a lower plant density or vegetation fraction observed from oblique position can produce similar mean values as those from a higher plant density or vegetation fraction from a nadir observation position. This is because from an oblique angle the sensing system cannot adequately sense the nonvegetated or poorly vegetated ground. Consequently, nadir-derived CSMs also can provide nonspectral information on

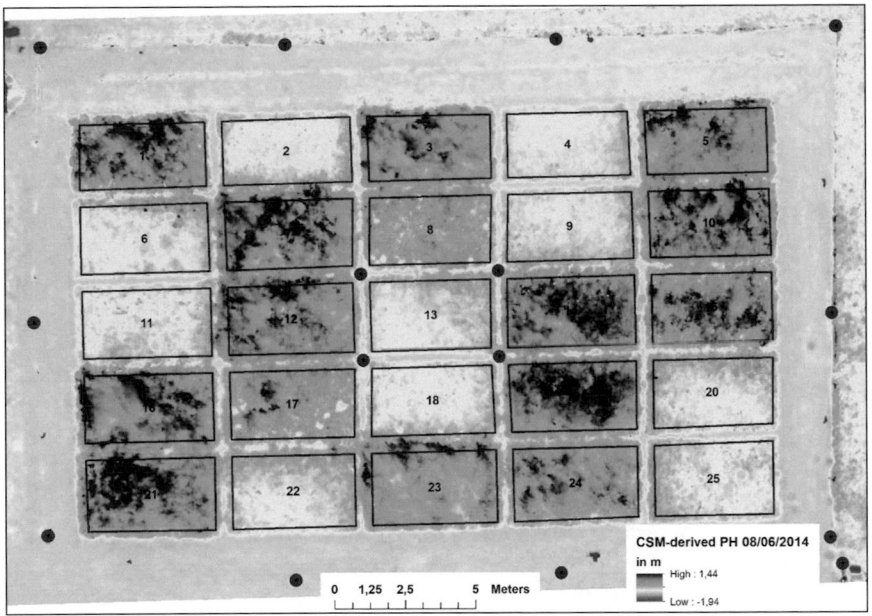

FIGURE 4.7   Pixel-wise plant height in meters for a grassland experiment computed from two digital surface models (DSMs) at 2 cm resolution (see Figure 4.4). Polygons represent the individual experiment plots and the red points are ground control points.

emergence, lodging, fractional vegetation cover, vegetation or sowing density, and vegetation surface roughness (Bareth et al., 2016; Jin et al., 2017; Madec et al., 2017).

Furthermore, a strong and robust relation between UAV/SfM- and TLS-derived crop heights is given and shown in Figure 4.10. The slightly higher crop height values of the TLS approach can be explained by the oblique observing position as shown and explained in Figure 4.9. Similar findings are reported by Malambo et al. (2018). Summarizing, crop height can be robustly determined by using the CSM approach on plot and field scale with both approaches.

Crop biomass is important in precision agriculture in the context of nitrogen management. It is a key parameter for the nitrogen nutrition index (Lemaire and Mcynart, 1997; Mistele and Schmidhalter, 2008), which is used for optimized fertilization. As is shown in Figure 4.1 from manual measurements in wheat, crop height serves as a good predictor for biomass. This was the motivation for investigating CSM-derived crop height. As is shown in Figure 4.11, the UAV

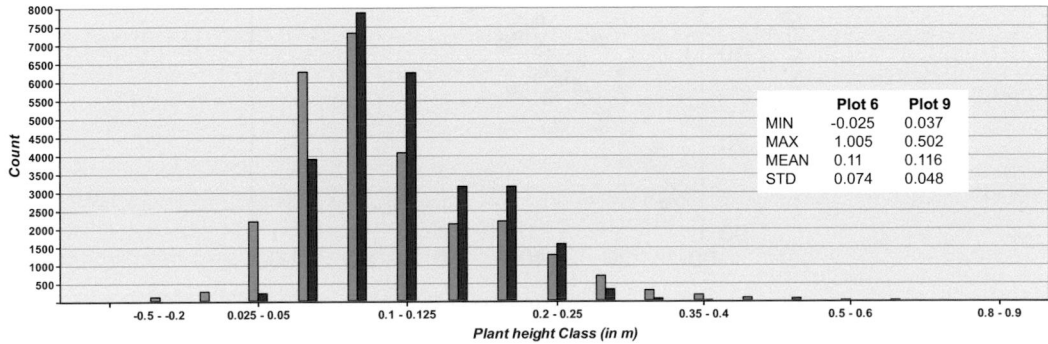

FIGURE 4.8   Histograms of DSM-derived plant height in meters for plot 6 (blue) and plot 9 (red) of Figure 4.7 of a grassland experiment. Computed with the ArcGIS function zonal histogram.

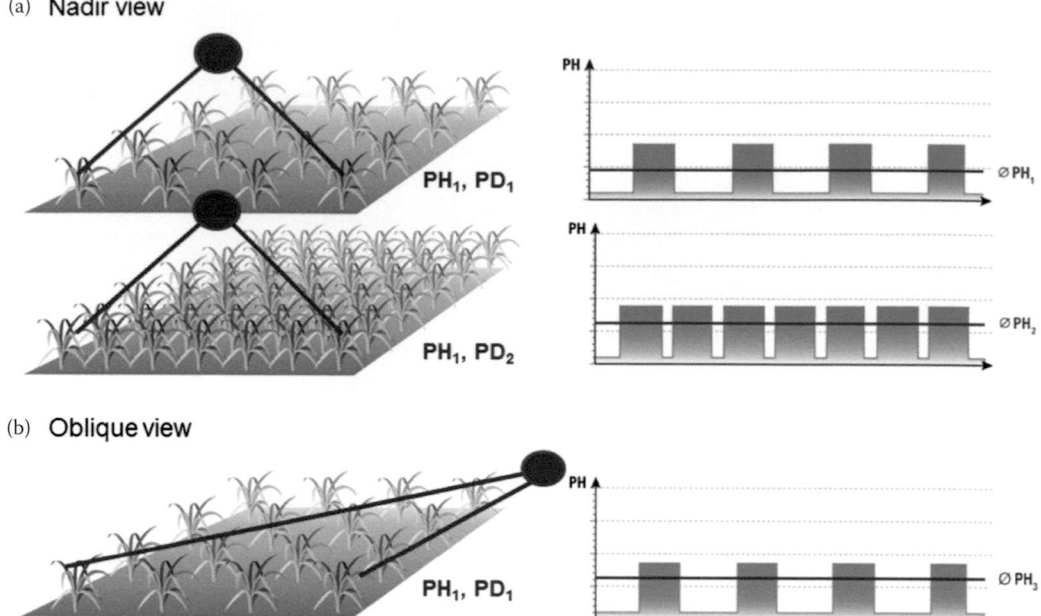

FIGURE 4.9 Influence of different viewing geometries on the mean plant height values of crop surface models (CSMs): (a) nadir view: equal plant height (PH_1) but different plant densities (PD_1, PD_2) result in different mean plant height values ($\varnothing PH_1 \neq \varnothing PH_2$); (b) Oblique view: equal plant height (PH_1) and equal plant densities (PD_1) result in different plant height values of nadir and oblique viewing angles ($\varnothing PH_1 \approx \varnothing PH_3$). (From Bareth et al., 2016. *Photogramm. Fernerkund. Geoinformation* 85–94. https://doi.org/10.1127/pfg/2016/0289. With permission.)

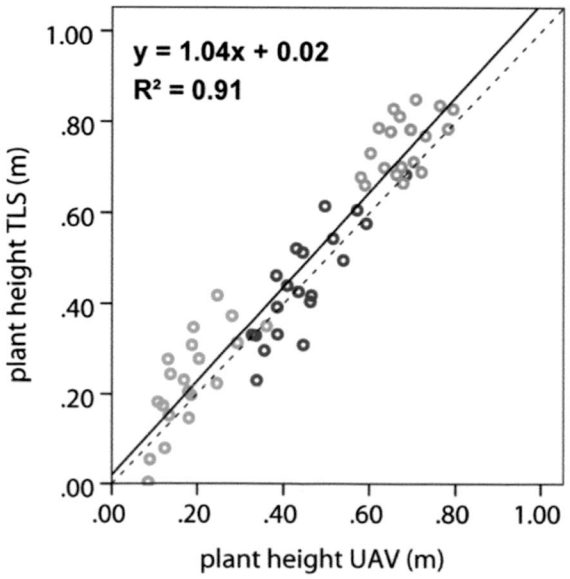

FIGURE 4.10 Trend between UAV- and TLS-derived crop height in a barley experiment. (From Bareth et al., 2016. *Photogramm. Fernerkund. Geoinformation* 85–94. https://doi.org/10.1127/pfg/2016/0289. With permission.)

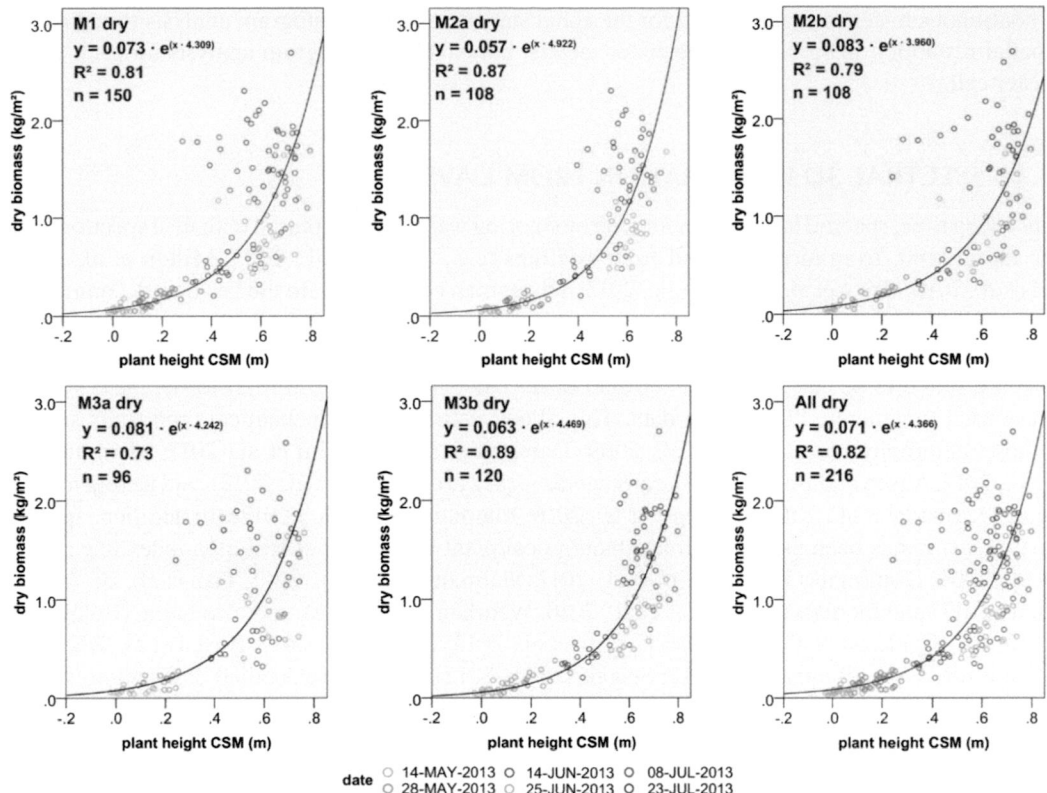

FIGURE 4.11 Cross-validation relationship of dry biomass and CSM-derived crop height in a barley experiment. Model 1 (M1): 70% calibration set; M2a: 40 kg N/m²; M2b: 80 kg N/m²; M3a: old cultivars; M3b: new cultivars, and all values; $p < 0.0001$ for all R2. (From Bendig et al., 2014. *Remote Sens.* 6, 10395–10412. https://doi.org/10.3390/rs61110395.)

derived CSM approach for determining crop height is a good nonspectral estimator for crop biomass. R^2 ranges between 0.79 and 0.89, apart from that for the old cultivars (M3a in Figure 4.11) for which crop height tends to be affected by lodging in later growth stages. Studies of TLS-derived CSMs by Tilly et al. (2015b) for rice and by Brocks and Bareth (2018) for barley confirmed these first findings of Bendig et al. (2014). Similar results were reported for yield or other important crop traits by Ehlert et al. (2008) for oilseed rape, winter rye, and winter wheat, by Geipel et al. (2014) for maize, by Maimaitijiang et al. (2017) for soybean, by Chang et al. (2017) for sorghum, and by Iqbal et al. (2017) for poppy crops.

The relationship of plant height to different crop traits is likely to depend on the cultivar, management techniques (e.g., sowing density and row spacing), and other parameters. More studies are needed to evaluate how generic these relationships are. Similarly, the influences of the viewing geometry still need investigation.

Resulting from that work the analysis of UAV/SfM- or TLS-derived multitemporal DSMs enabled the nondestructive and robust determination of several nonspectral crop traits. By using GIS analysis tools, the pixel-wise and spatially continuous determination of crop height at ultrahigh resolution (<2 cm) was possible with both approaches, UAV-based RGB imaging combined with SfM analysis as well as with TLS-based data acquisition (Bendig et al., 2013; Tilly et al., 2014; Hoffmeister et al., 2016). Additionally, the ultrahigh-resolution DSM data can be summarized for fields, zones, or plots using zonal statistics or zonal histograms. From the latter, additional information such as surface roughness or FVC can be extracted. Bareth et al. (2016) even suggest applying a polygonal grid as a continuous raster in vector format for the zonal statistic or zonal histogram analysis to reduce the spatial resolution to several decimeters or meters, but enabling histogram analysis for each vector raster cell.

4.3 SPECTRAL 3D INFORMATION FROM UAVS

For a long time, spectral information for crop monitoring was mostly captured with field spectrometers on the ground, from airplanes, and from satellites (e.g., Ungar et al., 2003; Milton et al., 2009; Li et al., 2010; Gnyp et al., 2013; Mulla, 2013; Schaepman et al., 2015). In the last decade, unmanned robotic vehicles have enabled a new era in remote sensing (Zarco-Tejada, 2008). During this time, a variety of small and lightweight point, push broom, and 2D spectral sensors have become available that are suited to be carried by UAVs (Adão et al., 2017; Aasen et al., 2018). Mostly, these systems were used to generate 2D spectral data. This allows estimates of biochemical crop traits such as canopy chlorophyll (e.g., Berni et al., 2009; Domingues Franceschini et al., 2017; Maimaitijiang et al., 2017; Aasen and Bolten, 2018), carotenoids (e.g., Zarco-Tejada et al., 2013), and nitrogen status (e.g., Quemada et al., 2014; Caturegli et al., 2016; Maimaitijiang et al., 2017). In addition, spectral UAV sensing has been used to estimate biophysical plant traits such as leaf area index (e.g., Berni et al., 2009; Domingues Franceschini et al., 2017; Maimaitijiang et al., 2017; Tian et al., 2017; Yuan et al., 2017) and biomass (e.g., Lukas et al., 2016; Wehrhan et al., 2016), and also biotic (Bock et al., 2010; e.g., Calderón et al., 2013; Garcia-Ruiz et al., 2013; Dash et al., 2017; Kuska et al., 2017) and abiotic stress (e.g., Baluja et al., 2012; Stagakis et al., 2012). For a more detailed description of these applications the interested reader is referred to other reviews (Zhang and Kovacs, 2012; Adão et al., 2017; Hunt and Daughtry, 2017; Yang et al., 2017; Aasen et al., 2018). For spectral 3D mapping, data from different sensing systems have to be combined. Mostly, airplanes were used to carry (hyper-) spectral and separate LiDAR systems (e.g., Cook et al., 2013), since the sensors were rather big and heavy. The same strategy has also been used with UAVs. 3D information was either gathered by LiDAR (Jaakkola et al., 2010; Sankey et al., 2017) or RGB SfM (Suomalainen et al., 2014), while the spectral information being captured with a separate device. Subsequently, the spectral and 3D data had to be co-registered (e.g., Clark et al., 2011; Swatantran et al., 2011; Hladik et al., 2013; Alonzo et al., 2014). For a few years now, lightweight 2D spectral imagers have been available that can be used with UAVs to derive spectral and 3D data at the same time.

4.3.1 2D SPECTRAL IMAGERS

Two-dimensional spectral imagers allow the collection of spectral 2D images from platforms such as UAVs. Today, several spectral 2D imaging sensors are available. They can be distinguished by the technique used to collect the different spectral bands (Aasen et al., 2018), including multi-camera systems such as the Tetracam mini-MCA (TETRACAM Inc, 2015), the Micasense Parrot sequoia (MicaSense, 2017a) or RedEdge (MicaSense, 2017b), which record each spectral band with separate sensors that are combined into a common housing. The cameras have a limited number of spectral bands (mostly up to six) but a high spatial resolution. Sequential 2D imagers such as the Fabry–Pérot interferometer camera Rikola FPI (SENOP, 2017) sequentially record bands or band packages through the same lens system by tuning a filter to allow light for a specific bandpass to transit through. With this technique, many spectral bands could be recorded in a rather high spatial resolution but not at the exact same time. Snapshot 2D imagers recorded all spectral bands at the same time.

Different techniques exist to collect and combine data sets. Multipoint spectrometers such as the Cubert UHD-185 Firefly (Cubert GmbH, 2016) use a beam splitter to divide the 2D image into sections and the signal of these sections is then spread in the spectral domain. These cameras record many spectral bands but their spatial resolution is rather limited.

Another type are filter-on-chip cameras. Here the spectral filters are directly built onto the chip, similarly to the Bayern pattern of RGB cameras. The pixels are arranged in tiles, such that every tile contains the filters for all bands. This information is then compiled to derive the spectral signature of an object. Several companies (Cubert GmbH, 2017; photon focus AG, 2017; XIMEA, 2018) sell these cameras, which are based on chips manufactured by imec (imec, 2017). These cameras have several bands (usually 16 bands from 470 to 630 nm and 25 bands from 600 to 950 nm) and a high spatial resolution, but technically every pixel on the ground is sampled in just one band due to the tiles.

Different postprocessing steps are needed for the different cameras. For multicamera and sequential-band 2D imagers the different spectral bands need to be spatially aligned (Laliberte et al., 2011; Torres-Sánchez et al., 2013; Jhan et al., 2016; Honkavaara et al., 2017). To use the low-resolution spectral images of multipoint spectrometers one has to attach the spectral information to a high-resolution grayscale image (Aasen et al., 2015). For filter-on-chip sensors, the data needs to be interpolated to get a spectral signature for each spot on the ground.

4.3.2 3D Spectral Information from 2D Imagers

Since 2D spectral imagers record 2D spatial information, the SfM approach (cf. Section 4.2.2) can be used (i) to estimate the relative position and orientation of the spectral 2D images, (ii) to generate a 3D spectral point cloud of the scene, and (iii) to reconstruct the 3D geometry of the scene (Honkavaara et al., 2013; Aasen et al., 2015). With this information (hyper-)spectral digital surface models (SDSMs) can be created. SDSMs are a representation of the surface in 3D space linked with (hyper-)spectral information emitted and reflected by the objects covered by the surface (Aasen et al., 2015). Figure 4.12 shows an SDSM of a barley field phenotyping experiment at 96 days after sowing (early senescence) derived from a 2D spectral imager. Both the canopy height, symbolized by the vertical extent, and the spectral data, visualized as the red edge inflection point, were derived from the same system (2D spectral snapshot camera) at the same time. The SDSM is embedded into a larger digital surface model generated with an RGB camera. From these data, 3D and spectral information can be extracted to derive crop traits.

FIGURE 4.12 Spectral digital surface model (SDSM) of a barley field phenotyping experiment at 96 days after sowing (early senescence) derived from a 2D spectral imager embedded into a larger digital surface model generated with an RGB camera (vertical exaggeration: 2). The color of the SDSM corresponds to the red edge inflection point (REIP) VI. The logging in some plots is visible in the height as well as the spectral data. In the back, a tower of a monitoring station can be seen.

4.4 CROP TRAIT ESTIMATION FROM 3D AND SPECTRAL INFORMATION

Different approaches have been used to benefit from having 3D and spectral data together. They can be categorized and distinguished into, first, a segmentation, complementation, and combination approach using spectral and 3D data together. A second approach uses both types of data to derive different parameters. The third approach is to combine both types of information to estimate traits.

The next sections describe the different approaches together with some examples from crop and forest applications.

4.4.1 SEGMENTATION APPROACH

The segmentation approach uses either spectral (or color) information or 3D information for a presegmentation or classification to then extract relevant features from the other. The segmentation approach is used by Geipel et al. (2014) and others. They used the excess green index (Woebbecke et al., 1995), vegetation green index (Meyer et al., 1998; Gitelson et al., 2002), and an adapted broad-band variant of the plant pigment ratio (Metternicht, 2003) on RGB images to segment crop from noncrop pixels. Subsequently, the PH was calculated using the DSM height solely from crop-classified pixels. The authors concluded that preclassification significantly improved the results. With complete canopy closure, the preclassification did not improve the results anymore. Näsi et al. (2015) used 3D and spectral data from a 2D spectral imager to apply an object-based method for bark beetle damage detection in Norway spruce. First, the height information was used to segment individual trees. The tree crowns were delineated from the CHMs with a watershed segmentation approach (Hyyppa et al., 2001; Tanhuanpää et al., 2014). Second, spectral data was extracted from a circular window of approximately 1 m diameter centered on the tree. This information was then used to classify bark beetle damage. The authors pointed out that the NIR area of the spectrum in particular showed promising classification results. In a follow-up study, the authors investigated the effect of the spatial resolution by operating the same camera on an aircraft at 500 m with a ground sampling distance of 0.5 m and a UAV at 90 m with a ground sampling distance of 0.1 m. They found that the classification accuracy for the infested trees was better in almost every case with the high-resolution UAV data (Näsi et al., 2018). They concluded that the individual tree-based approach, made possible by the combination of 3D and spectral data, provides a new level of precision and efficiency for forest health management practices.

4.4.2 COMPLEMENTATION APPROACH

The complementation approach uses spectral and 3D data as complementary data to estimate different traits from both types of data. Honkavaara et al. (2013) used the height and spectral information by means of NDVI separately to derive dry biomass of a wheat field. In their study that looked at one individual date, the NDVI performed better that the height. Additionally, the study showed that normalizing for changing illumination conditions during the flight improved the results.

Aasen et al. (2015) derived PH from the 3D and chlorophyll from the spectral data for one flight in spring barely in the growing stage heading (BBCH 52–59). Their results showed good results for PH estimation ($R^2 = 0.7$) and chlorophyll ($R^2 = 0.5$) with the VI blue green pigment index 2 (Zarco-Tejada et al., 2005). In a follow up study (Aasen and Bolten, 2018), they derived chlorophyll and PH for different growth stages. The estimation of chlorophyll varied with growth stage and results were good when multiple growth stages were combined. The PH information helped to explain within-plot variability of the spectral data. Additionally, they found and explained that upscaling of results from nonimaging field spectrometers to imaging data was potentially challenging due to the different viewing geometries of the data (Aasen and Bolten, 2018). The estimation of PH was consistently good, as long as the prerequisites for a good DSM reconstruction with SfM were met (Aasen, 2016). Honkavaara et al. (2016) investigated the topography and moisture status of peatlands based on a visible and near-infrared (VNIR) and SWIR 2D spectral imager. They found potential deficiencies during the reconstruction of the DSM from hyperspectral 2D imagers in comparison with good-quality customer digital color cameras due to smaller numbers of pixels (poorer spatial resolution), smaller image size, and lower signal-to-noise ratio (because of measuring narrower spectral bands). For the moisture estimation the best accuracy was obtained when using the reflectance difference of SWIR (1246 nm) and VNIR band (859 nm), which gave an RMSE of 5.21 mass fraction in percentage points and a normalized RMSE of 7.61%. van der Meij et al. (2017) investigated plant trait response of oats (*Avena sativa*) to six different cover-crop species and species combinations from three different plant families (*Poaceae, Fabaceae, Brassicaceae*) by means of height, as was retrieved with UAV/SfM, fresh biomass, nitrogen content, and leaf chlorophyll content estimated by VIs and partial least-squares regression. The spectral data with partial least-squares modeling showed a good correlation to chlorophyll content ($R^2 = 0.80$), N content ($R^2 = 0.68$), and fresh biomass ($R^2 = 0.56$) and PH ($R^2 = 0.81$). Biomass was better estimated based on the SfM-derived PH ($R^2 = 0.74$). The segmentation approach has also been used in forest applications from airborne measurements. Lately, the approach has also been transferred to data from 2D spectral imagers. Saarinen et al. (2018) used both a segmentation and a complementation approach. They estimated biodiversity parameters in a boreal forest based on 3D and spectral data of a 2D spectral imager. They used the 3D information to delineate the trees and generate the structural tree metrics including maximum height, mean height, and different height percentiles. Additionally, the spectral information was extracted for each tree (which was previously delineated) to predict the health status and tree species as well as species group. A random forest (Breiman, 2001) was used to select the most important 3D and spectral metrics for different traits. Often, the complementary approach was also used in the combination approach, for example, to evaluate whether a combination of 3D and spectral data was beneficial over just using one type of data for a specific trait.

4.4.3   Combination Approach

In the combination approach, 3D and spectral data are combined to estimate one trait. So far, only a few studies have investigated such an approach. Marshall and Thenkabail (2015) investigated different combined models derived by *in situ* field measurements of nonspectral traits (crop height, fraction of absorbed photosynthetically active radiation, LAI, and fractional vegetation cover) with hyperspectral narrow bands (HNBs). Tilly et al. (2015a) combined TLS-derived crop height and hyperspectral *in situ* measurements to estimate barley biomass. And Bendig et al. (2015) investigated the UAV-derived PH in combination with the hyperspectral *in situ* measurements also for barley biomass. In these three studies on combined analysis of nonspectral and hyperspectral data, (i) PH performed as a robust predictor for crop biomass and partly outperformed spectral predictors, (ii) the combined analysis slightly improved the prediction performance, and (iii) the consideration of broadband RGB-based VIs improved the prediction performance by a similar magnitude as the spectral indices. Similarly, a study by Schaefer and Lamb (2016) found that a combination of LiDAR height and NDVI information was beneficial for pasture biomass estimation. In forest applications, the fusion of LiDAR and spectral information is already widely used, for example, for the classification of tree species (Dalponte et al., 2012). As recently shown by Bareth et al. (2015b), similar approaches (with RGB data) can also be used to classify different plant communities, resulting from different nutrient availability, in grasslands. Such approaches should be further explored.

The key results of combining spectral and nonspectral analysis for barley biomass of Tilly et al. (2015a) are given in Tables 4.1 and 4.2. Table 4.1 shows the regression analysis for bivariate and multivariate regression. Multivariate regression combined crop height and one VI. In both cases, linear and exponential bivariate biomass regression models (BRM), PH was outperforming the selected VIs. Only the NIR/SWIR-based VIs, the NRI and GnyLi VIs (Koppe et al., 2012; Gnyp et al., 2014) were competitive with PH. In contrast, the multivariate BRMs did not show such a significant difference, all models performed more or less slightly better than PH, NRI, or GnyLi

TABLE 4.1

Statistics for Model Calibration of a Barley Experiment as Mean Values of the Four Subset Combinations

| | | Bivariate BRMs | | | | | Multivariate BRMs | | | |
| | | Whole Period | | Pre-Anthesis | | | Whole Period | | Pre-Anthesis | |
	Estimator	R^2	SE_E[a]	R^2	SE_E[a]	Estimator[b]	R^2	SE_E[a]	R^2	SE_E[a]
Dry biomass — Linear	PH	0.65	10.03	0.76	5.73					
	GnyLi	0.52	11.75	0.68	6.67	GnyLi	0.65	34.63	0.77	25.41
	NDVI	0.07	16.38	0.34	9.56	NDVI	0.69	21.49	0.76	20.73
	NRI	0.54	11.58	0.70	6.40	NRI	0.65	35.04	0.77	24.86
	RDVI	0.13	15.87	0.39	9.21	RDVI	0.69	19.18	0.76	21.40
	REIP	0.12	15.92	0.58	7.60	REIP	0.73	1933.86	0.76	258.29
	RGBVI	0.05	16.55	0.26	10.10	RGBVI	0.68	22.28	0.76	23.23
Dry biomass — Exponential	PH	0.84	0.37	0.84	0.34					
	GnyLi	0.80	0.42	0.85	0.32	GnyLi	0.86	2.43	0.88	2.14
	NDVI	0.30	0.77	0.61	0.53	NDVI	0.85	2.84	0.88	3.99
	NRI	0.81	0.40	0.87	0.30	NRI	0.87	2.29	0.89	1.96
	RDVI	0.41	0.71	0.68	0.48	RDVI	0.85	2.52	0.88	2.84
	REIP	0.37	0.73	0.77	0.40	REIP	0.84	30.37	0.86	48.49
	RGBVI	0.23	0.81	0.48	0.60	RGBVI	0.85	2.51	0.87	2.73
Fresh biomass — Linear	PH	0.59	901.99	0.60	843.32					
	GnyLi	0.58	913.81	0.62	829.48	GnyLi	0.62	3295.30	0.64	2968.91
	NDVI	0.25	1222.39	0.42	1022.79	NDVI	0.60	4561.69	0.63	5008.60
	NRI	0.59	909.94	0.62	821.35	NRI	0.62	3056.34	0.64	2718.09
	RDVI	0.35	1143.49	0.50	945.26	RDVI	0.61	3813.94	0.64	3955.80
	REIP	0.30	1180.82	0.55	894.62	REIP	0.60	14599.87	0.63	59169.39
	RGBVI	0.22	1243.84	0.37	1066.53	RGBVI	0.61	4007.93	0.64	3881.46
Fresh biomass — Exponential	PH	0.70	0.37	0.68	0.39	PH				
	GnyLi	0.76	0.33	0.76	0.34	GnyLi	0.77	1.87	0.77	1.77
	NDVI	0.46	0.50	0.65	0.41	NDVI	0.77	3.74	0.79	4.30
	NRI	0.77	0.33	0.77	0.33	NRI	0.77	1.67	0.77	1.56
	RDVI	0.59	0.43	0.74	0.35	RDVI	0.79	2.69	0.82	2.89
	REIP	0.47	0.49	0.71	0.37	REIP	0.72	22.27	0.74	73.05
	RGBVI	0.38	0.53	0.55	0.47	RGBVI	0.77	2.58	0.78	2.68

Source: Tilly, N. et al. 2015a. *Remote Sens.* 7, 11449–11480. https://doi.org/10.3390/rs70911449.

[a] The SE_E for exponential models is calculated from natural log-transformed biomass values.

[b] Each fused with PH.

alone. This was a rather interesting result and indicated an increase in robustness by combining spectral information with PH, even if it was broadband RGB information.

This finding is also visible when applying and validating these BRMs (Table 4.2). Again, PH was outperforming the VIs for biomass. Only the NIR/SWIR VIs, NRI and GnyLi, showed similar performance to the PH. The validation of the multivariate BRMs also showed a similar calibration pattern: the combination of the nonspectral PH data with almost any given spectral data, like RGB broadband data, increased the robustness of the model performance. Similar results were also found by Marshall and Thenkabail (2015): "the nonspectral predictors were the most important, while the

TABLE 4.2

Statistics for Model Validation of a Barley Experiment as Mean Values of Four Subset Combinations

| | | Bivariate BRMs | | | | | | | Multivariate BRMs | | | | | | |
| | | Whole Period | | | Pre-Anthesis | | | | Whole Period | | | Pre-Anthesis | | |
	Estimator	R^2	RMSE[a]	d	R^2	RMSE[a]	d	Estimator[b]	R^2	RMSE[a]	d	R^2	RMSE[a]	d
Dry biomass — Linear	PH	0.66	257.57	0.88	0.80	147.75	0.92							
	GnyLi	0.54	299.67	0.81	0.72	173.31	0.88	GnyLi	0.65	262.19	0.88	0.79	148.20	0.92
	NDVI	0.07	412.70	0.33	0.38	244.47	0.64	NDVI	0.71	250.35	0.89	0.80	148.32	0.92
	NRI	0.55	295.41	0.82	0.74	166.41	0.89	NRI	0.66	261.77	0.88	0.80	147.67	0.92
	RDVI	0.13	400.36	0.44	0.41	233.53	0.71	RDVI	0.72	247.16	0.89	0.80	148.27	0.92
	REIP	0.15	404.95	0.46	0.68	197.50	0.83	REIP	0.73	228.46	0.91	0.80	147.88	0.92
	RGBVI	0.04	416.42	0.26	0.28	254.41	0.58	RGBVI	0.70	261.30	0.88	0.80	149.33	0.92
Dry biomass — Exponential	PH	0.85	0.39	0.95	0.85	0.36	0.95							
	GnyLi	0.80	0.42	0.94	0.86	0.33	0.95	GnyLi	0.87	0.36	0.96	0.89	0.31	0.96
	NDVI	0.29	0.77	0.63	0.59	0.54	0.81	NDVI	0.85	0.38	0.95	0.87	0.30	0.96
	NRI	0.81	0.40	0.94	0.87	0.31	0.96	NRI	0.87	0.36	0.96	0.89	0.29	0.96
	RDVI	0.40	0.71	0.73	0.66	0.48	0.87	RDVI	0.85	0.38	0.95	0.88	0.30	0.96
	REIP	0.40	0.75	0.72	0.82	0.43	0.90	REIP	0.85	0.39	0.95	0.89	0.34	0.95
	RGBVI	0.22	0.82	0.55	0.48	0.62	0.75	RGBVI	0.85	0.38	0.95	0.86	0.31	0.96
Fresh biomass — Linear	PH	0.67	963.45	0.84	0.70	892.55	0.85							
	GnyLi	0.65	970.70	0.83	0.72	886.24	0.84	GnyLi	0.69	939.84	0.85	0.74	861.73	0.86
	NDVI	0.27	1254.02	0.58	0.51	1053.83	0.70	NDVI	0.67	952.58	0.84	0.73	862.84	0.85
	NRI	0.65	962.49	0.83	0.72	873.75	0.85	NRI	0.69	938.46	0.85	0.74	857.99	0.86
	RDVI	0.38	1175.32	0.67	0.59	964.42	0.77	RDVI	0.68	943.96	0.85	0.74	841.36	0.86
	REIP	0.41	1244.11	0.66	0.77	951.74	0.81	REIP	0.67	966.67	0.84	0.77	908.74	0.84
	RGBVI	0.21	1260.32	0.53	0.41	1066.26	0.67	RGBVI	0.66	948.90	0.85	0.71	852.97	0.86
Fresh biomass — Exponential	PH	0.73	0.40	0.89	0.71	0.42	0.88							
	GnyLi	0.78	0.35	0.92	0.79	0.36	0.91	GnyLi	0.79	0.34	0.92	0.80	0.36	0.92
	NDVI	0.44	0.51	0.73	0.64	0.42	0.83	NDVI	0.78	0.34	0.92	0.79	0.34	0.92
	NRI	0.77	0.34	0.92	0.79	0.35	0.92	NRI	0.79	0.34	0.92	0.79	0.35	0.92
	RDVI	0.57	0.44	0.82	0.73	0.36	0.89	RDVI	0.80	0.33	0.93	0.83	0.31	0.93
	REIP	0.54	0.53	0.77	0.82	0.42	0.87	REIP	0.77	0.39	0.90	0.82	0.40	0.88
	RGBVI	0.36	0.54	0.68	0.53	0.47	0.78	RGBVI	0.76	0.34	0.92	0.76	0.34	0.92

Source: Tilly, N. et al. 2015a. *Remote Sens.* 7, 11449–11480. https://doi.org/10.3390/rs70911449.

[a] The RMSE for exponential models is calculated from natural log-transformed biomass values.

[b] Each fused with PH.

HNBs [hyperspectral narrow bands] explained additional and statistically significant predictors, but with lower variance."

Also UAVs have been used for a combined approach. Yue et al. (2017) estimated above-ground biomass of winter wheat with a 2D spectral imager by combining crop height and spectral data. Similarly to Aasen et al. (2015) they generated an SDSM and extracted crop height and spectral data from the SDSM. From the spectral data, they used individual wavelengths as well as different VIs in linear models. Additionally, they used partial least-squares regression to combine more than two parameters. Similarly to Tilly et al. (2015a), their results indicated that incorporating crop height into the models improved the accuracy of above-ground biomass estimations. 3D and spectral data have also been used in forest studies. Nevalainen et al. (2017) combined 3D and spectral data from a 2D spectral imager to classify species in a boreal forest. They tested several classifiers to combine both types of data. The best classification results were obtained with random forest and multilayer perceptron. The feature selection results indicated that spectral features were more significant in discriminating different tree species than the structural features derived from the photogrammetric point clouds.

4.5 OVERALL DISCUSSION

4.5.1 GROUND- AND UAV-BASED SENSING SYSTEMS

Mobile and fast data acquisition technologies that allow capture of crop traits are of great interest in field phenotyping and precision agriculture applications (Furbank and Tester, 2011; Fiorani and Schurr, 2013; Mulla, 2013; Araus and Cairns, 2014; Walter et al., 2015). Additionally, they are one key element to implement sustainable agriculture though smart farming (Walter et al., 2017). Spectral and 3D-based approaches allow information on crop traits to be derived and data can be captured in different ways. Ground-based observations such as ruler or field spectrometer measurements are rather slow, time consuming, and thus not feasible for practical applications. In the CROP.SENSe. net (https://www.cropsense.uni-bonn.de/) phenotyping experiment, it took two people about 1.5 h to screen 36 plots of 3 m by 7 m with six field spectrometer measurements in optimal conditions. As soon as the weather became more unstable, it took much longer because of the need to recalibrate more frequently (Tilly et al., 2015a). The spectral UAV flights took about 15–20 minutes and several hundreds of spectra were captured per plot (Aasen et al., 2015; Bareth et al., 2015a; Aasen and Bolten, 2018). The RGB flights took only a couple of minutes. However, UAV sensing has its limitations. Very strong winds (Beaufort scale >4), precipitation, and temperatures below 0°C and above 40°C limit the use of standard UAVs. Still, from our own experiences in Germany, UAV-based RGB image acquisition around ±2 days of a planned date are possible in 90% of the time of scheduled dates. Besides, even with state-of-the-art spectral fixed-wing UAV systems the area that can be covered in one UAV campaign is usually limited to approx. 100 ha due the trade-off between flying height, speed, ground sampling distance and coverage. Additionally, aviation regulations restrict UAV operations. Multiple campaigns can cover up to 500–1000 ha per day, and even more when out-of-sight operations are possible. Additionally, sensing technology is advancing rapidly. Compared to 3 years ago, sensors today are able to capture data faster and with much higher resolution, which allows flying higher and faster and covering a much larger area (Bareth et al., 2011; Aasen et al.,2018). These advances are key if technology is to be implemented for large-scale crop trait estimation, for example, in precision agriculture and high-throughput field phenotyping applications, where thousands of crop lines or plots need to be screened at multiple locations. In addition, with this performance these technologies will become economically important as soon as their suitability and reliability has been proven.

4.5.2 CROP TRAIT ESTIMATION WITH 3D DATA

The good and robust performance of nonspectral crop traits from TLS (Tilly et al., 2014) or UAV/SfM (Bendig et al., 2014) is of great economic importance. TLS and UAV/SfM have been cross-validated (Bareth et al., 2016; Cooper et al., 2017; Madec et al., 2017) and shown to reliably estimate crop height, biomass, emergence, vegetation density in early phenological stages, and lodging (Bendig et al., 2014, 2015; Tilly et al., 2014, 2015a; Bareth et al., 2015b, 2016; Aasen, 2016; Crommelinck and Höfle, 2016; Friedli et al., 2016). First approaches have also used the PH information to detect the response of plant growth to different fertilizers (Bareth et al., 2015b; Holman et al., 2016) and investigated plant–soil interactions (van der Meij et al., 2017). Laser scanning is increasingly implemented in field phenotyping platforms that are able to measure 3D information at high spatial and temporal resolution (Kirchgessner et al., 2017; Madec et al., 2017; Virlet et al., 2017; Jimenez-Berni et al.,

2018). UAV/SfM, on the other hand, is more flexible, less expensive, and thus rapidly adopted in the scientific community and beyond.

The key advantage of nonspectral crop monitoring with UAVs is the (near) independence from good and/or sunny weather, since—in contrast to spectral remote sensing—cloud cover and varying irradiation conditions do not affect the UAV-based image acquisition. Although very recently promising approaches have been developed to normalize spectral data for changing illumination conditions (Honkavaara and Khoramshahi, 2018; Miyoshi et al., 2018), these have not yet been implemented in common data processing procedures. For deriving 3D data from UAV/SfM, cloudy conditions can even be favorable because each image is more homogenously illuminated due to the absence of shadows. The higher flexibility of nonspectral crop monitoring is of considerable importance when precise timing of the measurements is necessary, for example, in characteristic phenological stages. For these applications, UAV-based remote sensing is unmatched in terms of acquiring data in a certain time window and making multitemporal campaigns possible and affordable.

When it comes to supraregional or continental-scale mapping, only satellite approaches are feasible. So far, only first approaches for crop height extraction from space have been undertaken with SAR (synthetic-aperture radar) interferometry. On the potential of Polarimetric SAR Interferometry to characterize the biomass, moisture and structure of agricultural crops at L-, C- and X-Bands (https://doi.org/10.1016/j.rse.2017.09.039). (Erten et al., 2016; Hütt et al., 2016; Lopez-Sanchez et al., 2017). But new missions such as the Global Ecosystem Dynamics Investigation (GEDI) LiDAR mission are coming up. GEDI is planned to be mounted on the international space station in 2018. It records ground tracks of 25 m diameter footprints providing the most complete set of measurements of canopy structure and bare earth (under canopy) topography yet achieved (Qi and Dubayah, 2016). In future, such missions could take the methods outlined in this chapter from local to global application.

While most work has focused on using DSMs for 3D plant trait mapping, recently the analysis of point clouds has also attracted more attention (e.g., Friedli et al., 2016; Jimenez-Berni et al., 2018). We expect this trend to continue. Additionally, we think that object-based analysis based on super-high-resolution UAV imagery (e.g., Jin et al., 2017; Chen et al., 2018; Roth et al., 2018) is an upcoming trend in crop trait extraction (Aasen et al., 2018). Finally, and importantly, we think that the increasing availability of 3D and spectral data will open up new approaches for crop monitoring.

4.5.3 Approaches for Crop Trait Estimation with Spectral and 3D Data

(Hyper-) spectral data allows one to derive a variety of crop traits, as many works have demonstrated (e.g., Thenkabail et al., 2012; Adão et al., 2017; Yang et al., 2017; Aasen et al., 2018). Today, UAV sensors that can capture a few spectral bands can weigh only a few hundred grams and can be bought off-the-shelf for a couple of thousand euro. And although UAV sensing systems are still behind their ground-based and airborne counterparts when it comes to data quality and spectral range, sensor development is fast progressing (Aasen et al., 2018).

Spectral UAV sensing can complement ground- or UAV-based 3D data collection because 3D techniques are not able to derive biochemical crop traits. With the advent of 2D spectral imagers, it is moreover possible to derive 3D and spectral information at the same time. This complements traditional approaches in which both types of information had to be captured separately, and brings exciting new opportunities for trait extraction. Only a few papers have been published that use 3D and spectral data, eventually captured by separate sensors (field/field; field/UAV; UAV/UAV), for crop trait mapping. Spectral 3D data can be used in a segmentation, complementation, and combination approach and several studies have found promising results by using these approaches. All approaches build on the complementary information of 3D and spectral data. With the increased use of 2D spectral imagers, and consequently the increasing availability of 3D and spectral data, we expect this trend to continue. Additionally, the results of studies that have shown the benefit of one of both types of information can theoretically be transferred to cases where both types of data are available. However, in practice one needs to be careful since comparing results from different devices might not be straightforward (Bareth et al., 2015a; von Bueren et al., 2015; Domingues Franceschini et al., 2017) and transferring results from one device to another might be challenging, for example, because of different angular properties of the data (Aasen and Bolten, 2018). In addition, data standardization is still an issue for UAV spectral sensing. Compared to satellite systems, were most data products are standardized, the variety of UAV sensors and protocols makes this issue more complex (Aasen et al., 2018).

Combining 3D and spectral data to estimate one trait makes sense in particular if it can be assumed that a plant trait can be better modeled by a combination of information. One example is biomass, which can be described as volume × density. 3D data allows precise derivation a DSM of a plant canopy and, thus, its volume. However, 3D data can primarily be captured from the topmost parts of the canopy due to occlusion (although multiple return LiDAR systems can also penetrate deeper into the canopy). Since light penetrates into the canopy and may be scattered multiple times, spectral data contains information from deeper inside the canopy (although the proportion rapidly decreases with leaf layers). Thus, it can give information about, for example, the canopy density. In theory, a combination of 3D and spectral data should perform better than just one type alone. In practice however, PH alone already performs very well, in particular, when looking at the several dates during the growing season. Here, studies should investigate the benefit when only one specific date and different varieties are taken into account.

In the future, novel data analysis approaches to combine 3D and spectral data are needed. While both types of data are becoming increasingly available, the development of algorithms to explore the data falls behind. This is particularly interesting, since combined spectral and 3D information will soon also become available globally. GEDI will be mounted together with the Ecosystem Spaceborne Thermal Radiometer Experiment on Space Station (ECOSTRESS) and the spectral instruments Orbiting Carbon Observatory 3 (OCO-3) and Hyperspectral Imager Suite (HISUI). Coordinating the spatial and temporal coincidence of measurements from GEDI, ECOSTRESS, OCO-3, and HISUI would be an opportunity to address ecosystem dynamics questions that cannot be answered from any one instrument (Stavros et al., 2017). Such instrument suites would also allow testing of the capabilities of the discussed data fusion approaches for whole landscapes. But it still has to be shown whether spaceborne instruments will have a suitable resolution for 3D crop growth monitoring.

4.6   CONCLUSIONS

Three developments in the last decade revolutionized remote sensing-based high-resolution crop monitoring: (i) the capability to create ultrahigh-resolution 3D data with terrestrial or airborne laser scanning, (ii) the software developments around the structure for motion (SfM) approach in combination with affordable UAV systems carrying standard digital cameras for 3D data acquisition, and (iii) miniaturization of multi- and hyperspectral sensors to fit on UAVs, making it possible to capture spectral information for a wide audience.

Nonspectral 3D data from laser scanning or UAV/SfM allows estimation of crop traits such as height, biomass, emergence, crop density, and lodging. The accuracies in x, y, and z coordinates are in the centimeter range and enable precise monitoring of crop growth and development. The key advantage of nonspectral crop monitoring with UAVs is the (near) independence from clear sky weather. This significantly increases the flexibility of 3D approaches in comparison with spectral approaches.

Nevertheless, spectral data are needed and complement 3D data when it comes to biochemical parameters. Today, spectral UAV systems can be bought off-the-shelf and allow the capture of spectral information of several hectares in a couple of minutes. Additionally, UAV-based remote sensing with 2D spectral imagers allows the capturing of spectral and 3D information at the same time. This supports rather new and promising approaches that (i) uses spectral (or color) information or 3D information for a presegmentation or classification, (ii) uses spectral and 3D data as complementary data to estimate different traits from both type of data, and (iii) combines 3D and spectral data to estimate one trait. While so far only some studies on the use of 3D and spectral data have been published, we think that these approaches will be explored to a greater extent in the future.

ACKNOWLEDGMENTS

We thank all current and former members of the GIS & RS Group (University of Cologne), especially Juliane Bendig, Andreas Bolten, Sebastian Brocks, Martin Leon Gnyp, Dirk Hoffmeister, Christoph Hütt, Rainer Laudien, Victoria Lenz-Wiedemann, Nora Tilly, Ulrike Lussem, and Kang Yu, for continuous support and research efforts on this topic in the last 10 years.

REFERENCES

Aasen, H., 2016. High-resolution 3D hyperspectral digital surface models from lightweight UAV snapshot cameras – potentials for precision agriculture applications. Presented at the *International Conference on Precision Agriculture*, St. Louis, Missouri USA.

Aasen, H., Bolten, A., 2018. Multi-temporal high-resolution imaging spectroscopy with hyperspectral 2D imagers – From theory to application. *Remote Sens. Environ.* 205, 374–389. https://doi.org/10.1016/j.rse.2017.10.043

Aasen, H., Burkart, A., Bolten, A., Bareth, G., 2015. Generating 3D hyperspectral information with lightweight UAV snapshot cameras for vegetation monitoring: From camera calibration to quality assurance. *ISPRS J. Photogramm. Remote Sens.* 108, 245–259. https://doi.org/10.1016/j.isprsjprs.2015.08.002

Aasen, H., Honkavaara, E., Lucieer, A., Zarco-Tejada, P.J., 2018. Quantitative remote sensing at ultra-high resolution with UAV spectroscopy: A review of sensor technology, measurement procedures, and data correction workflows. Remote Sens. 10, 1091, https://doi.org/10.3390/rs10071091

Adão, T., Hruška, J., Pádua, L., Bessa, J., Peres, E., Morais, R., Sousa, J., 2017. Hyperspectral imaging: A review on UAV-based sensors, data processing and applications for agriculture and forestry. *Remote Sens.* 9, 1110. https://doi.org/10.3390/rs9111110

Alchanatis, V., Cohan, Y., 2012. Spectral and spatial methods for hyperspectral image analysis for estimation of biophysical and biochemical properties of agricultural crops, in: Thenkabail, P.S., Lyon, J.G., Huete, A. (Eds.), *Hyperspectral Remote Sensing of Vegetation*. CRC Press, Boca Raton, FL, pp. 289–308.

Alexandridis, T., Tamouridou, A.A., Pantazi, X.E., Lagopodi, A., Kashefi, J., Ovakoglou, G., Polychronos, V., Moshou, D., 2017. Novelty detection classifiers in weed mapping: Silybum marianum detection on UAV multispectral images. *Sensors* 17, 2007. https://doi.org/10.3390/s17092007

Alonzo, M., Bookhagen, B., Roberts, D.A., 2014. Urban tree species mapping using hyperspectral and lidar data fusion. *Remote Sens. Environ.* 148, 70–83. https://doi.org/10.1016/j.rse.2014.03.018

Alshawabkeh, Y., Haala, N., Fritsch, D., 2007. Registrierung terrestrischer Bild- und LIDAR-Daten für die Dokumentation von Kulturdenkmälern. *Photogramm. Fernerkund. Geoinformation* 197–206.

Anderson, K., Gaston, K.J., 2013. Lightweight unmanned aerial vehicles will revolutionize spatial ecology. *Front. Ecol. Environ.* 11, 138–146. https://doi.org/10.1890/120150

Araus, J.L., Cairns, J.E., 2014. Field high-throughput phenotyping: The new crop breeding frontier. *Trends Plant Sci.* 19, 52–61. https://doi.org/10.1016/j.tplants.2013.09.008

Baluja, J., Diago, M.P., Balda, P., Zorer, R., Meggio, F., Morales, F., Tardaguila, J., 2012. Assessment of vineyard water status variability by thermal and multispectral imagery using an unmanned aerial vehicle (UAV). *Irrig. Sci.* 30, 511–522. https://doi.org/10.1007/s00271-012-0382-9

Bareth, G., Aasen, H., Bendig, J., Gnyp, M.L., Bolten, A., Jung, A., Michels, R., Soukkamäki, J., 2015a. Low-weight and UAV-based hyperspectral full-frame cameras for monitoring crops: Spectral comparison with portable spectroradiometer measurements. *Photogramm. Fernerkund. Geoinformation* 2015, 69–79. https://doi.org/10.1127/pfg/2015/0256

Bareth, G., Bendig, J., Tilly, N., Hoffmeister, D., Aasen, H., Bolten, A., 2016. A comparison of UAV- and TLS-derived plant height for crop monitoring: Using polygon grids for the analysis of crop surface models (CSMs). *Photogramm. Fernerkund. Geoinformation* 85–94. https://doi.org/10.1127/pfg/2016/0289

Bareth, G., Bolten, A., Bendig, J., Lenz-Wiedemann, V., Bareth, G., 2011. Potentials of Low-cost Mini-UAVs. *Geographisches Institut der Universität zu Köln - Kölner Geographische Arbeiten*. https://doi.org/10.5880/TR32DB.KGA92.2

Bareth, G., Bolten, A., Hollberg, J., Aasen, H., Burkart, A., Schellberg, J., 2015b. Feasibility study of using non-calibrated UAV-based RGB imagery for grassland monitoring: Case study at the Rengen Long-term Grassland Experiment (RGE), Germany. DGPF-Proceedings. Presented at the *DGPF Annual Conference'15*, Cologne, Germany, pp. 55–62.

Bendig, J., Bolten, A., Bareth, G., 2013. UAV-based Imaging for Multi-Temporal, very high resolution crop surface models to monitor crop growth variability. *Photogramm. - Fernerkund. - Geoinformation* 2013, 551–562. https://doi.org/10.1127/1432-8364/2013/0200

Bendig, J., Bolten, A., Bennertz, S., Broscheit, J., Eichfuss, S., Bareth, G., 2014. Estimating biomass of barley using crop surface models (CSMs) derived from UAV-based RGB imaging. *Remote Sens.* 6, 10395–10412. https://doi.org/10.3390/rs61110395

Bendig, J., Yu, K., Aasen, H., Bolten, A., Bennertz, S., Broscheit, J., Gnyp, M.L., Bareth, G., 2015. Combining UAV-based plant height from crop surface models, visible, and near infrared vegetation indices for biomass monitoring in barley. *Int. J. Appl. Earth Obs. Geoinformation* 39, 79–87. https://doi.org/10.1016/j.jag.2015.02.012

Berni, J., Zarco-Tejada, P.J., Suarez, L., Fereres, E., 2009. Thermal and narrowband multispectral remote sensing for vegetation monitoring from an unmanned aerial vehicle. *IEEE Trans. Geosci. Remote Sens.* 47, 722–738. https://doi.org/10.1109/TGRS.2008.2010457

Berra, E.F., Gaulton, R., Barr, S., 2017. Commercial off-the-shelf digital cameras on unmanned aerial vehicles for multitemporal monitoring of vegetation reflectance and NDVI. *IEEE Trans. Geosci. Remote Sens.* 55, 4878–4886. https://doi.org/10.1109/TGRS.2017.2655365

Bock, C.H., Poole, G.H., Parker, P.E., Gottwald, T.R., 2010. Plant disease severity estimated visually, by digital photography and image analysis, and by hyperspectral imaging. *Crit. Rev. Plant Sci.* 29, 59–107. https://doi.org/10.1080/07352681003617285

Breiman, L., 2001. Random forests. *Mach. Learn.* 45, 5–32. https://doi.org/10.1023/A:1010933404324

Brocks, S., Bareth, G., 2018. Estimating barley biomass with crop surface models from oblique RGB imagery. *Remote Sens.* 10, 268. https://doi.org/10.3390/rs10020268

Calderón, R., Navas-Cortés, J.A., Lucena, C., Zarco-Tejada, P.J., 2013. High-resolution airborne hyperspectral and thermal imagery for early detection of Verticillium wilt of olive using fluorescence, temperature and narrow-band spectral indices. *Remote Sens. Environ.* 139, 231–245. https://doi.org/10.1016/j.rse.2013.07.031

Caturegli, L., Corniglia, M., Gaetani, M., Grossi, N., Magni, S., Migliazzi, M., Angelini, L. et al. 2016. Unmanned aerial vehicle to estimate nitrogen status of turfgrasses. *PLoS ONE* 11, e0158268. https://doi.org/10.1371/journal.pone.0158268

Chang, A., Jung, J., Maeda, M.M., Landivar, J., 2017. Crop height monitoring with digital imagery from Unmanned Aerial System (UAS). *Comput. Electron. Agric.* 141, 232–237. https://doi.org/10.1016/j.compag.2017.07.008

Chen, R., Chu, T., Landivar, J.A., Yang, C., Maeda, M.M., 2018. Monitoring cotton (Gossypium hirsutum L.) germination using ultrahigh-resolution UAS images. *Precis. Agric.* 19, 161–177. https://doi.org/10.1007/s11119-017-9508-7

Clark, M.L., Roberts, D.A., Ewel, J.J., Clark, D.B., 2011. Estimation of tropical rain forest aboveground biomass with small-footprint lidar and hyperspectral sensors. *Remote Sens. Environ.* 115, 2931–2942. https://doi.org/10.1016/j.rse.2010.08.029

Colomina, I., Molina, P., 2014. Unmanned aerial systems for photogrammetry and remote sensing: A review. *ISPRS J. Photogramm. Remote Sens.* 92, 79–97. https://doi.org/10.1016/j.isprsjprs.2014.02.013

Cook, B., Corp, L., Nelson, R., Middleton, E., Morton, D., McCorkel, J., Masek, J., Ranson, K., Ly, V., Montesano, P., 2013. NASA Goddard's LiDAR, Hyperspectral and Thermal (G-LiHT) Airborne Imager. *Remote Sens.* 5, 4045–4066. https://doi.org/10.3390/rs5084045

Cooper, S., Roy, D., Schaaf, C., Paynter, I., 2017. Examination of the potential of terrestrial laser scanning and structure-from-motion photogrammetry for rapid nondestructive field measurement of grass biomass. *Remote Sens.* 9, 531. https://doi.org/10.3390/rs9060531

Crommelinck, S., Höfle, B., 2016. Simulating an autonomously operating low-cost static terrestrial LiDAR for multitemporal maize crop height measurements. *Remote Sens.* 8, 205. https://doi.org/10.3390/rs8030205

Cubert GmbH, 2016. UHD 185 – Firefly [WWW Document]. Hyperspectral Imaging UHD 185 - Firefly Cubert-GmbH. URL http://cubert-gmbh.de/uhd-185-firefly/ (accessed 4.21.16).

Cubert GmbH, 2017. ButterflEYE [WWW Document]. URL http://cubert-gmbh.com/product-category/spectral-cameras/butterfleye/ (accessed 10.11.17).

Dalponte, M., Bruzzone, L., Gianelle, D., 2012. Tree species classification in the Southern Alps based on the fusion of very high geometrical resolution multispectral/hyperspectral images and LiDAR data. *Remote Sens. Environ.* 123, 258–270. https://doi.org/10.1016/j.rse.2012.03.013

Dandois, J.P., Ellis, E.C., 2013. High spatial resolution three-dimensional mapping of vegetation spectral dynamics using computer vision. *Remote Sens. Environ.* 136, 259–276. https://doi.org/10.1016/j.rse.2013.04.005

Dash, J.P., Watt, M.S., Pearse, G.D., Heaphy, M., Dungey, H.S., 2017. Assessing very high resolution UAV imagery for monitoring forest health during a simulated disease outbreak. *ISPRS J. Photogramm. Remote Sens.* 131, 1–14. https://doi.org/10.1016/j.isprsjprs.2017.07.007

DJI, 2017. Phantom 4 Pro - User Manual V1.4, https://dl.djicdn.com/downloads/phantom_4_pro/20171017/Phantom_4_Pro_Pro_Plus_User_Manual_EN.pdf.

Domingues Franceschini, M., Bartholomeus, H., van Apeldoorn, D., Suomalainen, J., Kooistra, L., 2017. Intercomparison of unmanned aerial vehicle and ground-based narrow band spectrometers applied to crop trait monitoring in organic potato production. *Sensors* 17, 1428. https://doi.org/10.3390/s17061428

Ehlert, D., Horn, H.-J., Adamek, R., 2008. Measuring crop biomass density by laser triangulation. *Comput. Electron. Agric.* 61, 117–125. https://doi.org/10.1016/j.compag.2007.09.013

Eitel, J.U.H., Vierling, L.A., Long, D.S., Hunt, E.R., 2011. Early season remote sensing of wheat nitrogen status using a green scanning laser. *Agric. For. Meteorol.* 151, 1338–1345. https://doi.org/10.1016/j.agrformet.2011.05.015

Erten, E., Lopez-Sanchez, J.M., Yuzugullu, O., Hajnsek, I., 2016. Retrieval of agricultural crop height from space: A comparison of SAR techniques. *Remote Sens. Environ.* 187, 130–144. https://doi.org/10.1016/j.rse.2016.10.007

Fiorani, F., Schurr, U., 2013. Future scenarios for plant phenotyping. *Annu. Rev. Plant Biol.* 64, 267–291. https://doi.org/10.1146/annurev-arplant-050312-120137

Friedli, M., Kirchgessner, N., Grieder, C., Liebisch, F., Mannale, M., Walter, A., 2016. Terrestrial 3D laser scanning to track the increase in canopy height of both monocot and dicot crop species under field conditions. *Plant Methods* 12. https://doi.org/10.1186/s13007-016-0109-7

Furbank, R.T., Tester, M., 2011. Phenomics – technologies to relieve the phenotyping bottleneck. *Trends Plant Sci.* 16, 635–644. https://doi.org/10.1016/j.tplants.2011.09.005

Gago, J., Douthe, C., Coopman, R.E., Gallego, P.P., Ribas-Carbo, M., Flexas, J., Escalona, J., Medrano, H., 2015. UAVs challenge to assess water stress for sustainable agriculture. *Agric. Water Manag.* 153, 9–19. https://doi.org/10.1016/j.agwat.2015.01.020

Garcia-Ruiz, F., Sankaran, S., Maja, J.M., Lee, W.S., Rasmussen, J., Ehsani, R., 2013. Comparison of two aerial imaging platforms for identification of Huanglongbing-infected citrus trees. *Comput. Electron. Agric.* 91, 106–115. https://doi.org/10.1016/j.compag.2012.12.002

Geipel, J., Link, J., Claupein, W., 2014. Combined spectral and spatial modeling of corn yield based on aerial images and crop surface models acquired with an unmanned aircraft system. *Remote Sens.* 6, 10335–10355. https://doi.org/10.3390/rs61110335

Girardeau-Montaut, D., Roux, M., Marc, R., Thibault, G., 2005. Change detection on points cloud DATA acquired with a ground laser scanner.

Gitelson, A.A., Kaufman, Y.J., Stark, R., Rundquist, D., 2002. Novel algorithms for remote estimation of vegetation fraction. *Remote Sens. Environ.* 80, 76–87. https://doi.org/10.1016/S0034-4257(01)00289-9

Gnyp, M.L., Bareth, G., Li, F., Lenz-Wiedemann, V.I.S., Koppe, W., Miao, Y., Hennig, S.D. et al. 2014. Development and implementation of a multiscale biomass model using hyperspectral vegetation indices for winter wheat in the North China Plain. *Int. J. Appl. Earth Obs. Geoinformation* 33, 232–242. https://doi.org/10.1016/j.jag.2014.05.006

Gnyp, M.L., Yu, K., Aasen, H., Yao, Y., Huang, S., Miao, Y., Bareth, G., 2013. Analysis of crop reflectance for estimating biomass in rice canopies at different phenological stages. *Photogramm. - Fernerkund. - Geoinformation* 2013, 351–365. https://doi.org/10.1127/1432-8364/2013/0182

Gower, S.T., Kucharik, C.J., Norman, J.M., 1999. Direct and indirect estimation of leaf area index, fAPAR, and net primary production of terrestrial ecosystems. *Remote Sens. Environ.* 70, 29–51. https://doi.org/10.1016/S0034-4257(99)00056-5

Haala, N., 2013. The landscape of dense image matching algorithms. Presented at the *Photogrammetric Week'13*, pp. 271–284.

Haboudane, D., Miller, J.R., Pattey, E., Zarco-Tejada, P.J., Strachan, I.B., 2004. Hyperspectral vegetation indices and novel algorithms for predicting green LAI of crop canopies: Modeling and validation in the context of precision agriculture. *Remote Sens. Environ.* 90, 337–352. https://doi.org/10.1016/j.rse.2003.12.013

Hämmerle, M., Höfle, B., 2014. Effects of reduced terrestrial LiDAR point density on high-resolution grain crop surface models in precision agriculture. *Sensors* 14, 24212–24230. https://doi.org/10.3390/s141224212

Herwitz, S.R., Johnson, L.F., Dunagan, S.E., Higgins, R.G., Sullivan, D.V., Zheng, J., Lobitz, B.M., Leung, J.G., Gallmeyer, B.A., Aoyagi, M., Slye, R.E., Brass, J.A., 2004. Imaging from an unmanned aerial vehicle: Agricultural surveillance and decision support. *Computers and Electronics in Agriculture* 44: 49-61.

Hladik, C., Schalles, J., Alber, M., 2013. Salt marsh elevation and habitat mapping using hyperspectral and LIDAR data. *Remote Sens. Environ.* 139, 318–330. https://doi.org/10.1016/j.rse.2013.08.003

Hoffmeister, D., 2014. Feasibility studies of terrestrial laser scanning in Coastal Geomorphology, *Agronomy, and Geoarchaeology (PhD thesis)*. Universität zu Köln, Cologne, Germany.

Hoffmeister, D., 2016. Laser scanning approaches for crop Monitoring. In: V. Scognamiglio, G. Rea, F. Arduini, and G. Palleschi (eds.): *Biosensors for Sustainable Food - New Opportunities and Technical Challenges, Comprehensive Analytical Chemistry.* 74, 343–361. doi: 10.1016/bs.coac.2016.02.018

Hoffmeister, D., Bolten, A., Curdt, C., Waldhoff, G., Bareth, G., 2010. High-resolution Crop Surface Models (CSM) and Crop Volume Models (CVM) on field level by terrestrial laser scanning, *The Sixth International Symposium on Digital Earth. International Society for Optics and Photonics,* p. 6.

Hoffmeister, D., Waldhoff, G., Korres, W., Curdt, C., Bareth, G., 2016. Crop height variability detection in a single field by multi-temporal terrestrial laser scanning. *Precis. Agric.* 17, 296–312. https://doi.org/10.1007/s11119-015-9420-y

Holman, F., Riche, A., Michalski, A., Castle, M., Wooster, M., Hawkesford, M., 2016. High throughput field phenotyping of wheat plant height and growth rate in field plot trials using UAV based remote sensing. *Remote Sens.* 8, 1031. https://doi.org/10.3390/rs8121031

Honkavaara, E., Eskelinen, M.A., Polonen, I., Saari, H., Ojanen, H., Mannila, R., Holmlund, C. et al. 2016. Remote sensing of 3-D geometry and surface moisture of a peat production area using hyperspectral frame cameras in visible to short-wave infrared spectral ranges onboard a small unmanned airborne vehicle (UAV). *IEEE Trans. Geosci. Remote Sens.* 1–15. https://doi.org/10.1109/TGRS.2016.2565471

Honkavaara, E., Khoramshahi, E., 2018. Radiometric correction of close-range spectral image blocks captured using an unmanned aerial vehicle with a radiometric block adjustment. *Remote Sens.* 10, 256. https://doi.org/10.3390/rs10020256

Honkavaara, E., Rosnell, T., Oliveira, R., Tommaselli, A., 2017. Band registration of tuneable frame format hyperspectral UAV imagers in complex scenes. *ISPRS J. Photogramm. Remote Sens.* 134, 96–109. https://doi.org/10.1016/j.isprsjprs.2017.10.014

Honkavaara, E., Saari, H., Kaivosoja, J., Pölönen, I., Hakala, T., Litkey, P., Mäkynen, J., Pesonen, L., 2013. Processing and assessment of spectrometric, stereoscopic imagery collected using a lightweight UAV spectral camera for precision agriculture. *Remote Sens.* 5, 5006–5039. https://doi.org/10.3390/rs5105006

Hunt, E.R., Daughtry, C.S.T., 2017. What good are unmanned aircraft systems for agricultural remote sensing and precision agriculture? *Int. J. Remote Sens.* 1–32. https://doi.org/10.1080/01431161.2017.1410300

Hütt, C., Tilly, N., Schiedung, H., Bareth, G., 2016. Potential of multitemporal tandem-X derived crop surface models for maize growth monitoring. *ISPRS - Int. Arch. Photogramm. Remote Sens. Spat. Inf. Sci.* XLI-B7, 803–808. https://doi.org/10.5194/isprs-archives-XLI-B7-803-2016

Hyyppa, J., Kelle, O., Lehikoinen, M., Inkinen, M., 2001. A segmentation-based method to retrieve stem volume estimates from 3-D tree height models produced by laser scanners. *IEEE Trans. Geosci. Remote Sens.* 39, 969–975. https://doi.org/10.1109/36.921414

imec, 2017. Hyperspectral imaging [WWW Document]. Hyperspectral Imaging. URL https://www.imec-int.com/en/hyperspectral-imaging (accessed 10.11.17).

Iqbal, F., Lucieer, A., Barry, K., Wells, R., 2017. Poppy crop height and capsule volume estimation from a single UAS flight. *Remote Sens.* 9, 647. https://doi.org/10.3390/rs9070647

Jaakkola, A., Hyyppä, J., Kukko, A., Yu, X., Kaartinen, H., Lehtomäki, M., Lin, Y., 2010. A low-cost multi-sensoral mobile mapping system and its feasibility for tree measurements. *ISPRS J. Photogramm. Remote Sens.* 65, 514–522. https://doi.org/10.1016/j.isprsjprs.2010.08.002

Jagt, B., Lucieer, A., Wallace, L., Turner, D., Durand, M., 2015. Snow depth retrieval with UAS using photogrammetric techniques. *Geosciences* 5, 264–285. https://doi.org/10.3390/geosciences5030264

Jhan, J.-P., Rau, J.-Y., Huang, C.-Y., 2016. Band-to-band registration and ortho-rectification of multilens/multispectral imagery: A case study of MiniMCA-12 acquired by a fixed-wing UAS. *ISPRS J. Photogramm. Remote Sens.* 114, 66–77. https://doi.org/10.1016/j.isprsjprs.2016.01.008

Jimenez-Berni, J.A., Deery, D.M., Rozas-Larraondo, P., Condon, A.T.G., Rebetzke, G.J., James, R.A., Bovill, W.D., Furbank, R.T., Sirault, X.R.R., 2018. High throughput determination of plant height, ground cover, and above-ground biomass in wheat with LiDAR. *Front. Plant Sci.* 9. https://doi.org/10.3389/fpls.2018.00237

Jin, X., Liu, S., Baret, F., Hemerlé, M., Comar, A., 2017. Estimates of plant density of wheat crops at emergence from very low altitude UAV imagery. *Remote Sens. Environ.* 198, 105–114. https://doi.org/10.1016/j.rse.2017.06.007

Kirchgessner, N., Liebisch, F., Yu, K., Pfeifer, J., Friedli, M., Hund, A., Walter, A., 2017. The ETH field phenotyping platform FIP: A cable-suspended multi-sensor system. *Funct. Plant Biol.* 44, 154. https://doi.org/10.1071/FP16165

Koppe, W., Gnyp, M.L., Hennig, S.D., Li, F., Miao, Y., Chen, X., Jia, L., Bareth, G., 2012. Multi-temporal hyperspectral and radar remote sensing for estimating winter wheat biomass in the North China Plain. *Photogramm. - Fernerkund. - Geoinformation* 2012, 281–298. https://doi.org/10.1127/1432-8364/2012/0117

Koppe, W., Li, F., Gnyp, M.L., Miao, Y., Jia, L., Chen, X., Zhang, F., Bareth, G., 2010. Evaluating multispectral and hyperspectral satellite remote sensing data for estimating winter wheat growth parameters at regional scale in the North China Plain. *Photogramm. - Fernerkund. - Geoinformation* 2010, 167–178. https://doi.org/10.1127/1432-8364/2010/0047

Kronenberg, L., Yu, K., Walter, A., Hund, A., 2017. Monitoring the dynamics of wheat stem elongation: Genotypes differ at critical stages. *Euphytica* 213. https://doi.org/10.1007/s10681-017-1940-2

Kuska, M.T., Brugger, A., Thomas, S., Wahabzada, M., Kersting, K., Oerke, E.-C., Steiner, U., Mahlein, A.-K., 2017. Spectral patterns reveal early resistance reactions of barley against Blumeria graminis f. sp. hordei. *Phytopathology* 107, 1388–1398. https://doi.org/10.1094/PHYTO-04-17-0128-R

Laliberte, A.S., Goforth, M.A., Steele, C.M., Rango, A., 2011. Multispectral remote Sensing from Unmanned Aircraft: Image Processing Workflows and Applications for Rangeland Environments. *Remote Sens.* 3, 2529–2551. https://doi.org/10.3390/rs3112529

Lemaire, G., Gastal, F., 1997. N uptake and distribution in plant canopies. in: Lemaire, G. (Ed.), *Diagnosis of the Nitrogen Status in Crops*. Springer Berlin Heidelberg, pp. 3–44. https://doi.org/10.1007/978-3-642-60684-7

Lemaire, G., Meynart, J., 1997. Use of the nitrogen nutrition index for the analysis of agronomic data. in: Lemaire, G. (Ed.), *Diagnosis of the Nitrogen Status in Crops*. Springer Berlin Heidelberg, pp. 45–56. https://doi.org/10.1007/978-3-642-60684-7

Li, F., Gnyp, M.L., Jia, L., Miao, Y., Yu, Z., Koppe, W., Bareth, G., Chen, X., Zhang, F., 2008. Estimating N status of winter wheat using a handheld spectrometer in the North China Plain. *Field Crops Res.* 106, 77–85. https://doi.org/10.1016/j.fcr.2007.11.001

Li, F., Miao, Y., Hennig, S.D., Gnyp, M.L., Chen, X., Jia, L., Bareth, G., 2010. Evaluating hyperspectral vegetation indices for estimating nitrogen concentration of winter wheat at different growth stages. *Precis. Agric.* 11, 335–357. https://doi.org/10.1007/s11119-010-9165-6

Lopez-Sanchez, J.M., Vicente-Guijalba, F., Erten, E., Campos-Taberner, M., Garcia-Haro, F.J., 2017. Retrieval of vegetation height in rice fields using polarimetric SAR interferometry with TanDEM-X data. *Remote Sens. Environ.* 192, 30–44. https://doi.org/10.1016/j.rse.2017.02.004

Lowe, D.G., 2004. Distinctive image features from scale-invariant keypoints. *Int. J. Comput. Vis.* 60, 91–110. https://doi.org/10.1023/B:VISI.0000029664.99615.94

Lu, D., 2006. The potential and challenge of remote sensing-based biomass estimation. *Int. J. Remote Sens.* 27, 1297–1328. https://doi.org/10.1080/01431160500486732

Lucieer, A., Malenovský, Z., Veness, T., Wallace, L., 2014a. HyperUAS imaging spectroscopy from a multirotor unmanned aircraft system: HyperUAS-imaging spectroscopy from a multirotor unmanned. *J. Field Robot.* 31, 571–590. https://doi.org/10.1002/rob.21508

Lucieer, A., Turner, D., King, D.H., Robinson, S.A., 2014b. Using an Unmanned Aerial Vehicle (UAV) to capture micro-topography of Antarctic moss beds. *Int. J. Appl. Earth Obs. Geoinformation* 27, 53–62. https://doi.org/10.1016/j.jag.2013.05.011

Luhmann, T., Robson, S., Kyle, S., Boehm, J., 2014. *Close-Range Photogrammetry and 3D Imaging*, 2nd edition. ed, De Gruyter textbook. De Gruyter, Berlin.

Lukas, V., Novák, J., Neudert, L., Svobodova, I., Rodriguez-Moreno, F., Edrees, M., Kren, J., 2016. The combination of UAV survey and landsat imagery for monitoring of crop vigor in precision agriculture, *ISPRS - International Archives of the Photogrammetry, Remote Sensing and Spatial Information Sciences*. pp. 953–957. https://doi.org/10.5194/isprs-archives-XLI-B8-953-2016

Madec, S., Baret, F., de Solan, B., Thomas, S., Dutartre, D., Jezequel, S., Hemmerlé, M., Colombeau, G., Comar, A., 2017. High-throughput phenotyping of plant height: Comparing unmanned aerial vehicles and ground LiDAR estimates. *Front. Plant Sci.* 8. https://doi.org/10.3389/fpls.2017.02002

Maimaitijiang, M., Ghulam, A., Sidike, P., Hartling, S., Maimaitiyiming, M., Peterson, K., Shavers, E. et al. 2017. Unmanned Aerial System (UAS)-based phenotyping of soybean using multi-sensor data fusion and extreme learning machine. *ISPRS J. Photogramm. Remote Sens.* 134, 43–58. https://doi.org/10.1016/j.isprsjprs.2017.10.011

Malambo, L., Popescu, S.C., Murray, S.C., Putman, E., Pugh, N.A., Horne, D.W., Richardson, G. et al. 2018. Multitemporal field-based plant height estimation using 3D point clouds generated from small unmanned aerial systems high-resolution imagery. *Int. J. Appl. Earth Obs. Geoinformation* 64, 31–42. https://doi.org/10.1016/j.jag.2017.08.014

Malenovský, Z., Lucieer, A., King, D.H., Turnbull, J.D., Robinson, S.A., 2017. Unmanned aircraft system advances health mapping of fragile polar vegetation. *Methods Ecol. Evol.* 8, 1842–1857. https://doi.org/10.1111/2041-210X.12833

Marshall, M., Thenkabail, P., 2015. Developing in situ non-destructive estimates of crop biomass to address issues of scale in remote sensing. *Remote Sens.* 7, 808–835. https://doi.org/10.3390/rs70100808

Metternicht, G., 2003. Vegetation indices derived from high-resolution airborne videography for precision crop management. *Int. J. Remote Sens.* 24, 2855–2877. https://doi.org/10.1080/01431160210163074

Meyer, G.E., Mehta, T., Kocher, M.F., Mortensen, D.A., Samal, A., 1998. Textural imaging and discriminant analysis for distinguishing weeds for spot spraying. *Trans. ASAE* 41, 1189–1197. https://doi.org/10.13031/2013.17244

MicaSense, 2017a. Parrot Sequoia [WWW Document]. MicaSense. URL https://www.micasense.com/parrotsequoia/ (accessed 12.14.17).

MicaSense, 2017b. RedEdge-M [WWW Document]. MicaSense. URL https://www.micasense.com/rededge-m/ (accessed 12.14.17).

Milton, E.J., Schaepman, M.E., Anderson, K., Kneubühler, M., Fox, N., 2009. Progress in field spectroscopy. *Remote Sens. Environ.* 113, S92–S109. https://doi.org/10.1016/j.rse.2007.08.001

Mistele, B., Schmidhalter, U., 2008. Estimating the nitrogen nutrition index using spectral canopy reflectance measurements. *Eur. J. Agron.* 29, 184–190. https://doi.org/10.1016/j.eja.2008.05.007

Miyoshi, G.T., Imai, N.N., Tommaselli, A.M.G., Honkavaara, E., Näsi, R., Moriya, É.A.S., 2018. Radiometric block adjustment of hyperspectral image blocks in the Brazilian environment. *Int. J. Remote Sens.* 1–21. https://doi.org/10.1080/01431161.2018.1425570

Mulla, D.J., 2013. Twenty five years of remote sensing in precision agriculture: Key advances and remaining knowledge gaps. *Biosyst. Eng.* 114, 358–371. https://doi.org/10.1016/j.biosystemseng.2012.08.009

Mutanga, O., Skidmore, A.K., 2004. Narrow band vegetation indices overcome the saturation problem in biomass estimation. *Int. J. Remote Sens.* 25, 3999–4014. https://doi.org/10.1080/01431160310001654923

Näsi, R., Honkavaara, E., Blomqvist, M., Lyytikäinen-Saarenmaa, P., Hakala, T., Viljanen, N., Kantola, T., Holopainen, M., 2018. Remote sensing of bark beetle damage in urban forests at individual tree level using a novel hyperspectral camera from UAV and aircraft. *Urban For. Urban Green.* 30, 72–83. https://doi.org/10.1016/j.ufug.2018.01.010

Näsi, R., Honkavaara, E., Lyytikäinen-Saarenmaa, P., Blomqvist, M., Litkey, P., Hakala, T., Viljanen, N., Kantola, T., Tanhuanpää, T., Holopainen, M., 2015. Using UAV-Based photogrammetry and hyperspectral imaging for mapping bark beetle damage at tree-level. *Remote Sens.* 7, 15467–15493. https://doi.org/10.3390/rs71115467

Nevalainen, O., Honkavaara, E., Tuominen, S., Viljanen, N., Hakala, T., Yu, X., Hyyppä, J. et al. 2017. Individual tree detection and classification with UAV-based photogrammetric point clouds and hyperspectral imaging. *Remote Sens.* 9, 185. https://doi.org/10.3390/rs9030185

Omasa, K., Hosoi, F., Konishi, A., 2007. 3D lidar imaging for detecting and understanding plant responses and canopy structure. *J. Exp. Bot.* 58, 881–898. https://doi.org/10.1093/jxb/erl142

Pajares, G., 2015. Overview and current status of remote sensing applications based on unmanned aerial vehicles (UAVs). *Photogramm. Eng. Remote Sens.* 81, 281–330. https://doi.org/10.14358/PERS.81.4.281

photon focus AG, 2017. Hyperspectral cameras [WWW Document]. URL http://www.photonfocus.com/de/produkte/kamerafinder/?no_cache=1&cid=9&pfid=2 (accessed 10.11.17).

Pittman, J., Arnall, D., Interrante, S., Moffet, C., Butler, T., 2015. Estimation of biomass and canopy height in bermudagrass, alfalfa, and wheat using ultrasonic, laser, and spectral sensors. *Sensors* 15, 2920–2943. https://doi.org/10.3390/s150202920

Pohl, C., Van Genderen, J.L., 1998. Multisensor image fusion in remote sensing: Concepts, methods and applications. *Int. J. Remote Sens.* 19, 823–854. https://doi.org/10.1080/014311698215748

Qi, W., Dubayah, R.O., 2016. Combining Tandem-X InSAR and simulated GEDI lidar observations for forest structure mapping. *Remote Sens. Environ.* 187, 253–266. https://doi.org/10.1016/j.rse.2016.10.018

Quemada, M., Gabriel, J., Zarco-Tejada, P., 2014. Airborne hyperspectral images and ground-level optical sensors as assessment tools for maize nitrogen fertilization. *Remote Sens.* 6, 2940–2962. https://doi.org/10.3390/rs6042940

Rasmussen, J., Ntakos, G., Nielsen, J., Svensgaard, J., Poulsen, R.N., Christensen, S., 2016. Are vegetation indices derived from consumer-grade cameras mounted on UAVs sufficiently reliable for assessing experimental plots? *Eur. J. Agron.* 74, 75–92. https://doi.org/10.1016/j.eja.2015.11.026

Reed, D.C., Rassweiler, A., Arkema, K.K., 2008. Biomass rather than growth rate determines variation in net primary production by giant kelp. *Ecology* 89, 2493–2505. https://doi.org/10.1890/07-1106.1

Remondino, F., Spera, M.G., Nocerino, E., Menna, F., Nex, F., 2014. State of the art in high density image matching. *Photogramm. Rec.* 29, 144–166. https://doi.org/10.1111/phor.12063

Roth, L., Aasen, H., Walter, A., Liebisch, F., 2018. Extracting leaf area index using viewing geometry effects—A new perspective on high-resolution unmanned aerial system photography. *ISPRS J. Photogramm. Remote Sens.* 141, 161–175. https://doi.org/10.1016/j.isprsjprs.2018.04.012

Roth, L., Streit, B., 2017. Predicting cover crop biomass by lightweight UAS-based RGB and NIR photography: An applied photogrammetric approach. *Precis. Agric.* https://doi.org/10.1007/s11119-017-9501-1

Saarinen, N., Vastaranta, M., Näsi, R., Rosnell, T., Hakala, T., Honkavaara, E., Wulder, M. et al. 2018. Assessing Biodiversity in Boreal Forests with UAV-Based Photogrammetric Point Clouds and Hyperspectral Imaging. *Remote Sens.* 10, 338. https://doi.org/10.3390/rs10020338

Salamí, E., Barrado, C., Pastor, E., 2014. UAV flight experiments applied to the remote sensing of vegetated areas. *Remote Sens.* 6, 11051–11081. https://doi.org/10.3390/rs61111051

Sanchez-Azofeifa, A., Antonio Guzmán, J., Campos, C.A., Castro, S., Garcia-Millan, V., Nightingale, J., Rankine, C., 2017. Twenty-first century remote sensing technologies are revolutionizing the study of tropical forests. *Biotropica* 49, 604–619. https://doi.org/10.1111/btp.12454

Sankey, T., Donager, J., McVay, J., Sankey, J.B., 2017. UAV lidar and hyperspectral fusion for forest monitoring in the southwestern USA. *Remote Sens. Environ.* 195, 30–43. https://doi.org/10.1016/j.rse.2017.04.007

Schaefer, M., Lamb, D., 2016. A Combination of Plant NDVI and LiDAR Measurements Improve the Estimation of Pasture Biomass in Tall Fescue (Festuca arundinacea var. Fletcher). *Remote Sens.* 8, 10. https://doi.org/10.3390/rs8020109

Schaepman, M.E., Jehle, M., Hueni, A., D'Odorico, P., Damm, A., Weyermann, J., Schneider, F.D. et al. 2015. Advanced radiometry measurements and Earth science applications with the Airborne Prism Experiment (APEX). *Remote Sens. Environ.* 158, 207–219. https://doi.org/10.1016/j.rse.2014.11.014

SENOP, 2017. Optronics Hyperspectral [WWW Document]. URL http://senop.fi/en/optronics-hyperspectral (accessed 10.11.17).

Skovsgaard, J.P., Vanclay, J.K., 2008. Forest site productivity: A review of the evolution of dendrometric concepts for even-aged stands. *Forestry* 81, 13–31. https://doi.org/10.1093/forestry/cpm041

Snavely, N., 2016. Bundler: Structure from Motion (SfM) for Unordered Image Collections [WWW Document]. Bundler - Struct. Motion SfM Unordered Image Collect. URL http://www.cs.cornell.edu/~snavely/bundler/ (accessed 4.3.16).

Snavely, N., Seitz, S.M., Szeliski, R., 2008. Modeling the world from internet photo collections. *Int. J. Comput. Vis.* 80, 189–210. https://doi.org/10.1007/s11263-007-0107-3

Stagakis, S., González-Dugo, V., Cid, P., Guillén-Climent, M.L., Zarco-Tejada, P.J., 2012. Monitoring water stress and fruit quality in an orange orchard under regulated deficit irrigation using narrow-band structural and physiological remote sensing indices. *ISPRS J. Photogramm. Remote Sens.* 71, 47–61. https://doi.org/10.1016/j.isprsjprs.2012.05.003

Stavros, E.N., Schimel, D., Pavlick, R., Serbin, S., Swann, A., Duncanson, L., Fisher, J.B. et al. 2017. ISS observations offer insights into plant function. *Nat. Ecol. Evol.* 1, 0194. https://doi.org/10.1038/s41559-017-0194

Suomalainen, J., Anders, N., Iqbal, S., Roerink, G., Franke, J., Wenting, P., Hünniger, D., Bartholomeus, H., Becker, R., Kooistra, L., 2014. A lightweight hyperspectral mapping system and photogrammetric processing chain for unmanned aerial vehicles. *Remote Sens.* 6, 11013–11030. https://doi.org/10.3390/rs61111013

Swatantran, A., Dubayah, R., Roberts, D., Hofton, M., Blair, J.B., 2011. Mapping biomass and stress in the Sierra Nevada using lidar and hyperspectral data fusion. *Remote Sens. Environ.* 115, 2917–2930. https://doi.org/10.1016/j.rse.2010.08.027

Tanhuanpää, T., Vastaranta, M., Kankare, V., Holopainen, M., Hyyppä, J., Hyyppä, H., Alho, P., Raisio, J., 2014. Mapping of urban roadside trees – A case study in the tree register update process in Helsinki City. *Urban For. Urban Green.* 13, 562–570. https://doi.org/10.1016/j.ufug.2014.03.005

TETRACAM Inc, 2015. Tetracam's Miniature Multiple Camera Array [WWW Document]. Tetracam Mini-MCA. URL http://www.tetracam.com/Products-Mini_MCA.htm (accessed 4.16.16).

Thenkabail, P.S., Lyon, J.G., Huete, A. (Eds.), 2012. *Hyperspectral Remote Sensing of Vegetation*. CRC Press, Boca Raton, FL.

Tian, J., Wang, L., Li, X., Gong, H., Shi, C., Zhong, R., Liu, X., 2017. Comparison of UAV and WorldView-2 imagery for mapping leaf area index of mangrove forest. *Int. J. Appl. Earth Obs. Geoinformation* 61, 22–31. https://doi.org/10.1016/j.jag.2017.05.002

Tilly, N., Aasen, H., Bareth, G., 2015a. Fusion of plant height and vegetation indices for the estimation of barley biomass. *Remote Sens.* 7, 11449–11480. https://doi.org/10.3390/rs70911449

Tilly, N., Hoffmeister, D., Cao, Q., Huang, S., Lenz-Wiedemann, V., Miao, Y., Bareth, G., 2014. Multitemporal crop surface models: Accurate plant height measurement and biomass estimation with terrestrial laser scanning in paddy rice. *J. Appl. Remote Sens.* 8, 083671. https://doi.org/10.1117/1.JRS.8.083671

Tilly, N., Hoffmeister, D., Cao, Q., Lenz-Wiedemann, V., Miao, Y., Bareth, G., 2015b. Transferability of models for estimating paddy rice biomass from spatial plant height data. *Agriculture* 5, 538–560. https://doi.org/10.3390/agriculture5030538

Torres-Sánchez, J., López-Granados, F., De Castro, A.I., Peña-Barragán, J.M., 2013. Configuration and specifications of an unmanned aerial vehicle (UAV) for early site specific weed management. *PLoS ONE* 8, e58210. https://doi.org/10.1371/journal.pone.0058210

Toth, C., Jóźków, G., 2016. Remote sensing platforms and sensors: A survey. *ISPRS J. Photogramm. Remote Sens.* 115, 22–36. https://doi.org/10.1016/j.isprsjprs.2015.10.004

Triggs, B., McLauchlan, P.F., Hartley, R.I., Fitzgibbon, A.W., 2000. Bundle adjustment — A modern synthesis, in: Triggs, B., Zisserman, A., Szeliski, R. (Eds.), *Vision Algorithms: Theory and Practice*. Springer Berlin Heidelberg, Berlin, Heidelberg, pp. 298–372.

Turner, D., Lucieer, A., Wallace, L., 2014. Direct georeferencing of ultrahigh-resolution UAV imagery. *IEEE Trans. Geosci. Remote Sens.* 52, 2738–2745. https://doi.org/10.1109/TGRS.2013.2265295

Ullman, S., 1979. The interpretation of structure from motion. *Proc. R. Soc. B Biol. Sci.* 203, 405–426. https://doi.org/10.1098/rspb.1979.0006

Ungar, S.G., Pearlman, J.S., Mendenhall, J.A., Reuter, D., 2003. Overview of the earth observing one (eo-1) mission. *IEEE Trans. Geosci. Remote Sens.* 41, 1149–1159. https://doi.org/10.1109/TGRS.2003.815999

van der Meij, B., Kooistra, L., Suomalainen, J., Barel, J.M., De Deyn, G.B., 2017. Remote sensing of plant trait responses to field-based plant–soil feedback using UAV-based optical sensors. *Biogeosciences* 14, 733–749. https://doi.org/10.5194/bg-14-733-2017

van Leeuwen, M., Hilker, T., Coops, N.C., Frazer, G., Wulder, M.A., Newnham, G.J., Culvenor, D.S., 2011. Assessment of standing wood and fiber quality using ground and airborne laser scanning: A review. *For. Ecol. Manag.* 261, 1467–1478. https://doi.org/10.1016/j.foreco.2011.01.032

Virlet, N., Sabermanesh, K., Sadeghi-Tehran, P., Hawkesford, M.J., 2017. Field Scanalyzer: An automated robotic field phenotyping platform for detailed crop monitoring. *Funct. Plant Biol.* 44, 143. https://doi.org/10.1071/FP16163

Vivoni, E.R., Rango, A., Anderson, C.A., Pierini, N.A., Schreiner-McGraw, A.P., Saripalli, S., Laliberte, A.S., 2014. Ecohydrology with unmanned aerial vehicles. *Ecosphere* 5, art130. https://doi.org/10.1890/ES14-00217.1

von Bueren, S.K., Burkart, A., Hueni, A., Rascher, U., Tuohy, M.P., Yule, I.J., 2015. Deploying four optical UAV-based sensors over grassland: Challenges and limitations. *Biogeosciences* 12, 163–175. https://doi.org/10.5194/bg-12-163-2015

Walter, A., Finger, R., Huber, R., Buchmann, N., 2017. Opinion: Smart farming is key to developing sustainable agriculture. *Proc. Natl. Acad. Sci.* 114, 6148–6150. https://doi.org/10.1073/pnas.1707462114

Walter, A., Liebisch, F., Hund, A., 2015. Plant phenotyping: From bean weighing to image analysis. *Plant Methods* 11, 14. https://doi.org/10.1186/s13007-015-0056-8

Warner, T.A., Cracknell, A.P., 2017. Unmanned aerial vehicles for environmental applications. *Int. J. Remote Sens.* 1–8. https://doi.org/10.1080/01431161.2017.1301705

Wehrhan, M., Rauneker, P., Sommer, M., 2016. UAV-based estimation of carbon exports from heterogeneous soil landscapes—A case study from the CarboZALF experimental area. *Sensors* 16, 255. https://doi.org/10.3390/s16020255

Wiedemann, A., 2001. Kombination von Laserscanner-Systemen und photogrammetrischen Methoden im Nahbereich. *Photogramm. Fernerkund. Geoinformation* 261–270.

Woebbcke, D.M., Meyer, G.E., Bargen, K., Mortensen, D.A., 1995. Color indices for weed identification under various soil, residue, and lighting conditions. *Trans. ASAE* 38, 259–269. https://doi.org/10.13031/2013.27838

XIMEA, 2018. XIMEA - Hyperspectral cameras based on USB3 - xiSpec [WWW Document]. URL https://www.ximea.com/en/products/xilab-application-specific-oem-custom/hyperspectral-cameras-based-on-usb3-xispec (accessed 1.23.18).

Yang, G., Liu, J., Zhao, C., Li, Z., Huang, Y., Yu, H., Xu, B. et al. 2017. Unmanned Aerial Vehicle Remote Sensing for Field-Based Crop Phenotyping: Current Status and Perspectives. *Front. Plant Sci.* 8. https://doi.org/10.3389/fpls.2017.01111

Yuan, H., Yang, G., Li, C., Wang, Y., Liu, J., Yu, H., Feng, H., Xu, B., Zhao, X., Yang, X., 2017. Retrieving soybean leaf area index from unmanned aerial vehicle hyperspectral remote sensing: Analysis of RF, ANN, and SVM regression models. *Remote Sens.* 9, 309. https://doi.org/10.3390/rs9040309

Yue, J., Yang, G., Li, C., Li, Z., Wang, Y., Feng, H., Xu, B., 2017. Estimation of winter wheat above-ground biomass using unmanned aerial vehicle-based snapshot hyperspectral sensor and crop height improved models. *Remote Sens.* 9, 708. https://doi.org/10.3390/rs9070708

Zarco-Tejada, P.J., 2008. A new era in remote sensing of crops with unmanned robots. *SPIE Newsroom*. https://doi.org/10.1117/2.1200812.1438

Zarco-Tejada, P.J., Berjon, A., Lopezlozano, R., Miller, J., Martin, P., Cachorro, V., Gonzalez, M., Defrutos, A., 2005. Assessing vineyard condition with hyperspectral indices: Leaf and canopy reflectance simulation in a row-structured discontinuous canopy. *Remote Sens. Environ.* 99, 271–287. https://doi.org/10.1016/j.rse.2005.09.002

Zarco-Tejada, P.J., Diaz-Varela, R., Angileri, V., Loudjani, P., 2014. Tree height quantification using very high resolution imagery acquired from an unmanned aerial vehicle (UAV) and automatic 3D photo-reconstruction methods. *Eur. J. Agron.* 55, 89–99. https://doi.org/10.1016/j.eja.2014.01.004

Zarco-Tejada, P.J., Guillén-Climent, M.L., Hernández-Clemente, R., Catalina, A., González, M.R., Martín, P., 2013. Estimating leaf carotenoid content in vineyards using high resolution hyperspectral imagery acquired from an unmanned aerial vehicle (UAV). *Agric. For. Meteorol.* 171–172, 281–294. https://doi.org/10.1016/j.agrformet.2012.12.013

Zhang, C., Kovacs, J.M., 2012. The application of small unmanned aerial systems for precision agriculture: A review. *Precis. Agric.* 13, 693–712. https://doi.org/10.1007/s11119-012-9274-5

5 Photosynthetic Efficiency and Vegetation Stress

Elizabeth M. Middleton, K. Fred Huemmrich,
Qingyuan Zhang, Petya K.E. Campbell, and David R. Landis

CONTENTS

5.1 INTRODUCTION

The term "spectral bioindicators" refers to a developing scientific field that uses remote sensing technologies to assess the health, physiology, and biodiversity of vegetation. In this chapter, we place emphasis on detecting and tracking responses to environmentally imposed stresses that affect photosynthetic efficiency, and relationships to photosynthesis, at the scale of canopies and ecosystems (Jones and Vaughan, 2010). We define "plant stress" as the condition of performing below expected, or optimal, levels for basic biological processes such as growth, maintenance, and reproduction—due to limiting or unfavorable environmental or physiological factors. Spectral bioindicators utilize narrow spectral bands that most typically can be obtained from hyperspectral data. The most useful narrow-band hyperspectral data (e.g., ≤10 nm spectral resolution) are obtained from the full spectrum, visible through short-wave infrared wavelengths. This is often referred to as the VSWIR spectrum and is used to detect and monitor plant physiological responses to stresses.

Indices derived from continuous and/or highly sampled reflectance or radiance spectra show us new information about plant physiological processes, such as those that are obtained from remote sensing observations at different scales, whether at the land surface or from aircraft or satellite sensors. These types of spectral indices were not available from broad-band satellite data (such as Landsat data) that have been collected for more than three decades. Historical datasets, primarily utilizing spectral vegetation indices such as the normalized difference vegetation index (NDVI), have been foundational in demonstrating the potential of satellite remote sensing for studying ecosystem processes (Rouse and Schell, 1973; Tucker, 1979; Prince, 1991; Tucker et al., 2005). Based on the reflectance characteristics of green vegetation, these satellite broad-band (≥20 nm) observations have made it possible to track interannual trends and to infer increased photosynthetic activity and longer growing seasons in northern hemisphere land regions (Goetz and Curtiss, 1996; Myneni et al., 1997b; Forkel, 2016). However, current and future narrow-band sensors and spectrometers give us potent new tools (e.g., Rascher et al., 2010a,b; Guanter et al., 2015b; Lee et al., 2015; Schaepman et al., 2015; Kerr et al., 2016; Pinto et al., 2016) to examine ecosystem function, health, composition, and biodiversity (Yu et al., 2014; Jetz et al., 2016; Alonso et al., 2017; Maimaitiyiming et al., 2017; Springer et al., 2017) remotely and from space, supplementing the existing capability for monitoring seasonality and phenology of terrestrial ecosystems. Furthermore, these new measurements push us beyond *inference* of photosynthetic activity into more direct observations closely related to *measurement* of photosynthetic function, with less reliance on supporting meteorological data for interpretation. This is critical to enable a global capability for detecting and monitoring the dynamics of our Earth's ecosystems in time and space. In the face of global climate change, increasing human populations, and decline in natural biodiversity, there is a pressing need to bring the best technologies to bear on monitoring our home planet, Earth (IPCC, 2014).

5.2 DESCRIBING AND MEASURING ECOSYSTEM PHOTOSYNTHESIS AND RELATED PROCESSES

(Terms used in this section are defined or explained in Table 5.1.)

Vegetation assemblages grow under a specific range of environmental conditions, characterized as canopies and biomes, thriving under optimal conditions. However, vegetation is subjected to "stress" when environmental conditions differ significantly from optimal, such as when drought or extreme (high/low) temperatures occur, or insufficient or excess nutrients are present (Mooney et al., 1991; Cech et al., 1998). When stress factors impair physiological function through biochemical mechanisms, this impacts the exchange of carbon, water, and energy between vegetation and the atmosphere. Since nonoptimal conditions commonly occur, assumptions based on optimal conditions are typically not representative or reasonable, either in interpreting observations or for model retrievals. Nevertheless, models typically either presume optimal conditions or include simplified assumptions about types of stresses and stress responses.

It is imperative to understand how, where, and when environmental stresses invoke limiting conditions on terrestrial ecosystems. Although there are multiple ways that plants manage their exposure to environmental stresses, we focus on the subset of physiological responses that affect photosynthetic function and efficiency that are captured in spectral responses. Consequently, we will discuss photosynthetic efficiency as it is connected to two related, and physiologically mediated, leaf energy dissipation pathways—heat loss via nonphotochemical quenching and chlorophyll fluorescence. We will also examine spectral bioindicators that capture these processes and related spectral parameters and methods, at observation scales from leaf to satellite.

TABLE 5.1
Definition of Terms and Parameters Used in Section 5.2

Parameter	Definition
APAR	Absorbed photosynthetically active radiation, the amount of photosynthetically active radiation absorbed by plants
Carbon source or sink	A component of the Earth's system (forests, soil, oceans, the atmosphere, and fossil fuels) that contains carbon. Movement of carbon into the system is categorized as a sink, while movement out of the system is considered a source
CLM	Community land model, a model widely used by the vegetation community to simulate the interaction between the atmosphere and the terrestrial biosphere
CO_2	Carbon dioxide
EC	Eddy covariance, a technique to measure and calculate vertical turbulent fluxes within the atmosphere
fAPAR	The fraction of incident PAR absorbed by plants
GEP	Gross ecosystem production (GEP), the carbon taken up by photosynthesis
GPP	Gross primary production, the carbon taken up by ecosystems through the process of photosynthesis
LUE	Light use efficiency, the overall ability of plants/canopies/ecosystems to convert solar energy into usable carbohydrates and biomass, the ratio of GEP to the APAR
NEP	Net ecosystem production (NEP), the net carbon flux of the ecosystem, the balance between the carbon taken up by photosynthesis (GEP) vs. that released due to metabolic activity by microbial and plant cells due to respiration (R_{eco})
PAR	Photosynthetically active radiation, radiation between 400 and 700 nm
R_{eco}	Ecosystem respiration, the carbon exiting the ecosystem due to metabolic activity by microbial and plant cells
RUE	Radiation use efficiency, same as LUE
VPD	Vapor pressure deficit, the difference (deficit) between the amount of moisture in the air and the maximum amount of moisture the air can hold when it is saturated

5.2.1 Relevance of Ecosystem Photosynthetic Processes to Climate Change

Detecting vegetation stress and photosynthetic function at synoptic landscape scales is important for assessing the physiological status of terrestrial ecosystems, in order to quantify impacts and uncertainties associated with climate change and human activities (Figure 5.1), and to predict impacts on future human societies (IPCC, 2014). Photosynthesis represents the largest carbon flux between the biosphere and the atmosphere and is of critical importance to atmospheric CO_2 concentrations and, thus, climate change (Bousquet et al., 2000; Nemani et al., 2003; Canadell et al., 2007; Piao et al., 2009; Bloom et al., 2016). Based on direct observations, gross global terrestrial photosynthesis has been estimated to be 123 ± 8 Pg C y^{-1} (1 Pg = petagram = 10^{15} g) (Denman et al., 2007; Beer et al., 2010; Piao et al., 2013; Le Quéré et al., 2015), while photosynthesis in oceans was estimated over a decade ago to be 93 ± 19 Pg C yr^{-1} (Falkowski et al., 1998; Keeling et al., 2005). As of January 2018, the atmospheric CO_2 concentration reported by the Scripps CO_2 Program was 408 ppm whereas it was ~315 ppm in 1958 (Scrippts, 2018), nearing a 100 ppm increase in only 60 years.

Present approaches for assessing ecosystem photosynthetic function and productivity depend on models driven by meteorological data to predict vegetation responses to environmental conditions. Uncertainties are introduced to model predictions when influencing factors are not well described (e.g., nitrogen deposition) or known (e.g., photosynthetic efficiency) (Baker et al., 2008), and because responses of terrestrial systems to climate–ecosystem feedback factors are not well understood. Spectral bioindicators observe wavelength-related radiance or reflectance (or apparent reflectance) changes that are directly related to vegetation physiological status and stress responses, thus providing an independent check on the models.

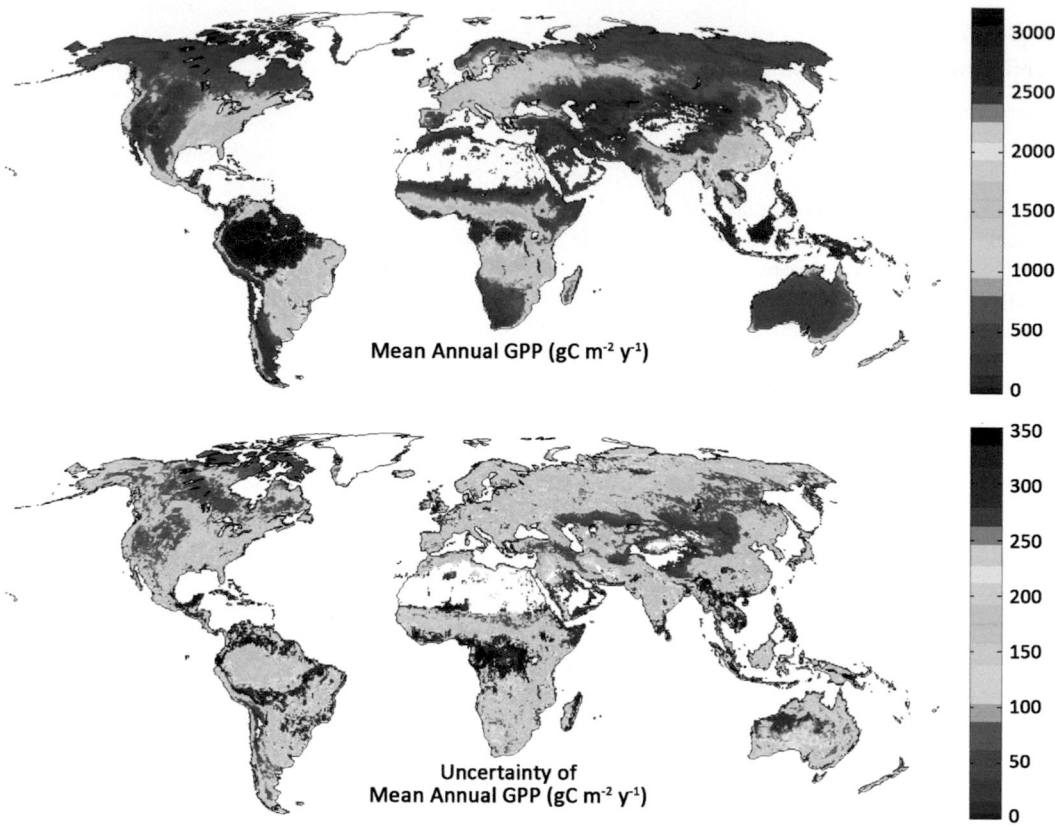

FIGURE 5.1 The top map shows the estimated global mean annual gross primary productivity (GPP, gC m^{-2} y^{-1}) from 2012. The lower map shows the uncertainty of the mean values, >10% in some regions. (From IPCC. 2014. In: Core Writing Team, R.K. Pachauri and L.A. Meyer (eds.). *Contribution of Working Groups I, II and III to the Fifth Assessment Report of the Intergovernmental Panel on Climate Change.* Geneva, Switzerland: IPCC.)

5.2.2 Ecosystem Photosynthesis, Photosynthetic Efficiency, and Plant Stress

Photosynthesis in plants is a fundamental biochemical process that occurs within chloroplasts, primarily in cells of leaf mesophyll tissue (Larcher et al., 2003) but sometimes also in stem and bark tissue (Kharouk et al., 1995). Photosynthesis requires light energy, water, nutrients, moderate temperatures, and CO_2. During photosynthesis, CO_2 is removed from the atmosphere and O_2 is released back; CO_2 is also returned as a by-product of respiratory processes by plants or soil microbes. With photosynthesis, CO_2 is fixed into carbohydrates through a complex series of enzyme-regulated steps to provide metabolic energy that fuels biochemical processes for growth, maintenance, reproduction, and biomass accumulation (Morton, 2009; Wohlfahrt and Gu, 2015). An ecosystem is referred to as a net carbon "sink" if the CO_2 influx from photosynthesis exceeds the efflux from respiration, but is referred to as a net carbon "source" if more carbon is lost than is gained (Goulden et al., 2011). In the aggregate across terrestrial ecosystems, small-scale physiological processes within plants can affect CO_2 exchange, thus potentially having large-scale global impacts on the Earth's climate, atmosphere, and biogeochemical cycles (Beer et al., 2010; Jung and Reichstein, 2010; Piao et al., 2013; IPCC, 2014).

The photosynthetic efficiency of terrestrial ecosystems, and associated ecosystem services (e.g., carbon sequestration, climate regulation, hydrological function, and wildlife habitat) (Costanza et al.,

1997; De Groot et al., 2002), can be negatively impacted by stress factors such as climate change (e.g., those related to temperature, irradiance levels, and atmospheric CO_2 concentrations), land use change, chemical pollution, biodiversity loss, and an overabundance or imbalance of nitrogen, phosphorus, and other essential minerals. Rockström et al. (2009) made a first attempt to define the boundary conditions beyond which various stress factors could push terrestrial ecosystems to the point where significant negative impacts accrue on ecosystems and human well-being. While the response to these stress factors can be abrupt and irreversible, the responses can also be gradual and thus more difficult to detect in their initial stages. Rockström has been actively advocating for technology interventions to reduce CO_2 levels (Rockström et al., 2017) and he and colleagues (Figueres et al., 2017) insist this must be initiated before 2020. New methods based on space observations will surely be critical for tracking the state of ecosystems globally in the face of predicted climate changes and potential recovery.

At the terrestrial ecosystem and landscape scales, three primary approaches have been utilized to examine the dynamics of photosynthesis: measurements from instrumented towers, remote sensing observations from various platforms, and process models. By any of these methods, it should be borne in mind that ecosystem health and physiological status are evaluated *indirectly*, and inferred based on responses over time.

5.2.3 How Do We Measure Canopy/Ecosystem Photosynthesis?

The eddy covariance (EC) flux tower technique, developed in the 1980s–1990s, has been adopted for networks around the world to measure the net exchange of carbon, water, and energy between an ecosystem and the atmosphere. This offers a powerful means for obtaining near-continuous time series of ecosystem-level photosynthesis over multiple years (Baldocchi et al., 2001; FLUXNET, 2015). Hundreds of EC flux towers are now distributed around the world, collecting data in every major biome (http://fluxnet.fluxdata.org/sites/site-summary/). Data from EC tower sites supply direct, 30-minute measurements of the net CO_2 flux for ecosystems, representing a large range of different climatic zones, plant functional types, and ecological disturbances (Baldocchi et al., 2001; Baldocchi, 2008; Beer et al., 2010; FLUXNET, 2015). The net flux is determined from the covariance of the vertical wind speed and the CO_2 concentration, measured using sonic anemometers and infrared gas analyzers, respectively. For towers of different heights, the measured flux footprint typically averages between 0.5 and 1.0 km^2 (Kljun et al., 2004) over a period of several weeks or months, varying with wind direction, turbulence, and surface characteristics. To interpret and analyze the net flux, it is important to make precise simultaneous meteorological measurements for air and soil temperatures, soil moisture, atmospheric vapor pressure deficit (VPD), and precipitation (Falge et al., 2001; Desai et al., 2008). The net CO_2 flux is referred to as net ecosystem productivity (NEP), which can be partitioned into two streams: gross ecosystem productivity (GEP) or gross primary production (GPP)—the carbon taken up by photosynthesis; and ecosystem respiration (R_{eco})—the carbon flux exiting the ecosystem due to metabolic activity by microbial and plant cells (Reichstein et al., 2005; Mahecha et al., 2010).

Recently, there has been increasing interest in the use of spectral bioindicators to help make predictions about biological processes with direct measurements, and to provide inputs to dynamic models for large geographical areas so as to gain understanding of those underlying processes. Flux towers provide the data with which to develop, calibrate, and, especially, validate either spectral bioindicator measurements or predictions from dynamic models driven by them.

5.2.4 The Photosynthetic Light Use Efficiency Model

A widely used approach for modeling canopy or ecosystem photosynthesis is based on the simple photosynthetic light use efficiency (LUE) concept, also referred to as radiation use efficiency (RUE). An LUE model, rather than modeling complex biochemical processes that depend on ancillary and meteorological information, attempts to describe the overall ability of plants/canopies/ecosystems to convert solar energy into usable carbohydrates and biomass (Monteith, 1972, 1977; Kumar and Monteith, 1981). This LUE model assumes that gross maximum photosynthesis (under ideal conditions) is proportional to the available photosynthetically active radiation (PAR, $\sim$400–700 nm) that is absorbed by vegetation (or APAR), adjusted by an efficiency term (ε). Typically, the efficiency term attempts to account for different vegetation types as well as all nonoptimal or limiting environmental stress factors (e.g., air and soil temperature, soil moisture, nutrient availability, VPD, etc.) that affect the absorption and utilization of PAR. Therefore, the basic LUE model describes GEP in terms of the physiologically active APAR fraction (fAPAR) and a photosynthetic efficiency term (Monteith, 1977):

$$\int \text{GEP}(t)\, dt = \int \varepsilon(t) * \text{fAPAR}(t) * Q_{in}(t)\, dt \qquad (5.1)$$

where Q_{in} is the incident PAR, APAR is the product of fAPAR and $Q_{in,}$ and ε is the LUE parameter at time t. The GEP accumulated over time, as for an hour, a day, or a growing season, represents the integral of the instantaneous GEP.

As originally conceived, the LUE model was used for crop dry biomass production at harvest (Monteith, 1977) and for wheat yields (Kumar and Monteith, 1981), thus the start and end times for the integration in Equation 5.1 were the start and end of the growing season. Such determinations for natural ecosystems (e.g., forests) cannot practically be done using the agriculture harvest approach, but the approach can be applied to ecosystems for any defined time period (Russell et al., 1989). However, a noninvasive method is needed, and with the advent of instrumented towers to measure carbon exchange, ecosystem LUE can be determined from EC tower-based measurements through incorporation of measured fluxes such as NEP and meteorological information into models. It should be realized that net ecosystem CO_2 and LUE determinations from a tower include not only the amount of carbon fixed by the overstory foliage, but also the net storage by understory plants and the respiratory contributions from nonphotosynthetic material (i.e., bark, standing dead foliage), litter fall, and soil components. Techniques have been developed by the EC flux community to partition the measured net CO_2 flux into ecosystem respiration (the emission of CO_2 from the ecosystem) and GPP (e.g., Baldocchi, 2003, 2008; Reichstein et al., 2005).

LUE is computed as the ratio of GPP/APAR from EC towers or can be estimated with spectral bioindicators at chosen time periods (e.g., midday, at the time of satellite overpass observation) or as a temporal average. These two methods are compatible and provide a straightforward approach for describing ecosystem carbon uptake using both flux and spectral information. When the original LUE concept was formulated, it was suggested that the full growing season LUE could be a conservative value, not varying widely (Daughtry et al., 1992; Goetz and Prince, 1999). However, further field and remote sensing studies revealed that seasonally averaged LUE varies substantially among and within vegetation types (Figure 5.2), and in response to environmental conditions (e.g., Field, 1991; Gower et al., 1999; Anderson et al., 2000; Nichol et al., 2000; Green et al., 2003; Middleton et al., 2009a; Coops et al., 2010; Goerner et al., 2011; Cheng et al., 2014; Soudani et al., 2014; Gamon et al., 2016; Middleton et al., 2016). It has also been observed that LUE can be more variable over short time periods (e.g., within days), since it is affected by many factors: temperature, illumination intensity and quality, amount of diffuse light, atmospheric humidity, water availability, nutrient availability, plant type, plant age, disease, and soil type.

This new ability to track photosynthetic activity and LUE from a dense temporal dataset (using tower and/or remote sensing information) has enabled researchers to appreciate and quantify the daily, seasonal, and interannual LUE variability across ecosystems. Consequently, variations on the basic LUE model have been developed that incorporate environmental information. Typically, these draw upon values from a look-up table for maximum *unstressed* LUE (ε^*) or LUEmax (Heinsch et al., 2003), and apply modifying factors ranging between 0 and 1 that explicitly describe the

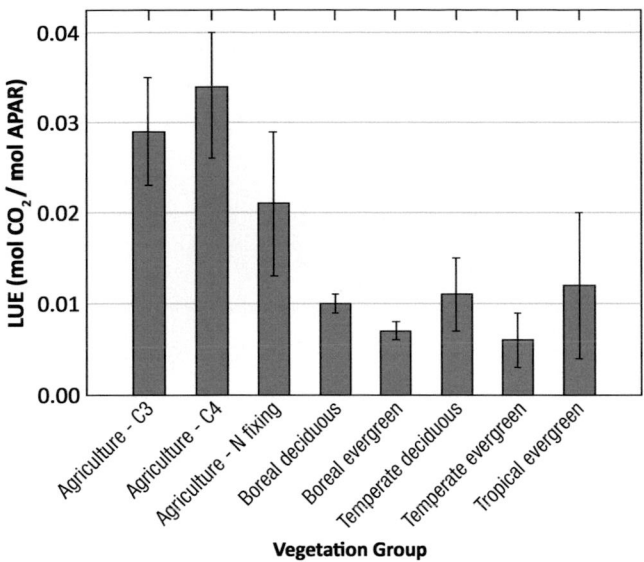

FIGURE 5.2 The variability in annual values for photosynthetic light use efficiency (LUE, mol CO_2/ mol APAR) is shown for a range of vegetation groups. Sample sizes vary, $n \geq 50$. (From Gower, S.T. et al., 1999. *Remote Sensing of Environment*, 70(1), 29–51; Anderson, M.C. et al., 2000. *Agricultural and Forest Meteorology*, 101(4), 265–289.)

expected effects of environmental variables such as soil water deficits, low temperatures, and VPD (Law and Waring, 1994; Prince and Goward, 1995; Matross et al., 2006). One such LUE based regional model is the vegetation photosynthesis and respiration model (VPRM, Mahadevan et al., 2008) that has been used to calculate GEP, and which can explain 60%–80% of hourly variability in fluxes at test sites. Matross et al. (2006) used atmospheric CO_2 concentrations in an inverse modeling framework to derive the LUE efficiency parameters in VPRM. Furthermore, Lin et al. (2011) demonstrated that biases can accumulate in estimating LUE from fluxes alone, potentially causing extremely large biases at seasonal to annual time scales.

LUE estimates based on EC tower data vary greatly within a day (due in part to measurement methods) and day to day, as shown for a grassland (Figure 5.3), but weekly average values are fairly stable. While average monthly values track midseason LUE, they overestimate or underestimate at early and late phenology periods. Further, in this grassland ecosystem, the assumed LUEmax rarely agreed with observations.

It is clear that additional means are required to constrain modeled estimates of LUE that are developed from a photosynthetic efficiency term, which rely on assumed maximum values and meteorological measurements that often represent the average for a large area rather than the local conditions to which a plant canopy/ecosystem is subjected. Furthermore, some variables affecting LUE, such as soil moisture, can have significant spatial variations and are difficult to measure over large areas. Developing approaches that use spectral information to directly determine LUE, such as using a composite efficiency term from remote sensing (Gamon et al., 2001), can reduce uncertainties in this key ecosystem descriptor as it varies over the landscape and over time (e.g., Lobell et al., 2002; Middleton et al., 2016; Springer et al., 2017).

One such approach draws on a model widely used by the vegetation community, the Community Land Model (CLM) developed by the National Center for Atmospheric Research (NCAR CLM4.5), which can utilize satellite spectral observations such as those available from the Moderate Resolution Spectroradiometer (MODIS) (Bonan et al., 2011; Hudiburg et al., 2013; Oleson et al., 2013).

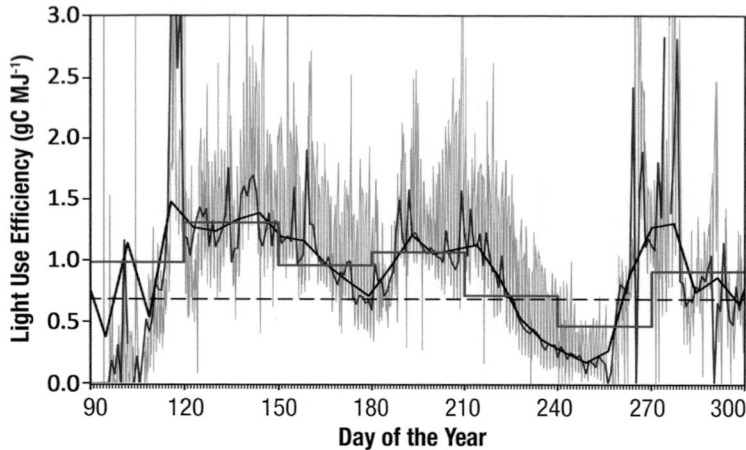

FIGURE 5.3 Variations in LUE at different temporal resolutions for a Shindler, OK, C4 grassland, from green-up in April through senescence in October in 1998. LUE calculated using GEP and incident PAR from flux tower, with green fPAR estimated from broad band NDVI using PAR and shortwave radiation sensors. The blue line is half hourly LUE during daylight periods, the red line is daily LUE, the solid black line is weekly LUE, and the green line is monthly LUE. Flux data from S. Verma (University of Nebraska). The dashed line is LUEmax, an assumed constant, maximum value for C4 agriculture (Heinsch et al., 2003). (From co-author K.F. Huemmrich.)

5.3 PHYSIOLOGICAL BASIS FOR SPECTRAL RESPONSES OF PHOTOSYNTHETIC FUNCTION AND STRESS

(Terms used in this section are defined or explained in Table 5.2.)

Plants often absorb more energy from the sun than they can use for photosynthesis. Plants continually adjust energy flow to support photosynthesis, while avoiding photo-oxidation under stressful conditions. Excess light is a special problem of overwintering evergreen plants (Öquist and Huner, 2003; Štroch et al., 2008). Various protection mechanisms have been developed so that plants can utilize sunlight while disposing of dangerous excess energy. These include biochemical mechanisms related to the light reactions of Photosystem II (PSII) and Photosystem I (PSI) and the production of screening pigments in the external epidermal leaf layers. Without these protective processes, excess energy could result in lethal photo-oxidations, endangering the organism (Demmig-Adams and Adams, 1998, 2000). In addition to these physiological mechanisms, plants have a number of other ways to adjust to environmental stress, such as altering canopy structural components (e.g., leaf angle distributions, leaf area, plant form, root depth), epidermal screening pigments, or phenology, but those will not be addressed here.

The photosynthetic process begins with APAR, or more specifically APAR$_{chl}$, the PAR absorbed by chlorophyll, as discussed below (Section 5.4.1). Energized chlorophyll molecules release photochemical energy as they return to the ground state in the light reactions associated with PSII and PSI. Excess energy is discarded from the PSII light reactions as heat via nonphotochemical quenching (NPQ), as chlorophyll fluorescence emissions (ChlF), or as metabolic heat. The remaining energy is passed along the electron transport chain to PSI and channeled to dark cycle processes for photochemistry leading to carbon fixation. Thus, there are four interrelated energy pathways for APAR$_{chl}$, but only one path that leads to photosynthesis, as diagrammed in Figure 5.4. A description of photosynthesis for the general reader can be found in Morton (2009) and Larcher et al. (2003).

Physical and chemical changes occur within leaves when they are subjected to stress conditions. The developing science of spectral bioindicators endeavors to extract information about the health and physiology of ecosystems, putting emphasis on spectroscopic techniques requiring continuous

TABLE 5.2
Definition of Terms and Parameters Used in Section 5.3

Parameter	Definition
$APAR_{chl}$	The APAR absorbed by chlorophyll
ChlF	Chlorophyll fluorescence emissions
Chlorophyll fluorescence	The passive emission of photons from the photosystems of foliage of absorbed APAR not utilized in photosythesis
Epoxidation/de-epoxidation	A reversible conformational change of the xanthophyll violaxanthin to zeaxanthin and back
Fluorescence	The property of absorbing light at shorter wavelengths (higher energy photons) and emitting light at longer wavelengths (lower energy photons)
NPQ	Nonphotochemical quenching, a mechanism to regulate and protect photosynthesis in environments where light energy absorption exceeds the capacity for light utilization in photosynthesis. It is a method of harmlessly dissipating excess excitation energy as heat through molecular vibrations
PPFD	Photosynthetic photon flux density, a measure of the number of photons in the 400–700 nm range that fall on a square meter of target area per second
PRI	Photochemical reflectance index, a normalized difference ratio of two narrow visible wavelength band reflectances, expressed as: $PRI = (\rho_{531} - \rho_{ref})/(\rho_{531} + \rho_{ref})$, where ρ_{531} is the reflectance at 531 nm, the band sensitive to the epoxidation state of the xanthophyll cycle pigments and ρ_{ref} is a second reference band
PSII and PSI	Photosystems II and I are where light reactions take place in plants. They consist of chlorophyll molecules, accessory pigments, proteins and other molecules that are involved in the release and transfer of excited electrons to the electron transport chain.
SIF	Solar-induced fluorescence, steady-state chlorophyll fluorescence driven by solar radiation
Xanthophyll cycle	This physical cycle consists of two reactions: de-epoxidation of the violaxanthin to the zeaxanthin and the reverse process of epoxidation
Xanthophylls	A category of carotenoid plant pigments that have a role in nonphotochemical quenching. Three types of xanthophyll pigments involved in NPQ are zeaxanthin (Z), antheraxanthin (A), and violaxanthin (V) where V→A→Z by the sequential biochemical removal of oxygen (de-epoxidation), or the reverse recovery process (epoxidation, Z→A→V).

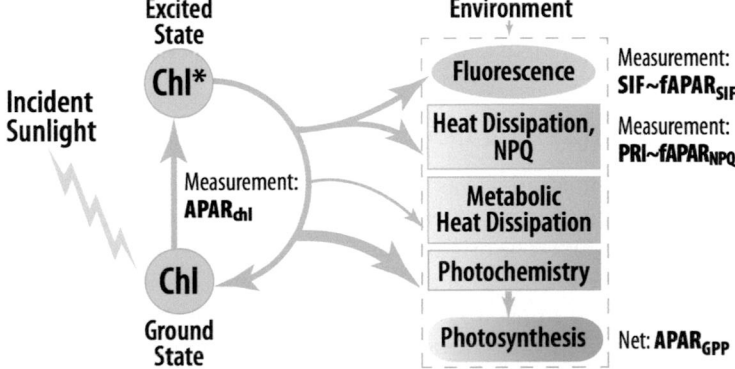

FIGURE 5.4 The fate of light energy captured by photosynthetic pigments, especially chlorophyll ($APAR_{chl}$) is shown. The excited chlorophyll (Chl) molecules release energy along several pathways. The primary pathway is toward photochemistry and photosynthesis (thick arrow) but energy is also discarded as fluorescence from PSII and PSI and as NPQ (PSII only). Sensible heat is a small dissipation path. The parameters that can be measured with remote sensing variables are indicated ($APAR_{chl}$, SIF, PRI, canopy temperature, and $APAR_{GPP}$). (From co-author E.M. Middleton.)

narrow spectral bands (i.e., spectrometers) or a collection of well-chosen physiologically responsive narrow bands (i.e., with hyperspectral sensors). Reflectance-based indices typically benefit from having spectral resolution between 1 and 10 nm, whereas fluorescence retrievals require much higher spectral resolution (<0.5 nm) and higher instrument fidelity. Spectroscopic techniques based on optical properties (reflectance, transmittance, absorptance) of leaf tissues portray light interaction properties of foliage constituents (Goetz and Curtiss, 1996). On the other hand, spectroscopic methods based on fluorescence emissions derive information primarily about function. This understanding of vegetation reflectance and fluorescence properties at the leaf level also applies to integrated values representing a canopy or ecosystem, and is the basis of successful remote sensing methods.

5.3.1 The Xanthophyll Pigment Cycle in Leaves

The xanthophyll cycle operates as a light- and pH-induced interconversion of specialized carotenoid pigments that ultimately sheds excess energy through heat dissipation (Pfündel and Dilley, 1993; Pfündel and Bilger, 1994; Gilmore and Yamasaki, 1998; Havaux and Niyogi, 1999; Müller et al., 2001). The xanthophyll-driven NPQ process is cyclic and reversible. Its primary function is the prevention of oxidative damage to the photosynthetic apparatus imbedded in the chloroplasts' thylakoid membranes (Bilger and Björkman, 1990; Demmig-Adams and Adams, 2000; Li et al., 2000; Demmig-Adams, 2003). This nonphotochemical energy reduction mechanism involves the interconversion of the xanthophylls violaxanthin (V) to zeaxanthin (Z) under alternating light and dark regimes, as described by Müller et al. (2001). The V form predominates in the dark, but during daylight it is converted first to antheraxanthin (A) and then to Z in high light (or in response to other stressors). This reversible biochemical cycle begins when sufficiently high light intensities induce a lower chloroplast lumen pH, initiating the sequential removal from V of double-bonded oxygen groups (de-epoxidation), first to form A and then to form Z. Z binds to PSII proteins causing a conformational change that releases excess excitation energy.

The expression of xanthophyll de-epoxidation state depends on lumen pH and the fraction of PSII antennae complexes bound to Z. The xanthophyll cycle is typically activated sometime between midmorning to midafternoon, depending on the combined strength of the environmental factors (high light and temperatures, drought). In the dark (overnight), Z reverts back to V through two epoxidation steps. This protective mechanism can be observed because the underlying physiological responses induce spectral changes at 531 nm, captured by a spectral bioindicator referred to as the photochemical reflectance index (PRI) (Section 5.3.2). Since A and Z have higher absorption coefficients than V at ~531 nm, reflectance decreases and a lower PRI value is produced when A and Z accumulate in the thylakoid membranes, as under high light conditions.

5.3.2 Photochemical Reflectance Index Measurements at Leaf Level

The use of the PRI for LUE monitoring was first documented at the leaf level by Gamon et al. (1990, 1992, 1997) and substantiated by many others (e.g., Filella et al., 1996; Peñuelas et al., 1997; Meroni et al., 2008b; Suárez et al., 2008). The original demonstration of the PRI and its relationship to LUE was made in sunflower leaves (Peñuelas et al., 1994; Gamon et al., 2001), linking the PRI response to the xanthophyll cycle's role in photosynthetic down-regulation and to photosynthetic function (Gamon et al., 2007) (Figure 5.5). Subsequently, several groups investigated the capability of the PRI under ambient conditions to track variations in plant physiological status and photosynthetic activities induced by various environmental stressors including excessive sunlight, water limitations, and nitrogen availability; soybean plants were studied under different soil moisture conditions (Peñuelas et al., 1994, 1995, 1997, 2011; Inoue and Peñuelas, 2006; Filella et al., 2009). From measurements of individual plants, Trotter and colleagues (2002) reported a linear relationship between PRI and LUE among eight different species that exhibited a range of physiological properties and nitrogen content. These early studies were based on *in situ* leaf or plant observations acquired using hand-held

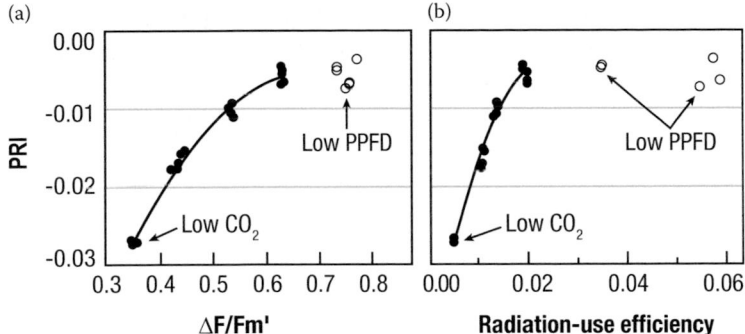

FIGURE 5.5 Measurements on leaves of *G. barbadense* (cotton) exposed to light and CO_2 treatments. PRI is plotted against (a) the photochemical efficiency, expressed by the relative change in chlorophyll fluorescence in the light, or $\Delta F/F_m'$ ($r^2 = 0.93$ for ●); and (b) photosynthetic radiation use efficiency (RUE or LUE, $r^2 = 0.94$ for ●). Low-light points (○; 0–100 μmol m^{-2} s^{-1}); medium to high-light points (●; 500–1500 μmol m^{-2} s^{-1}). (From Gamon, J.A. et al., 1997. *Oecologia*, 112(4), 492–501.)

spectroradiometers and were supported by other recent research (e.g., Atherton et al., 2016). Remote sensing examinations of the PRI have subsequently been made for ecosystems and landscapes and from instruments on towers, aircraft, and satellites, as discussed below (Section 5.4).

The PRI is a normalized difference spectral index, and its original formulation was based on two narrow (3–10 nm) wavelengths in the green spectrum to capture physiological responses (Gamon et al., 1992, 1997, 2001). The PRI expresses the relative down-regulation of photosynthesis induced primarily by high light intensities via the xanthophyll pigment cycle (Gamon et al., 2001), but it also is affected by secondary or compounding factors such as drought and senescence. The general formulation for the PRI is:

$$PRI_{REF\lambda} = [\rho_{531} - \rho_{REF\lambda}]/[\rho_{531} + \rho_{REF\lambda}] \tag{5.2}$$

where ρ_{531} is the reflectance for the essential physiologically active 531 nm spectral band and $\rho_{REF\lambda}$ is the reference reflectance band for a second nonresponsive wavelength, which in the original formulation of the PRI used a REF λ at 570 nm, or PRI_{570}. It is essential to include a narrow band (optimally ~5 nm) centered at ~531 nm to capture the xanthophyll cycle expression/progression, but the choice of a reference band is more flexible. Several different reference bands (~10 nm) centered on wavelengths at 551, 488, and 670 nm have been used. Conceptually, values for the PRI can vary between +1.0 and −1.0, although they commonly fall between +0.5 and −0.5. For PRI_{570}, PRI_{551}, and PRI_{670}, short term (i.e., daily) responses to high light conditions are indicated by relatively lower and typically negative values, so that the greatest stress-induced relative photosynthetic down-regulation is indicated by the most negative values observed. The reverse situation occurs with PRI_{488}, for which increasingly greater down-regulation is indicated with larger positive values. We note that there are many two-band and three-band spectral indices for plant traits and pigment content estimates (Thenkabail et al., 2000) which can assist interpretation of plant status along with the PRI.

5.3.3 Chlorophyll Fluorescence of Leaves

The other primary dynamic pathway that plants use to dissipate excess energy is through chlorophyll fluorescence emissions (ChlF), emitted from chlorophyll *a* molecules associated with the photosystem's antennae pigment complexes in chloroplasts (e.g., Lichtenthaler et al., 1986; Genty et al., 1989). The emission spectrum of ChlF (650–800 nm) is characterized by two broad peaks: the first centered at ∼685 nm (red, F685); and the second at ∼740 nm (far red, F740) (Figure 5.6a–c), although the wavelengths at the peaks may shift ≤5 nm due to various biochemical or morphological conditions (e.g., N content, leaf thickness), which serves as an important secondary diagnostic characteristic. This shift is shown in Figure 5.6c, where the far-red peak for pine needles (*Pinus virginiana* Mill.) occurred at ∼735 rather than 740 nm.

ChlF is emitted from both sides of a leaf, as shown for emission spectra of corn leaves (*Zea mays* L.) obtained with a unique leaf clip/spectrometer system (Van Wittenberghe et al., 2015) under ambient conditions Figure 5.6a. The ChlF emitted within the first red peak results almost exclusively from excitation of PSII by red light and shorter wavelengths (i.e., ultraviolet to red, λ < ∼660 nm) through absorption and energy transfer among the accessory photosynthetic pigments (the carotenoids and

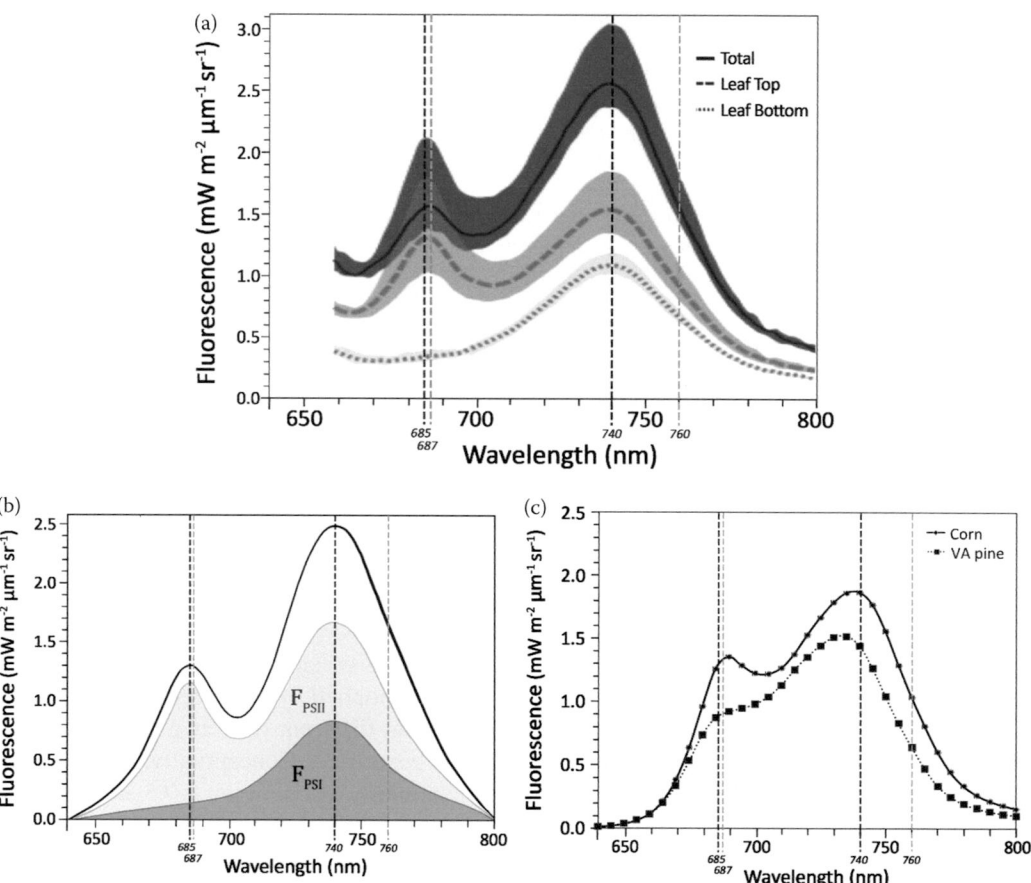

FIGURE 5.6 Leaf chlorophyll fluorescence spectra: (a) Average (±SD) fluorescence spectra for corn leaves acquired under ambient field conditions with a FluoWat leaf clip (Van Wittenberghe et al., 2015) and spectroradiometer system, in 2014 at the USDA/ARS research site in Beltsville, MD, showing the upward emissions measured above the adaxial surface (light blue curve), the downward emissions measured below the leaf (abaxial) undersides (dotted curve), and the total emissions produced (dark blue curve). (b) Conceptual figure showing the relative contributions of ChlF originating from PSII (blue region) and PSI (green) to the total measured emissions (maroon curve). (c) Average fluorescence spectra (mW m^{-2} mm^{-1} sr^{-1}) for upper (adaxial) surfaces of corn leaves (*Zea mays* L.) and pine needles (*Pinus virginiana* L.) acquired with a laboratory spectrofluorometer with an excitation wavelength of 532 nm. Note that the adaxial curves in (a) and (c) are similar. The left peak (red light) is at 685 mn and the right peak (far-red light) is at 740 nm, both marked with black vertical dashed lines. There are Fraunhoffer lines at 687 and 760 nm, both marked with gray vertical dashed lines. (From co-authors, E.M. Middleton and Q. Zhang.)

chlorophyll *b*), whereas ChlF produced from PSI in the second peak is stimulated primarily by red and far-red light ($\lambda \sim$ 670–720 nm). For corn leaves, approximately 20%–30% of the ChlF produced in the (first) red peak was emitted through the leaf undersides, providing additional red photons available to be absorbed by chlorophyll *a* molecules in PSI. This is a process referred to as "reabsorption," in which photons originating from PSII are fluoresced, and are captured again in PSI either within the same or an adjacent leaf, whereby they then contribute to the significant ChlF fraction (e.g., $\sim$50%) subsequently emitted from the second peak. The relative contributions to ChlF across the spectrum are shown conceptually, with the photon source indicated as coming from PSII (blue region, F_{PSII}) or PSI (green, F_{PSI}) in Figure 5.6b.

It is important to recognize that the primary signal available for remote sensing above a vegetation canopy is the ChlF associated with the illuminated, typically the upper (adaxial), surfaces of the leaves (the dark blue shaded region in Figure 5.6a) (Rascher et al., 2000). The shape of this upward flux emission spectrum obtained in the field displays comparable peaks (with red < far-red), and matches the emission shape obtained from illuminated adaxial surfaces of corn leaves in the laboratory (Figure 5.6c). Furthermore, the magnitudes of the ChlF obtained in the laboratory are similar to those obtained *in situ*: red peak $\sim$1.4 mW m^{-2} μm^{-1} sr^{-1}; far-red peak, $\sim$1.8 mW m^{-2} μm^{-1} sr^{-1}. Lower emissions were obtained from the pine needles at the red and far-red peaks ($\sim$0.8 and 1.5 mW m^{-2} μm^{-1} sr^{-1}, respectively).

The total ChlF produced, however, is significantly higher, especially in the far-red peak ($\sim$2.5 mW m^{-2} μm^{-1} sr^{-1}), than ChlF obtained only from adaxial (upward) emissions. In particular, the adaxial (upward) far-red ChlF is roughly half of the total emission in this wavelength region, because almost half is transmitted downward into the canopy. Finally, the emissions obtained under field conditions in the O_2-A band region near 760 nm (Figure 5.6, shown with a vertical dotted line) are only slightly more than half the adaxial far-red 740 nm peak intensity, and only $\sim$40% of the total (upward + downward) ChlF at the 740 nm peak. In contrast, the red peak is well represented at the O_2-B wavelength (Figure 5.6, dotted line at 687 nm) and the surface emissions at 685–687 nm underestimate the total by a small amount.

In general, ChlF (especially in the red peak) and photosynthesis rate are negatively correlated— at a fixed chlorophyll content and in the absence of stress conditions strong enough to induce significant photoprotective ChlF and NPQ responses. This is because under optimal conditions, the available APAR$_{chl}$ is channeled to the dark reactions where it is utilized for photochemistry, with only a baseline amount of photons discarded as ChlF emissions. These processes are described in several physiological models (Cerovic et al., 1999; Maier et al., 2003; van der Tol et al., 2009; Porcar-Castell et al., 2014; Verrelst et al., 2015; Atherton et al., 2016; Vilfan et al., 2016). Also under optimal conditions, ChlF in the second peak is believed by many in the community to be correlated with APAR$_{chl}$. Therefore, ChlF serves as an important indicator of plant photosynthetic function. However, the interpretation of ChlF as it relates to stress factors could appear ambiguous since emissions can drop lower, or be augmented, when the system shifts toward increasing dissipation of excess energy as heat (Zarco-Tejada et al., 2003) by invoking the xanthophyll cycle. The red/far-red ratio was identified as an important indicator of chlorophyll content (and energy transfer between the photosystems) (Lichtenthaler et al., 1986; Lichtenthaler et al., 1990; D'Ambrosio et al., 1992; Agati et al., 1995; Buschmann, 2007), and another ratio (far-red/F700) enabled accurate chlorophyll content determination (Gitelson et al., 1999).

A stress indicator developed by Ac et al. (2015) compares the stressed and unstressed values as a ratio, drawing on leaf and plant measurements made in controlled experiments. This ratio equals 1.0 in the absence of stress (e.g., F740$_{observed}$/F740$_{control}$ = 1.0) or deviates from 1.0 if stress occurs. For F740, this index <1.0 for most stress-inducing conditions, such as insufficient water or low nitrogen. In contrast, the comparable index for F685 may be augmented (>1.0) under low nitrogen, but this index declines for both F740 and F685 when vegetation is exposed to unfavorably higher temperatures. This study (Ac et al., 2015) also showed that the red/far-red ChlF ratio (F685/F740) provided the clearest indicator of temperature stress (declining, due to higher red ChlF) and nitrogen

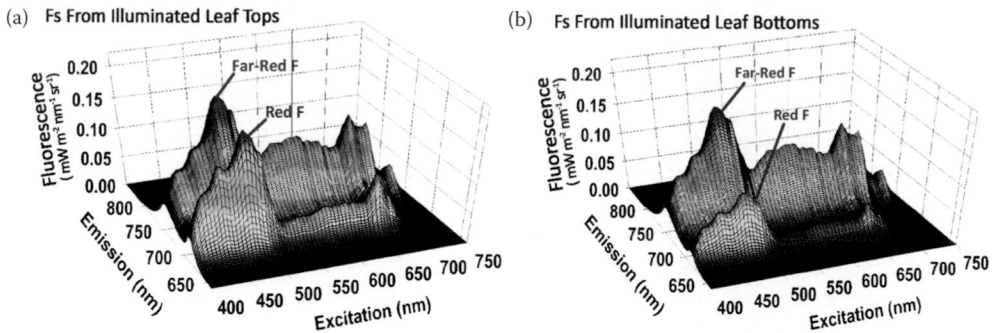

FIGURE 5.7 Fluorescence excitation–emission matrices (EMM) for corn, grown under optimal nitrogen levels in Beltsville, MD. (a) Steady-state (Fs) upward emissions from the leaf tops; (b) Steady-state downward emissions from the leaf bottoms (direct illumination of the leaf bottom). The intensity data are in mW m^{-2} nm^{-1} sr^{-1}. (Courtesy of Lawrence A. Corp.)

deficiency (increasing, due to higher far-red ChlF). Consequently, stress detection with ChlF requires measurements of both the red and far-red ChlF.

Most of the foundational research on ChlF was accomplished with active systems, including lasers and other illumination sources in laboratory and field studies (Chappelle and Williams, 1987; Lichtenthaler and Rinderle, 1988; Agati et al., 1995; Mohammed et al., 1995; Cerovic et al., 1999; Corp et al., 2003; Meroni et al., 2009). Laboratory research originally relied upon dark-adapted leaves that were given an ultraviolet or visible radiation laser pulse to stimulate fluorescence. An example of a laboratory dataset comprised of ChlF intensities as a function of excitation and emission wavelengths is shown as a 3D excitation–emission matrix (EEM) (Figure 5.7), acquired with sequential monochromatic light pulses from a spectroradiometer. This figure shows the upward steady-state ChlF flux (Figure 5.7a) of illuminated adaxial surfaces of corn leaves, which displays the highest and near similar red and far-red peaks when excited by blue-green light. However, when the undersides (i.e., abaxial surfaces) are similarly illuminated (Figure 5.7b), the red peak is much lower than the far-red peak (which maintains values similar to adaxial illumination), due to reabsorption within the leaf. This agrees with measurements shown in a previous figure (Figure 5.6a) acquired under field conditions that naturally experience enriched blue-green wavelengths in solar light. This type of dataset can be used to determine the excitation wavelengths that contribute to the emissions, as shown for leaves of three agricultural and three deciduous tree species (Figure 5.8a,b). The relative contributions to ChlF of "sunlight" wavelengths are more variable for the red (685 nm) than for the far-red (740 nm) peaks.

The advent of new field fluorometers that can measure variable and steady-state ChlF for individual leaves or conifer shoots under ambient light, especially those coupled with photosynthesis systems, has broadened the application of ChlF measurements in crops, trees, and ecosystems (Schreiber et al., 1986; Maxwell et al., 1998). Leaf and intercellular ChlF serves as an important tool in basic and applied plant physiology research to evaluate the function and efficiency of the photosynthetic apparatus at the leaf level (Krause and Weis, 1991; Maxwell and Johnson, 2000; Baker, 2008). For example, Bernacchi et al. (2002), using a gas exchange system coupled with a fluorimeter, investigated the limitations to photosynthesis due to temperature responses of the mesophyll layer. ChlF measurements transitioned to the field, for assessing photosynthetic function with the development of the pulse amplitude-modulated (PAM, Bolhàr-Nordenkampf et al., 1989). Recently Porcar-Castell et al. (2014) investigated the linkage between chlorophyll fluorescence, as measured by a PAM fluorimeter, to photosynthesis and to solar induced fluorescence (SIF), which can be used to detect stress in remote sensing applications at larger canopy and landscape scales. In addition, we also offer a few of our leaf-level laboratory and field plant fluorescence studies,

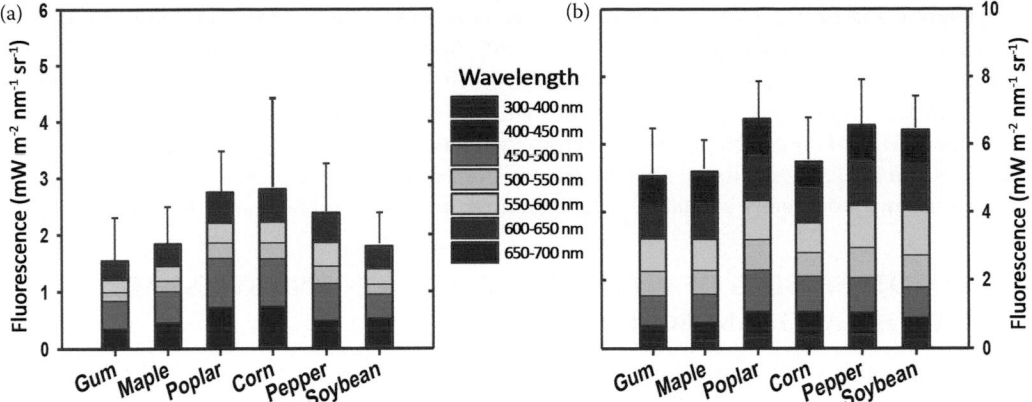

FIGURE 5.8 Leaf-level SIF simulated from laboratory excitation–emission matrices with solar correction. The spectral profile of the lamp was adjusted to the solar spectrum, yielding a photosynthetic photon flux density (PPFD) of 1660 μmol m^{-2} s^{-1} or 400 W m^{-2} nm^{-1}. This provides "adjusted" SIF intensities reported in mW m^{-2} nm^{-1} sr^{-1}. (a) is red fluorescence and (b) is far-red fluorescence. (Courtesy of Lawrence A. Corp.)

researching the effects of seasonality, diurnal variations, nitrogen and water availability on the ChlF emissions: Chappelle and Williams, 1987; Middleton et al., 1996, 1997, 2003, 2005, 2008, 2015a,b; Krizek et al., 2001; Campbell et al., 2007, 2008, 2014; Julitta et al., 2016. Several leaf-level studies are investigating the commonalities between actively and passively induced fluorescence (e.g., Magney et al., 2017), the role of light exposure (Wyber et al., 2017; Peng et al., 2017), and combined reflectance, fluorescence, and photosynthesis (Springer et al., 2017).

5.3.4 Influence of PAR Levels on Vegetation Responses

Optimal PAR levels for photosynthesis in foliage of most species of higher plants typically range from 1200–2000 μmol m^{-2} s^{-1} for "sun" leaves that grow in the exposed ambient light environment experienced by upper and outer canopy foliage (Jones, 1983; Thayer and Björkman, 1990). Optimal PAR levels are lower for "shade" leaves that grow in the lower or inner portions of a canopy, where transmitted PAR dominates and shaded foliage receives a higher proportion of PAR as diffuse radiation.

In addition to high PAR intensity levels, it has been realized that LUE is strongly affected by the direct/diffuse irradiance ratio, increasing under diffuse solar radiation (Gu et al., 2002; Alton et al., 2007; Damm et al., 2015a). At the canopy level, this direct/diffuse ratio influences the instantaneous fraction of shaded vs. sunlit foliage, thus altering available PAR. For scaling up photosynthetic activities from leaf to canopy level, the importance of separating sunlit and shaded leaves has been recognized because of the nonlinear response of leaf carbon assimilation to light intensity (Hollinger et al., 1994; De Pury and Farquhar, 1997; Cheng et al., 2009). Carbon uptake rates of shaded leaves are likely to have a linear response to irradiance, whereas this response often saturates for leaves exposed to direct sunlight (Chen et al., 1999). Moreover, since sunlit leaves might be much warmer than shaded leaves on a clear and sunny day, ignoring the temperature difference could also bias estimates among carbon, water, and heat fluxes (Boardman, 1977; Zhang et al., 1995; Demmig-Adams, 1998; Alton et al., 2007; Sarijeva et al., 2007).

The physiological and morphological differences of sun and shade foliage have been well documented, especially concerning leaf anatomy and pigment concentrations (Lichtenthaler, 1987; Thayer and Björkman, 1990; Middleton et al., 1997; Demmig-Adams, 1998; De Lucia, 1991; Dias, 2007; Dörken and Lepetit, 2018; Stylinski et al., 2002; Lichtenthaler et al., 2007; Sarijeva et al.,

2007; Gamon and Berry, 2012). As a result, several modeling methods separating canopy into sunlit and shaded fractions were developed (Spitters, 1986; Spitters et al., 1986; Sellers et al., 1992; De Pury and Farquhar, 1997; Wang and Leuning, 1998). However, it should be noted that unfavorable environmental conditions such as insufficient water and nutrient levels, and low/high temperatures also reduce LUE, likely contributing to responses mediated through the xanthophyll cycle. High light intensities can be damaging to the photosystems of both sun and shade leaves, especially if water or nutrients are not sufficiently available, or temperatures are unfavorable (Zarco-Tejada et al., 2012).

5.4 REMOTE SENSING OF VEGETATION FUNCTION AND STRESS AT CANOPY AND LANDSCAPE SCALES

(Terms used in this section are defined or explained in Table 5.3.)

Our research team at NASA/GSFC has uniquely addressed the topic of plant stress spectral bioindices at all scales: leaf, canopy, aircraft, and satellite observations. Having already addressed the leaf level topic (Section 5.3), in the remaining two sections of this chapter we provide pertinent examples of our research findings at the field and landscape scales—from pole-mounted, tower-mounted, aircraft, and satellites sensors. We will also summarize the current state of remote sensing for vegetation dynamics related to photosynthetic function, and point to future developments. However, a comprehensive treatment is beyond the scope of this chapter.

TABLE 5.3
Definition of Terms and Parameters Used in Section 5.4

Parameter	Definition
3FLD	A version of the Fraunhofer line depth method of measuring SIF that uses three spectral bands, two outside and one inside the Fraunhofer absorption line
BRDF	Bi-directional reflectance distribution function, a function of four real variables that defines how light is reflected at an opaque surface
EVI	Enhanced vegetation index, $EVI = G(\rho_{NIR} - \rho_{red})/(\rho_{NIR} + C_1\rho_{red} + C_2\rho_{blue} + L)$, where ρ_{NIR} is reflectance in a near infrared wavelength band, ρ_{red} is reflectance in a red band, and ρ_{blue} is reflectance in a blue band. The coefficients for MODIS-EVI are $L = 1$, $C_1 = 6$, $C_2 = 7.5$, and G (gain factor) $= 2.5$.
MTCI	MERIS terrestrial chlorophyll index, $MTCI = (\rho_{753.75} - \rho_{708.75})/(\rho_{708.75} + \rho_{681.25})$, where ρ_{xxx} is reflectance
NDVI	Normalized difference vegetation index, $NDVI = (\rho_{NIR} - \rho_{red})/(\rho_{NIR} + \rho_{red})$, where ρ_{NIR} is reflectance in a near infrared wavelength band and ρ_{red} is reflectance in a red or visible wavelength band
NIR	Near-infrared radiation, ~700–1000 nm
O_2-A	Oxygen band A, telluric oxygen (O_2) feature centered at 760.4 nm
O_2-B	Oxygen band B, telluric oxygen (O_2) feature centered at 687.0 nm
Red/far-red fluorescence ratio	The ratio of chlorophyll fluorescence measured in either the red (685 nm) and far-red (~740 nm) peaks, or with the SIF values obtained in the O_2-B (~687 nm) and O_2-A (~760 nm) oxygen absorption features
SCOPE	Soil canopy observation, photochemistry and energy fluxes, a SVAT model incorporating leaf and canopy radiation and biochemical processes
SVAT	Soil-vegetation-atmosphere transfer, a vegetation canopy energy balance model
SWIR	Short-wave infrared radiation, ~1000–3000 nm
VAA	View azimuth angle, the horizontal view angle (as a compass bearing) relative to true (geographic) north, increasing in the clockwise direction (VAA 0° is north, 90° is east, 180° is south, 270° is west)
VZA	View zenith angle, the vertical view angle, in degrees offset from straight down (VZA 0°), and VZA 90° is viewing the horizon

5.4.1 Remote Sensing Estimates of APAR

APAR is a key variable in LUE models, and is assumed to represent the integrated influences from all physiologically active absorbing pigments present in a vegetated canopy. The rationale for remote sensing measurements of the components of APAR, incident PAR, and fAPAR, is well established (Daughtry et al., 1982; Hipps et al., 1983; Wiegand and Richardson, 1984; Myneni et al., 1997a; White et al., 2001; Pinker et al., 2003; Cheng et al., 2014) and its estimation is affected by the way it is defined and measured (Gitelson and Gamon, 2015). Incident PAR is variable over time periods from minutes to days and months, due to predictable daily and seasonal changes in solar elevation angle as well as less predictable variations in atmospheric scattering and clouds. A number of approaches have been developed to estimate atmospheric scattering from satellite altitudes above the Earth's atmosphere to determine incident PAR at the surface (Frouin and Pinker, 1995; Liang et al., 2006) and atmospheric corrections (Berk et al., 2014), and these approaches are evolving as new sensor capabilities come into play.

The fAPAR parameter is also temporally variable, but as it is related to vegetation structure and composition (as well as solar angle), its change is generally smoother over time. The most commonly employed method utilizes the NDVI (Rouse et al., 1974; Tucker, 1979) as a surrogate for fAPAR, drawing on relatively broad-band ($\sim$20–50 nm) observations in the red ($\sim$620–670 nm) and NIR ($\sim$840–875 nm) spectrum:

$$\text{NDVI} = [(\rho_{\text{NIR}} - \rho_{\text{Red}})/(\rho_{\text{NIR}} + \rho_{\text{Red}})] \qquad (5.3)$$

For photosynthesis studies based on remote sensing, the quantity that is actually needed is APAR_{chl}, as discussed above in Section 5.3. Some spectral indices are more sensitive to the green vegetated fraction rather than to total fAPAR absorbed by the whole canopy (e.g., White et al., 2001), such as those developed to estimate the canopy chlorophyll content. Examples include the MERIS (Medium Resolution Imaging Spectrometer) terrestrial chlorophyll index (MTCI, Dash and Curran, 2007; Dash et al., 2010), and chlorophyll and carotenoid equations empirically derived by Gitelson et al. (2002, 2005, 2012). In addition, some research teams use the enhanced vegetation index (EVI, Huete et al., 1997, 2002) to estimate $\text{fAPAR}_{\text{chl}}$ and simply assume $\text{fAPAR}_{\text{chl}} = \text{EVI}$, replacing the denominator in Equation 5.3 with an adjusted reference quantity that includes a blue band ($\sim$470 nm), as:

$$\text{EVI} = 2.5 * [(\rho_{\text{NIR}} - \rho_{\text{Red}})/(1 + \rho_{\text{NIR}} + (6 * \rho_{\text{Red}}) - (7.5 * \rho_{\text{Blue}}))] \qquad (5.4)$$

All of these indices provide a relative estimate of the canopy green fraction, but not a physically based quantity. A radiative transfer model, ProSAIL-2 (Zhang et al., 2005, 2009), has been developed to quantitatively retrieve $\text{fAPAR}_{\text{chl}}$ and $\text{fAPAR}_{\text{non-chl}}$. The physically based ProSAIL-2 model utilizes seven 10–20 nm wide reflectance bands distributed across the full shortwave (visible, NIR, SWIR) spectrum as model inputs. APAR_{chl} at the canopy/landscape scale is obtained as the product of $\text{fAPAR}_{\text{chl}}$ and PAR.

5.4.2 Field Studies: The Photochemical Reflectance Index Light Use Efficiency

Successful leaf-level investigations have led to canopy-level studies of PRI by many research teams (Filella et al., 1996, 2009; Barton and North, 2001; Gamon et al., 2001; Garbulsky et al., 2011; Garrity et al., 2011; Jenkins et al., 2007; Hilker et al., 2008, 2009; Cheng et al., 2009, 2010, 2013; Middleton et al., 2009a,b, 2014a,b, 2015a,b; Rossini et al., 2013, 2015; Stagakis et al., 2014; Magney et al., 2016; Nakaji et al., 2016; Zhang et al., 2016; Chou et al., 2017; Zhang et al., 2017). Although the PRI signal is relatively weak compared with spectral indices using red vs. NIR wavelengths (e.g., NDVI, EVI), it has successfully been observed remotely over crops and forests from aircraft and towers and from satellites (Section 5.4). For crops, PRI measurements made from hand-held and pole-mounted spectrometers have shown clear diurnal and seasonal dynamics that can be related to variations in LUE or GEP obtained at an adjacent EC tower (Huemmrich et al., 2008; Strachan et al., 2008; Middleton et al., 2009b, 2014; Campbell et al., 2014). However, this success depends upon collecting measurements in a systematic way that accounts for geometric effects on measurements, as well as canopy structure influences on the canopy light environment, given different responses of sun and shade canopy sectors (Hilker et al., 2008, 2009; Middleton et al., 2009a). Some studies support the hypothesis that PRI is an indicator of PSII photochemical efficiency at the canopy scale, indicated for example, with controlled studies for elevated CO_2 and low temperatures (Guo and Trotter, 2006) and studies showing strong relationships to drought stress (e.g., Zarco-Tejada et al., 2012, 2013a; Rossini et al., 2013). However, there is currently no community consensus as to whether the PRI can be considered a reliable proxy for NPQ, or whether it serves as a more general indicator of plant stress limitations on photosynthesis.

The PRI has reliably tracked crop LUE, as determined from continuous EC tower measurements at an USDA-ARS research cornfield fertilized at the best management practices rate (100% N) in Beltsville, MD, USA (Huemmrich et al., 2008; Cheng et al., 2009, 2013; Middleton et al., 2009b, 2014b, 2015b; Campbell et al., 2014, 2016a,b; Sabater et al., 2017). An example is given (Figure 5.9) for data in two summers (2007, 2008) acquired with a nadir view from pole-mounted spectrometers, which were substantiated by similar pole-mounted and tower-mounted measurements for different varieties and environmental conditions in this cornfield for nine more years, all indicating that the PRI accounts for between 65% and 70% of the variation in daily or seasonal LUE.

The impact of sun and shade illumination differences across a canopy affects the PRI responses, (Takata and Mottus, 2016) contributing to canopy anisotropy, as demonstrated for midday measurements at the fully developed vegetated growth stage in this cornfield. Eight view azimuth angles (VAAs) were acquired in a circular fashion around an imaginary polar plane at 1 m above the crop from a pole-mounted spectrometer, to systematically sample the bidirectional reflectance distribution function (BRDF) (Cheng et al., 2010) for three view zenith angle (VZA) sets (Figure 5.10). During each VZA set, the PRI changed smoothly from the lowest values in the sunlit hotspot (VAA = 0; the starting points) associated with backscattered (reflected) radiation to the highest values (at top of curves, VAA = 180°) in the shaded (cool spot) sector associated with the transmitted forward scattered radiation. The PRI values per azimuth measurement set increased as the VZA

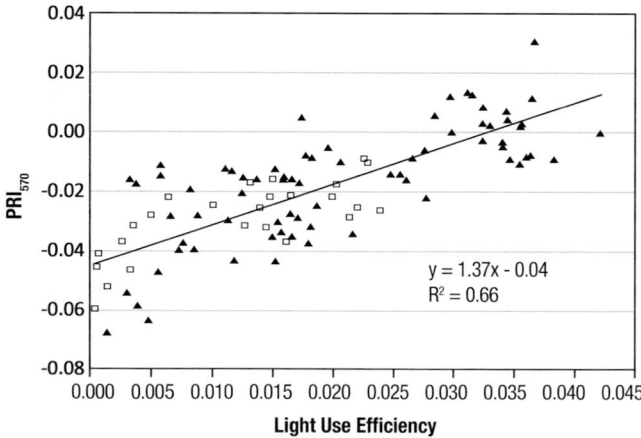

FIGURE 5.9 Relationship between light use efficiency (LUE) and PRI_{570} for cornfield in Beltsville, MD, collected at multiple times during selected clear days during the 2007 (□) and 2008 (▲) growing seasons. PRI_{570} values from averages of nadir spectral reflectance collected along a 100 m transect in field. LUE (with units of mol C mol^{-1} APAR) are hourly values calculated using GEP and incident PAR from flux tower and green fPAR estimated from NDVI from reflectance data. Flux data from W.P. Kustas (USDA/Beltsville Agricultural Research Service). (From co-author, Huemmrich, K.F. et al. 2008. *IEEE International Geoscience & Remote Sensing Symposium (IGARSS)*, Boston, MA, USA.)

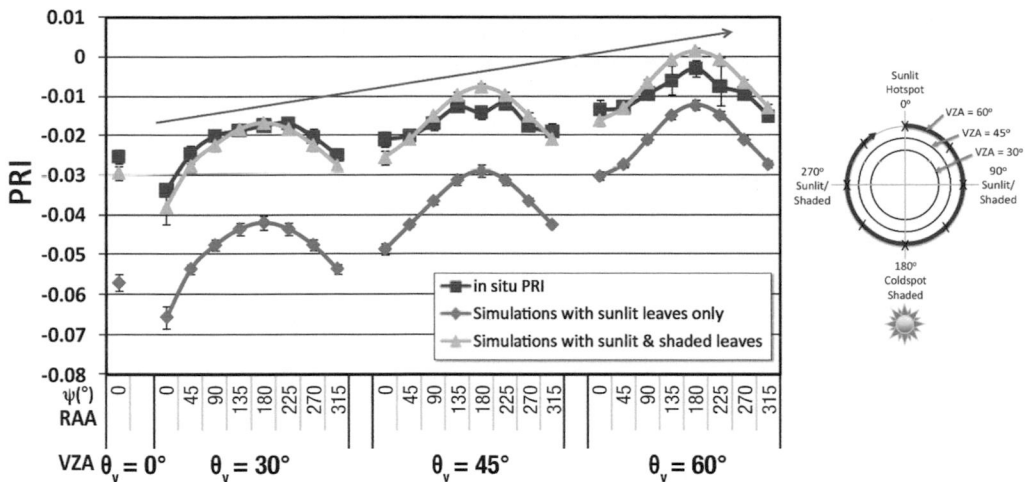

FIGURE 5.10 Effects of viewing geometry on photochemical reflectance index (PRI) values for canopy measurements in a USDA cornfield in Beltsville, MD. Comparisons show *in situ* observations (■) and radiative transfer modeling simulations using sunlit leaves only (◆) versus both sunlit and shaded leaves (△). The data are from July 1, 2008. (From Cheng, Y.-B. et al. 2010. *Ecological Informatics*, 5(5), 330–338.)

increased from 30° to 60° (shown with red arrow), since a greater fraction of shaded foliage was viewed. Successful modeling of these observations required the addition of shaded leaves in a two-layer radiative transfer simulation. Since shaded foliage receives mostly diffuse light and does not experience the high light stress of sunlit leaves, it is less physiologically stressed and attains higher photosynthetic efficiency.

A further study of the BRDF responses of cornfields was conducted in controlled plots that had been provided one of four variable nitrogen (N) fertilization augmentation rates (0%, 50%, 100%, 150% of the optimal amount), and watering regimes from drip irrigation (0, 1), forming eight treatments (watered and nonwatered, at each of four N levels) (Middleton et al., 2014a). BRDF observations of PRI at 1 m above the canopies were derived from pole-mounted spectroradiometer data collected at seven positions (nadir, VZA = 0°; ±30°; ±45°; ±60°) within the solar principal plane. Measurements were made in mornings and afternoons throughout a full summer (Figure 5.11). In midseason, when the site experienced prolonged high temperatures and low precipitation, the crop was visibly stressed. During that period, BRDFs could be vertically stacked, with the watered 100% or 150% N plots having the highest (near zero) PRI responses, indicating the lowest stress responses under these conditions. In contrast, the nonwatered 50% and 0% drought plots exhibited the lowest (most stressed) PRI values (<-0.04). In the watered plots, a "J" shape emerged along the solar principal plane (blue and green curves), with highest values (describing lower stress) attained in the forward (cool spot) direction. Plants in the drought treatments (yellow and brown curves) exhibited PRI BRDFs having a V or W shape. This resulted because water-deficient plants invoked protective structural strategies, in addition to photoprotective physiological processes (e.g., NPQ; and ChlF, not shown), to reduce the high light load. These structural changes included moving leaves to more vertical positions combined with leaf roll, exposing the waxy leaf undersides when viewed from above. This underscores the relevance of ChlF from leaf bottoms shown in Figure 5.7b. Note that the vertical stacking of the eight treatment groups from least to most stressed (top to bottom of the stack, watered plots >drought plots) is clearly apparent on July 17, 2012 (marked with dotted vertical line), just before an ameliorating rainfall.

After a major rainfall in the reproductive growth stage, all stands recovered from drought stress, and treatment differences in terms of PRI values and BRDF shapes disappeared. With onset of senescence, the PRI values gradually declined through the remaining season. The watered high-N

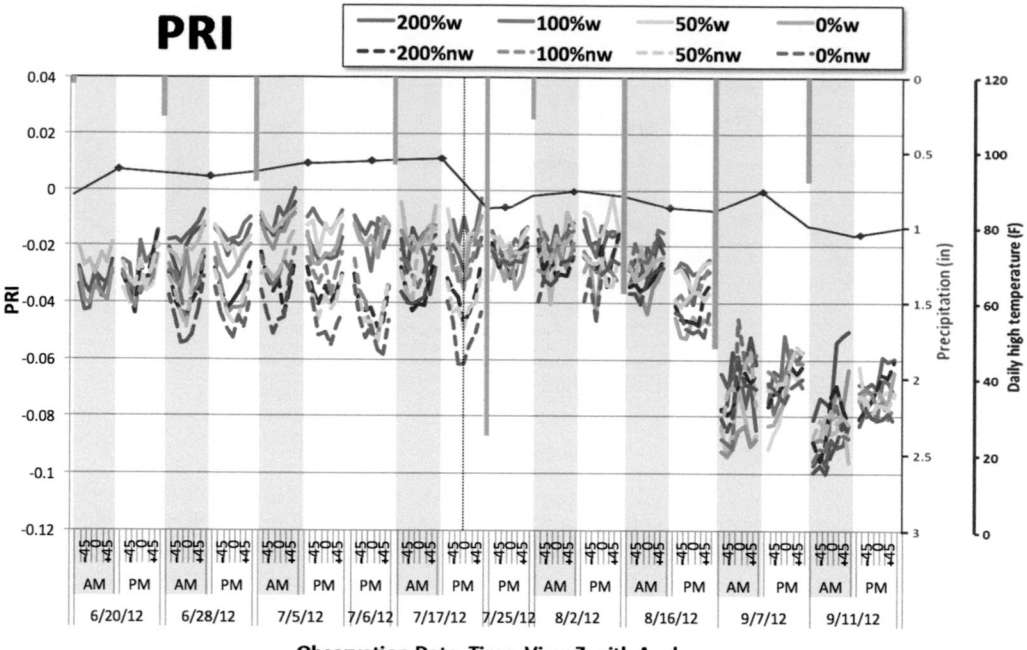

FIGURE 5.11 Seven BRDF measurements ($0° \pm 45°$, $90°$, $135°$) in the solar principal plane were made for the PRI during midmorning and midafternoons across a corn crop's growing season, for eight treatment plots supplied experimental nitrogen and/or water augmentations (N: 0%, 50%, 100%, 200%; water: 0, 1). From 6/28/2012 until 7/25/2012, the PRI for watered plots (blue and green curves) > unwatered plots (red, orange, yellow, brown curves). The PRI was similar in all plots when sufficient water was available (7/25/2012 through senescence). Watered plots also exhibited a different BRDF shape than unwatered plots. The daily high temperature ($°C$) is tracked with the red curve, and precipitation events are shown with aqua vertical bars. (From co-author, Middleton, E.M. et al. 2014a. *Proceedings, 5th International Workshop on Remote Sensing of Vegetation Fluorescence*, Paris, France, April 22–24, 2014.)

plots fared best throughout the summer, attaining the highest PRI values (between 0 and -0.02), and highest leaf-level photosynthetic rates and photosynthetic efficiencies (Middleton et al., 2014a). In addition to capturing responses to experimental treatments (N, water) over a range of phenological stages, enhanced by harsh environmental conditions, this example illustrates the benefit provided by the PRI for stress monitoring. It also illustrates the complexity of drawing upon spectral reflectance information (only) to unravel functional status.

The development of automated near surface instruments (e.g., Hilker et al., 2007; Corp et al., 2010) allows frequent observations of a canopy at the temporal and spatial scales relevant to flux measurements (Leuning et al., 2006). The challenge for these systems is the development of methods to acquire observations representative of a flux tower footprint. Multiple approaches have been tried including the following: mounting a spectroradiometer on a tram that moves on a track above the canopy (e.g., Gamon et al., 2006a); mounting a spectroradiometer or an imaging spectrometer from a tower or fixed platform (Leuning et al., 2006; Hilker et al., 2007; Corp et al., 2010); or placing multiple narrow-band sensors (e.g., NDVI and/or PRI bands) to observe different parts of the canopy (Garrity et al., 2010). By frequently sampling a number of smaller areas within a flux tower footprint, near surface measurements provide rich datasets for the study of spectral bioindicators. The dynamics of PRI variability within canopies have been addressed using automated tower-mounted spectrometers that collect measurements for known angular sectors of a canopy over time and over a range of illumination conditions, as was first undertaken for a conifer forest in 2006 (Hilker et al., 2008), clearly showing different PRI behavior associated with sunlit vs. shaded canopy sectors (Cheng et al., 2009, 2010; Middleton et al., 2009b, 2014a).

5.4.3 Field Studies: Solar-Induced Chlorophyll Fluorescence

5.4.3.1 How Is Chlorophyll Fluorescence Measured in Ambient Light?

New methods have emerged in the past decade to retrieve ChlF as a vegetation canopy remote sensing measurement, derived from hyperspectral radiance (Meroni et al., 2009), especially very high-resolution spectra (<0.5 nm) using various Fraunhofer line depth (FLD) approaches (Plascyk, 1975; Maier et al., 2003; Alonso et al., 2008; Damm et al., 2014) and spectral fitting methods (Mazzoni et al., 2010; Meroni et al., 2010; Cogliati et al., 2015). These typically utilize the atmospheric absorption features located in the vicinity of the two telluric oxygen (O_2) features centered at 687.0 nm (O_2-B) and 760.4 nm (O_2-A) (e.g., Alonso et al., 2008). The quantity retrieved is the solar-induced fluorescence (SIF) that is passively induced by sunlight and assumed to represent a steady-state condition. The SIF retrieval methods enable separation of upwelling radiances into two components: the weak ChlF emission signal (e.g., <5% of the total upwelling spectral signal) and the much greater reflected contribution (Zarco-Tejada et al., 2003). Since atmospheric "windows" such as Fraunhofer lines and absorption features are narrow, continuous high-resolution spectra (i.e., 0.025–1 nm) must be used. The original FLD method utilized two narrow bands inside and outside of the Fraunhofer feature, while other methods require additional narrow bands. For example, the 3FLD approach uses three wavelengths and linear interpolation—the wavelength at the deepest absorption (e.g., inside the feature) and wavelengths on both shoulders of the feature (outside) (Alonso et al., 2008; Damm et al., 2014). Nonlinear spectral fitting methods require all of the wavelengths within the absorption feature (Mazzoni et al., 2010; Meroni et al., 2010; Cogliati et al., 2015; Verhoef et al., 2018). These radiance-based methods have the advantage of retrieving SIF with actual physical units (e.g., $mW\ m^{-2}\ nm^{-2}\ sr^{-2}$) and are less sensitive to other changes in leaf biochemistry.

Spectral reflectance obtained above green vegetation in nature is contaminated with the ChlF contribution, from which it cannot be isolated, and thus is referred to as "apparent reflectance" (Campbell et al., 2008). Reflectance-based approaches to identify ChlF contributions examine small distortions in apparent reflectance of plants in the red edge spectral region, and have been developed from either derivative reflectance spectra or the original reflectance spectra. These indices are usually combinations of narrow bands located at the ChlF signal peaks (685 nm or 740 nm) coupled with unaffected reference bands. Reflectance-based methods do not yield ChlF in physical units, but rather provide a relative influence of ChlF on apparent reflectance, and may be contaminated by changes in other biochemical properties. Nonetheless, these reflectance-based methods provide an easier alternative approach to capturing relative ChlF information and open a way for further investigations and progress (Zarco-Tejada et al., 2003; Dobrowski et al., 2005).

5.4.3.2 What Can Be Learned from SIF Field Studies?

Only a few research groups have accomplished quality field research on SIF, due to the expense of quality sensors that are capable of acquiring consistent measurements from which fluorescence can be reliably retrieved and the experience required to achieve useable measurements under field conditions. Measurements and instrumentation protocols are currently being developed and evolved to support remote sensing fluorescence observations from aircraft and satellites by the OPTIMISE coalition (ESSEM COST Action ES1309; http://optimise.dcs.aber.ac.uk/), endorsed by European Space Agency (ESA). These protocols primarily draw on the collective experiences of the research groups that support the ESA Fluorescence Explorer (FLEX) mission (refer to Section 5.3.5.2), and others. Published works generally address the evaluation of midday far-red SIF (e.g., SIF760) in the context of GPP and canopy types (e.g., crop, growth stage) at a tower site (e.g., Yang et al., 2015), platform-based observations, or other *in situ* measurements, such as with pole-mounted instruments (e.g., Meroni et al., 2008a; Damm et al., 2010; Rascher et al., 2010a,b; Guanter et al., 2013; Sabater et al., 2017, 2018). Several of these studies examined far-red SIF and PRI together (Schickling et al., 2016). A few studies have examined both far-red SIF and red SIF (SIF687) and/or their ratio (Meroni et al., 2008a; Middleton et al., 2014a,b; Rossini et al., 2016; Campbell et al., 2016b). Measurements over the full day are also being made by a number of research groups (e.g., Suess et al., 2016).

Our research team has extensively studied SIF responses at the USDA-ARS cornfield in Beltsville, MD, typically in conjunction with PRI determinations (Middleton et al., 2009b, 2014a,b, 2015b; Cheng et al., 2013; Campbell et al., 2014, 2016b). We provide three SIF examples. The first two examples document anisotropy of the SIF energy flux with BRDF responses for SIF acquired from our automated tower-mounted FUSION system that surveys a cornfield (Corp et al., 2010; Sabater et al., 2017) (Figure 5.12); and an index describing the relative anisotropy from BRDF measurements made in the N and water treatment plots described above in Section 5.4.2

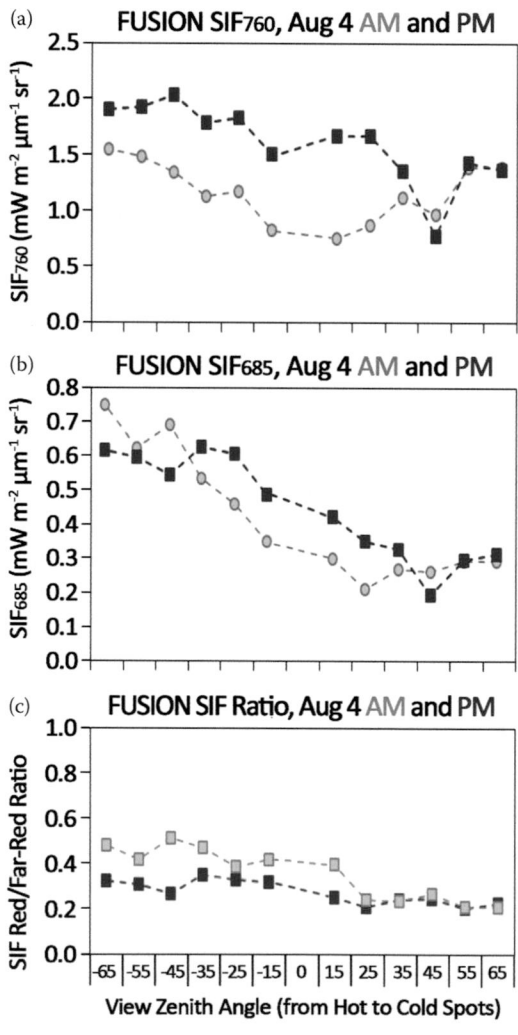

FIGURE 5.12 Changes in canopy solar induced fluorescence (SIF) throughout the day and with viewing geometry, measured on August 4, 2014 with GSFC/FUSION tower system at a cornfield in Beltsville, MD (100%N, rainfed). Morning measurements are shown with the blue curves, afternoons with red curves. (From co-author, Middleton, E.M. et al. 2015b. *Proceedings, (on CD) International Geoscience and Remote Sensing Symposium (IGARSS 2015)*, Milan, Italy, 27–31 July, 2015.)

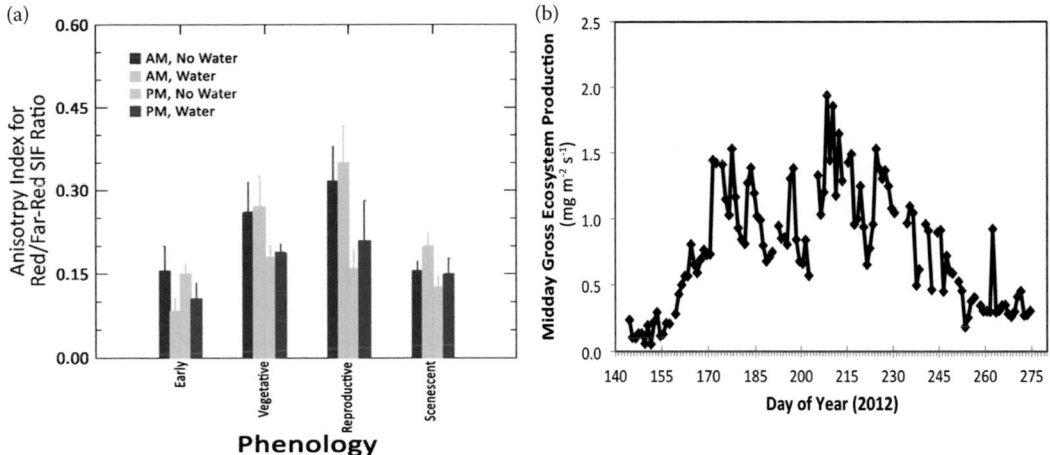

FIGURE 5.13 (a) Canopy solar-induced fluorescence (SIF) exhibits significant changes with vegetation phenology, as represented for corn by the differences in the SIF red/far-red ratio. The SIF ratio increased significantly from the early to the vegetative and reproductive stages, with the increase in canopy leaf area and chlorophyll content, and decreased with the offset of leaf senescence. These differences were most pronounced earlier in the day (AM) and in the middle of the growing season (e.g., during vegetative and reproductive stages). (b) The seasonal trend for the SIF ratio followed the 2012 GEP seasonal profile. Figure (b) shows midday GEP from eddy covariance across the 2012 season for the 100%N field. Flux data from W.P. Kustas (USDA/Beltsville Agricultural Research Service). (From co-author, Middleton, E.M. et al. 2015a. Multi-angle hyperspectral observations with SIF and PRI to detect plant stress & GPP in a cornfield. *Proceedings, 9th EARSeL SIG Workshop on Imaging Spectroscopy, CD-ROM*, Luxembourg City, Luxembourg, April 2015, p. 10.)

(Figure 5.13). The automated FUSION tower system was mounted on a 10 m tall tower located in a large cornfield that was annually provided an optimal 100% N fertilization rate. This system consists of several spectroradiometers, including dedicated sensors for SIF measurements in the O_2-A and O_2-B regions centered on 760 and 687 nm (Corp et al., 2010). The instruments view the canopy repetitively throughout the day, collecting spectra at fixed azimuth intervals in an almost full circular azimuth plane, repeated at select VZAs, within a 15 minute cycle, daily.

We selected a clear midsummer date and present the solar principal plane BRDF from morning and afternoon for the far-red SIF, the red SIF (calculated using the 3FLD method), and their red/far-red SIF ratio (Figure 5.12). The fluorescence responses for both red and far-red SIF were higher, as expected, in the backscatter hot-spot region (denoted as negative VZAs), and were higher in the afternoon than in morning, and the morning to afternoon differences for SIF radiances were greater in the far-red on most days examined. On August 4, 2014, the far-red SIF was 3–4 times greater than red SIF, and varied between a morning low (0.8 mW m^{-2} μm^{-1} sr^{-1}) and an afternoon high (2.2 mW m^{-2} μm^{-1} sr^{-1}). Over the day, red SIF varied between 0.2 and 0.7 mW m^{-2} μm^{-1} sr^{-1}. Shaded and sunlit canopy sectors changed through the day, since the solar principal plane rotates with the sun's azimuth, so that morning and afternoon observations are for different canopy sectors. The BRDFs for the SIF ratio were much flatter and more similar for morning and afternoon; however, the ratio was lower in the instantaneously shaded forward scatter sector (i.e., less stressed) than in the sunlit backscatter sector all day.

BRDFs in the solar principal plane for SIF were acquired, in addition to those for the PRI, in the N/water treatment plots described above (Section 5.4.2). These SIF determinations revealed that the water availability (i.e., rainfed vs. a drip line), significantly influenced the degree of anisotropy expressed, which changed over the crop's development at different growth stages. Here, an index of anisotropy was used, based on the difference between red/far-red SIF ratio values in the cold spot (CS) and the hotspot (HS) SIF values along each BRDF (Middleton et al., 2015b), where larger values indicate greater deviation from isotropy. Anisotropy was significantly expressed in both far-red and red SIF, and for the SIF ratio (Figure 5.12). For the young corn plants, lack of water was captured with the SIF ratio in the "no water" plots (red and green bars) throughout the day (Figure 5.13). During vegetative and reproductive growth stages, when high temperatures prevailed in that year, plants exhibited substantially higher anisotropy during the mornings (red and aqua bars), and generally higher anisotropy in all plots for mature canopy stands, as compared to young and senescing stands. We note, however, that with this red/far-red SIF anisotropy index, senescent stands were not discriminated from young corn stands.

Many of our field observations included measurements that enabled us to derive values and trends for both PRI and SIF. When combined, we found that the red/far-red SIF ratio was highly correlated ($r = 0.75$) with the PRI measured in these plots over a growing season (Figure 5.14). This figure shows average daily values collected from the nadir view for the well-watered plots provided optimal N levels. With the lowest PRI and SIF ratio values, senescent vegetation (blue symbols and ellipse) was clearly distinguished from earlier growth stages, which populated different ranges within the overall dataset. These phenology groups are displayed with different symbols and ellipse outline colors (red = young stands; aqua = vegetative stands, and green = reproductive stands). Both the PRI and red/far-red SIF ratio generally increased from midrange values in young stands to highest values during reproduction, before crashing to low senescent values. We also found that the combination of the PRI and the red SIF was well correlated with the LUE determined from the EC flux tower in the USDA-ARS rainfed cornfield with optimal N fertilization level (Cheng et al., 2013).

The field and laboratory research over the past several decades has provided extensive datasets and advanced understanding about fluorescence emissions as a function of vegetation properties and environmental conditions, and relationship to photosynthesis. The first important model in support of fluorescence was FluorMOD (Zarco-Tejada et al., 2006). A major development was attained with a vertical 1D integrated radiative transfer and energy balance soil-vegetation-atmosphere transfer (SVAT) model, incorporating leaf and canopy processes (van der Tol et al., 2009), which is referred to as the SCOPE model (soil canopy observation, photochemistry and energy fluxes). SCOPE has

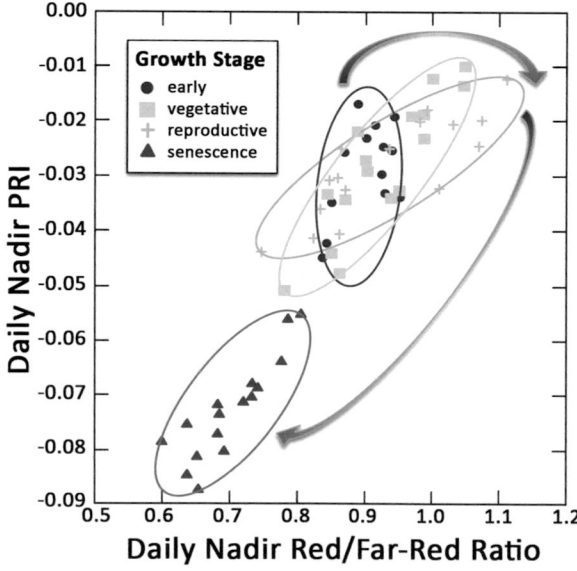

FIGURE 5.14 The daily average nadir values for the PRI and the red/far-red SIF ratio changed in concert over the 2012 growing season ($r = 0.75$) in the USD/ARS cornfield. These data represent the low-stress treatment plots (100% N, drip line). (From co-author E.M. Middleton.)

undergone several upgrades over the past half-decade (van der Tol et al., 2014; Zhao et al., 2016; Timmermans et al., 2017; Yang et al., 2017) and has been used for simulating and interpreting SIF and photosynthesis in two- and three-dimensional canopies (van der Tol et al., 2014). Another critical modeling piece, Fluspect-B, was recently achieved (Vilfan et al., 2016) to accurately describe fluorescence and reflectance properties within leaves, adding a powerful module to link with SCOPE. Evaluations of leaf biochemical and morphological factors as well as environmental factors in predicting fluorescence and photosynthesis was undertaken with the SCOPE model (Verrelst et al., 2016). A modeling capability to link SIF retrieval with atmospheric correction in a single processing chain, with application to the ESA/FLEX mission, was recently published (Verhoef et al., 2018). The SCOPE model and process-based model performances were examined in recent studies (van der Tol et al., 2016; Cui et al., 2017; Hu et al., 2018).

5.4.4 Aircraft Studies: PRI, SIF, and Hyperspectral Observations

The need to capture vegetation under a range of conditions presents a challenge for studying spectral bioindicators, accomplished either by viewing multiple sites and/or by capturing varying conditions at different times (Gamon et al., 2006b). This is only possible from remote platforms such as aircraft or satellites. The first airborne collection of the PRI was made during the 1994 BOReal Ecosystem and Atmosphere Study (BOREAS) (Sellers et al., 1995), where airborne hyperspectral VNIR sensors (10 nm spectral resolution) observed multiple flux towers sites (Nichol et al., 2000) and demonstrated a significant linear correlation across four types of boreal forests between LUE and PRI_{570}. Additionally, Rahman et al. (2001) used AVIRIS aircraft data and a scaled version of PRI (combined with the NDVI) to map carbon fluxes over the BOREAS study area. Since then, AVIRIS aircraft data (Green et al., 1998) have also been used in PRI studies of chaparral (Fuentes et al., 2006). Inoue and Peñuelas (2006) found that the correlation to LUE appeared stronger when PRI was derived from narrower spectroradiometer measurements (~3 nm bandwidth) than from simulated AVIRIS bandwidth of ~10 nm. Huemmrich et al. (2009) identified significant differences in the PRI:LUE relationships for deciduous versus conifer forests. Strachan et al. (2008) used PRI from aircraft to identify stressed areas that were correlated with lower yield in corn and wheat fields. Airborne-derived PRI was found sensitive to diurnal changes in water stress in orchards (Suárez et al., 2008), and to a thermal index (Zarco-Tejada et al., 2013a). Relationships between net photosynthesis and steady-state chlorophyll fluorescence (Zarco-Tejada et al., 2013b) and spatiotemporal patterns of ChlF and physiological were retrieved from airborne hyperspectral imagery (Zarco-Tejada et al., 2013c).

Far-red SIF was captured by an airborne spectrometer (Rossini et al., 2013; Damm et al., 2015b; Wieneke et al., 2016) and the first maps were obtained from the HyPlant fluorescence imager (Rascher et al., 2015; Rossini et al., 2015), the prototype system for the FLEX orbital mission. Another airborne system designed to capture SIF is the NASA/JPL Chlorophyll Fluorescence Imaging System (CFIS, Drewry et al., 2015), and supports the Orbiting Carbon Observatory (OCO-2) mission. Airborne based spectroscopy of both red and far-red SIF fluorescence were reported (Wieneke et al., 2016).

An airborne study conducted in 2013 as the first joint campaign between our NASA team and the ESA/FLEX) mission (Middleton et al., 2017a), collected both SIF and PRI observations with the HyPlant airborne spectrometers at four times of day over pine stands of known different ages in a pine plantation (Figure 5.15). Red SIF and the PRI were negatively correlated and exhibited diurnal hysteresis (indicated by arrows). Mature (right cluster) and young (left) forest stands showed similar behavior, but were clearly discriminated based on the PRI range of responses, with greater stress (i.e., lower PRI) indicated for the young stands. For the mature stands, relatively low morning stress was indicated by high PRI coupled with low red SIF values. Diurnal hysteresis was exhibited (in both groups) as stress increased in midday, indicated by lower PRI and higher SIF, returning to original values in late afternoon, in concert with illumination and temperature levels. This study also documented a strong relationship between the red/far-red SIF ratio and the LUE from the tower EC located within a mature stand (Middleton et al., 2017a).

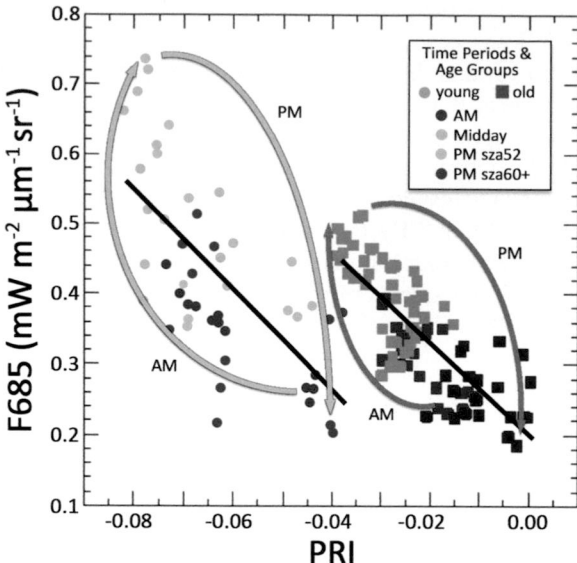

FIGURE 5.15 Hysteresis over the diurnal cycle is expressed in the combined responses acquired from the HyPlant airborne spectrometers for red SIF and the PRI from the mature stand at the NC2 tower site (right group) and the young stands (left group) at the Parker Tract Loblolly Pine (*Pinus taeda* L.) Plantation located on the lower coastal plain near Plymouth, NC (35°48′N 76°40′W). The diurnal cycling is indicated with arrows. (From co-author, Middleton, E.M. et al. 2017a. *Remote Sensing*, 9, 612, 30pp. doi: 10.3390/rs9060612.)

5.4.5 SATELLITE STUDIES: VEGETATION PHOTOSYNTHESIS AND STRESS

(Terms used in this section are defined or explained in Table 5.4.)

A flux tower can provide a continuous temporal profile of whole ecosystem carbon exchange, but only for a single location, and field studies are unavoidably constrained in scope by manpower and resources. In contrast, aircraft and satellite remote sensing provide information on spatial and temporal variability, thereby extending observations over the entire landscape. Remote sensing does not disturb the vegetation being measured, and offers efficient, repeated tools for monitoring ecosystem responses to environmental stresses for large land areas over time (Running et al., 2004; Grace et al., 2007). However, this information is predominantly associated with the exposed, upper layer of the canopy, not the whole system sampled by towers or the intensive measurements possible from *in situ* research.

Most commonly, remote sensing observations from aircraft or satellite sensors provide snapshots of vegetation functioning at variable time intervals across a growing season. And these snapshots, when acquired from sun-synchronous polar orbiting satellites, are limited to a single observation per day at a fixed time of day, spaced in intervals of days to multiples of weeks, depending on the footprint. From these data, a typical measurement series represents samples acquired on clear days composited over periods of several days or weeks, although limited daily observations are possible from some orbital platforms (e.g., the International Space Station, ISS). Nevertheless, this information is extremely useful for examining the spatial and temporal patterns of ecosystem health and physiological status, including LUE and vegetation stress.

The developing field of spectral bioindicators benefits greatly from the valuable GEP time series acquired from flux towers, providing an essential way to validate remote sensing estimates of photosynthesis. Moreover, the *in situ* and space-based observations can be used together to develop and test different methods of analyzing and modeling the spectral data against the "ground truth" of the flux tower measurements. For example, time series from flux towers have been extensively used to validate the MODIS satellite sensor algorithm that computes global GEP (Turner et al., 2003), and the various satellite-retrieved SIF observations (e.g., Frankenberg et al., 2011; Sanders et al., 2016).

TABLE 5.4

Definition of Terms and Parameters Used in Section 5.4.5

Parameter	Definition
EO-1	NASA's Earth Observing 1 satellite, a pointable multispectral and hyperspectral satellite (2000–2017)
FLEX	FLuorescence Explorer, the ESA Earth Explorer satellite mission planned for launch in 2022/23
GOME	ESA's Global Ozone Monitoring Experiment global coverage satellites sponsored by ESA—GOME (1995) and GOME-2 (2006)
GOSAT	Japan's Greenhouse gases Observing SATellite (2009)
HICO	Hyperspectral Imager for the Coastal Ocean, a hyperspectral sensor developed by the US Navy and operated on the ISS (2009–2014)
Hyperion	The first spaceborne high spectral resolution imaging spectrometer, onboard EO-1
HyspIRI	Hyperspectral Infrared Imager, a pre-Phase A NASA satellite concept based on a full-spectrum spectrometer (visible through shortwave infrared wavelengths) and a multiband thermal sensor
ISS	International Space Station
MAIAC	Multi-Angle Implementation of Atmospheric Correction, an atmospherically corrected collection of MODIS surface reflectance data
MERIS	Medium Resolution Imaging Spectrometer, an instrument on the ESA ENVISAT satellite (2002)
MODIS	Moderate Resolution Imaging Spectroradiometer, the multispectral instruments on NASA's Terra (2000) and Aqua (2002) global coverage satellites
PLSR	Partial least-squares regression, a statistical method utilized in spectroscopy to find the fundamental relations between two matrices

Civilian US satellite instruments with adequate spectral characteristics to track LUE and vegetation stress include MODIS on Terra (since 2000) and Aqua (since 2003), using ocean bands having 1 km+ coarse spatial resolution. We note that Terra and Aqua have already greatly exceeded their life expectancies, so cannot be expected to be available very far into the future. Between 2000 and 2017, EO-1/Hyperion collected time series of VSWIR images, from which the PRI and the fAPAR$_{chl}$ products can still be derived as reference datasets with relatively high spatial resolution sensor (30 m). Unfortunately, none of the US existing or future instruments planned, such as Landsats 8 & 9, NPOESS, NPP, or NOAA-R & S have included the physiologically essential band at ~530 nm in their multiband sensor designs. To follow after HICO, a second VNIR spectrometer, the DLR Earth Sensing Imaging Spectrometer, DESIS, sponsored by Teledyne and DLR (the German Space Agency) will be placed on the ISS in early 2018. Notably, the US National Research Council's 2017 Decadal Survey has recently recommended (January 2018, ESAS, 2017) that NASA continue development of global hyperspectral and thermal capability, which has already been designed into the HyspIRI VSWIR spectrometer/thermal global mission concept (first identified in the 2007 Decadal Survey, NRC, 2007). Meanwhile in 2015, ESA selected the FLuorescence EXplorer (FLEX) as its Explorer 8 mission, to be launched in 2023 in tandem with Sentinel-3C, carrying high–spectral resolution imaging spectrometers to measure fluorescence and reflectance. FLEX mission was specifically designed to retrieve and monitor SIF globally, at a 300 m ecologically relevant spatial scale. Unfortunately, the advancements made in the remote sensing of vegetation function over the past decades that utilize spectral bioindicators for photosynthetic efficiency and stress indicators have not yet been widely incorporated into satellite remote sensing sensor design plans.

5.4.5.1 Satellite Observations: PRI Studies

Opportunities to obtain the PRI from space have been limited to the EO-1/Hyperion (30 m) (Section 5.4.5.3), MODIS (1000 m ocean band) investigations (e.g., Gamon et al., 2016; Middleton et al., 2016), and ESA's programmable VNIR hyperspectral CHRIS-PROBA (<30 m) smallsat.

The MODIS instruments flying on the NASA Aqua and Terra satellites (Justice et al., 1998) have several relatively narrow (10 nm) spectral bands intended for ocean studies in the region needed for PRI, in particular band 11 centered at 531 nm (and reference band 12, 550 nm), but these MODIS bands were not routinely used over land (Rahman et al., 2004) until the recent Collection 6, which processed all the land and ocean bands together, and provided atmospherically corrected products. Since the MODIS sensor lacks the commonly used PRI reference band at 570 nm, investigators used a number of different reference bands, including narrow (10 nm) "ocean" bands centered at 488, 551, 667, and 678 nm (all at 1 km spatial scale), as well as the broad (50 nm) land band centered at 645 nm (500 m resolution) (Rahman et al., 2004; Drolet et al., 2005, 2008; Goerner et al., 2009; Gamon et al., 2016).

Over a decade ago, Drolet et al. (2005, 2008) demonstrated that a MODIS PRI using the ocean bands could track LUE from flux towers in Canadian forests. Several other successful studies have since been conducted using a MODIS PRI based on ocean/land band combinations for ecosystem surveys of LUE in Mediterranean forests (Goerner et al., 2009, 2011; Garbulsky et al., 2014), deciduous and evergreen broadleaf forests (Soudani et al., 2014), a Mediterranean deciduous forest (Guarini et al., 2014), photosynthetic phenology in evergreen conifers (Gamon et al., 2016), and growing season length in northern forest (Ulsig et al., 2017). In the earlier studies that applied an atmospheric correction routine, the 6S model was used. (Vermote et al., 1997). Since 2011, investigators of most studies have applied the multiangle implementation of atmospheric correction (MAIAC) algorithms (Lyapustin et al., 2011), and this processing is now applied operationally.

The consensus from these studies is that the PRI from MODIS can be used to determine ecosystem LUE, although an optimum approach for calculating MODIS PRI has not yet emerged. However, some of these studies relied on top-of-atmosphere (non–atmospherically corrected) imagery or did not adequately account for confounding factors, such as illumination changes, viewing geometry, and time of day. These factors, and the combination of Terra with Aqua observations, were rigorously examined statistically in our multiyear and multisite study of Canadian forests (Middleton et al., 2016) which fused morning Terra and afternoon Aqua MODIS observations ($r^2 = 0.62$) (Figure 5.16). The linear relationship obtained for four forests was similar to the one obtained in our multiyear

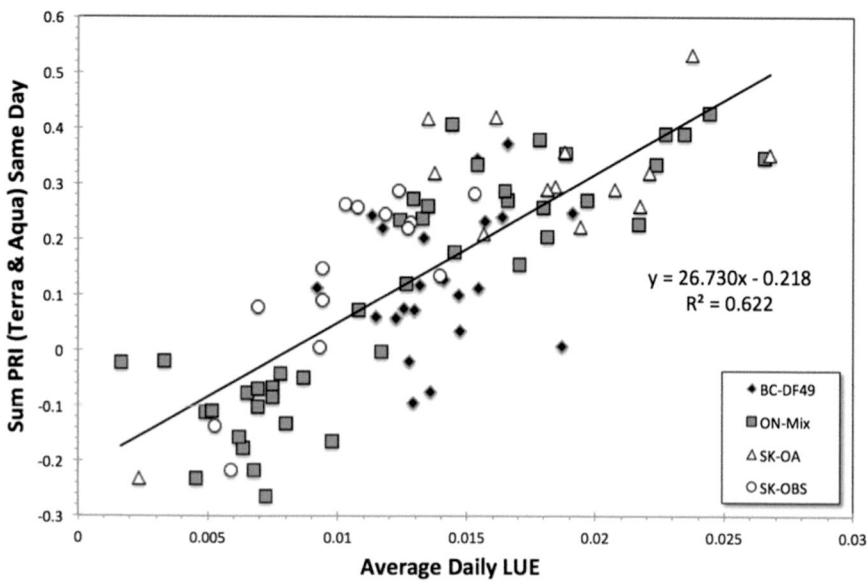

FIGURE 5.16 MODIS narrow ocean bands are used over land to monitor stress responses that inhibit carbon uptake in Canadian forest ecosystems. This study highlights the additional value of off-nadir directional reflectance observations along with the pairing of morning and afternoon satellite observations to improve retrievals of photosynthesis. (From co-author E.M. Middleton.)

variable time of day cornfield study ($r^2 = 0.68$) (Figure 5.9). The view angle and canopy architecture factors were initially examined by Drolet et al. (2005, 2008), who found that viewing direction (back vs. forward scattering) was an important factor influencing the PRI value obtained from MODIS imagery, with sunlit canopy displaying lower PRI than shadowed canopy. Sunlit foliage has lower LUE, and thus lower PRI values, due to increased light-induced stress at high light intensities. This finding revealed that canopy three-dimensional structure was affecting the canopy PRI–LUE relationships, due to sunlit and shaded components. This is consistent with earlier modeling studies (Barton and North, 2001), as well as aircraft or tower measurements that are influenced by canopy structure due to varying amounts of shadowed versus sunlit foliage in the instantaneous instrument field of view (Cheng et al., 2009, 2013; Middleton et al., 2009a).

Other satellite studies were conducted using the Earth Observing 1 (EO-1) hyperspectral spaceborne imager, Hyperion (Middleton et al., 2013), to calculate PRI, which successfully detected drought stress responses in Amazon forests (Asner et al., 2004). Further, data from the ESA CHRIS/PROBA (Compact High Resolution Imaging Spectrometer) satellite sensor were also successfully used to examine LUE (Hilker et al., 2011) with multiangle views.

We provide a new example of the use of EO-1/Hyperion to derive the PRI_{670} and relate it to tower-based NEP at the Duke Forest, NC, USA, which has tree stands comprised primarily of loblolly pine (LP) or mixed deciduous hardwood (HW) (Figure 5.17). Using equations generated from summer

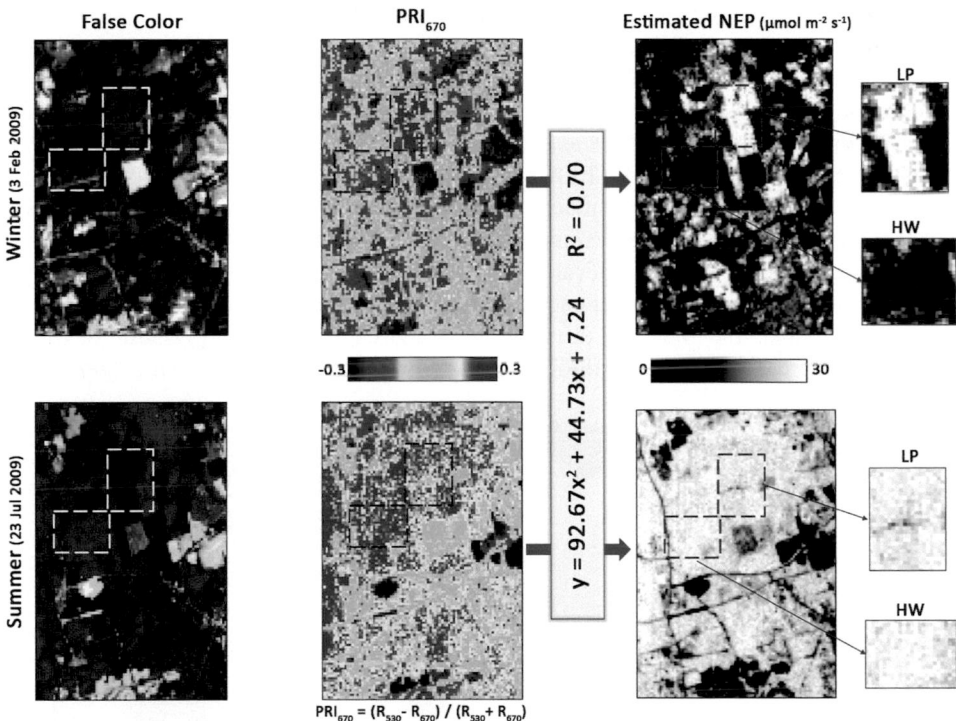

FIGURE 5.17 Difference in ecosystem productivity for deciduous and coniferous forests at Duke Forest, NC, based on the relationship between the narrow band vegetation index (VI) PRI_{670} to net ecosystem production (NEP). Satellite EO-1 Hyperion reflectance time series, covering VSWIR region at 10 nm, were used to derive VIs and determine their ability to capture the seasonal dynamics in ecosystem productivity, at sites equipped with eddy covariance towers (Campbell et al., 2013). At Duke Forest, the modified photochemical reflectance index PRI_{670} using reflectance at 531 and 670, successfully captured the dynamics in NEP for loblolly pine (LP) and hardwood (HW) forests (r^2 to NEP: HW 0.95, LP 0.85, HW + LP 0.74). The study demonstrates the potential of spaceborne spectrometers for providing a generalized approach for tracing the seasonal dynamics in ecosystem function and scaling across multiple ecosystems (Campbell et al., 2013). (From co-author P.K.E. Campbell.)

and winter scenes, maps for NEP were generated ($r^2 = 0.70$). The NEP maps show that multiple pine stands had high winter photosynthetic function (NEP > 20 μmol m^{-2} s^{-1}), and that all forest stands operated at this or higher function during the summer. The PRI maps show that the least stressed stands (red pixels) had the highest NEP, summer and winter. The study demonstrates the potential of space-borne spectrometers for providing a generalized approach for tracing the seasonal dynamics in ecosystem function and scaling across multiple ecosystems (Campbell et al., 2013, 2016a).

5.4.5.2 Satellite Observations: SIF Studies

In 2007, Grace et al. (2007) asked this question: Can we measure terrestrial photosynthesis from space directly, using spectral reflectance and fluorescence? By 2011, this was answered with the demonstration that far-red SIF can be retrieved from space using the FLD approach for Fraunhofer lines in the vicinity of the O_2-A 760 nm feature from several atmospheric chemistry satellites (Frankenberg et al., 2011, 2012; Joiner et al., 2011, 2012; Guanter et al., 2012, 2014). This was shown for global terrestrial maps derived from the GOSAT, Japan's Greenhouse Gases Observing Satellite, launched in January 2009 with a 10.5 km nadir footprint. SIF was also retrieved using Fraunhofer lines surrounding the 740 nm far-red ChlF peak from two ESA atmospheric chemistry missions, SCIAMACHY (32 km $\times$ 215 km), followed by GOME-1/GOME-2 (Global Ozone Monitoring Experiment, with a 40 km $\times$ 80 km nominal footprint) (Joiner et al., 2012, 2013, 2016; Köhler et al., 2015). The GOME-2 instrument carries on operational European meteorological satellites (the EUMETSAT MetOp series) has a revisit time of approximately 1.5 days and an equitorial overpass time of 09:30a.m.

An example of global far-red SIF maps is provided in Figure 5.18, which shows the annually averaged values from 2009 GOME-2 data (Joiner et al., 2013). The NASA satellite developed to obtain precision atmospheric CO_2 concentrations in and around the O_2-A feature with systematically acquired 10 km samples, OCO-2, has also been used to retrieve the far-red SIF near $\sim$760 nm (Frankenberg et al., 2014). Another ESA satellite, the TROPOspheric Monitoring Instrument (TROPOMI), was launched on the Sentinel-5P satellite in late 2017 and will measure CO_2 and SIF at a higher spatial resolution ($\sim$7 km^2) (Gaunter et al., 2015a). None of these atmospheric monitoring satellites were designed with the retrieval of SIF in mind, and therefore, have severe limitations. Validation of the coarse spatial resolution SIF retrievals has not been possible in comparison with ground measurements, and Magnani et al. (2014) urged caution in interpretation of the signal. Recently, modeling SIF and photosynthesis with these satellite inputs has been addressed (Zhang

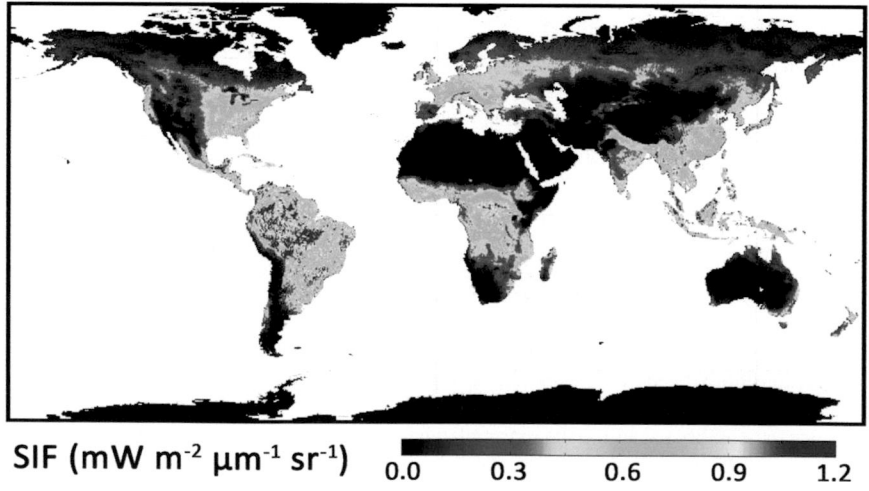

FIGURE 5.18 Global map showing the 2009 annual average of observations for far-red SIF derived from the GOME-2 satellite. (From Y. Yoshida, Poster, *American Geophysical Union 2016 Annual Meeting*, San Francisco, CA.)

et al., 2016b; Thum et al., 2017). The regional and global far-red SIF datasets have been successfully used to determine the onset and length of the growing season in several biomes (Sun et al., 2015), and far-red SIF in northern latitudes has been explored (Jeong et al., 2017; Luus et al., 2017).

In late 2015, the European Space Agency (ESA) selected the FLuorescence EXplorer (FLEX) as the Earth Explorer 8 mission for launch in 2022 (Drusch et al., 2016). FLEX will be the first satellite mission to be specifically designed to retrieve chlorophyll fluorescence across the full fluorescence emission spectrum (650–800 nm) along with hyperspectral reflectance, and the observations will be made globally over land at an ecologically relevant spatial scale of 300 m, approximately monthly. FLEX will be a tandem mission, paired with ESA's Sentinel-3B, which will provide additional reflectance bands and thermal information.

5.4.5.3 Satellite Observations: Hyperspectral Full-Spectrum Studies

Spectroscopy is the only remote sensing remote sensing technology that allows us to analyze the observations made from spaceborne sensors by applying laboratory-like methods, especially VSWIR spectroscopy. Since 2000, there have been several hyperspectral imagers placed in space. Notable are two experimental ESA imagers: CHRIS-Proba with 200 narrow bands that can be combined into wider (e.g., 18) bands at varying spatial coverage $\leq$30 m; and the MERIS (Medium Resolution Imaging Spectrometer) with 15 programmable bands (2.5–30 nm) which flew from 2002 to 2012. The U.S. Navy built the VNIR (6 nm resolution) Hyperspectral Imager for the Coastal Ocean (HICO) and deployed it on the International Space Station for five years (2009–2014). Although designed for coastal observations, it was successfully used to study GEP in forests at five flux towers in a study conducted by Huemmrich et al. (2017). A second VNIR spectrometer (DESIS) sponsored by Teledyne and DLR will be placed on the ISS in early 2018. In addition to these science-driven orbital instruments, there are several commercial civilian ventures that are routinely providing high spatial resolution imagery for a few spectral bands, a topic out of scope here.

The only civilian VSWIR spectrometer in space to date, Hyperion, was carried on NASA's recently decommissioned Earth Observing-1 (EO-1) satellite sampling mission (2000–2017) (Middleton et al., 2013, 2017b). Hyperion spectra have been used for numerous ecosystem studies (e.g., Asner et al., 2005, 2006), including GPP and LUE, which have been evaluated with the PLSR approach (Huemmrich et al., 2017), and PRI (e.g., Asner et al., 2005; Zhang et al., 2016a,b). EO-1/Hyperion has served as the orbital forerunner to the Hyperspectral Infrared Imager (HyspIRI) mission concept that was recently recommended to NASA as a future Designated Mission by the US National Academies of Sciences, Engineering and Medicine 2017 Decadal Survey of Earth Satellites (ESAS, 2017). HyspIRI would combine a full-spectrum VSWIR spectrometer (380–2500 nm) with a multiband IR sensor, collecting global reflectance imagery with high spectral ($\sim$5 nm), high spatial ($\sim$30 m), and a typical two-week repeat cycle, and continuing the previous MODIS, EO-1, and Landsat morning equatorial overpass time of 10–10:30 a.m. (Lee et al., 2015). Other orbital VNIR and VSWIR spectrometers are being developed by several countries including DLR (Germany) (Dotzler et al., 2015; Guanter et al., 2015b), PRISMA (Italy), and a VSWIR imaging spectrometer is under consideration by ESA for its future Sentinel-10 satellite.

The primary source of full-spectrum (hyperspectral) imagery (with discontinuous 10–50 nm wide bands) has been Terra and Aqua MODIS for regional and global studies (https://modis.gsfc.nasa.gov/about/) (Running et al., 2004). We include two examples of using spectral reflectance imagery from space: the first from MODIS data and the second from the HICO instrument which was flown on the International Space Station for four years.

The example drawn from MODIS data (Figure 5.19) features the fAPAR$_{chl}$ and related parameters (Zhang et al., 2005, 2009), and summarizes the 2004 growing season in Nebraska centered on a University of Nebraska EC flux tower site (NE1). The fAPAR$_{chl}$ and fAPAR$_{non-chl}$ parameters are determined as physically based quantities from forward radiative transfer modeling (Zhang et al., 2005, 2009), and the retrieval algorithm does not require ancillary meteorological data or biome type (or plant functional type) information as input. Here, the fAPAR$_{chl}$ and fAPAR$_{non-chl}$

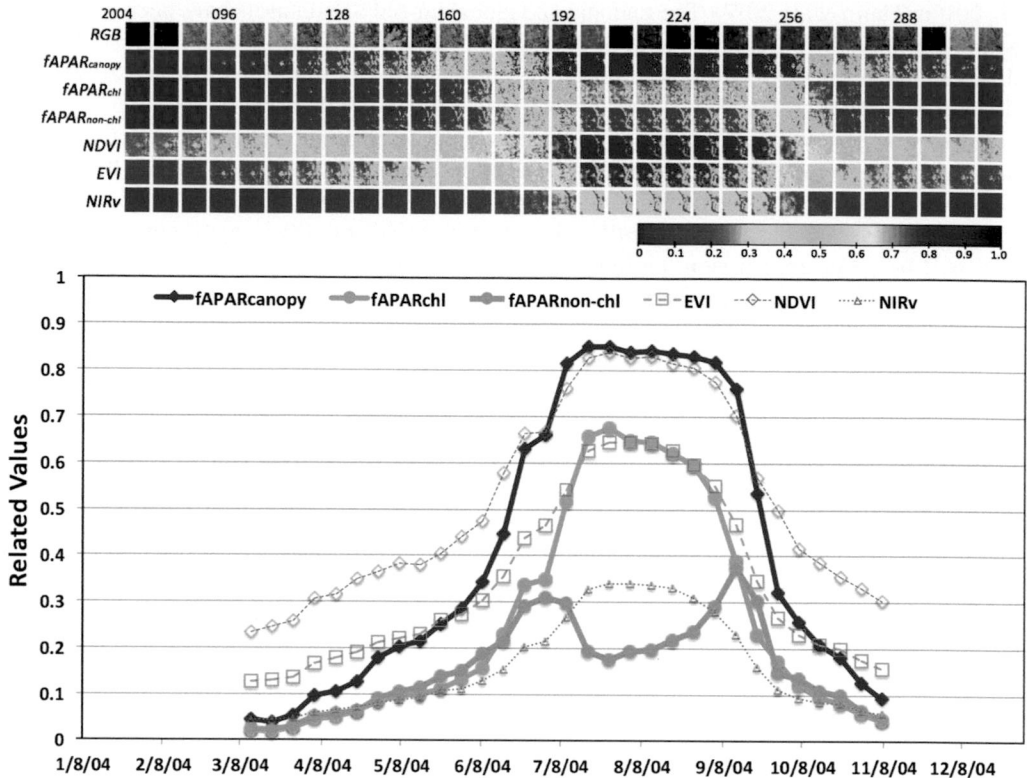

FIGURE 5.19 Satellite data from the NASA Moderate-Resolution Imaging Spectroradiometer (MODIS) centered on a Nebraska cornfield (NE1) were obtained at 8-day intervals throughout the 2004 crop growing season (31 images). The MODIS images of a 50 km by 50 km area were used to derive several products shown in the 7 rows of images in the top section (with a color scale for rows 2–7), and as a graph in the bottom section. Top section—the rows of images are in this order, from top: the red-green-blue band (RGB) composites; fAPAR$_{canopy}$; fAPAR$_{chl}$; fAPAR$_{non-chl}$; NDVI; EVI; and NIRv. Bottom section—the graph displays the seasonal change in fAPAR$_{chl}$ (green curve); fAPAR$_{non-chl}$ (orange curve); total canopy fAPAR$_{canopy}$ (red curve); the EVI (dashed green curve with squares); the NDVI (dotted blue curve with diamonds) and another index, the NIRv (dotted blue curve with triangles, a scaled version of the NDVI (Badgley et al., 2017)). The three fAPAR parameters were derived with the radiative transfer model PROSAIL2 (Zhang et al., 2009). (From co-author, Q. Zhang.)

parameters derived as physically based quantities are summed to obtain fAPAR$_{canopy}$, which yields values comparable to the NDVI in the midseason (∼0.85). The NDVI obviously overestimates the green vegetation fraction estimated more realistically by fAPAR$_{chl}$ and the EVI (∼0.65) but matches quite well the total canopy parameter, fAPAR$_{canopy}$, during the summertime. The two variables describing the total canopy (NDVI, fAPAR$_{canopy}$) are more constant during summer than fAPAR$_{chl}$ and EVI, and NDVI describes a longer growing season than indicated by fAPAR$_{chl}$. Only the variable fAPAR$_{chl}$ successfully describes both the seasonal change and the magnitude of the relative fraction of chlorophyll-containing canopies, and does not overestimate the early and late seasonal periods, as typically occurs with the two reflectance indices (NDVI, EVI). Furthermore, fAPAR$_{chl}$ can be combined with MODIS PRI or other satellite sources to track LUE (Zhang et al., 2016a,b). The variable fAPAR$_{non-chl}$ varies seasonally. As expected, the magnitude of NIRv (a scaled version of NDVI, Badgley et al., 2017) is lower than the magnitude of NDVI.

Clearly, estimates of GPP based on NDVI (which includes green and nongreen contributions) will be overestimated, as compared with estimates derived from fAPAR$_{chl}$ and the EVI. The

APAR needed for LUE estimates can be determined directly from $fAPAR_{chl}$ and the incident PAR (i.e., LUE = GPP/[$fAPAR_{chl}$*PAR]). Because they can be computed from any VSWIR remote sensing imagery, the $fAPAR_{chl}$ parameters should prove essential in future global mapping and modeling for ecosystem carbon studies, such as those accomplished with the Community Land Model.

The second example (Huemmrich et al., 2017) uses imagery from the Hyperspectral Imager for the Coastal Ocean (HICO) deployed for five years on the International Space Station (2009–2014). The oblique ISS orbit enables data collections at variable time intervals and times of day. By employing advanced statistical methods, such as the partial least-squares regression (PLSR) approach, the capability to monitor GEP from space from hyperspectral data is demonstrated (Figure 5.20). Here, the GEP was derived from PLSR analyses, which determines a composite spectral quantity from each image collected at four EC flux sites viewed from the ISS, and matched with the EC data collected in the nearest half-hour interval. An excellent retrieval was achieved ($r^2 = 0.87$) for sites representing grassland, shrub, and forest covers.

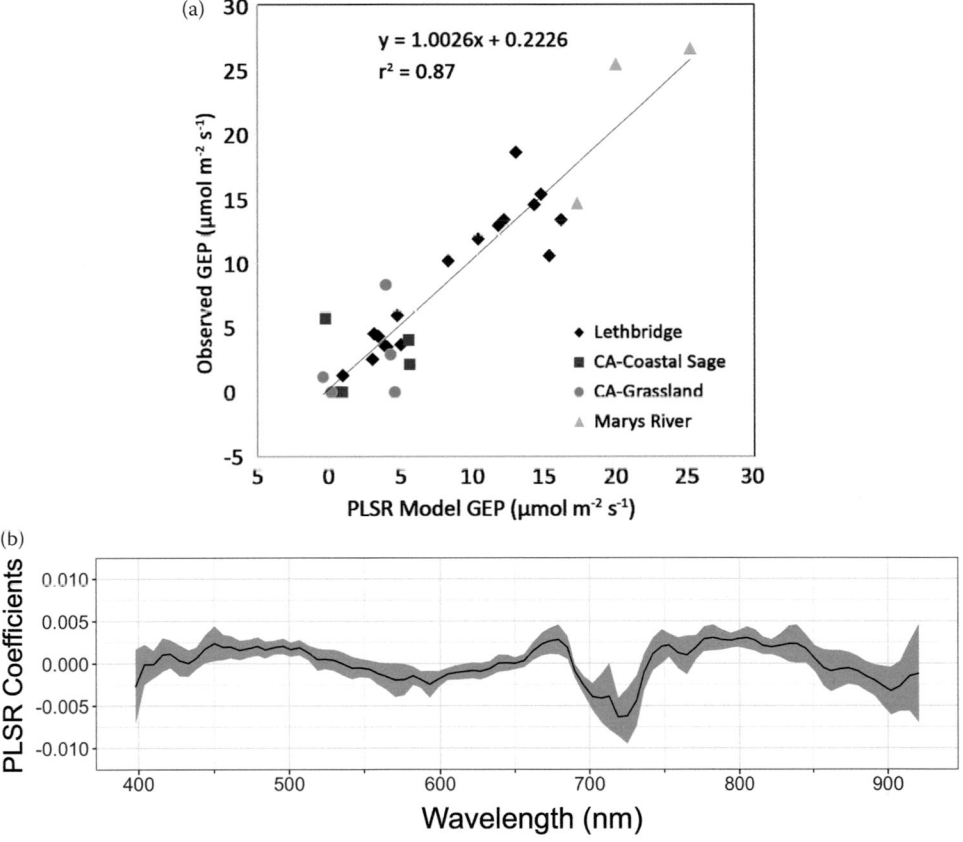

FIGURE 5.20   (a) Half-hourly GEP at time of overpass from EC compared to GEP estimated from HICO spectral reflectance using partial least-squares regression (PLSR) on 92 spectral bands between 398 and 920 nm. The different point shapes indicate data from different sites. Observations were collected from multiple sites with different vegetation types, at different times during the growing season and at different times of the day. (b) PLSR coefficients for each wavelength band for a regression on half-hourly GEP at the time of the overpass for all sites combined. The line indicates magnitude and sign of the PLSR coefficients for each wavelength band (with ±1 standard deviation from random cross-validation in gray). (From co-author, Huemmrich, K.F. et al. 2017. *IEEE Journal of Selected Topics in Applied Earth Observations and Remote Sensing*, 10(10), 4360–4375. doi 10.1109/JSTARS.2017.2725825.)

5.5 CONCLUSIONS

Environmental pressures on terrestrial vegetation, both from natural and human causes, induce physiological stress responses that result in changes in ecosystem processes, such as vegetation functioning and carbon exchange. These stress responses also affect the optical properties of plants. The new field of study addressing spectral bioindicators links observations of plant optical properties to ecosystem physiology. This approach can nondestructively monitor ecosystem health at a range of scales, providing a method that can connect surface, aircraft and satellite observations. Spectral bioindicators have been shown to provide unique information about ecosystem health and physiology that can be incorporated into process models and coupled climate carbon models to better assess the feedbacks between ecosystems and the climate system.

Our expectation is that improved and newly implemented spectral bioindicators will be able to detect photosynthetic stress at regional and global scales from satellites, aircraft, and ground-based sensors with greater accuracy over the coming decades, which will lead to lower uncertainties in the global carbon dynamics, including LUE and GEP. Narrow and/or continuous high spectral resolution data, including a narrow band centered at 531 nm along with appropriate reference bands such as 570, 551, 488, and 670 nm, are essential for obtaining viable instantaneous estimates of ecosystem stress responses. Interpretation of these remotely determined responses will be even more powerful when combined with other spectral indices (e.g., Thenkabail et al., 2000) and by using rigorous statistical and analytical methods that utilize the full VSWIR spectrum. Nevertheless, viewing geometry, soil background, and image preprocessing can significantly affect the retrieved measurements, and generalized approaches for mitigating these effects have yet to be implemented. However, these challenges can be overcome through sophisticated data processing methods such as cloud computing and machine learning, as well as with new observations from future orbital sensors that will provide greater spatial, spectral, radiometric, and temporal resolutions (including diurnal collections).

Advances in instruments that decrease costs and increase availability for ground-based and aircraft operation will make it easier to utilize spectral bioindicators locally, including SIF. While satellite imaging produces regular synoptic views, imaging spectrometers on aircraft can provide observations over specific areas at high spatial resolutions on demand. Utilizing unmanned aerial vehicles (UAV) will further decrease the cost and increase availability of spectral bioindicator data. Further, radiometers mounted on towers or within canopies can be integrated into sensor networks for continuous detailed vegetation monitoring.

There is a strong practical need to develop approaches to monitor plant physiological responses in a world with increasing population intensifying demands on forest and agricultural systems, coupled with heightened environmental stresses from a changing climate. It is presently difficult to acquire timely information on ecosystem productivity patterns in space and time, which leads to a critical need for new methods to monitor the ongoing changes in both human-managed and natural ecosystems. The spectral bioindicators approach may be a part of a solution to this problem by providing scalable temporal and spatial observations of vegetation production along with diagnostic information on the nature of and response to environmental stresses.

The 2017–2027 Decadal Survey for Earth Science and Applications from Space (ESAS, 2017) recommended that NASA designate an orbital mission as one of the key priorities for the 2017–2027 period. That mission should quantify the distribution of global ecosystem's functional traits, functional types, and composition, to better understand how and why they are changing spatially over time. There is a need to accurately monitor the key parameters governing vegetation function at a temporal scale, relevant to their dynamics (10–16 days), and map them at a spatial scale that allows practical assessment and management of their impacts on the hydrological and biological cycles. Long-term observations are required for environmental monitoring and change detection as ecosystems cycle through seasonal changes and respond to environmental conditions, climate change, and human alterations. Currently, no single sensor can provide data characterizing vegetation function at the desired spectral, temporal and spatial scales. Fusing spectral bioindicators from multiple sources from hyperspectral and multispectral orbital sources will help to quantify the key parameters describing ecosystem function and productivity, and their temporal trends. Better information on ecosystem composition, functioning and fluxes will support improved scientific understanding, applications, and decision-making.

ACKNOWLEDGMENTS

We thank NASA's Terrestrial Ecology Program for support of E.M.'s spectral bioindicator project. We thank several colleagues for assistance with data processing and presentation materials, including Dr. Sergio Bernardes (U. Georgia, formerly NASA post-doc) and Lawrence Corp (Science Systems and Analysis, Inc.). We are also grateful to Drs. Hank Margolis and Yen-Ben Cheng for their insights and work on the earlier version of this chapter.

REFERENCES

Ac, A. et al. 2015. Meta-analysis assessing potential of steady-state chlorophyll fluorescence for remote sensing detection of plant water, temperature and nitrogen stress. *Remote Sensing of Environment*, 168, 420–436.

Agati, G. et al. 1995. The F685/F730 chlorophyll fluorescence ratio as a tool in plant physiology: Response to physiological and environmental factors. *Journal of Plant Physiology*, 145, 228–238.

Alonso, L. et al. 2008. Improved Fraunhofer line discrimination method for vegetation fluorescence quantification. *Geoscience and Remote Sensing Letters, IEEE*, 5(4), 620–624.

Alonso, L. et al. 2017. Diurnal cycle relationships between passive fluorescence, PRI and NPQ of vegetation in a controlled stress experiment. *Remote Sensing*, 9(8), 770. doi: 10.3390/rs9080770

Alton, P.B., North, P.R., and Los, S.O. 2007. The impact of diffuse sunlight on canopy light use efficiency, gross photosynthetic product and net ecosystem exchange in three forest biomes. *Global Change Biology*, 13(4), 776–787.

Anderson, M.C. et al. 2000. An analytical model for estimating canopy transpiration and carbon assimilation fluxes based on canopy light-use efficiency. *Agricultural and Forest Meteorology*, 101(4), 265–289.

Asner, G.P. 2005. Substrate age and precipitation effects on Hawaiian forest canopies from spaceborne imaging spectroscopy. *Remote Sensing of Environment*, 98(4), 457–467.

Asner, G.P. et al. 2004. Drought stress and carbon uptake in an Amazon forest measured with spaceborne imaging spectroscopy. *Proceedings of the National Academy of Sciences*, 101(16), 6039–6044.

Asner, G.P. et al. 2006. Vegetation–climate interactions among native and invasive species in Hawaiian rainforest. *Ecosystems*, 9(7), 1106–1117.

Atherton, J. et al. 2016. Using spectral chlorophyll fluorescence and the photochemical reflectance index to predict physiological dynamics. *Remote Sensing of Environment*, 176, 17–30.

Badgley, G. et al. 2017. Canopy near-infrared reflectance and terrestrial photosynthesis. *Science Advances*, 3, e1602244.

Baker, I.T. et al. 2008. Seasonal drought stress in the Amazon: Reconciling models and observations. *Journal of Geophysical Research*, 113(G00B011). http://dx.doi.org/10.1029/2007JG000644

Baker, N.R. 2008. Chlorophyll fluorescence: A probe of photosynthesis in vivo. *Annual Review of Plant Biology*, 59, 89–113.

Baldocchi, D. 2003. Assessing the eddy covariance technique for evaluating carbon dioxide exchange rates of ecosystems: Past, present and future. *Global Change Biology*, 9, 479–492.

Baldocchi, D. 2008. Turner Review No. 15. 'Breathing' of the terrestrial biosphere: Lessons learned from a global network of carbon dioxide flux measurement systems. *Australian Journal of Botany*, 56(1), 1–26.

Baldocchi, D. et al. 2001. FLUXNET: A new tool to study the temporal and spatial variability of ecosystem-scale carbon dioxide, water vapor, and energy flux densities. *Bulletin of the American Meteorological Society*, 82(11), 2415–2434.

Barton, C.V.M., and North, P.R.J. 2001. Remote sensing of canopy light use efficiency using the photochemical reflectance index: Model and sensitivity analysis. *Remote Sensing of Environment*, 78, 264–273.

Beer, C. et al. 2010. Terrestrial gross carbon dioxide uptake: Global distribution and covariation with climate. *Science*, 329(5993), 834–838.

Berk, A. et al. 2014. MODTRAN6: A major upgrade of the MODTRAN radiative transfer code. *Proceedings of the International Society for Optics and Photonics (SPIE)*. 13 June 2014.

Bernacchi, C.J. et al. 2002. Temperature response of mesophyll conductance. Implications for the determination of rubisco enzyme kinetics and for limitations to photosynthesis in vivo. *Plant Physiol*, 130(4): 1992–1998. doi: 10.1104/pp.008250

Bilger, W., and Björkman, O. 1990. Role of the xanthophyll cycle in photoprotection elucidated by measurements of light-induced absorbance changes, fluorescence and photosynthesis in leaves of Hedera canariensis. *Photosynthesis Research*, 25(3), 173–185.

Bloom, A.A. et al. 2016. Retrievals of terrestrial carbon allocation, pools, and residence times. *Proceedings of the National Academy of Sciences*, 113, 1285–1290.

Boardman, N.K. 1977. Comparative photosynthesis of sun and shade plants. *Annual Review of Plant Physiology*, 28(1), 355–377.

Bolhàr-Nordenkampf, H.R. et al. 1989. Chlorophyll fluorescence as a probe of the photosynthetic competence of leaves in the field: A review of current instrumentation. *Functional Ecology*, 3, 497–514.

Bonan, G.B. et al. 2011. Improving canopy processes in the Community Land Model version 4 (CLM4) using global flux fields empirically inferred from FLUXNET data. *Journal of Geophysical Research*, 116, G02014.

Bousquet, P. et al. 2000. Regional changes in carbon dioxide fluxes of land and oceans since 1980. *Science*, 290(5495), 1342–1346.

Buschmann, C. 2007. Variability and application of the chlorophyll fluorescence emission ratio red/far-red of leaves. *Photosynthesis Research*, 92(2), 261–271.

Campbell, P.K.E. et al. 2007. Assessment of vegetation stress using reflectance or fluorescence measurements. *Journal of Environmental Quality*, 36, 832–845.

Campbell, P.K.E. et al. 2008. Contribution of chlorophyll fluorescence to the apparent vegetation reflectance. *Science of the Total Environment*, 404, 433–439.

Campbell, P.K.E. et al. 2013. EO-1 Hyperion reflectance time series at calibration and validation sites: Stability and sensitivity to seasonal dynamics. *IEEE Journal of Selected Topics in Applied Earth Observations and Remote Sensing (JSTARS)*, 6(2), 276–290.

Campbell, P.K.E. et al. 2014. Diurnal and phenological changes in vegetation fluorescence and reflectance, indicative of vegetation photosynthetic properties and function. *Proceedings of the 5th International Workshop on Remote Sensing of Vegetation Fluorescence*, Paris, France, April 2014, 12pp.

Campbell, P.K.E. et al. 2016a. EO-1 Spectral Time Series for Ecosystem Function/Production. *6th Annual HyspIRI Data Product Symposium and Aquatic Forum, NASA/GSFC*, Greenbelt, MD.

Campbell, P.K.E. et al. 2016b. Diurnal and Seasonal trends in Chlorophyll Fluorescence and Xanthophyll Cycle, Associated with Photosynthetic Function and CO2 Assimilation in Corn. *SAIL35 Symposium*, Enschede, Netherlands, 26–28 September 2016.

Canadell, J.G. et al. 2007. Contributions to accelerating atmospheric CO_2 growth from economic activity, carbon intensity, and efficiency of natural sinks. *Proceedings of the National Academy of Sciences*, 104(47), 18866–18870.

Cech, J.J.J., Wilson, B.W., and Crosby, D.G. (Eds.). 1998. *Multiple Stresses in Ecosystems*. Boca Raton, Florida: CRC Press.

Cerovic, Z.G. et al. 1999. Ultraviolet-induced fluorescence for plant monitoring: Present state and prospects. *Agronomie*, 19(7), 543–578.

Chappelle, E.W., and Williams, D.L. 1987. Laser induced fluorescence from plant foliage. *IEEE Transactions in Geoscience and Remote Sensing*, GE-25, 726–736.

Chen, J.M. et al. 1999. Daily canopy photosynthesis model through temporal and spatial scaling for remote sensing applications. *Ecological Modelling*, 124(2–3), 99–119.

Cheng, Y.-B. et al. 2009. Dynamics of spectral bio-indicators and their correlations with light use efficiency using directional observations at a Douglas-fir forest. *Measurement Science and Technology*, 20(9), 095107.

Cheng, Y.-B. et al. 2010. Utilizing in situ directional hyperspectral measurements to validate bio-indicator simulations for a corn crop canopy. *Ecological Informatics*, 5(5), 330–338.

Cheng, Y.-B. et al. 2013. Integrating solar induced fluorescence and the photochemical reflectance index for estimating gross primary production in a cornfield. *Remote Sensing*, 5, 6857–6879.

Cheng, Y.-B. et al. 2014. Impacts of light use efficiency and fPAR parameterization on gross primary production modeling. *Agricultural and Forest Meteorology*, 189, 187–197.

Chou, S. et al. 2017. Canopy-level photochemical reflectance index from hyperspectral remote sensing and leaf-level non-photochemical quenching as early indicators of water stress in maize. *Remote Sensing*, 9(8), 794.

Cogliati, S. et al. 2015. Retrieval of sun-induced fluorescence using advanced spectral fitting methods. *Remote Sensing of Environment*, 169, 344–357.

Coops, N.C. et al. 2010. Estimation of light-use efficiency of terrestrial ecosystems from space: A status report. *Bioscience*, 60(10), 788–797.

Corp, L.A. et al. 2003. Fluorescence sensing systems: *In vivo* detection of biophysical variations in field corn due to nitrogen supply. *Remote Sensing Environment*, 86, 470–479.

Corp, L.A. et al. 2010. FUSION: A fully ultraportable system for imaging objects in nature. *IEEE International Geoscience & Remote Sensing Symposium (IGARSS)*, Honolulu, HI.

Costanza, R. et al. 1997. The value of the world's ecosystem services and natural capital. *Nature*, 387(6630), 253–260.

Cui, T. et al. 2017. Estimating diurnal courses of gross primary production for maize: A comparison of sun-induced chlorophyll fluorescence, light-use efficiency and process-based models. *Remote Sensing*, 9(12), 1267.

D'Ambrosio, N. et al. 1992. Increase of the chlorophyll fluorescence ratio F690/F735 during the autumnal chlorophyll breakdown. *Radiation and Environmental Biophysics*, 31(1), 51–62.

Damm, A. et al. 2010. Remote sensing of sun-induced fluorescence to improve modeling of diurnal courses of gross primary production (GPP). *Global Change Biology*, 16, 171–186.

Damm, A. et al. 2014. FLD-based retrieval of sun-induced chlorophyll fluorescence from medium spectral resolution airborne spectroscopy data. *Remote Sensing of Environment*, 147, 256–266.

Damm, A. et al. 2015a. Impact of varying irradiance on vegetation indices and chlorophyll fluorescence derived from spectroscopy data. *Remote Sensing of Environment*, 156, 202–215.

Damm, A. et al. 2015b. Far-red sun-induced chlorophyll fluorescence shows ecosystem-specific relationships to gross primary production: An assessment based on observational and modeling approaches. *Remote Sensing of Environment*, 166, 91–105.

Dash, J., and Curran, P.J., 2007. Evaluation of the MERIS terrestrial chlorophyll index (MTCI). *Advances in Space Research*, 9(3), 100–104.

Dash, J. et al. 2010. Validating the MERIS Terrestrial Chlorophyll Index (MTCI) with ground chlorophyll content data at MERIS spatial resolution. *International Journal of Remote Sensing*, 31, 5513–5532.

Daughtry, C.S.T. et al. 1982. *Spectral estimates of solar radiation intercepted by corn canopies*. West Lafayette, Indiana: Purdue University. AgRISTARS Technical Report, sr-PZ-04236.

Daughtry, C.S.T. et al. 1992. Spectral estimates of absorbed radiation and phytomass production in corn and soybean canopies. *Remote Sensing of Environment*, 39(2), 141–152.

De Groot, R.S., Wilson, M.A., and Boumans, R.M.J. 2002. A typology for the classification, description and valuation of ecosystem functions, goods and services. *Ecological Economics*, 41(3), 393–408.

De Lucia, E.H. 1991. Photosynthetic symmetry of sun and shade leaves of different orientations. *Oecologia*, 87(1), 51–57. doi: 10.1007/BF00323779.

De Pury, D.G.G., and Farquhar, G.D. 1997. Simple scaling of photosynthesis from leaves to canopies without the errors of big-leaf models. *Plant Cell and Environment*, 20(5), 537 557.

Demmig-Adams, B. 1998. Survey of thermal energy dissipation and pigment composition in sun and shade leaves. *Plant and Cell Physiology*, 39(5), 474–482.

Demmig-Adams, B. 2003. Linking the xanthophyll cycle with thermal energy dissipation, *Photosynthesis Research*, 76: 73.

Demmig-Adams, B., and Adams, W.W. 2000. Photosynthesis: Harvesting sunlight safely. *Nature*, 403(6768), 371–374.

Denman, K.L. et al. 2007. Couplings between changes in the climate system and biogeochemistry. In S. Solomon, D. Qin, M. Manning, Z. Chen, M. Marquis, K.B. Avery, M. Tignor and H.L. Miller (Eds.), *Climate Change 2007: The Physical Science Basis. Contribution of Working Group I to the Fourth Assessment Report of the Intergovernmental Panel on Climate Change*. Cambridge, United Kingdom: Cambridge University Press.

Desai, A.R. et al. 2008. Cross-site evaluation of eddy covariance GPP and RE decomposition techniques. *Agricultural and Forest Meteorology*, 148(6–7), 821–838.

Dias, J. 2007. Physiological aspects of sun and shade leaves of Lithraea molleoides (Vell.) Engl. (Anacardiaceae). *Brazilian Archives of Biology and Technology*, 50(1), 91-99.

Dobrowski, S.Z. et al. 2005. Simple reflectance indices track heat and water stress-induced changes in steady-state chlorophyll fluorescence at the canopy scale. *Remote Sensing of Environment*, 97(3), 403–414.

Dörken, V.M. and Lepetit, B. 2018. Morpho-anatomical and physiological differences between sun and shade leaves in Abies alba Mill. (Pinaceae, Coniferales): A combined approach. *Plant Cell Environ.*, 41(7), 1683-1697. doi: 10.1111/pce.13213.

Dotzler, S. et al. 2015. The potential of EnMAP and Sentinel-2 for detecting drought stress phenomena in deciduous forest communities. *Remote Sensing*, 7(10), 14227–14258. http://dx.doi.org/10.3390/rs71014227

Drewry, D. et al. 2015. The chlorophyll fluorescence imaging spectrometer (CFIS): A new airborne instrument for quantifying solar-induced fluorescence. *American Geophysical Union, Fall Meeting 2015*, abstract id. B54D-04.

Drolet, G.G. et al. 2005. A MODIS-derived Photochemical Reflectance Index to detect interannual variations in the photosynthetic light-use efficiency of a boreal deciduous forest. *Remote Sensing of Environment*, 98, 212–224.

Drolet, G.G. et al. 2008. Regional mapping of gross light-use efficiency using MODIS spectral indices. *Remote Sensing of Environment*, 12, 3064–3078.

Drusch, M. et al. 2016. The Fluorescence EXplorer (FLEX) mission concept—ESA's Earth Explorer 8 (EE8). *IEEE Transactions on Geoscience and Remote Sensing*, 55, 1273–1284.

ESAS 2017. *National Academies of Sciences, Engineering, and Medicine. 2018. Thriving on Our Changing Planet: A Decadal Strategy for Earth Observation from Space*. Washington, DC: The National Academies Press. https://doi.org.10.17226/24938. http://sites.nationalacademies.org/DEPS/ESAS2017/index.htm

Falge, E. et al. 2001. Gap filling strategies for defensible annual sums of net ecosystem exchange. *Agricultural and Forest Meteorology*, 107(1), 43–69.

Falkowski et al. 1998. Biogeochemical controls and feedbacks on ocean primary production. *Science*, 281(5374), 200–206.

Field, C.B. 1991. Ecological scaling of carbon gain to stress and resource availability. In H.A. Mooney, W.E. Winner and E.J. Pell (Eds.), *Integrated Responses of Plants to Stress*. San Diego: Academic Press.

Figueres, C. et al., 2017. Three years to safeguard our climate. *Nature*, 546, 593–595. doi: 10.1038/546593a.

Filella, I. et al. 1996. Relationship between photosynthetic radiation-use efficiency of barley canopies and the photochemical reflectance index (PRI). *Physiologia Plantarum*, 96(2), 211–216.

Filella, I. et al. 2009. PRI assessment of long-term changes in carotenoids/chlorophyll ratio and short-term changes in de-epoxidation state of the xanthophyll cycle. *International Journal of Remote Sensing*, 30, 4443–4455.

FLUXNET 2015. *FLUXNET-ORNL*. https://fluxnet.ornl.gov/, hosted at the USA: Oak Ridge National Laboratory.

Forkel, M. 2016. Enhanced seasonal CO_2 exchange caused by amplified plant productivity in northern ecosystems. *Science*, 352, 696–699.

Frankenberg, C. et al. 2011. New global observations of the terrestrial carbon cycle from GOSAT: Patterns of plant fluorescence with gross primary productivity. *Geophysical Research Letters*, 38, L17706.

Frankenberg, C. et al. 2012. Remote sensing of near-infrared chlorophyll fluorescence from space in scattering atmospheres: Implications for its retrieval and interferences with atmospheric CO_2 retrievals. *Atmospheric Measurement Techniques*, 5, 2081–2094.

Frankenberg, C. et al. 2014. Prospects for chlorophyll fluorescence remote sensing from the Orbiting Carbon Observatory-2. *Remote Sensing of Environment*, 147, 1–12.

Frouin, R., and Pinker, R.T. 1995. Estimating photosynthetically active radiation (PAR) at the earth's surface from satellite observations. *Remote Sensing of Environment*, 51(1), 98–107.

Fuentes, D.A. et al. 2006. Mapping carbon and water vapor fluxes in a chaparral ecosystem using vegetation indices derived from AVIRIS. *Remote Sensing of Environment*, 103(3), 312–323.

Gamon, J.A., and Berry, J. 2012. Facultative and constitutive pigment effects on the Photochemical Reflectance Index (PRI) in sun and shade conifer needles. *Israel Journal of Plant Sciences*, 60, 85–95.

Gamon, J.A. et al. 1990. Remote sensing of the xanthophyll cycle and chlorophyll fluorescence in sunflower leaves and canopies. *Oecologia*, 85(1), 1–7.

Gamon, J.A. et al. 1992. A narrow-waveband spectral index that tracks diurnal changes in photosynthetic efficiency. *Remote Sensing of Environment*, 41(1), 35–44.

Gamon, J.A. et al. 1997. The photochemical reflectance index: An optical indicator of photosynthetic radiation use efficiency across species, functional types, and nutrient levels. *Oecologia*, 112(4), 492–501.

Gamon, J.A. et al. 2001. Assessing photosynthetic downregulation in sunflower stands with an optically-based model. *Photosynthesis Research*, 67(1), 113–125.

Gamon, J.A. et al. 2006a. A mobile tram system for systematic sampling of ecosystem optical properties. *Remote Sensing of Environment*, 103(3), 246–254.

Gamon, J.A. et al. 2006b. Spectral Network (SpecNet)–What is it and why do we need it?. *Remote Sensing of Environment*, 103(3), 227–235.

Gamon, J.A. et al. 2007. Ecological Applications of Remote Sensing at Multiple Scales. In: Pugnaire, F.I., Valladares, F. (Eds) *Plant Functional Ecology*, Second Edition. Boca Raton, FL: CRC Press, pp. 655–684.

Gamon, J.A. et al. 2016. A remotely sensed pigment index reveals photosynthetic phenology in evergreen conifers. *Proceedings of the National Academy of Sciences*, 113(46), 13087–13092.

Garbulsky, M.F. et al. 2011. The photochemical reflectance index (PRI) and the remote sensing of leaf, canopy and ecosystem radiation use efficiencies: A review and meta-analysis. *Remote Sensing of Environment*, 115(2), 281–297.

Garbulsky, M.F. et al. 2014. Photosynthetic light use efficiency from satellite sensors: From global to Mediterranean vegetation. *Environmental and Experimental Botany*, 103, 3–11.

Garrity, S.R. et al. 2010. A simple filtered photodiode instrument for continuous measurement of narrowband NDVI and PRI over vegetated canopies. *Agricultural and Forest Meteorology*, 150(3), 489–496.

Garrity, S.R. et al. 2011. Disentangling the relationships between plant pigments and the photochemical reflectance index reveals a new approach for remote estimation of carotenoid content. *Remote Sensing of Environment*, 115, 628–635.

Genty, B. et al. 1989. The relationship between the quantum yield of photosynthetic electron-transport and quenching of chlorophyll fluorescence. *Biochimica et Biophysica Acta*, 990, 87–92.

Gilmore, A.M., and Yamasaki, H. 1998. 9-Aminoacridine and dibucaine exhibit competitive interactions and complicated inhibitory effects that interfere with measurements of ΔpH and xanthophyll cycle-dependent Photosystem II energy dissipation. *Photosynthesis Research*, 57(2), 159–174.

Gitelson, A.A., and Gamon, J.A. 2015. The need for a common basis for defining light-use efficiency: Implications for productivity estimation. *Remote Sensing of Environment*, 156, 196–201.

Gitelson, A.A. et al. 1999. The chlorophyll fluorescence ratio F_{735}/F_{700} as an accurate measure of the chlorophyll content in plants. *Remote Sensing of Environment*, 69, 296-302.

Gitelson, A.A. et al. 2002. Assessing carotenoid content in plant leaves with reflectance spectroscopy. *Photochemistry and Photobiology*, 75, 272–281.

Gitelson, A.A. et al. 2005. Remote estimation of canopy chlorophyll content in crops. *Geophysical Research Letters*, 32(8), L08403.

Gitelson, A.A. et al. 2012. Remote estimation of crop gross primary production with Landsat data. *Remote Sensing of Environment*, 121, 404–414.

Goerner, A. et al. 2009. Tracking seasonal drought effects on ecosystem light use efficiency with satellite-based PRI in a Mediterranean forest. *Remote Sensing of Environment*, 113(5), 1101–1111.

Goerner, A. et al. 2011. Remote sensing of ecosystem light use efficiency with MODIS based PRI. *Biogeosciences*, 8, 189–202.

Goetz, A.F.H., and Curtiss, B. 1996. Hyperspectral imaging of the earth: Remote analytical chemistry in an uncontrolled environment. *Field Analytical Chemistry & Technology*, 1(2), 67–76.

Goetz, S.J., and Prince, S.D. 1999. Modelling terrestrial carbon exchange and storage: evidence and implications of functional convergence in light-use efficiency. In A.H. Fitter and D. Raffaelli (Eds.), *Advances in Ecological Research* (pp. 57–92): Academic Press.

Goulden, M.L. et al. 2011. Patterns of NPP, GPP, respiration, and NEP during boreal forest succession. *Global Change Biology*, 17(2), 855–871.

Gower, S.T. et al. 1999. Direct and indirect estimation of leaf area index, *f*APAR, and net primary production of terrestrial ecosystems. *Remote Sensing of Environment*, 70(1), 29–51.

Grace, J. et al. 2007. Can we measure terrestrial photosynthesis from space directly, using spectral reflectance and fluorescence? *Global Change Biology*, 13, 1484–1497. https://doi.org/10.1111/j.1365-2486.2007.01352.x

Green, D.S. et al. 2003. Foliar morphology and canopy nitrogen as predictors of light-use efficiency in terrestrial vegetation. *Agricultural and Forest Meteorology*, 115(3–4), 163–171.

Green, R.O. et al. 1998. Imaging spectroscopy and the airborne visible/infrared imaging spectrometer (AVIRIS). *Remote Sensing of Environment*, 65(3), 227–248.

Gu, L. et al. 2002. Advantages of diffuse radiation for terrestrial ecosystem productivity. *Journal of Geophysical Research*, 107(D6), 4050.

Guanter, L. et al. 2012. Retrieval and global assessment of terrestrial chlorophyll fluorescence from GOSAT space measurements. *Remote Sensing of Environment*, 121, 236–251.

Guanter, L. et al. 2013. Using field spectroscopy to assess the potential of statistical approaches for the retrieval of sun-induced chlorophyll fluorescence from space. *Remote Sensing of Environment*, 133, 52–61.

Guanter, L. et al. 2014. Global and time-resolved monitoring of crop photosynthesis with chlorophyll fluorescence. *Proceedings of the National Academy of Sciences*, 111, E1327–E1333.

Guanter, L. et al. 2015a. Potential of the TROPOspheric Monitoring Instrument (TROPOMI) onboard the Sentinel-5 Precursor for the monitoring of terrestrial chlorophyll fluorescence. *Atmospheric Measurement Techniques*, 8, 1337–1352.

Guanter, L. et al. 2015b. The EnMAP spaceborne imaging spectroscopy mission for earth observation. *Remote Sensing*, 7(7), 8830–8857.

Guarini, R. et al. 2014. The utility of MODIS-sPRI for investigating the photosynthetic light-use efficiency in a Mediterranean deciduous forest. *International Journal of Remote Sensing*, 35(16), 6157–6172.

Guo, J.M., and Trotter, C.M. 2006. Estimating photosynthetic light-use efficiency using the photochemical reflectance index: The effects of short-term exposure to elevated CO_2 and low temperature. *International Journal of Remote Sensing*, 27(20), 4677–4684.

Havaux, M., and Niyogi, K.K. 1999. The violaxanthin cycle protects plants from photooxidative damage by more than one mechanism. *Proceedings of the National Academy of Sciences*, 96(15), 8762–8767.

Heinsch, F.A. et al. 2003. *User's Guide: GPP and NPP (MOD17A2/A3) Products. NASA MODIS Land Algorithm. Version 1.3.* http://www.ntsg.umt.edu/project/modis/mod17.php

Hilker, T. et al. 2007. Instrumentation and approach for unattended year round tower based measurements of spectral reflectance. *Computers and Electronics in Agriculture*, 56(1), 72–84.

Hilker, T. et al. 2008. A modeling approach for up-scaling gross ecosystem production to the landscape scale using remote sensing data. *Journal of Geophysical Research: Biogeosciences*, 113(G3), (Art. No. G03006.).

Hilker, T. et al. 2009. An assessment of photosynthetic light use efficiency from space: Modeling the atmospheric and directional impacts on PRI reflectance. *Remote Sensing of Environment*, 113(11), 2463–2475.

Hilker, T. et al. 2011. Inferring terrestrial photosynthetic light use efficiency of temperate ecosystems from space. *Journal of Geophysical Research: Biogeosciences*, 116, 668–675. http://dx.doi.org/10.1029/2011JG001692

Hipps, L.E. et al. 1983. Assessing the interception of photosynthetically active radiation in winter wheat. *Agriculture and Forest Meteorology*, 28, 253–259.

Hollinger, D.Y. et al. 1994. Carbon dioxide exchange between an undisturbed old-growth temperate forest and the atmosphere. *Ecology*, 75(1), 134–150.

Hu, J. et al. 2018. Evaluating the performance of the SCOPE Model in simulating canopy solar-induced chlorophyll fluorescence. *Remote Sensing*, 10(2), 250.

Hudiburg, T.W., Law, B.E., and Thornton, P.E. 2013. Evaluation and improvement of the Community Land Model (CLM 4.0) in Oregon forests. *Biogeosciences*, 10(1), 453–470.

Huemmrich, K.F. et al. 2008. Using reflectance measurements to determine light use efficiency in corn. *IEEE International Geoscience & Remote Sensing Symposium (IGARSS)*, Boston, MA, USA.

Huemmrich, K.F. et al. 2009. Remote sensing of light use efficiency. *Proceedings of the 30th Canadian Symposium on Remote Sensing*, Lethbridge, Alberta, Canada.

Huemmrich, K.F. et al. 2017. ISS as a platform for optical remote sensing of ecosystem carbon fluxes: A case study using HICO. *IEEE Journal of Selected Topics in Applied Earth Observations and Remote Sensing*, 10(10), 4360–4375. doi: 10.1109/JSTARS.2017.2725825.

Huete, A.R. et al. 1997. A comparison of vegetation indices global set of TM images for EOS-MODIS. *Remote Sensing of Environment*, 59, 440–451.

Huete, A.R. et al. 2002. Overview of the radiometric and biophysical performance of the MODIS vegetation indices. *Remote Sensing of Environment*, 83, 195–213.

Inoue, Y., and Peñuelas, J. 2006. Relationship between light use efficiency and photochemical reflectance index in soybean leaves as affected by soil water content. *International Journal of Remote Sensing*, 27(22), 5109–5114.

IPCC. 2014. Climate change 2014: Synthesis report. In: Core Writing Team, R.K. Pachauri and L.A. Meyer (eds.). *Contribution of Working Groups I, II and III to the Fifth Assessment Report of the Intergovernmental Panel on Climate Change*. Geneva, Switzerland: IPCC, 151pp.

Jenkins, J. et al. 2007. Refining light-use efficiency calculations for a deciduous forest canopy using simultaneous tower-based carbon flux and radiometric measurements. *Agriculture and Forest Meteorology*, 143, 64–79.

Jeong, S.-J. et al. 2017. Application of satellite solar-induced chlorophyll fluorescence to understanding large-scale variations in vegetation phenology and function over northern high latitude forests. *Remote Sensing of Environment*, 190, 178–187.

Jetz, W. et al. 2016. Monitoring plant functional diversity from space. *Nature Plants*, 2, 16024.

Joiner, J. et al. 2011. First observations of global and seasonal terrestrial chlorophyll fluorescence from space. *Biogeosciences*, 8, 637–651.

Joiner, J. et al. 2012. Filling-in of near-infrared solar lines by terrestrial fluorescence and other geophysical effects: Simulations and space-based observations from SCIAMACHY and GOSAT. *Atmospheric Measurement Techniques*, 5, 809–829.

Joiner, J. et al. 2013. Global monitoring of terrestrial chlorophyll fluorescence from moderate-spectral-resolution near-infrared satellite measurements: Methodology, simulations, and application to GOME-2. *Atmospheric Measurement Techniques*, 6, 2803–2823.

Joiner, J. et al. 2016. New methods for retrieval of chlorophyll red fluorescence from hyperspectral satellite instruments: Simulations and application to GOME-2 and SCIAMACHY. *Atmospheric Measurement Techniques*, 9, 3939–3967.

Jones, H.G. 1983. *Plants and Microclimate, a Quantitative Approach to Environmental Plant Physiology*. Cambridge, U.K.: Cambridge Univeristy Press.

Jones, H.G., and Vaughan, R.A. 2010. *Remote Sensing of Vegetation: Principles, Techniques, and Applications*. Oxford and New York: Oxford University Press. ISBN-13: 978-0199207794.

Julitta, T. et al. 2016. Comparison of sun-induced fluorescence estimates obtained from four portable field spectroradiometers. *Remote Sensing*, 8, 122. doi: 10.3390/rs8020122.

Jung, M., and Reichstein, M. 2010. Recent decline in the global land evapotranspiration trend due to limited moisture supply. *Nature*, 467(7318), 951–954.

Justice, C.O. et al. 1998. The Moderate Resolution Imaging Spectroradiometer (MODIS): Land remote sensing for global change research. *IEEE Transactions on Geoscience and Remote Sensing*, 36(4), 1228–1249.

Keeling, C.D. et al. 2005. Atmospheric CO_2 and $^{13}CO_2$ exchange with the terrestrial biosphere and oceans from 1978 to 2000: observations and carbon cycle implications. In J.R. Ehleringer, T.E. Cerling and M.D. Dearing (Eds.), *A History of Atmospheric CO_2 and Its Effects on Plants, Animals, and Ecosystems*. New York: Springer Verlag, pp. 83–113.

Kerr, G. et al. 2016. The hyperspectral sensor DESIS on MUSES: Processing and applications. *Proceedings of the IEEE IGARSS (International Geoscience and Remote Sensing Symposium) Conference*, Beijing, China, July 10-15, 2016.

Kharouk, V.I. et al. 1995. Aspen bark photosynthesis and its significance to remote sensing and carbon budget estimates in the boreal ecosystem. *Water, Air, & Soil Pollution*, 82(1), 483–497.

Kljun, N. et al. 2004. A simple parameterisation for flux footprint predictions. *Boundary-Layer Meteorology*, 112(3), 503–523.

Köhler, P. et al. 2015. A linear method for the retrieval of sun-induced chlorophyll fluorescence from GOME-2 and SCIAMACHY data. *Atmospheric Measurement Techniques*, 8, 2589–2608. doi: 10.5194/amt-8-2589-2015. http://www.atmos-meas-tech.net/8/2589/2015/

Krause, G., and Weis, E. 1991. Chlorophyll fluorescence and photosynthesis: The basics. *Annual Review of Plant Biology*, 42, 313–349.

Krizek, D.T. et al. 2001. Evaluating UV-B effects and EDU protection in cucumber leaves using fluorescence images and fluorescence emission spectra. *Journal of Plant Physiology*, 158(1), 41–53.

Kumar, M., and Monteith, J.L. 1981. Remote sensing of crop growth. In H. Smith (Ed.), *Plants and the Daylight Spectrum*. London: Academic Press, pp. 134–144.

Larcher, W. et al. 2003. *Physiological Plant Ecology*, 4th edition. New York and London. ISBN 13: 9783540435167.

Law, B.E., and Waring, R.H. 1994. Combining remote sensing and climatic data to estimate net primary production across Oregon. *Ecological Applications*, 4(4), 717–728.

Le Quéré, C. et al. 2015. Global carbon budget 2015. *Earth System Science Data*, 7(2), 349–396.

Lee, C.M. et al. 2015. An introduction to the NASA Hyperspectral InfraRed Imager (HyspIRI) mission and preparatory activities: Special Issue on HyspIRI. *Remote Sensing of Environment*, 167, 6–19.

Leuning, R. et al. 2006. A multi-angle spectrometer for automatic measurement of plant canopy reflectance spectra. *Remote Sensing of Environment*, 103(3), 236–245.

Li, X.-P. et al. 2000. A pigment-binding protein essential for regulation of photosynthetic light harvesting. *Nature*, 403(6768), 391–395.

Liang, S. et al. 2006. Estimation of incident photosynthetically active radiation from modcrate resolution imaging spectrometer data. *Journal of Geophysical Research*, 111(D15), D15208.

Lichtenthaler, H.K. 1987. Chlorophylls and carotenoids, the pigments of photosynthetic biomembranes. In: Douce, R. and Packer, L. (Eds.), *Methods Enzymol.*, 148, New York: Academic Press Inc., pp. 350–382.

Lichtenthaler, H.K., and Rinderle, U. 1988. The role of chlorophyll fluorescence in the detection of stress conditions in plants. *CRC Critical Reviews in Analytical Chemistry*, 19, S29–S85.

Lichtenthaler, H.K. et al. 1986. Application of chlorophyll fluorescence in ecophysiology. *Radiation and Environmental Biophysics*, 25, 297–308.

Lichtenthaler, H.K. et al. 1990. The chlorophyll fluorescence ratio F690/F730 in leaves of different chlorophyll content. *Photosynthesis Research*, 25(3), 295–298.

Lichtenthaler, H.K. et al. 2007. Differences in pigment composition, photosynthetic rates and chlorophyll fluorescence images of sun and shade leaves of four tree species. *Plant Physiology and Biochemistry*, 45(8), 577–588.

Lin, J.C. et al. 2011. Attributing uncertainties in simulated biospheric carbon fluxes to different error sources. *Global Biogeochemical Cycles*, 25, GB2018.

Lobell, D.B. et al. 2002. Satellite estimates of productivity and light use efficiency in United States agriculture, 1982-98. *Global Change Biology*, 8(8), 722–735.

Luus, K.A. et al. 2017. Tundra photosynthesis captured by satellite-observed solar-induced chlorophyll fluorescence. *Geophysical Research Letters*, 44. doi: 10.1002/2016GL070842.

Lyapustin, A. et al. 2011. Multi-angle implementation of atmospheric correction (MAIAC): 1. Radiative transfer basis and look-up tables. *Journal of Geophysical Research: Atmospheres*, 116. http://dx.doi.org/10.1029/2010jd014985

Magnani, F. et al. 2014. Let's exploit available knowledge on vegetation fluorescence. *Proceedings of the National Academy of Sciences*, 111, E2510.

Magney, T.S. et al. 2016. Response of high frequency Photochemical Reflectance Index (PRI) measurements to environmental conditions in wheat. *Remote Sensing of Environment*, 173, 84–97.

Magney, T.S. et al. 2017. Connecting active to passive fluorescence with photosynthesis: A method for evaluating remote sensing measurements of chlorophyll fluorescence. *New Phytologist*, 215, 1594–1608.

Mahadevan, P. et al. 2008. A satellite-based biosphere parameterization for net ecosystem CO_2 exchange: Vegetation photosynthesis and respiration model (VPRM). *Global Biogeochemical Cycles*, 22(2), GB2005.

Mahecha, M.D. et al. 2010. Global convergence in the temperature sensitivity of respiration at ecosystem level. *Science*, 329(5993), 838–840.

Maier, S.W. et al. 2003. Sun-induced fluorescence: A new tool for precision farming. In T. VanToai, D. Major, M. McDonald, J. Schepers and L. Tarpley (Eds.), *Digital Imaging and Spectral Techniques: Applications to Precision Agriculture and Crop Physiology*. Madison: American Society of Agronomy, pp. 209–222.

Maimaitiyiming, M. et al. 2017. Early detection of plant physiological responses to different levels of water stress using reflectance spectroscopy. *Remote Sensing*, 9(7), Article 745.

Matross, D.M. et al. 2006. Estimating regional carbon exchange in New England and Quebec by combining atmospheric, ground-based and satellite data. *Tellus*, 58B, 344–358.

Maxwell, K., and Johnson, G.N. 2000. Chlorophyll fluorescence—A practical guide. *Journal of Experimental Botany*, 51, 659–668.

Maxwell, K. et al. 1998. A comparison of CO_2 and O_2 exchange patterns and the relationship with chlorophyll fluorescence during photosynthesis in C3 and CAM plants. *Australian Journal of Plant Physiology*, 25, 45–52.

Mazzoni, et al. 2010. High-resolution methods for fluorescence retrieval from space. *Optics Express*, 18(15), 15649–15663.

Meroni, M. et al. 2008a. Assessing Steady-state fluorescence and PRI from hyperspectral proximal sensing as early indicators of plant stress: The case of ozone exposure. *Sensors*, 8, 1740–1754.

Meroni, M. et al. 2008b. Leaf level early assessment of ozone injuries by passive fluorescence and photochemical reflectance index. *International Journal of Remote Sensing*, 29(17), 5409–5422.

Meroni, M. et al. 2009. Remote sensing of solar-induced chlorophyll fluorescence: Review of methods and applications. *Remote Sensing of Environment*, 113(10), 2037–2051.

Meroni, M. et al. 2010. Performance of spectral fitting methods for vegetation fluorescence quantification. *Remote Sensing of Environment*, 114(2), 363–374.

Middleton, E.M. et al. 1996. Initial assessment of physiological response to UV-B irradiation using fluorescence measurements. *Journal of Plant Physiology*, 148, 68–77.

Middleton, E.M. et al. 1997. Seasonal variability in foliar characteristics and physiology for boreal forest species at the five Saskatchewan tower sites during the 1994 Boreal Ecosystem-Atmosphere Study (BOREAS). *Journal of Geophysical Research*, 102(D24), 28831–28844.

Middleton, E.M. et al. 2003. Optical and fluorescence properties of corn leaves from different nitrogen regimes. In: M. Owe, G. D'Urso, L. Toulios (Eds.), *Remote Sensing for Agriculture, Ecosystems, and Hydrology IV, Proceedings of the International Society for Optics and Photonics (SPIE)*, Vol 4879, March 2003, pp. 72–83.

Middleton, E.M. et al. 2005. Evaluating UV B effects and EDU protection in soybean leaves using fluorescence. *Photochemistry and Photobiology*, 81, 1075–1085. Symposium in Print on the Effects of Ultraviolet Radiation on Terrestrial Ecosystems.

Middleton, E.M. et al. 2008. Comparison of measurements and FluorMOD simulations for solar induced chlorophyll fluorescence and reflectance of a corn crop under nitrogen treatments. *International Journal of Remote Sensing*, Special Issue for 2nd Recent Advances in Quantitative Remote Sensing (RAQRSII), 29(17), 5193–5213.

Middleton, E.M. et al. 2009a. Linking foliage spectral responses to canopy level ecosystem photosynthetic light use efficiency at a Douglas-fir forest in Canada. *Canadian Journal of Remote Sensing*, 35(2), 166–188.

Middleton, E.M. et al. 2009b. Diurnal and seasonal dynamics of canopy-level solar-induced chlorophyll fluorescence and spectral reflectance indices in a cornfield. *Proceedings, 6th EARSeL SIG Workshop on Imaging Spectroscopy, CD-ROM*, Tel-Aviv, Israel, March 16–19, 2009. 12pp.

Middleton, E.M. et al. 2013. The Earth Observing One (EO-1) satellite mission: Over a decade in space. *IEEE Journal of Selected Topics in Applied Earth Observations and Remote Sensing (JSTARS)*, 6(2), 243–256.

Middleton, E.M. et al. 2014a. Daily light use efficiency in a corn field can be related to the canopy red/far-red fluorescence ratio and leaf light use efficiency across a growing season. In: *Proceedings, 5th International Workshop on Remote Sensing of Vegetation Fluorescence*, Paris, France, April 22–24, 2014, 12pp.

Middleton, E.M. et al. 2014b. Directional hyperspectral observations to detect plant stress with the PRI and SIF in a cornfield. *Proceedings, 4th Recent Advances in Quantitative Remote Sensing (RAQRS'IV)*, Valencia, Spain, September 22–26, 2014, 9pp.

Middleton, E.M. et al. 2015a. Multi-angle hyperspectral observations with SIF and PRI to detect plant stress & GPP in a cornfield. *Proceedings, 9th EARSeL SIG Workshop on Imaging Spectroscopy, CD-ROM*, Luxembourg City, Luxembourg, April 2015, p. 10.

Middleton, E.M. et al. 2015b. Novel Leaf-Level Measurements of Chlorophyll Fluorescence for Photosynthetic Efficiency. *Proceedings, (on CD) International Geoscience and Remote Sensing Symposium (IGARSS 2015)*, Milan, Italy, 27–31 July, 2015.

Middleton, E.M. et al. 2016. Remote sensing of ecosystem light use efficiency using MODIS. *Remote Sensing of Environment*, 187, 345–366.

Middleton, E.M. et al. 2017a. The 2013 FLEX—US Airborne Campaign at the Parker Tract Loblolly Pine Plantation in North Carolina, USA. *Remote Sensing*, 9, 612, 30pp. doi: 10.3390/rs9060612.

Middleton, E.M. et al. 2017b. Hyperion: The First Global Orbital Spectrometer, the Earth Observing-1 (EO-1) Satellite (2000–2017). *IEEE International Geoscience & Remote Sensing Symposium (IGARSS) 2017*, Ft. Worth, TX.

Mohammed, G.H. et al. 1995. Chlorophyll fluorescence: A review of its practical forestry applications and instrumentation. *Scandinavian Journal of Forest Research*, 10(1), 383–410.

Monteith, J. 1972. Solar-radiation and productivity in tropical ecosystems. *Journal of Applied Ecology*, 9, 747–766.

Monteith, J.L. 1977. Climate and efficiency of crop production in Britain. *Philosophical Transactions of the Royal Society of London Series B-Biological Sciences*, 281(980), 277–294.

Mooney, H.A. et al. (Eds.) 1991. *Response of Plants to Multiple Stresses*. New York, USA: Academic Press.

Morton, O. 2009. *Eating the Sun: How Plants Power the Planet*. London, U.K.: Harper Collins.

Myneni, R.B. et al. 1997a. Estimation of global leaf area index and absorbed PAR using radiative transfer models. *Geoscience and Remote Sensing, IEEE Transactions on*, 35(6), 1380–1393.

Myneni, R.B. et al. 1997b. Increased plant growth in the northern high latitudes from 1981 to 1991. *Nature*, 386(6626), 698–702.

Müller, P. et al. 2001. Non-photochemical quenching. A response to excess light energy. *Plant Physiology*, 125(4), 1558–1566.

Nakaji, T. et al. 2016. Estimation of light-use efficiency through a combinational use of the photochemical reflectance index and vapor pressure deficit in an evergreen tropical rainforest at Pasoh, Peninsular Malaysia. *Remote Sensing of Environment*, 150, 82–92.

NCAR CLM4.5, http://www.cesm.ucar.edu/models/cesm1.2/clm/models/lnd/clm/doc/UsersGuide/f101.html

Nemani, R.R. et al. 2003. Climate-driven increases in global terrestrial net primary production from 1982 to 1999. *Science*, 300(5625), 1560–1563.

Nichol, C.J. et al. 2000. Remote sensing of photosynthetic light-use efficiency of boreal forest. *Agriculture and. Forest Meteorology*, 101, 131–142.

NRC 2007. *Decadal Survey: National Research Council 2007. Earth Science and Applications from Space: National Imperatives for the Next Decade and Beyond*. Washington, DC: The National Academies Press. https://doi.org/10.17226/11820. https://www.nap.edu/catalog/11820/earth-science-and-applications-from-space-national-imperatives-for-the

Oleson, K.W. et al. 2013. *Technical Description of version 4.5 of the Community Land Model (CLM)*. NCAR Technical Note NCAR/TN-503+STR, Boulder, Colorado: National Center for Atmospheric Research, 420pp.

Öquist, G., and Huner, N.P.A. 2003. Photosynthesis of overwintering evergreen plants. *Annual Review of Plant Biology*, 54(1), 329–355.

Peng, Y. et al. 2017. Using remotely sensed spectral reflectance to indicate leaf photosynthetic efficiency derived from active fluorescence measurements. *Journal of Applied Remote Sensing*, 11, Article 026034.

Peñuelas, J. et al. 1994. Reflectance indices associated with physiological changes in nitrogen- and water-limited sunflower leaves. *Remote Sensing of Environment*, 48(2), 135–146.

Peñuelas, J. et al. 1995. Assessment of photosynthetic radiation-use efficiency with spectral reflectance. *New Phytologist*, 131(3), 291–296.

Peñuelas, J. et al. 1997. Photochemical reflectance index and leaf photosynthetic radiation-use-effeciency assessment in Mediterranean trees. *International Journal of Remote Sensing*, 18(13), 2863–2868.

Peñuelas, J. et al. 2011. Photochemical reflectance index (PRI) and remote sensing of plant CO_2 uptake. *New Phytologist*, 191(3), 596–599.

Pfündel, E.E., and Bilger, W. 1994. Regulation and possible function of the violaxanthin cycle. *Photosynthesis Research*, 42(2), 89–109.

Pfündel, E.E., and Dilley, R.A. 1993. The pH dependence of violaxanthin deepoxidation in isolated Pea chloroplasts. *Plant Physiology*, 101(1), 65–71.

Piao, S. et al. 2009. The carbon balance of terrestrial ecosystems in China. *Nature*, 458(7241), 1009–1013.

Piao, S. et al. 2013. Evaluation of terrestrial carbon cycle models for their response to climate variability and to CO_2 trends. *Global Change Biology*, 19(7), 2117–2132.

Pinker, R.T. et al. 2003. Surface radiation budgets in support of the GEWEX Continental-Scale International Project (GCIP) and the GEWEX Americas Prediction Project (GAPP), including the North American Land Data Assimilation System (NLDAS) project. *Journal of Geophysical Research*, 108(D22), 8844.

Pinto, F. et al. 2016. Sun-induced chlorophyll fluorescence from high-resolution imaging spectroscopy data to quantify spatio-temporal patterns of photosynthetic function in crop canopies. *Plant Cell and Environment*, 39(7), 1500–1512.

Plascyk, J.A. 1975. The MK II Fraunhofer line discriminator (FLD-II) for airborne and orbital remote sensing of solar-stimulated luminescence. *Optical Engineering*, 14(4), 339–346.

Porcar-Castell, A. et al. 2014. Linking chlorophyll a fluorescence to photosynthesis for remote sensing applications: Mechanisms and challenges. *Journal Experimental Botany*, 5, 4065–4095.

Prince, S.D. 1991. A model of regional primary production for use with coarse resolution satellite data. *International Journal of Remote Sensing*, 12(6), 1313–1330.

Prince, S.D., and Goward, S.N. 1995. Global primary production: A remote sensing approach. *Journal of Biogeography*, 22(4–5), 815–835.

Rahman, A.F. et al. 2001. Modeling spatially distributed ecosystem flux of boreal forest using hyperspectral indices from AVIRIS imagery. *Journal of Geophysical Research*, 106(D24), 33579–33591.

Rahman, A.F. et al. 2004. Potential of MODIS ocean bands for estimating CO_2 flux from terrestrial vegetation: A novel approach. *Geophysical Research Letters*, 31, L10503. http://dx.doi.org/10.1029/2004GL019778.

Rascher, U. et al. 2000. Evaluation of instant light-response curves of chlorophyll fluorescence parameters obtained with a portable chlorophyll fluorometer on site in the field. *Plant Cell and Environment*, 23, 1397–1405.

Rascher, U. et al. 2010a. Canopy fluorescence improves modeling of diurnal courses of GPP-correlation of GPP and Fs over a variety of crops. *Proceedings of the 4th International Workshop on Remote Sensing of Vegetation Fluorescence*, Valencia, Spain, 15–17 November 2010.

Rascher, U. et al. 2010b. Sensing of photosynthetic activity of crops. *Precision Crop Protection—the Challenge and Use of Heterogeneity*, 87–99.

Rascher, U. et al. 2015. Sun-induced fluorescence—A new probe of photosynthesis: First maps from the imaging spectrometer HyPlant. *Global Change Biology*, 21, 4673–4684.

Reichstein, M. et al. 2005. On the separation of net ecosystem exchange into assimilation and ecosystem respiration: Review and improved algorithm. *Global Change Biology*, 11(9), 1424–1439.

Rockström, J. et al. 2009. A safe operating space for humanity. *Nature*, 461(7263), 472–475.

Rockström, J. et al. 2017. A roadmap for rapid decarbonization. *Science*, 355(6331), 1269–1271. DOI: 10.1126/science.aah3443.

Rossini, M. et al. 2013. Assessing canopy PRI from airborne imagery to map water stress in maize. *ISPRS Journal of Photogrammetry and Remote Sensing*, 86, 168–177.

Rossini, M. et al. 2015. Red and far red sun-induced chlorophyll fluorescence as a measure of plant photosynthesis. *Geophysical Research Letters*, 42, 1632–1639.

Rossini, M. et al. 2016. Analysis of red and far-red sun-induced chlorophyll fluorescence and their ratio in different canopies based on observed and modeled data. *Remote Sensing*, 8, 412.

Rouse, J.W., and Schell, J.A. 1973. Systems approach to earth observations. *IEEE Transactions on Aerospace and Electronic Systems*, AES9(5), 804–804.

Rouse, J.W. et al. 1974. Monitoring vegetation systems in the Great Plains with ERTS. *Proceedings of the 3rd Earth Resource Technology Satellite (ERTS) Symposium*, Washington, DC, USA, 10–14 December 1973, pp. 309–317.

Running, S. et al. 2004. A continuous satellite-derived measure of global terrestrial primary production. *Bioscience*, 54(6), 547–560.

Russell, G. et al. 1989. Absorption of radiation by canopies and stand growth. In G. Russell, B. Marshall and P.G. Jarvis (Eds.), *Plant Canopies: Their Growth, Form and Function*. Cambridge: Cambridge University Press, pp. 21–40.

Sabater, N. et al. 2017. Oxygen transmittance correction for solar-induced chlorophyll fluorescence measured on proximal sensing: application to the NASA-GSFC Fusion tower. *Proceedings, 2017 International Geoscience and Remote Sensing Symposium (IGARSS 2017)*, Ft. Worth, TX, 4 pp.

Sabater, N. et al. 2018. Compensation of oxygen transmittance effects for proximal sensing retrieval of canopy-leaving sun-induced chlorophyll fluorescence. *Remote Sensing of Environment*, in review (submitted Dec. 2017).

Sanders, A. et al. 2016. Spaceborne sun-induced vegetation fluorescence time series from 2007 to 2015 evaluated with Australian flux tower measurements. *Remote Sensing*, 8, 895, doi:10.3390/rs8110895.

Sarijeva, G. et al. 2007. Differences in photosynthetic activity, chlorophyll and carotenoid levels, and in chlorophyll fluorescence parameters in green sun and shade leaves of Ginkgo and Fagus. *Journal of Plant Physiology*, 164(7), 950–955.

Schaepman, M.E. et al. 2015. Advanced radiometry measurements and Earth science applications with the Airborne Prism Experiment (APEX). *Remote Sensing of Environment*, 158, 207–219.

Schickling, A. et al. 2016. Combining sun-induced chlorophyll fluorescence and photochemical reflectance index improves diurnal modeling of gross primary productivity. *Remote Sensing*, 8(7), Article 574.

Schreiber, U. et al. 1986. Continuous recording of photochemical and non-photochemical chlorophyll fluorescence quenching with a new type of modulation fluorometer. *Photosynthesis Research*, 10, 51–62.

Scripps CO_2 Program, 2018. *Scripps Institute of Oceanography*. San Diego, CA, USA.

Sellers, P.J. et al. 1992. Canopy reflectance, photosynthesis, and transpiration. III. A reanalysis using improved leaf models and a new canopy integration scheme. *Remote Sensing of Environment*, 42(3), 187–216.

Sellers, P.J. et al. 1995. The boreal ecosystem-atmosphere study (BOREAS): An overview and early results from the 1994 field year. *Bulletin of the American Meteorological Society*, 76(9), 1549–1577.

Soudani, K. et al. 2014. Relationships between photochemical reflectance index and light use efficiency in deciduous and evergreen broadleaf forests. *Remote Sensing of Environment*, 144, 73–84.

Spitters, C.J.T. 1986. Separating the diffuse and direct component of global radiation and its implications for modeling canopy photosynthesis Part II. Calculation of canopy photosynthesis. *Agricultural and Forest Meteorology*, 38(1–3), 231–242.

Spitters, C.J.T. et al. 1986. Separating the diffuse and direct component of global radiation and its implications for modeling canopy photosynthesis Part I. Components of incoming radiation. *Agricultural and Forest Meteorology*, 38(1–3), 217–229.

Springer, K.R. et al. 2017. Parallel seasonal patterns of photosynthesis, fluorescence, and reflectance indices in boreal trees. *Remote Sensing*, 9(7), 691. doi: 10.3390/rs9070691

Stagakis, S. et al. 2014. Tracking seasonal changes of leaf and canopy light use efficiency in a *Phlomis fruticosa* Mediterranean ecosystem using field measurements and multi-angular satellite hyperspectral imagery. *ISPRS Journal of Photogrammetry and Remote Sensing*, 97, 138–151. http://dx.doi.org/10.1016/j.isprsjprs.2014.08.012

Strachan, I.B. et al. 2008. Use of hyperspectral remote sensing to estimate the gross photosynthesis of agricultural fields. *Canadian Journal of Remote Sensing*, 34(3), 333–341.

Štroch, M. et al. 2008. Dynamics of the xanthophyll cycle and non-radiative dissipation of absorbed light energy during exposure of Norway spruce to high irradiance. *Journal of Plant Physiology*, 165(6), 612–622.

Stylinski, C.D. et al. 2002. Seasonal patterns of reflectance indices, carotenoid pigments and photosynthesis of evergreen chaparral species. *Oecologia*, 131, 366–374.

Suess, A. et al. 2016. Deriving diurnal variations in sun-induced chlorophyll-a fluorescence in winter wheat canopies and maize leaves from ground-based hyperspectral measurements. *International Journal of Remote Sensing*, 37, 60–77.

Sun, Y. et al. 2015. Drought onset mechanisms revealed by satellite solar-induced chlorophyll fluorescence: Insights from two contrasting extreme events. *Journal of Geophysical Research: Biogeosciences*, 120, 2427–2440.

Suárez, L. et al. 2008. Assessing canopy PRI for water stress detection with diurnal airborne imagery. *Remote Sensing of Environment*, 112(2), 560–575.

Takala, T.L.H., Mottus, M. 2016. Spatial variation of canopy PRI with shadow fraction caused by leaf-level irradiation conditions. *Remote Sensing of Environment*, 182, 99–112.

Thayer, S.S., and Björkman, O. 1990. Leaf xanthophyll content and composition in sun and shade determined by HPLC. *Photosynthesis Research*, 23(3), 331–343.

Thenkabail, P.S. et al. 2000. Hyperspectral vegetation indices and their relationships with agricultural crop characteristics. *Remote Sensing of Environment*, 71(2), 158–182.

Thum, T. et al. 2017. Modelling sun-induced fluorescence and photosynthesis with a land surface model at local and regional scales in northern Europe. *Biogeosciences*, 14, 1969–1984. doi: 10.5194/bg-14-1969-2017.

Timmermans, J. et al. 2017. Directional radiative transfer by SCOPE, SLC and DART using laser scan derived structural forest parameters. *EGU General Assembly Conference Abstracts*, 19, 6501.

Trotter, G.M. et al. 2002. The photochemical reflectance index as a measure of photosynthetic light use efficiency for plants with varying foliar nitrogen contents. *International Journal of Remote Sensing*, 23(6), 1207–1212.

Tucker, C.J. 1979. Red and photographic infrared linear combinations for monitoring vegetation. *Remote Sensing of Environment*, 8, 127–150.

Tucker, C.J. et al. 2005. An extended AVHRR 8-km NDVI dataset compatible with MODIS and SPOT vegetation NDVI data. *International Journal of Remote Sensing*, 26(20), 4485–4498.

Turner, D.P. et al. 2003. A cross-biome comparison of daily light use efficiency for gross primary production. *Global Change Biology*, 9, 383–395.

Ulsig, L. et al. 2017. Detecting inter-annual variations in the phenology of evergreen conifers using long-term MODIS vegetation index time series. *Remote Sensing*, 9(1), Article 49, doi: 10.3390/rs9010049.

van der Tol, C. et al. 2009. An integrated model of soil-canopy spectral radiances, photosynthesis, fluorescence, temperature and energy balance. *Biogeosciences*, 6, 3109–3129.

van der Tol, C. et al. 2014. Models of fluorescence and photosynthesis for interpreting measurements of solar-induced chlorophyll fluorescence. *Journal of Geophysical Research: Biogeosciences*, 119, 2312–2327.

van der Tol, C. et al. 2016. A model and measurement comparison of diurnal cycles of sun-induced Chlorophyll fluorescence of crops. *Remote Sensing of Environment*, 186, 663–677.

Van Wittenberghe, S. et al. 2015. FluoWat bidirectional sun-induced chlorophyll fluorescence emission is influenced by leaf structure and light scattering properties: A bottom-up approach. *Remote Sensing of Environment*, 158, 169–179.

Verhoef, W. et al. 2018. Hyperspectral radiative transfer modeling to explore the combined retrieval of biophysical parameters and canopy fluorescence from FLEX—Sentinel-3 tandem mission multi-sensor data. *Remote Sensing of Environment*, 204, 942–963. http://dx.doi.org/10.1016/j.rse.2017.08.006

Vermote, E.F. et al. 1997. Second simulation of the satellite signal in the solar spectrum, 6S: An overview. *IEEE Transsactions on Geoscience and Remote Sensing* , 35, 675–686.

Verrelst, J. et al. 2015. Global sensitivity analysis of the SCOPE model: What drives simulated canopy-leaving sun-induced fluorescence? *Remote Sensing of Environment*, 166, 8–21.

Verrelst, J. et al. 2016. Evaluating the predictive power of sun-induced chlorophyll fluorescence to estimate net photosynthesis of vegetation canopies: A SCOPE modeling study. *Remote Sensing of Environment*, 176, 139–151.

Vilfan, N. et al. 2016. Fluspect-B: A model for fluorescence, reflectance and transmittance spectra. *Remote Sensing of Environment*, 596–615.

Wang, Y.P., and Leuning, R. 1998. A two-leaf model for canopy conductance, photosynthesis and partitioning of available energy I: Model description and comparison with a multi-layered model. *Agricultural and Forest Meteorology*, 91(1–2), 89–111.

White, H.P. et al. 2001. Four-Scale Linear Model for Anisotropic Reflectance (FLAIR) for plant canopies. I. Model description and partial validation. *Geoscience and Remote Sensing, IEEE Transactions on*, 39(5), 1072–1083.

Wiegand, C.L., and Richardson, A.J. 1984. Leaf area, light interception and yield estimates from spectral components analysis. *Agronomy Journal*, 76, 543–548.

Wieneke, S. et al. 2016. Airborne based spectroscopy of red and far-red sun-induced chlorophyll fluorescence: Implications for improved estimates of gross primary productivity. *Remote Sensing of Environment*, 184, 654–667.

Wohlfahrt, G., and Gu, L. 2015. The many meanings of gross photosynthesis and their implication for photosynthesis research from leaf to globe. *Plant Cell and Environment*, 38(12), 2500–2507. doi: 10.1111/pce.12569.

Wyber, R.A. et al. 2017. Do daily and seasonal trends in leaf solar induced fluorescence reflect changes in photosynthesis, growth or light exposure?. *Remote Sensing*, 9, 604.

Yang, P. et al. 2017. The mSCOPE model: A simple adaptation to the SCOPE model to describe reflectance, fluorescence and photosynthesis of vertically heterogeneous canopies. *Remote Sensing of Environment*, 201, 1–11.

Yang, X. et al. 2015. Solar-induced chlorophyll fluorescence that correlates with canopy photosynthesis on diurnal and seasonal scales in a temperate deciduous forest. *Geophysical Research Letters*, 42, 2977–2987.

Yu, Q. et al. 2014. Narrowband bio-indicator monitoring of temperate forest carbon fluxes in Northeastern China. *Remote Sensing*, 6(9), 8986–9013.

Zarco-Tejada, P.J. et al. 2003. Steady-state chlorophyll a fluorescence detection from canopy derivative reflectance and double-peak red-edge effects. *Remote Sensing of Environment*, 84(2), 283–294.

Zarco-Tejada, P.J. et al. 2006. FluorMODgui V3.1—A graphic user interface for the leaf and canopy simulation of chlorophyll fluorescence. *Computers & Geosciences*, 32(5), 577–591.

Zarco-Tejada, P.J. et al. 2012. Fluorescence, temperature and narrow-band indices acquired from a UAV platform for water stress detection using a micro-hyperspectral imager and a thermal camera. *Remote Sensing of Environment*, 117, 322–337.

Zarco-Tejada, P.J. et al. 2013a. A PRI-based water stress index combining structural and chlorophyll effects: Assessment using diurnal narrow-band airborne imagery and the CWSI thermal index. *Remote Sensing of Environment*, 138, 38–50.

Zarco-Tejada, P.J. et al. 2013b. Relationships between net photosynthesis and steady-state chlorophyll fluorescence retrieved from airborne hyperspectral imagery. *Remote Sensing of Environment*, 136, 247–258.

Zarco-Tejada, P.J. et al. 2013c. Spatio temporal patterns of chlorophyll fluorescence and physiological and structural indices acquired from hyperspectral imagery as compared with carbon fluxes measured with eddy covariance. *Remote Sensing of Environment*, 133, 102–115.

Zhang, C. et al. 2016. Affecting factors and recent improvements of the photochemical reflectance index (PRI) for remotely sensing foliar, canopy and ecosystemic radiation-use efficiencies. *Remote Sensing*, 8(9), Article 677.

Zhang, C. et al. 2017. Photochemical Reflectance Index (PRI) for detecting responses of diurnal and seasonal photosynthetic activity to experimental drought and warming in a Mediterranean Shrubland. *Remote Sensing*, 9, 1189.

Zhang, H. et al. 1995. Photosynthetic characteristics of sun versus shade plants of *Encelia farinosa* as affected by photosynthetic photon flux density, intercellular CO_2 concentration, leaf water potential, and leaf temperature. *Functional Plant Biology*, 22(5), 833–841.

Zhang, Q. et al. 2005. Estimating light absorption by chlorophyll, leaf and canopy in a deciduous broadleaf forest using MODIS data and a radiative transfer model. *Remote Sensing of Environment*, 99, 357–371.

Zhang, Q. et al. 2009. Can a MODIS-derived estimate of the fraction of PAR absorbed by chlorophyll ($fAPAR_{chl}$) improve predictions of light-use efficiency and ecosystem photosynthesis for a boreal aspen forest?. *Remote Sensing of Environment*, 113, 880–888.

Zhang, Q. et al. 2016a. Remote estimation of corn daily gross primary production (GPP): Integration of $fAPAR_{chl}$ and PRI. *Remote Sensing of Environment*, 186, 311–321.

Zhang, Y. et al. 2016b. Model-based analysis of the relationship between sun-induced chlorophyll fluorescence and gross primary production for remote sensing applications. *Remote Sensing of Environment*, doi: 10.1016/j.rse.2016.10.016.

Zhao, F. et al. 2016. FluorWPS: A Monte Carlo ray-tracing model to compute sun-induced chlorophyll fluorescence of three-dimensional canopy. *Remote Sensing of Environment*, 187, 385–399.

Section II

Plant Species Identification and Discrimination

6 Crop Type Discrimination Using Hyperspectral Data
Advances and Perspectives

Lênio Soares Galvão, José Carlos Neves Epiphanio,
Fábio Marcelo Breunig, and Antônio Roberto Formaggio

CONTENTS

6.1 INTRODUCTION

The original definition of hyperspectral remote sensing, also known as imaging spectrometry or imaging spectroscopy, refers to the acquisition of images in hundreds of contiguous spectral bands to obtain high-spectral-resolution data for each pixel of the scene (Goetz et al., 1985). In reality, this concept is much more related to the ability of the sensors to measure narrow absorption bands rather than the number of bands. For example, despite the difference in the number of bands, the Airborne Visible/Infrared Imaging Spectrometer (AVIRIS) with 224 bands (400–2500 nm) and the Compact High Resolution Imaging Spectrometer (CHRIS)/PROBA with 62 bands (410–1000 nm) are both hyperspectral sensors because they have contiguous bands in each spectral range with bandwidths of approximately 10 nm. This bandwidth allows adequate measurement of most of the narrow spectral features that appear in land cover spectra.

Hyperspectral sensors enable the calculation of several narrow-band vegetation indices (NVIs) and absorption band parameters (e.g., depth, width, area, and asymmetry) on a per-pixel basis. The resultant images are useful for agriculture. For crop type discrimination, the use of narrow bands usually results in increased classification accuracy when compared to broad bands of the multispectral sensors (Thenkabail et al., 2002, 2004). However, signal-to-noise ratio (SNR) is an important factor in this comparison because it affects classification accuracy (Lobell and Asner, 2003). Unfortunately, several hyperspectral sensors have poor SNR, especially in the short-wave infrared (SWIR). Classification of hyperspectral data requires also feature selection to avoid the Hughes effect: the loss of accuracy when data dimensionality increases while the training sample size remains fixed (Hughes, 1968; Pal and Foody, 2010). The selected set of narrow bands or spectral attributes for crop type discrimination is not universally applicable and should be determined for each study area. NVIs are useful for crop type discrimination as well as for estimating biophysical (e.g., leaf area index—LAI, biomass) and biochemical (e.g., chlorophyll, leaf/canopy nitrogen, and water) crop attributes (Bannari et al., 2008; Haboudane et al., 2008; Li et al., 2008; Smith et al., 2008). At local scale, bands and derived vegetation indices can be correlated with crop yield (Zarco-Tejada et al., 2005; Galvão et al., 2009; Yang et al., 2009). Using images showing variation in absorption band parameters, one can detect changes in canopy constituents such as chlorophyll and leaf water content within and between agricultural fields. For a given crop type and reproductive stage, information on the causes (e.g., water stress, diseases, or nutrient deficiency) of such variation can be obtained.

Examples of crop type discrimination studies using imaging spectrometers include the following: (1) discrimination of sugarcane varieties with Hyperion/Earth Observing-1 (EO-1) data (Galvão et al., 2005, 2006; Everingham et al., 2007); (2) use of a spectral library generated from field spectroradiometric and Hyperion data for spectral angle mapper (SAM) classification of different crop varieties (Rao, 2008); (3) use of support vector machine (SVM) and linear discriminant analysis for crop classification using different hyperspectral datasets (Camps-Valls et al., 2003; Bandos et al., 2009); (4) SAM classification of crops using an SVM-based algorithm for reference spectra extraction (Filippi et al., 2009); (5) evaluation of multiangular information to improve crop classification accuracy when compared to a single nadir acquisition, using hyperspectral CHRIS/PROBA data (Duca and Del Frate, 2009); (6) use of ground-based hyperspectral imagery for crop–weed species discrimination and oriented herbicide application (Eddy et al., 2008); (7) combination of field and airborne hyperspectral Hymap data acquired in consecutive growing seasons for crop type discrimination (Nidamanuri and Zbell, 2011); (8) test and identification of optimal NVIs to map the world's main agricultural crops using Hyperion/EO-1 data (Thenkabail et al., 2013); (9) crop classification in a highly heterogeneous area of Kenya using airborne hyperspectral data and SVM (Piiroinen et al., 2015); and (10) application of sparse graph regularization for mapping eight crop types using airborne hyperspectral data acquired in China (Xue et al., 2017).

Hyperion, on board the EO-1 spacecraft, launched in 2000, was a pushbroom hyperspectral instrument with 242 spectral bands. However, it acquired images in 196 calibrated bands (10 nm of bandwidth) positioned in the visible, near-infrared (NIR), and SWIR (400–2400 nm range) (Pearlman et al., 2003). The instrument had a spatial resolution of 30 m and a swath width of 7.7 km. After more than 16 years of science observations, the EO-1 satellite was decommissioned in 2017. The 16-day revisit time of the sensor could be reduced by off-nadir pointing. Despite the poor SNR in the SWIR, the historical Hyperion data still provide an excellent opportunity to evaluate the hyperspectral technology from orbital level for crop type discrimination.

In this chapter, we first discuss the factors that affect crop type discrimination by remote sensing. Then, we use different Hyperion datasets in Brazil to show the spectral discrimination between flooded rice, coffee, sugarcane, bean, corn, cultivated pasture, and soybean. The role played by reflectance, reflectance ratios, NVIs, and absorption band parameters to differentiate crops is also demonstrated. Using Hyperion data to simulate the spectral response of selected multispectral sensors, we demonstrate the spectral resolution influence on cultivar discrimination from multispectral to hyperspectral remote sensing. Finally, recent advances in feature selection and classification techniques for crop type discrimination are discussed.

6.2 FACTORS AFFECTING CROP TYPE DISCRIMINATION BY REMOTE SENSING

Measurements of crop area are important for government and economic players. However, the first step for such measurements using remote sensing—the crop identification—is not necessarily an easy task. When compared to field observations, such difficulties increase when measuring the reflectance of crops from satellites because the spectral similarities between the crops can be even higher.

Remote sensing scientists are continually developing methods to improve crop type discrimination. Factors affecting discrimination may be grouped into five general categories: biophysical, crop development, management, crop calendar, and regional aspects.

In terms of biophysical aspects, each crop belongs to a family, species, and so forth. In the lowest scale, each crop belongs to a variety or cultivar. At this scale, plants are expected to be homogeneous concerning their genotypes and phenotypes. As we generalize from cultivar and variety to species and family, it is expected that plant characteristics depart from homogeneity and impose on crops increasing differences. With regard to remote sensing, the differences are related to many aspects such as leaf pigments, leaf structure, duration of life cycle, plant structure, height, and physiology (Gausman, 1985). In a classical review, Gausman (1985) showed how some of these factors could affect plant reflectance. At the variety or cultivar scale, leaf pigments affect spectral reflectance and are supposed to be very similar for some specific crops. However, even at this level, we can find differences between the cultivars (Yanqun et al., 2003; Falqueto et al., 2009). Because some cultivars have different canopy structure (e.g., leaf thickness, leaf angle), this parameter affects the way that light interacts with the canopy (Gonçalves et al., 2008). When we look at higher levels of species or classes, differences between the crops are even larger. For instance, leaf mesophyll layers of dicotyledonous plants have more well-defined palisade and spongy cells than those of the monocotyledonous plants. For crop type discrimination, it is well known that reflectance is influenced by the mesophyll structure (Bauer, 1975).

Crops are very distinct in their development aspects or phenological stages. In general, major cultivated crops can be classified into three main groups according to the duration of the life cycle: annual, perennial, and semiperennial crops. Annual crops are planted once or even three times a year. A perennial crop can stay in field for many years, while a semiperennial crop remains in field only for a few years. The duration of the crop cycle impacts on the chances of acquiring cloud-free images using optical remote sensing, which are obviously higher for perennial crops. Especially for annual crops, due to their short cycle of life, another important aspect for remote sensing is how they are split into development stages. Different crops present distinct phenological characteristics and timings according to their nature: germination, tillering, flowering, ball formation (e.g., cotton), ripening, and so forth. Even for the same crop and growing season, the duration and magnitude of each phenological stage can vary between the varieties, which introduce data variability for crop type discrimination with imaging systems.

Crop management is another important factor. The farmers use distinct management practices depending on the cultivated crop. For instance, tillage (e.g., soybean and corn), burning prior to harvest (sugarcane), type of pruning (coffee and orchards), and harvest timing (e.g., sugarcane can be harvested almost all year long) can introduce strong variability in spectral reflectance between agricultural fields of the same crop type. In addition, technological improvements (e.g., mechanical harvesting of sugarcane) are not adopted by farmers at the same time and place. Management practices in perennial crops have a special impact on crop type discrimination. In general, the row or tree spacing is larger in perennial crops than in annual crops, thus exposing the soil background, which influences the canopy reflectance (Heilman and Boyd, 1986; Gilabert et al., 1994).

Dry-land crops, which predominate in global production, depend on rain to develop. As a result, cloud cover in the rainy season is an important issue for their discrimination using optical remote sensing. For annual crops, the crop calendar is determined mainly by the water availability for dry-land crops. Irrigated crops have a complement of water supplied by mechanized systems. For irrigated cultivations, the crop calendar is not as fixed as it is for dry-land crops, since the water availability is a minor problem for them. The main limitations are temperature and light availability. If irrigation is used, farmers can cultivate more crop types at more times during the year. In general, there is a preference for some crops to be irrigated, and there is a dependence on the irrigation method. For instance, rice is cultivated by flooding rather than by use of sprinkler systems. Therefore, irrigation management can help crop discrimination in some cases (e.g., rice paddies) or can pose difficulties in others (e.g., various crops under the same center pivot system).

Finally, regional aspects can control the crop calendar. For instance, low, humid, and flat lands are suited to growing flooded rice. Intermediate temperatures are favorable for coffee. Gentle topography and fertile soils are preferred for corn, cotton, oats, soybeans, sugarcane, and wheat. Sugarcane, vineyards, apple tree, and coffee can be cultivated also in undulated topography.

When using hyperspectral remote sensing in agriculture, one should consider the influence of these factors and take advantage of them to facilitate crop type discrimination. For example, the water background influence on canopy reflectance can facilitate discrimination of flooded rice if two images are acquired in distinct reproductive stages or in flooded and nonflooded phases of crop development.

6.3 CROP TYPE DISCRIMINATION USING HYPERION DATA

Here, we discuss crop type discrimination using Hyperion data acquired over different agricultural areas in Brazil. We use also multispectral sensor-simulated images from this instrument. Obviously, this discussion is a simplification of the "real world," where the spectral variability between and within crops is much stronger due to the factors mentioned in the previous section. For instance, the Hyperion scenes registered a limited number of growth stages per crop, which facilitates the discrimination for some of them. However, the current approach serves to illustrate major differences in reflectance, reflectance ratios, NVIs, and absorption band parameters between the crops. To facilitate the discussion, we use the following nomenclature for the spectral ranges: visible (400–700 nm), red edge (701–760 nm), NIR-1 (761–900 nm), NIR-2 (901–1400 nm), SWIR-1 (1401–1900 nm), and SWIR-2 (1901–2500 nm).

6.3.1 SELECTED CROPS

Six crop types were selected for analysis: coffee (*Coffea arabica* L.), sugarcane (*Saccharum* spp.), flooded rice (*Oryza sativa* L.), common bean (*Phaseolus vulgaris* L.), corn (*Zea mays* L.), and soybean (*Glycine max* (L.) Merrill). Coffee is a perennial crop and sugarcane a semiperennial crop. The others are annual crops. Cultivated pasture was also included. It is important in Brazil for animal production as the main feed resource for cattle. Brazil is one of the largest world producers of these crops, especially of sugarcane and coffee (world leader), and soybean (second in rank). Sugarcane has an important environmental role providing ethanol—a renewable fuel for the great number of Brazilian flexible-fuel vehicles. Coffee has had an historical participation in the Brazilian economy, and soybeans have experienced an impressive rise in production over the last few decades.

From a remote sensing perspective, the selected crops represent completely distinct canopies. Leguminous plants like soybean and bean have broader leaves and a more planophile canopy architecture than grass plants like rice, sugarcane, and corn with a more erectophile architecture. Among the erectophile crops, rice has the largest number of leaves per linear meter (around 20). Naturally, structural or physiognomic characteristic and differences between the crops influence their reflectance. For instance, in contrast to other crops that have less than 1 m of maximum height, sugarcane and corn are the tallest crops and usually reach more than 2 m height. Such characteristics affect the scattering and absorption of radiation by the canopy components, especially the leaves. They affect also the ability to estimate foliar chemistry from hyperspectral remote sensing using radiative transfer models (e.g., PROSAIL described in Jacquemoud et al., 2009). The chances of doing this are higher for crops with planophile architecture than for crops with erectophile architecture due to reduced influence of shadows for the sensors. Therefore, crop chemistry estimates from imaging spectroscopy depends on the ability to reduce confounding effects of vegetation structure on the hyperspectral signal.

Furthermore, annual crops like rice and soybean are generally planted in October–November and harvested in March–April in southern and central Brazil, respectively. Because the peak of crop development is also coincident with the peak of cloud cover in Brazil, optical remote sensing of these crops is sometimes difficult, as demonstrated by Sugawara et al. (2008) when mapping soybean in southern Brazil with Landsat instruments. Much better revisit times than that provided by Landsat (16 days) would be recommended to increase the chances of acquiring cloud-free data for these crops.

6.3.2 Hyperion Datasets and Preprocessing

In this study, we used six Hyperion images from distinct study areas acquired with solar elevation angles that ranged from 36° to 54° (Table 6.1). Due to the narrow swath width of Hyperion (7.7 km), we did not find a single Hyperion scene with multiple crops at variable phenological stages to test their classification. Data acquisition for pasture and soybean was performed under off-nadir viewing and in the backscattering direction with pointing angles of −12° and −26°, respectively. Off-nadir pointing was an alternative to reduce the 16 day revisit time of the Hyperion, as mentioned before. However, it may result in strong reflectance variation for a given crop type, when compared to nadir data acquisition, depending on the view angle and view direction used in data collection (Galvão et al., 2009).

The 196 radiometrically calibrated Hyperion bands, sampled at approximately 10 nm intervals in the 426–2395 nm range, were used in the analysis. An algorithm to identify bad pixels and to reduce striping effects was applied to the Hyperion images replacing abnormal vertical lines by the average response of adjacent columns. The Hyperion radiance values from the 196 bands were converted into surface reflectance images using the fast line-of-sight atmospheric analysis of spectral hypercubes (FLAASH) algorithm, a MODTRAN4-based approach to remove atmospheric scattering and absorption effects (Harris Geospatial Solutions, Inc., Melbourne, Florida). The two-band (K-T) method, which uses an approach based on the dark pixel reflectance ratio with bands placed around 660 and 2100 nm (Kaufman et al., 1997), was selected to estimate the amount of aerosols and the scene average visibility. Precipitable water vapor was calculated on a per-pixel basis using the 1140 nm absorption band. A correction for adjacency effects was also applied to data.

Model parameters included tropical (soybean dataset) and midlatitude summer (remaining datasets) atmospheres with a rural aerosol model. Bands around 1400 and 1900 nm were not useful, even after atmospheric correction, due to the strong atmospheric water vapor absorption. For crop type discrimination, only 150 bands were considered in the data analysis after the removal of the noisy bands, including those affected by the atmosphere at 1400 and 1900 nm.

TABLE 6.1

Crops and Attributes of the Hyperion Datasets Used in the Data Analysis

Crop	Location in Brazil (State)	Latitude[a]	Longitude[a]	Acquisition Date (y/m/d)	Sun Azimuth	Sun Elevation	Pointing Angle
Rice	Rio G. do Sul	29°45'37"S	55°46'57"W	2003/04/10	47°	40°	+1°
Coffee	Minas Gerais	20°54'18"S	47°03'51"W	2008/08/21	48°	43°	+2°
Sugarcane	São Paulo	20°32'16"S	47°25'09"W	2002/07/16	40°	36°	+1°
Bean and Corn	São Paulo	23°06'00"S	49°20'23"W	2004/02/19	77°	54°	+2°
Pasture	São Paulo	21°40'41"S	52°22'19"W	2002/05/27	39°	36°	−12°
Soybean	Mato Grosso	12°45'06"S	52°22'19"W	2006/01/14	108°	54°	−26°

[a] Center coordinates of the study area.

6.3.3 Hyperion Color Composites and Agronomic Information

Hyperion color composites of the six study areas are illustrated in Figure 6.1. For annual crops like rice and soybean, knowledge of the phenological stages in the period of image acquisition is essential for the correct interpretation of the data.

In southern Brazil, rice is flooded and water-seeded in October–November, flowers in January–February, and is harvested in March–April. Thus, the Hyperion image of the rice fields (April 10, 2003) comprises the harvest period, as indicated by several harvested crop fields in cyan color (Figure 6.1a). In this period, the spectral influence of the background water to decrease the visible and especially NIR and SWIR reflectance is greatly reduced. The dominant response is from the senesced mature leaves associated with the ripening growth stage. Because the reflectance of the rice fields changes dramatically with crop development, or with background water modifications, inspection of images from two dates (flooded and nonflooded rice stages) is helpful in case of eventual classification problems between rice and other scene components (D'Arco et al., 2006). Even a single image from the establishment phase (vegetative stage), when the spectral response of the background water is stronger, can be used to discriminate flooded rice from other crops and to provide an early estimate of its cultivated area (Van Niel and McVicar, 2004).

Well-developed coffee canopies in the central and upper portions of the Hyperion scene are characterized by a low reflectance in the visible region (dark green leaves plus shadows within and between canopies) or by canopies in dark-green colors in the true color composite (Figure 6.1b). The observed spectral variability between the fields is probably introduced by factors such as row spacing, relief, age, cultivar, and management practices (Moreira et al., 2004). Emerging coffee canopies appear in dark reddish shades due to the strong soil background influence.

Most of the reddish shades in Figure 6.1c are associated with sugarcane. Color variability is the resultant of a combination of interrelated factors such as canopy closure, variety, age, successive ratoon harvests (shoot sprouting), leaf width and size, erectness or canopy architecture, degree of lodging, soil type and fertility, local water stress, pests, and diseases. Harvested fields with the predominance of nonphotosynthetic vegetation over the surface (crop residues) occur in cyan. Harvesting may be

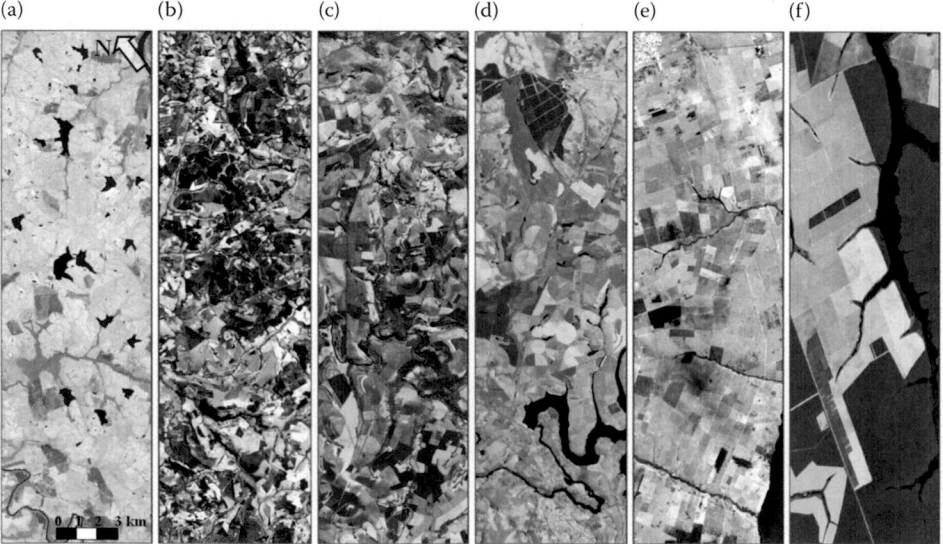

(a)   (b)   (c)   (d)   (e)   (f)

FIGURE 6.1 Hyperion color composites of different study areas showing the following crop types: (a) rice (reddish shades); (b) coffee (dark green color); (c) sugarcane (reddish shades); (d) corn and bean (pivots with dark and bright reddish shades, respectively); (e) pasture; and (f) soybean. In (a), (c), (d), and (f), bands centered at 864 nm (red), 1649 nm (green), and 671 nm (blue) were used in the false color composites. In (b) and (e), the true color composites refer to bands at 671 nm (red), 569 nm (green), and 487 nm (blue).

mechanical or manual. In manual harvesting, sugarcane is burned prior to cutting the stems in order to eliminate the straw and facilitate harvesting (Rudorff et al., 2010). The presence of ash over the soil surface produces dark blue shades in the false color composite (Figure 6.1c), which is, thus, an indicator of manual harvesting. Remote sensing is then used in a Brazilian agroenvironmental program to detect areas of biomass burning and to support a protocol that establishes deadlines to cease straw burning practice depending on the terrain slope (Rudorff et al., 2010).

Selected agricultural fields to represent bean and corn came from the same Hyperion image (Figure 6.1d). In the study area, they were cultivated under systems of center pivot irrigation. Center pivots with bean have brighter reddish shades than those with corn due to the larger amounts of radiation scattered in the NIR by the planophile canopies of bean. Pivots with nonphotosynthetic vegetation appear in cyan.

Pasture predominates in the true color composite of Figure 6.1e. Hyperion acquired the image at the end of the rainy season, and pastures with different vigor were observed. Variation in green color indicates changes in live to senescent biomass ratio. The magnitude of these changes is also dependent on the grass species under analysis, as pointed out by Numata et al. (2008) when using Hyperion data to estimate biophysical parameters of grazed pasture in the Amazon region.

Soybean is the highest-reflective studied crop, which facilitates its discrimination in false color composites (Figure 6.1f). Hyperion acquired data on January 14, 2006 over seven soybean varieties at reproductive stages that ranged from R1 (beginning bloom) to R3 (beginning pod). The observed color variation in the southern portion of the farm (brighter yellowish shades) is associated with one variety (Monsoy 9010), which was sensed by Hyperion at an earlier reproductive stage (Galvão et al., 2009).

6.3.4 REFLECTANCE AND BAND RATIO DIFFERENCES BETWEEN THE CROPS

We randomly selected 300 pixels per crop from different agricultural fields to illustrate reflectance differences between them. Average reflectance spectra (Figure 6.2) showed that the lowest reflectance was displayed by coffee due to the combined influence of its dark green leaves on the visible and of shadows within and between canopies over the entire wavelength region. Depending on the row orientation, the topographic position of the plantations (slope aspect and gradient), and the viewing–illumination geometry, shadow effects can be dominant in the spectral response of coffee. Such effects may affect the relationships between reflectance and biophysical parameters. On the other hand, the largest reflectance was observed for soybean and bean because of the efficiency of their planophile canopy architecture in scattering radiation toward the sensor. Furthermore, Hyperion data were acquired over the soybean farm (known as *Tanguro* farm) with off-nadir viewing in the backscattering direction, which resulted in increased reflectance due to the predominance of sunlit canopy components toward the sensor. Senesced pasture presented high reflectance in the red, NIR-2, SWIR-1, and SWIR-2 spectral intervals due to decreasing amounts of chlorophyll (live biomass) and leaf water and to the resultant predominance of nonphotosynthetic vegetation over live biomass. The lowest variability was observed for soybean, which presented the smallest standard deviation values in all Hyperion bands, whereas the highest variability was observed for sugarcane, corn, and coffee (results not shown).

The scatterplot of the relationship between the reflectance of the red (e.g., 660 nm; maximum chlorophyll absorption) and NIR (e.g., 864 nm; canopy scattering in a well-defined atmospheric window) bands can be thought of as a triangle whose vertices have different crop type positioning (Figure 6.3). Soybean is placed at the first corner of the triangle because of the chlorophyll absorption in the red and the strong scattering of radiation in the NIR by the planophile canopy. Coffee is located at the second corner (low NIR and red reflectance) because of the dark green leaves and shadow effects mentioned before. Finally, senesced pasture and sugarcane occurred close to the third corner due to decreasing amounts of chlorophyll (presence of nonphotosynthetic vegetation) over the substrate and within canopy, and the resultant higher red reflectance. Inside the triangle,

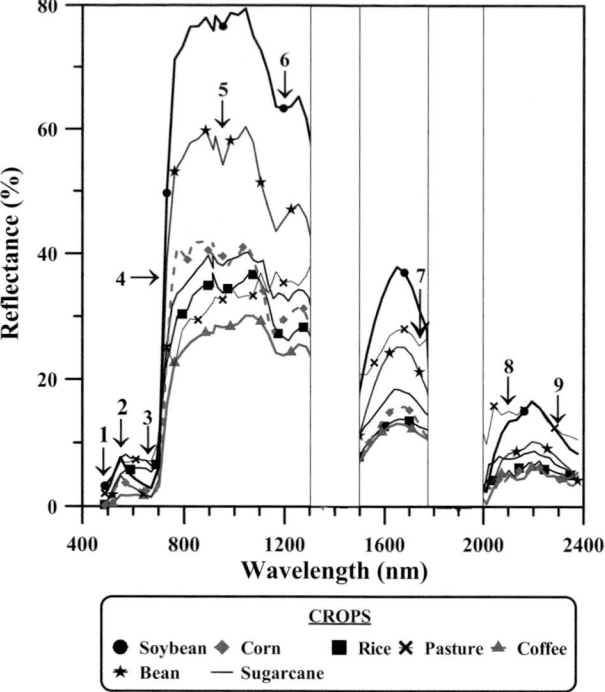

FIGURE 6.2 Average Hyperion reflectance spectra of the crop types under analysis obtained from 300 pixels per crop. The major spectral features are as follows: 1 and 3, blue–red chlorophyll absorption bands; 2, green reflectance peak; 4, red edge; 5 and 6, leaf water absorption bands; 7–9, lignin–cellulose spectral features. The two major intervals (1400 and 1900 nm) of strong atmospheric water vapor absorption are indicated. The growth stages of the crops are discussed in the text.

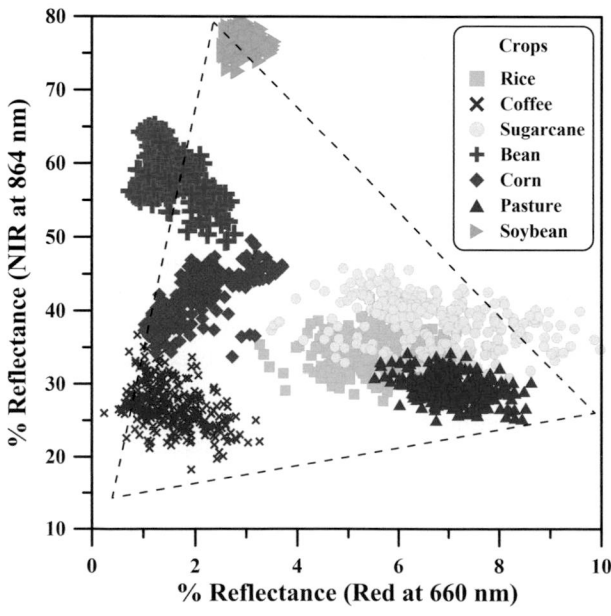

FIGURE 6.3 Relationships between the reflectance of the red and NIR Hyperion bands for the studied crop types. The triangle is discussed in the text.

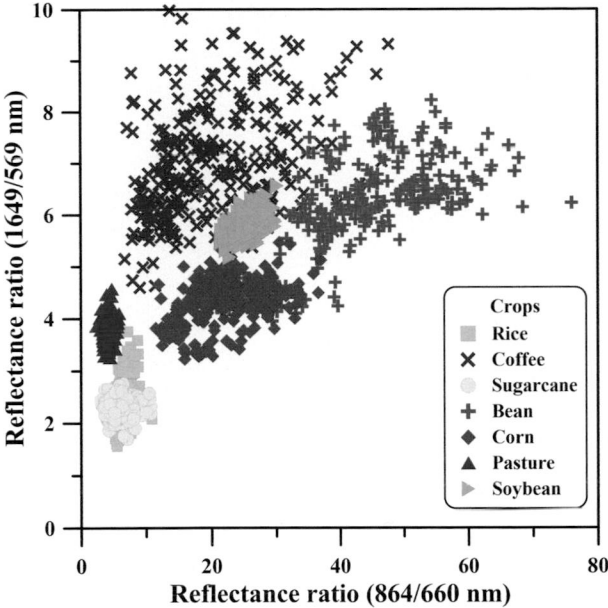

FIGURE 6.4 Variation in NIR-1/red and SWIR-1/green reflectance ratios for the crop types under study.

rice occurred together with pasture and sugarcane because Hyperion sensed this crop in the mature grain stage (ripening), when the red reflectance was increased due to senesced vegetation. In this stage, the filled spikelets change in color from green to yellow.

In addition to the differences in overall brightness, the crops presented also variation in reflectance ratios, especially for NIR/red and SWIR-1/green ratios. For example, pasture, sugarcane, and rice with senesced foliage had lower 864/660 and 1649/569 nm reflectance ratios than soybean, bean, corn, and coffee (Figure 6.4). The same behavior was verified for the NIR/SWIR-1 reflectance ratios (results not shown). Inspection of the Hyperion crop spectra (Figure 6.2) highlighted the importance of the NIR-1, SWIR-1, and red bands for crop type discrimination.

The importance of each band for discriminating crops can be also studied using multiple discriminant analysis (MDA). MDA performs linear discriminant analysis for multiple groups with the final objective of classifying samples into one of the groups. MDA deals simultaneously with the maximal separability between the groups, feature selection, and classification. Feature selection will be discussed later. Data distribution is generally assumed as normal, but statistical procedures (e.g., the Kolmogorov–Smirnov and Shapiro–Wilk tests) should be used to evaluate data normality, which is usually dependent on the wavelength and on the crop classes under analysis.

In this context, it is important to split the samples randomly into training and validation datasets to test the discriminatory power of the functions before classifying them. In the present example, the 2100 crop pixel spectra were used for training and a separate subset of 700 pixel spectra (100 per crop) was used for validation. We tested the reflectance of each of the 150 Hyperion bands placed between 426 and 2395 nm as input variables for MDA. Figure 6.5 confirms the importance of the NIR-1 and SWIR-1 bands for crop type discrimination. Other important bands were located in the red and green regions, in the NIR-2 interval, around the 1205 nm leaf water absorption, and in the red edge region around 720 nm.

6.3.5 Narrow-Band Vegetation Indices

Using hyperspectral remote sensing, one can calculate many narrow-band vegetation indices (NVIs) (Galvão et al., 2009; Roberts et al., 2012; Thenkabail et al., 2014). Care is necessary when using

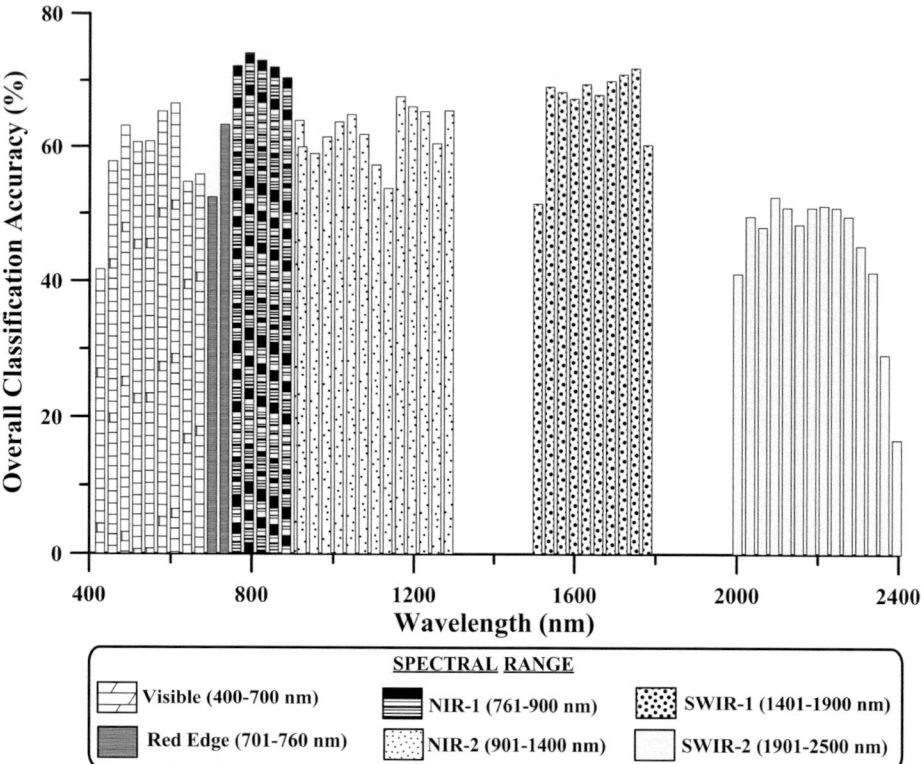

FIGURE 6.5 MDA-derived classification accuracy of the Hyperion bands for a subset of 700 validation pixels representing seven crop types from different study areas listed in Table 6.1. Spectral intervals are indicated to facilitate discussion.

them because some indices are simply reflectance ratios. Others have local applicability and were proposed empirically to solve specific problems. Several indices are also correlated with each other. In fact, only a few indices have been elaborated on a rigorous scientific basis.

Table 6.2 shows 21 NVIs that can be calculated with Hyperion data. Equations and references are indicated there. We generated images for the following indices: (1) Atmospherically Resistant Vegetation Index (ARVI); (2) Enhanced Vegetation Index (EVI); (3) Normalized Difference Vegetation Index (NDVI); (4) Simple Ratio (SR); (5) Sum Green Index (SGI); (6) Normalized Difference Infrared Index (NDII); (7) Normalized Difference Water Index (NDWI); (8) Water Band Index (WBI); (9) Leaf Water Vegetation Index (LWVI-2); (10) Disease Water Stress Index (DWSI); (11) Moisture Stress Index (MSI); (12) Plant Senescence Reflectance Index (PSRI); (13) Carotenoid Reflectance Index (CRI); (14) Anthocyanin Reflectance Index (ARI); (15) Photochemical Reflectance Index (PRI); (16) Structure Insensitive Pigment Index (SIPI); (17) Red Edge Normalized Difference Vegetation Index (RENDVI); (18) Red Edge Position (REP); (19) Vogelmann red edge index 1 (VOG-1); (20) Visible Atmospherically Resistant Index (VARI); and (21) Visible Green Index (VIg).

In general, ARVI, EVI, NDVI, SR, ARI, CRI, SGI, VARI, and VIg are indices closely related to greenness and leaf pigments (e.g., chlorophyll and carotenoids). MSI, NDII, NDWI, LWVI-2, WBI, and DWSI are much more associated with leaf/canopy water content. PSRI indicates canopy stress (pigment changes), whereas PRI and SIPI express light use efficiency. RENDVI, REP, and VOG-1 show spectral variations associated with the red edge wavelength position, which may be affected by changes in chlorophyll concentration or water stress. A comprehensive overview on hyperspectral vegetation indices was provided by Roberts et al. (2012), who classified the NVIs according to

TABLE 6.2

Narrow-Band Vegetation Indices (NVIs) Calculated from Hyperion Data

Vegetation Index	Formula[a]	References
ARVI	$[\rho_{864} - (2*\rho_{671} - \rho_{467})]/[\rho_{864} + (2*\rho_{671} - \rho_{467})]$	Kaufman and Tanré (1992)
EVI	$2.5*(\rho_{864} - \rho_{671})/(\rho_{864} + 6*\rho_{671} - 7.5*\rho_{467} + 1)$	Huete et al. (2002)
NDVI	$(\rho_{864} - \rho_{671})/(\rho_{864} + \rho_{671})$	Rouse et al. (1973)
SR	ρ_{864}/ρ_{671}	Rouse et al. (1973)
SGI	$(\rho_{508} + \rho_{518} + \rho_{528} + \rho_{538} + \rho_{549} + \rho_{559} + \rho_{569} + \rho_{579} + \rho_{590} + \rho_{600})/10$	Lobell and Asner (2003)
NDII	$(\rho_{823} - \rho_{1649})/(\rho_{823} + \rho_{1649})$	Hunt Jr. and Rock (1989)
NDWI	$(\rho_{854} - \rho_{1245})/(\rho_{854} + \rho_{1245})$	Gao (1996)
WBI	ρ_{905}/ρ_{973}	Penuelas et al. (1997)
LWVI-2	$(\rho_{1094} - \rho_{1205})/(\rho_{1094} + \rho_{1205})$	Galvão et al. (2005)
DWSI	ρ_{803}/ρ_{1598}	Apan et al. (2004)
MSI	ρ_{1598}/ρ_{823}	Hunt Jr. and Rock (1989)
PSRI	$(\rho_{681} - \rho_{498})/\rho_{752}$	Merzlyak et al. (1999)
CRI	$(1/\rho_{508}) - (1/\rho_{701})$	Gitelson et al. (2002b)
ARI	$(1/\rho_{549}) - (1/\rho_{701})$	Gitelson et al. (2001)
PRI	$(\rho_{529} - \rho_{569})/(\rho_{529} + \rho_{569})$	Gamon et al. (1997)
SIPI	$(\rho_{803} - \rho_{467})/(\rho_{803} + \rho_{681})$	Penuelas et al. (1995)
RENDVI	$(\rho_{752} - \rho_{701})/(\rho_{752} + \rho_{701})$	Gitelson et al. (1996)
REP	$(\rho_{n+1} - \rho_n)/10$ in the 690–750 nm interval	Curran et al. (1995)
VOG-1	ρ_{742}/ρ_{722}	Vogelmann et al. (1993)
VARI	$(\rho_{559} - \rho_{640})/(\rho_{559} + \rho_{640} - \rho_{467})$	Gitelson et al. (2002a)
VIg	$(\rho_{559} - \rho_{640})/(\rho_{559} + \rho_{640})$	Gitelson et al. (2002a)

[a] ρ is the reflectance of the closest Hyperion bands (n, center in nanometers) to the original wavelength formulations.

their predominant association with vegetation structure (e.g., LAI and biomass), biochemistry (e.g., pigments, water, lignin and cellulose) and physiology (e.g., light use efficiency and stress).

Narrow-band vegetation indices can be used as potential variables for crop type discrimination. Classification of the training dataset using discriminant analysis showed that the 10 best NVIs to discriminate the seven crop types were ARVI, EVI, NDVI, and SGI (greenness/leaf pigment indices); RENDVI and VOG-1 (chlorophyll red edge); SIPI and PRI (light use efficiency); and DWSI and NDWI (leaf water) (Figure 6.6). Examples of the relationships between pairs of indices are shown in Figures 6.7 and 6.8. Pasture presented the lowest NDWI value, as expected. Pasture, sugarcane, and rice displayed lower NDVI than the other crops because of the influence of nonphotosynthetic vegetation over the substrate and within the canopies (Figure 6.7). In agreement with these results, decreasing amounts of chlorophyll (presence of senesced canopy foliage) resulted in larger SIPI and lower RENDVI values (canopy stress) for the three mentioned crops (Figure 6.8).

Although useful for crop type discrimination, the major value of the vegetation indices for agriculture is to provide information on the variability of biophysical and biochemical canopy attributes within a given crop rather than between crops. For example, some indices (e.g., NDVI and EVI) are usually correlated with LAI and others with leaf water content (e.g., NDWI). Variation in LAI and leaf water content for a given crop may be produced by several factors such as water stress, nutrient deficiency, pests, and diseases, or even by local problems of soil adaptation. Especially for annual crops, such variation may be also produced by differences in reproductive stages of the cultivars.

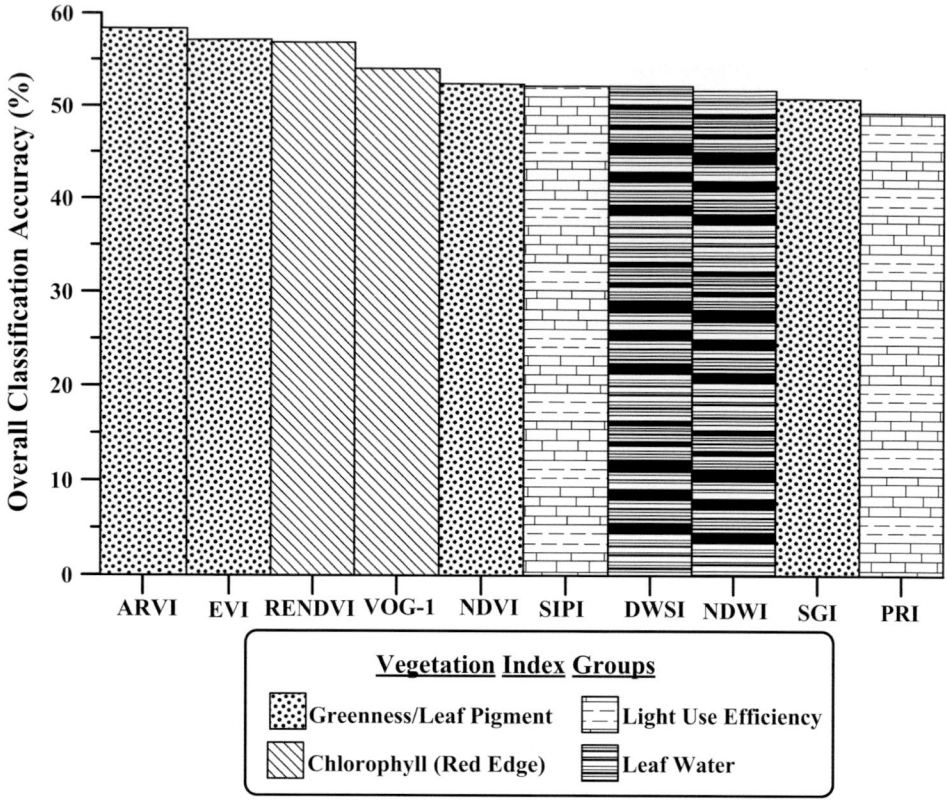

FIGURE 6.6   MDA-derived classification accuracy of the 10 best narrowband vegetation indices for a subset of 700 validation pixels representing the seven crop types. Indices were grouped according to their meaning.

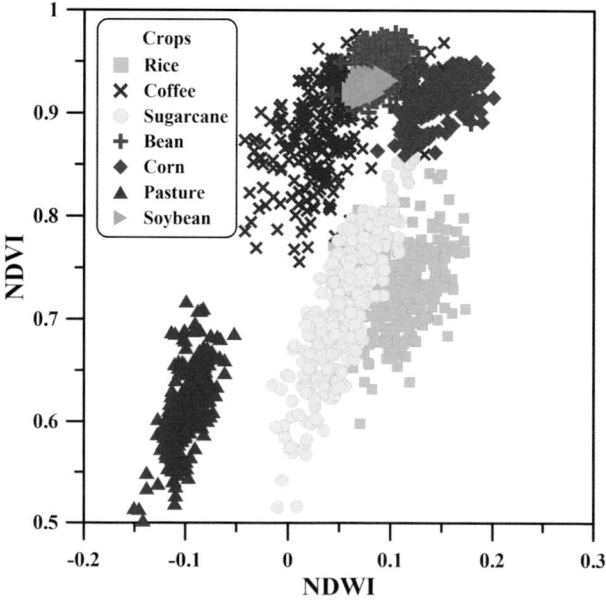

FIGURE 6.7   Relationships between normalized difference vegetation index (NDVI) and normalized difference water index (NDWI) for the studied crops.

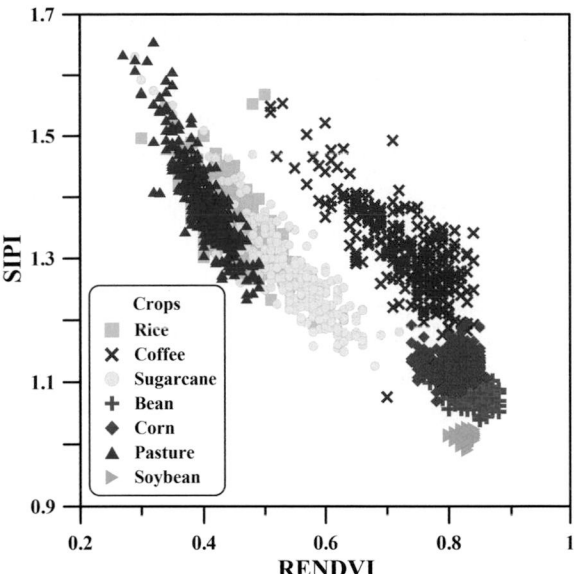

FIGURE 6.8 Relationships between structure-insensitive pigment index (SIPI) and red edge normalized difference vegetation index (RENDVI) for the studied crops.

An example is presented in Figure 6.9 for seven soybean varieties (300 pixels per variety extracted from different fields) planted in the same growing season and farm (*Tanguro* farm in Figure 6.1f) and for two indices (MSI and VOG-1). By definition, higher MSI and VOG-1 values indicate greater water stress and chlorophyll content, respectively. Using a single date (January 14, 2006) and without any agronomic information, one can wrongly conclude that the fields of Monsoy 9010 (brighter yellowish shades in the southern portion of the farm—Figure 6.1f) were under water stress during the Hyperion image acquisition. In reality, Monsoy 9010 was sensed by Hyperion in an earlier reproductive stage than the other varieties, which is the major cause of the observed higher MSI and

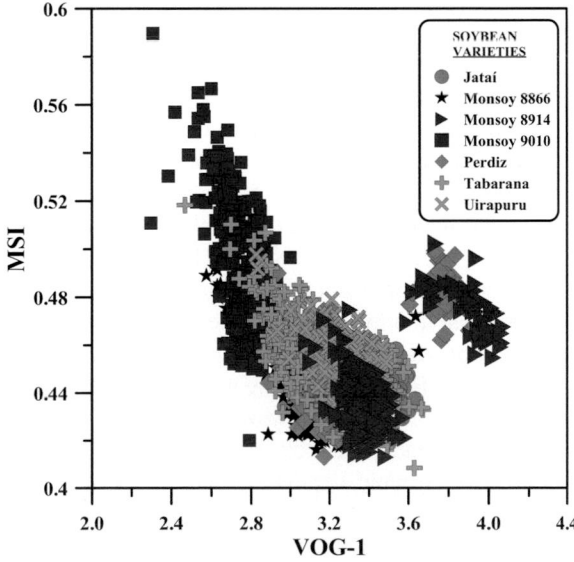

FIGURE 6.9 Relationships between moisture stress index (MSI) and Vogelmann red edge index-1 (VOG-1) for seven soybean varieties planted in the *Tanguro* farm (Figure 6.1f).

lower VOG-1 values (Figure 6.9). On the other hand, some pixels representing distinct agricultural fields of Monsoy 8914 and Perdiz displayed comparatively larger MSI values (right side of Figure 6.9), which would have required attention for potential plant stress.

Thus, interpretation of the causes of the vegetation index variability for annual crops should be performed relatively to crop development. It is greatly facilitated with image acquisition at different dates along the growing season.

6.3.6 SPECTRAL FEATURES

We can determine absorption band parameters and test them for crop type discrimination. In Figure 6.2, the major spectral features detected by Hyperion were indicated by numbers. Most of the crop spectra displayed well-defined chlorophyll absorption bands in the blue (447 nm) and red (671 nm) wavelengths (e.g., soybean) and the characteristic green reflectance peak at 569 nm. Exceptions were pasture, rice, and sugarcane, which had comparatively higher red reflectance due to the spectral influence of senesced canopy components.

Senesced pasture showed also lignin–cellulose spectral features in the SWIR-1 and SWIR-2 ranges (1750, 2103, and 2304 nm). Absorption bands produced by leaf water at 983 and 1205 nm, especially the second feature, were observed in all spectra, except for pasture. Unfortunately, the deepest leaf water absorption bands at 1400 and 1900 nm, usually detected in laboratory spectra, generally cannot be measured by airborne or orbital hyperspectral sensors because of the strong atmospheric water vapor absorption at these wavelengths. Even in field data acquisition, if the sun is used as the illumination source, measurements of these features are difficult. This occurs because of the small amounts of energy that can be detected by the sensor after it passes through the atmosphere and is reflected by the surface toward the instrument.

Another spectral feature indicated in Figure 6.2 is the red edge, which varies between the crops studied. This term has been used in the literature with two different meanings: (1) to characterize the spectral region of rapid reflectance change between the visible and NIR (690–750 nm); and (2) as the wavelength position (inflection point) equivalent to the maximum rate of reflectance change (slope). The position of the inflection point is usually associated with changes in chlorophyll content. The inflection point varies also for the same crop with canopy development (results not shown). At a fixed phenological stage of a given crop, shifts in red edge in agricultural fields may be associated with canopy stress.

Quantitatively, the detection of the red edge (inflection point) requires derivative analysis applied to the interval mentioned, preceded by some smoothing procedure to reduce noise effects on results. This is critical when using instruments with poor SNR. In the present example, we applied derivative analysis to the 2100 crop spectra in the 701–760 nm interval and used the Savitzky–Golay smoothing procedure to obtain the first-order derivative spectra and the wavelength position of the red edge. First-order derivative results (not presented) showed that senesced pasture, rice, and sugarcane had red edge values positioned at shorter wavelengths (lower than 725 nm) than those observed for the other crops, which was consistent with RENDVI results of Figure 6.8 and with the spectral influence of nonphotosynthetic vegetation. It is important to note that red edge shifting occurred over a 30 nm wavelength range, which is equivalent only to three Hyperion bandwidths.

Quantitatively, we can describe each absorption band by calculating several parameters (e.g., depth, width, area, and asymmetry). Band depth is the most frequently used parameter. It is usually correlated with the abundance of the absorber. However, there are cases in which this correlation is affected by the presence of other substances. For example, opaque minerals (magnetite and ilmenite) in tropical soils decrease absorption band depths of other constituents at different wavelengths (Galvão et al., 1997).

A simple way to calculate absorption band depth is to select bands at the edges and center of the feature and to compute the normalized difference between them. For example, LVWI-2 (Table 6.2) is an indirect measurement of the leaf water absorption band centred at 1205 nm. A more elaborate procedure is to use the continuum removal method to normalize the curves, to isolate and filter the

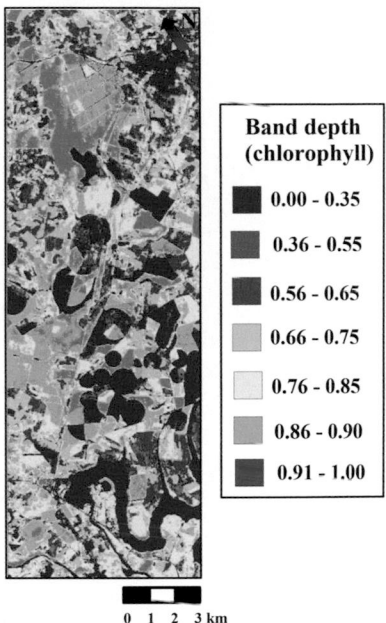

FIGURE 6.10 Band depth image for the 671 nm chlorophyll absorption band derived from the Hyperion image of Figure 6.1d and the continuum removal method. Pixels in sea-green indicate green vegetation with deep chlorophyll absorption bands in their spectra. Pixels in red refer to nonvegetated surfaces (e.g., bare soils or water).

features from the spectra, and to allow their comparison from a common baseline. Bands of a given hyperspectral sensor placed at the edges (reflectance maxima) of an absorption feature are selected to represent the limits of straight-line segments (continuum). The depth (D) of each absorption band is 1 minus R_b/R_c, where R_b is the reflectance at the center of the absorption band and R_c is the reflectance of the continuum at the same wavelength as R_b (Clark and Roush, 1984). In practice, user-friendly software can be applied to hyperspectral data to generate images of the absorption band parameters. For instance, the Processing Routines in IDL for Spectroscopic Measurements (PRISM) software (Kokaly and Skidmore, 2015), which is based on the continuum removal method, allows quantification on a per-pixel basis of the depth, area, width, and asymmetry of the major absorption bands observed in vegetation spectra.

Using the different Hyperion datasets and the continuum removal method, we generated band depth images for the following absorption bands: chlorophyll at 671 nm (edges at 569 and 763 nm); leaf water at 983 nm (edges at 933 and 1094 nm) and 1205 nm (edges at 1094 and 1286 nm); and lignin–cellulose at 2103 nm (edges at 2052 and 2214 nm) and 2304 nm (edges at 2214 and 2385 nm). Figure 6.10 shows an example of the 671 nm chlorophyll absorption band depth derived from the Hyperion image (corn and bean dataset of Figure 6.1d). Pixels in sea-green indicate green vegetation with deep 671 nm chlorophyll absorption bands in the corresponding Hyperion spectra. Pixels in red are related to nonvegetated surfaces such as bare soils or water.

In hyperspectral data, the canopy/leaf water 1205 nm absorption band is generally better defined than the corresponding 983 nm feature in vegetation spectra, especially for instruments with poor SNR like Hyperion. Figure 6.11 shows crop type discrimination based on the relationship between the 671 nm chlorophyll absorption band and the 1205 nm spectral feature for the 2100 crop spectra under study. Pasture, rice, and sugarcane presented the shallowest chlorophyll absorption bands due to senesced foliage, whereas bean showed the deepest chlorophyll feature. Flooded rice displayed deeper leaf water features when compared to sugarcane and pasture, which may reflect to some extent the influence of water over the leaf surface and the great number of individuals per linear meter.

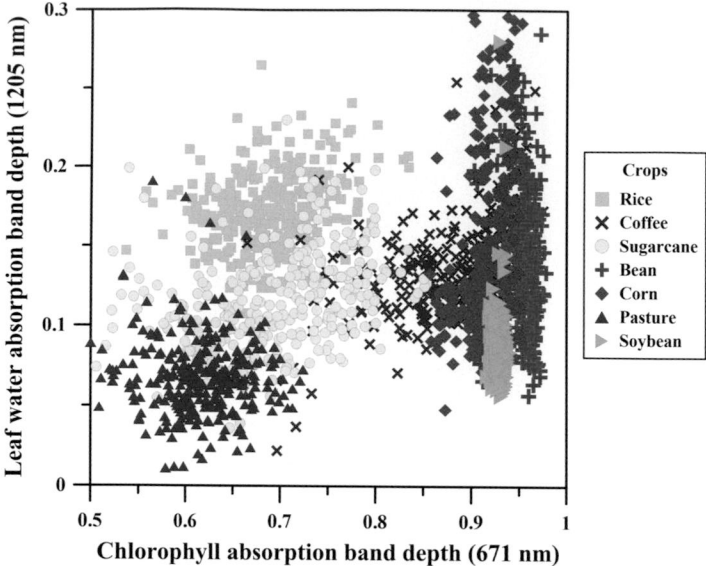

FIGURE 6.11 Relationship between band depth values of the chlorophyll and leaf water absorption bands (2100 training pixels).

6.3.7 CULTIVAR DISCRIMINATION

Because of the greater number of narrow bands and contiguous bands, hyperspectral sensors provide an enhanced level of information that increases the chances to discriminate materials with very small reflectance differences. The example presented here is based on the work by Galvão et al. (2005), who used Hyperion data to discriminate five sugarcane varieties in Southeastern Brazil (image of Figure 6.1c). The main hypothesis was that cultivar-related agronomical differences produced different spectral responses that could be detected by hyperspectral sensors. Detailed sugarcane information was provided by an owner company. The agronomic data included variety, date of planting, frequency and date of cutting, soil types, and farms. Sugarcane age ranged from 1 to 4 years.

Reflectance in the Hyperion bands, narrow-band ratios, NVIs, and spectral features parameters, which were discussed in the previous items, were considered as potential variables for MDA. The objective was to look for the optimum discriminant function to differentiate the varieties or to maximize the Mahalanobis distance for the two most similar groups. A training dataset of 200 pixels and a subset of 100 pixels were used to obtain the functions and test their discriminatory power. Final validation was performed by comparing the MDA classification results with the available ground information. Galvão et al. (2005) used a stepwise method to select the best variables (feature selection). The probability of F was used as a criterion to include (0.05) and to remove (0.10) variables in forward and backward steps. As mentioned before, in classification of hyperspectral data, feature selection is necessary to avoid the Hughes effect or phenomenon (Hughes, 1968; Pal and Foody, 2010). In other words, depending on the classifier, the use of the whole set of highly correlated Hyperion bands may actually produce worse classification results than the use of a selected set of bands with good discriminatory power. In the literature, several procedures have been proposed for feature selection (Bajcsy and Groves, 2004; Lu et al., 2007; Pal, 2009) and recent advances will be reported later in this chapter. Nonparametric classification techniques, such as SVM, are pointed out by some researchers as insensitive to the Hughes effect. However, according to Pal and Foody (2010), even SVM is affected by the Hughes phenomenon, and feature selection is still recommended before SVM classification.

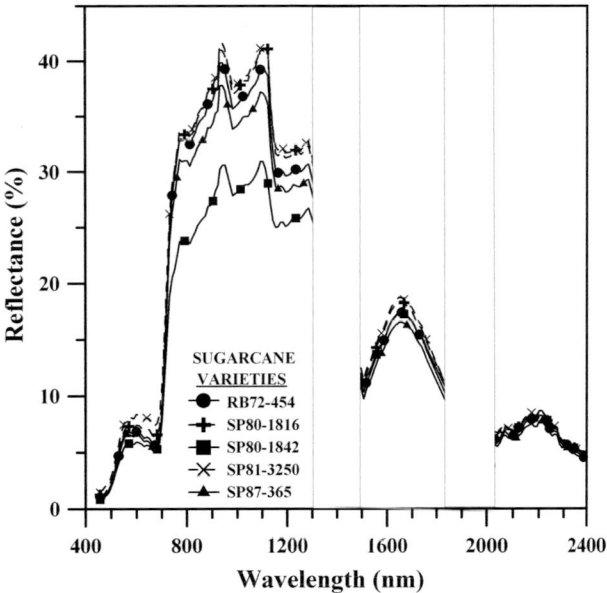

FIGURE 6.12   Average Hyperion surface reflectance spectra of the five studied sugarcane varieties. (Adapted from Galvão, L.S. et al. 2005. *Remote Sensing of Environment*, 94, 523–534.)

Figure 6.12 shows the average Hyperion reflectance spectra of the five sugarcane varieties. The transition from the low (SP80-1842) to the high reflectance (SP81-3250) varieties represented the change from erect (SP80-1842) to medium arch foliage (SP81-3250). Such differences in canopy structure affect sunlight penetration and reflectance, resulting in a higher reflectance for planophile than erectophile plants (Jackson and Pinter, 1986). In comparison with the other sugarcane varieties, SP80-1842 presented also deeper lignin–cellulose absorption bands at 2103 and 2304 nm and shallower leaf liquid water absorption bands at 983 and 1205 nm. The transition from the variety SP80-1842 to SP81-3250 was characterized by an increase in the LWVI-2 values and by a decrease in depth of the 2304 nm absorption band, both due to the larger amounts of nonphotosynthetic constituents within the canopy viewed by the sensor in the SP80-1842 variety (Figure 6.13).

The variety SP80-1842 was easily discriminated from the other four varieties due to its lower NIR reflectance (Figure 6.12). Thus, the simplest way to perform such discrimination was to use a band threshold in the NIR interval (e.g., pixels with reflectance values lower than 30% at 864 nm). Discrimination between the remaining four varieties (RB72-454, SP80-1816, SP81-3250, and SP87-365) was much more difficult due to the similarity of their average reflectance spectra. It required discriminant analysis.

The predictive power of the reflectance of the Hyperion narrow bands, the reflectance ratios, and of some spectral parameters tested by Galvão et al. (2005) to discriminate the four sugarcane varieties is shown in Figures 6.14 through 6.16, respectively. The Hyperion bands were considered individually for the subsequent calculation of the average classification accuracy of the discriminant functions at different spectral intervals. A single-variable run procedure was used also for reflectance ratios and spectral parameters. Results refer to the training dataset of pixels. The best spectral intervals of Hyperion narrow-band positioning to discriminate the four Brazilian sugarcane varieties were the SWIR-1 (1498–1780 nm), the NIR-2 (915–1296 nm), and the green (508–600 nm) (Figure 6.14). Hyperion narrow bands located in these intervals showed the largest classification accuracy values.

For Hyperion narrow-band ratios, the best results (36%–40% of correct classification of the pixels) were obtained for (1) SWIR-1/green and SWIR-2/green ratios; (2) SWIR-1/NIR-1 and SWIR-1/NIR-2

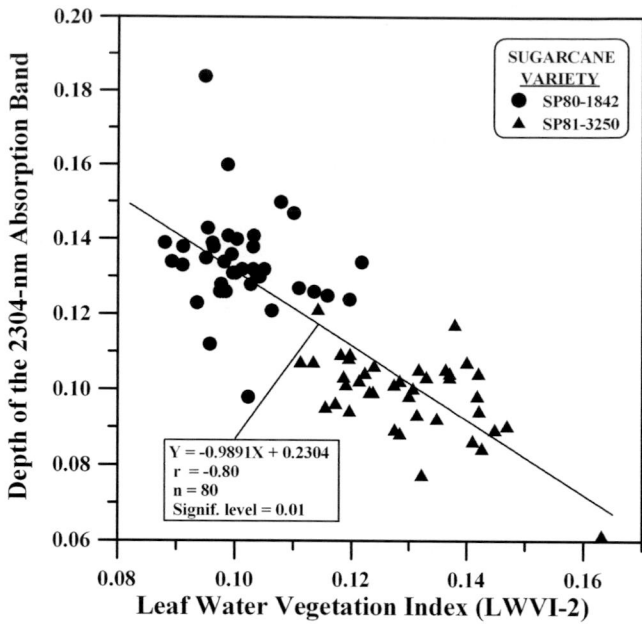

FIGURE 6.13 Relationship between the band depth of the 2304 nm absorption band (lignin–cellulose) and the LWVI-2 (leaf water) for the two sugarcane varieties with the lowest and highest reflectance. (Adapted from Galvão, L.S. et al. 2005. *Remote Sensing of Environment*, 94, 523–534.)

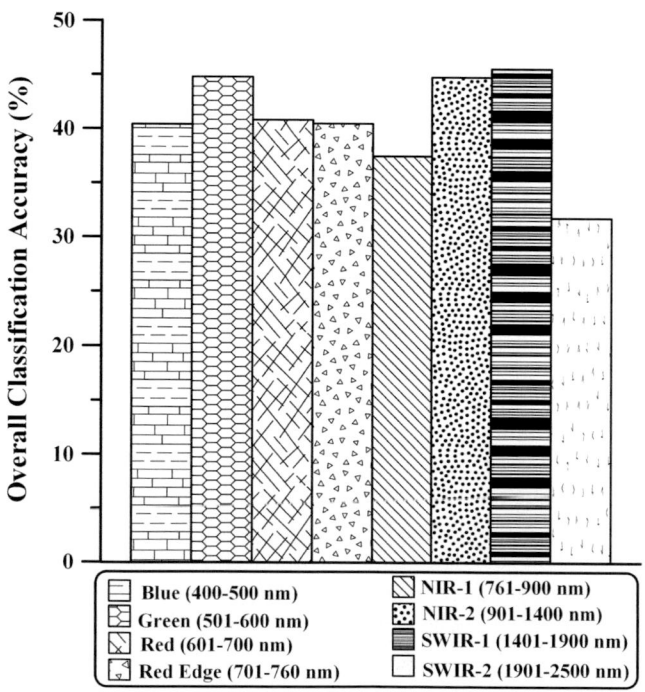

FIGURE 6.14 The potential of selected spectral intervals of Hyperion narrow-band positioning to discriminate four sugarcane varieties (RB72-454, SP80-1816, SP81-3250, and SP87-365). Results refer to the average classification accuracy values and the training dataset. (Adapted from Galvão, L.S. et al. 2005. *Remote Sensing of Environment*, 94, 523–534.)

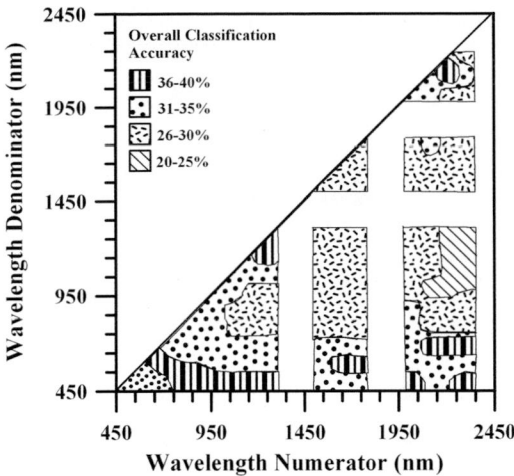

FIGURE 6.15 The potential of the Hyperion narrow-band ratios to discriminate sugarcane varieties. Results around 1400 and 1900 nm were omitted due to atmospheric water vapor absorption. (Adapted from Galvão, L.S. et al. 2005. *Remote Sensing of Environment*, 94, 523–534.)

ratios; and (3) NIR-2/NIR-1 and NIR-2/NIR-2 ratios (Figure 6.15). Finally, the spectral indices that exhibited the best discriminatory power were associated with the leaf liquid water (LWVI-2 and depth of the 1205 and 983 nm absorption bands) and lignin–cellulose (depth of the 2304 nm absorption band) spectral features (Figure 6.16).

Using the stepwise procedure, Galvão et al. (2005) selected the following set of variables to compose the final discriminant model: (1) the Hyperion bands placed at 651, 722, 813, 1084, 1124, 1649, and 2002 nm; (2) the reflectance ratios 2355/2052, 1750/478, 1750/569, and 1255/478 nm; (3) the depth of the absorption bands centered at 671 nm (chlorophyll), 983 nm (leaf liquid water),

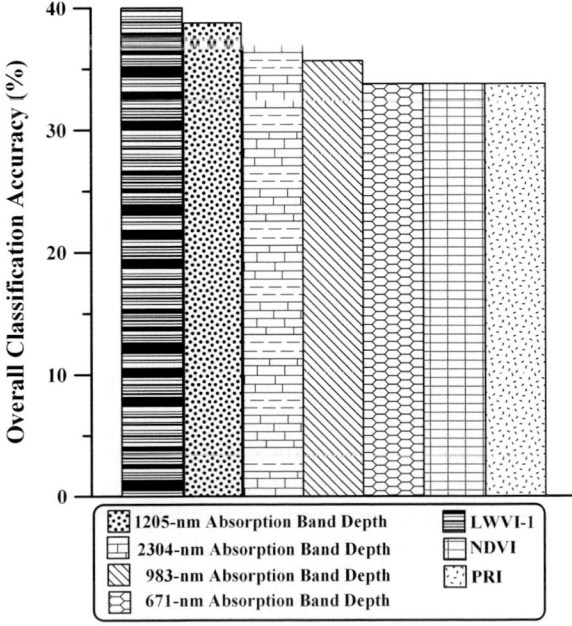

FIGURE 6.16 The potential of absorption band depth and some vegetation indices to discriminate sugarcane varieties. (Adapted from Galvão, L.S. et al. 2005. *Remote Sensing of Environment*, 94, 523–534.)

(a)   (b)

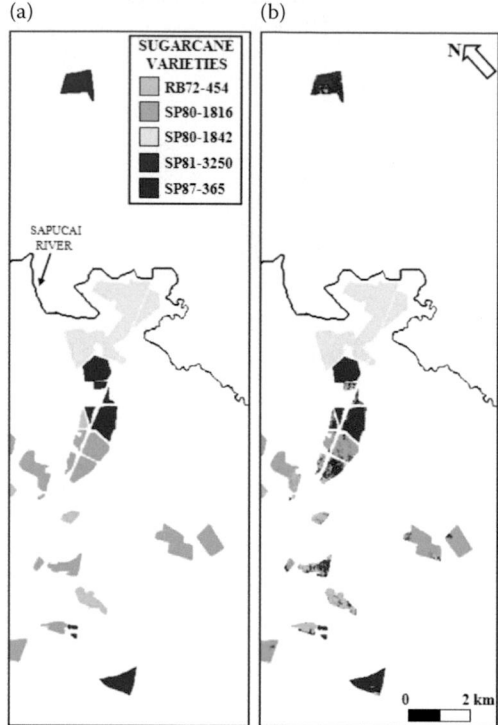

FIGURE 6.17 (a) Ground truth image showing the spatial distribution of the five sugarcane varieties under study in the central portion of the study area. (b) Hyperspectral classification image derived from MDA. (Adapted from Galvão, L.S. et al. 2005. *Remote Sensing of Environment*, 94, 523–534.)

and 2304 nm (lignin–cellulose); and (4) the NDWI and DSWI. When applied to the subset of pixels, this discriminant model reached 87.5% of classification accuracy. Comparison between the ground truth data (Figure 6.17a) and the MDA-derived classified image (Figure 6.17b) in the central portion of the study area confirmed the good performance of the discriminant model that presented the best results for the cultivars SP87-365 and RB72-454.

6.3.8 MULTISPECTRAL VERSUS HYPERSPECTRAL DISCRIMINATION

We used the Hyperion image showing planted soybeans (Figure 6.1f) to evaluate the spectral resolution influence on the discrimination of seven cultivars: Jataí; Monsoy 8866, 8914, and 9010; Perdiz; Tabarana; and Uirapuru. Filter functions were used to simulate the spectral response of the following multispectral instruments (Table 6.3): Advanced Very High Resolution Radiometer (AVHRR/NOAA-17); CCD Camera on Board the China–Brazil Earth Resources Satellite (CCD/CBERS-2); High Geometric Resolution Instrument (HRG/SPOT-5); Enhanced Thematic Mapper Plus (ETM+/Landsat-7); Moderate Resolution Imaging Spectroradiometer (MODIS/Terra); and Advanced Spaceborne Thermal Emission and Reflection Radiometer (ASTER/Terra). MODIS bands 8 (405–420 nm) and 20–36 (3660–14385 nm) were not simulated from Hyperion because they were out of the spectral range of operability. The same occurred for ASTER bands 9 (2360–2430 nm) and 10–14 (thermal). The spatial resolution was not simulated.

The discrimination between the soybean varieties was tested using reflectance values of the Hyperion and sensor-simulated images extracted from training (300 pixels per variety) and subset samples (100 pixels per cultivar) as input variables for MDA. The stepwise procedure described before was used for band selection.

TABLE 6.3

Simulated Multispectral Sensor Bands from Hyperion Data Using the Filter Functions

Sensor	Blue Band (nm)	Green Band (nm)	Red Band (nm)	Red Edge Band (nm)	NIR-1 Band (nm)	NIR-2 Band (nm)	SWIR-1 Band (nm)	SWIR-2 Band (nm)
AVHRR/NOAA-17	–	–	580–680	–	725–1000	–	1580–1640	–
CCD/CBERS-2	450–520	520–590	630–690	–	770–890	–	–	–
HRG/SPOT-5	–	500–590	610–680	–	780–890	–	1580–1750	–
ETM+/Landsat-7	450–515	525–605	630–690	–	775–900	–	1550–1750	2090–2350
MODIS/Terra	438–448	526–536	620–670	743–753	841–876	1230–1250	1628–1652	2105–2155
	459–479	545–565	662–672		862–877	931–941		
	483–493	546–556	673–683		890–920	915–965		
ASTER/Terra	–	520–600	630–690	–	760–860	–	1600–1700	2145–2185
								2185–2225
								2235–2285
								2295–2365

Inspection of the average surface reflectance spectra measured by Hyperion (not presented) showed a better discrimination for cultivar Monsoy 9010, which was sensed by Hyperion in an earlier reproductive stage. Considering only the spectral resolution influence on results, MODIS was the best multispectral sensor when compared to Hyperion (Figure 6.18). MODIS allows the measurement of the chlorophyll features in the visible and of the 1205 nm leaf water absorption in the NIR-2. It has also red edge (743–753 nm) and SWIR-1 and SWIR-2 bands. On the other hand, CCD/CBERS-2 does not acquire data in the SWIR-1 interval, and AVHRR/NOAA-17 acquires data in a very broad NIR band (Table 6.3). From an agricultural point of view, these are two undesirable specifications for crop type discrimination, as deduced from the previous sections.

Among the multispectral sensors, the lowest classification accuracies were observed for AVHRR (53.9%) and CCD (54.7%), and the highest accuracy was verified for MODIS (75.4%

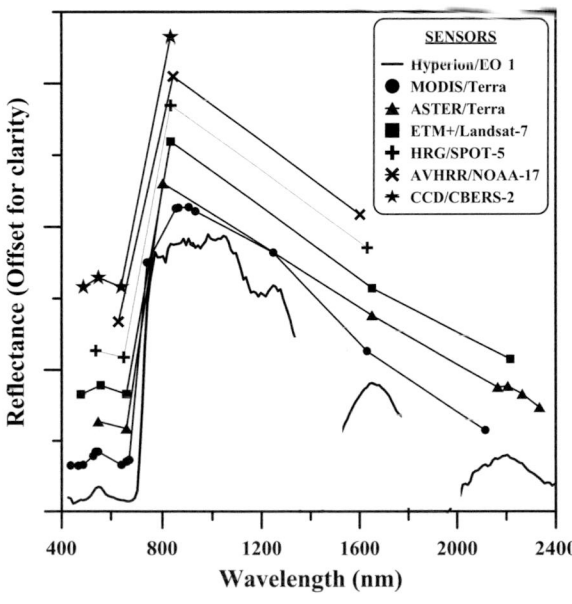

FIGURE 6.18 Spectral response of the cultivar Monsoy 8914 for different multispectral sensors simulated from Hyperion data.

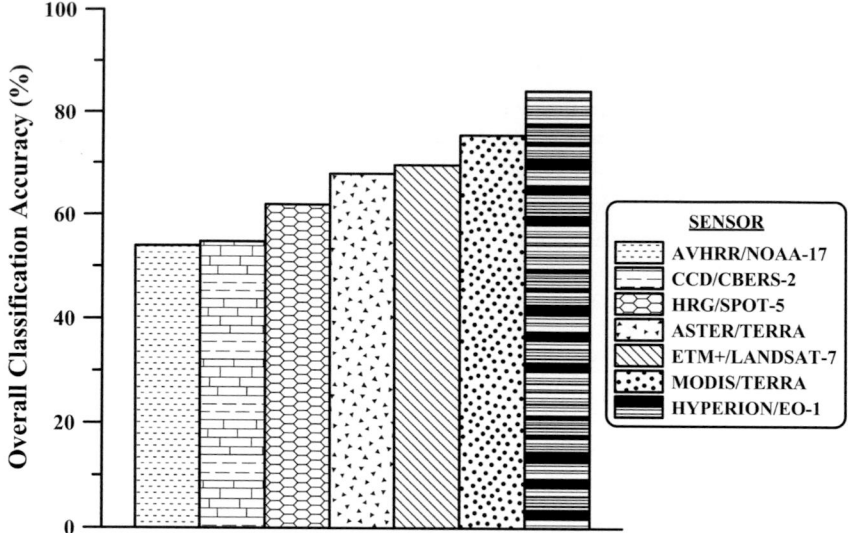

FIGURE 6.19 Overall classification accuracy for Hyperion and simulated multispectral sensors from this sensor. Results refer to the subset of 700 pixels and seven soybean varieties.

in Figure 6.19). In comparison with the simulated multispectral sensors, Hyperion had the best result. The overall classification accuracy was 84.1% for the subset of pixels using 36 narrow bands centered at the following wavelengths: 477, 528, 548, 569, 589, 609, 640, 660, 681, 701, 721, 742, 762, 793, 823, 864, 884, 925, 972, 1023, 1063, 1083, 1104, 1194, 1205, 1245, 1517, 1548, 1568, 1649, 1699, 1749, 2133, 2203, 2244, and 2284 nm. Results from the sensor simulation (Figure 6.19) are in agreement with those obtained by Galvão et al. (2006) when studying sugarcane varieties with Hyperion. Discriminant scores plotted for four of the seven soybean varieties confirmed the poor discrimination observed for AVHRR (Figure 6.20a) and the better discrimination provided by Hyperion (Figure 6.20b).

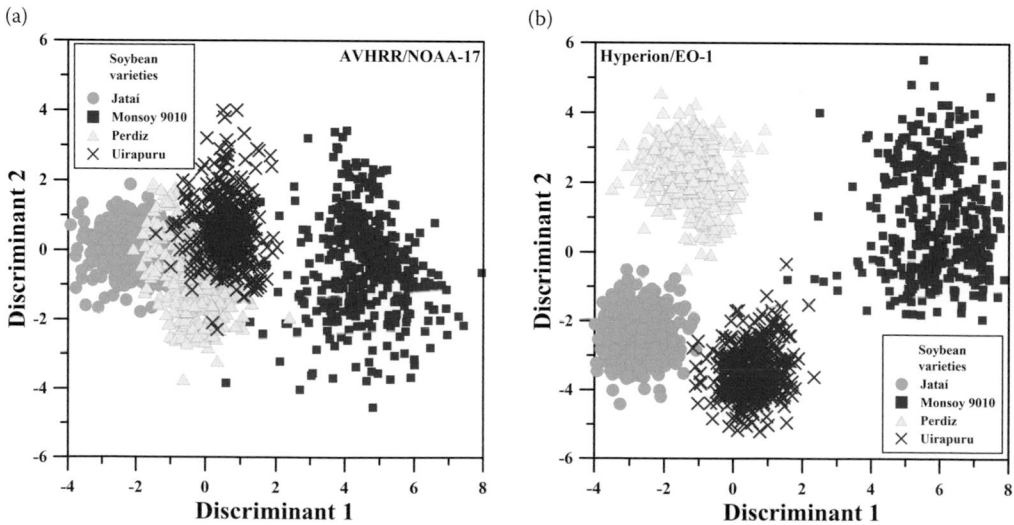

FIGURE 6.20 Projection of (a) the AVHRR/NOAA-17 and (b) the Hyperion/EO-1 discriminant scores for four of the seven studied soybean cultivars. Monsoy 9010 was sensed by the satellite at an earlier growth stage than the other varieties, which facilitated its discrimination.

6.4 ADVANCES IN FEATURE SELECTION AND CLASSIFICATION TECHNIQUES

In the previous sections, we used MDA preceded by a feature selection stepwise procedure to classify crops and their varieties. The increasing availability of hyperspectral data and the perspectives of new orbital missions have introduced advances in algorithms that deal with the high dimensionality of the data, the Hughes effect, and the reduction in computational time for classification. Feature extraction or feature selection approaches have been proposed in recent years. Feature extraction methods reduce the data dimensionality using, for instance, techniques similar to principal component analysis (PCA) applied to the reflectance of hyperspectral bands. On the other hand, feature selection methods pick the best subset of bands, NVIs, or absorption band parameters to maximize the discrimination between the classes tested for classification using a given criterion (e.g., distance, class separability, information, or dependence measurements). As discussed by Damodaran et al. (2017), feature selection is preferable over feature extraction because the selected spectral attributes for classification are more easily interpretable from a physical point of view than the statistically transformed data.

Damodaran et al. (2017) proposed a surrogate-kernel-based feature selection for hyperspectral image classification, which outperformed two other competitive methods (kernel dependence measure and constrained band selection). Taskin et al. (2017) developed a novel algorithm based on the method called high-dimensional model representation, which was very effective in selecting discriminative features with a relatively low computational cost. In a recent classification study, Ma et al. (2017) indicated the correlation-based feature selection (CFS) algorithm, from the Waikato Environment for Knowledge Analysis (WEKA) software, as the best feature selection procedure between eight filter, wrapper, and embedded methods. In general, comparison between these studies and other different methods reported recently in the literature did not allow identification of a single solution for feature selection in hyperspectral classification. By contrast, depending on the complexity of the crop classification scenario, different algorithms should be considered.

In addition to MDA, different classifiers can be tested for crop type discrimination. SVM is popular in hyperspectral image classification. Other useful techniques are Decision Trees (DT), Random Forests (RF), and Spectral Angle Mapper (SAM). Novel approaches (e.g., rotation forest, sparse representation, and different ensemble classifiers) have been proposed for classifying hyperspectral data (Chen et al., 2016; Xia et al., 2017; Zhang et al., 2017). However, most of these techniques have not been analyzed and compared over crops. For other land cover classes, including urban areas, the RF classifier generally outperforms the DT and SAM classifiers but produces similar results compared to the SVM technique (Ghimire et al., 2012; Dalponte et al., 2013; Ghosh et al., 2014; Toniol et al., 2017).

A review of the relative performance of advanced spectral classifiers applied to different hyperspectral datasets was provided by Ghamisi et al. (2017). After comparing SVM, RF, neural networks, deep learning approaches, logistic regression-based techniques, and sparse representation-based classifiers in different study areas, they observed that their classification performance varied between the studied hyperspectral datasets. Therefore, the choice of an appropriate classifier varies with the land cover types under analyses. A single classifier that consistently provides the best classification accuracy for all crop scenarios does not exist. Different classifiers should be therefore tested whenever necessary.

For crop type discrimination using hyperspectral images, an interesting strategy is to combine different metrics (e.g., reflectance, NVIs, absorption band parameters) and different classifiers (e.g., SVM, RF and DT) (Jacon et al., 2017). The consistency between the classification results from different classifiers can be therefore inspected on a per-pixel basis, as done by Toniol et al. (2017) over the savannas in Brazil.

6.5 CONCLUDING REMARKS

In general, hyperspectral sensors provide significantly better classification results than multispectral sensors, but their classification performance depends on other factors such as SNR, spectral range of data acquisition, phenological stages of multiple crops and cultivars, and the adequate use of feature selection and classification techniques. Depending on the classifier, the use of the whole set of highly correlated hyperspectral metrics may actually produce worse classification results than the use of a selected set of metrics with good discriminatory power.

Examples in this chapter demonstrated the discriminatory power of reflectance of the 10 nm Hyperion bands placed in the NIR (e.g., 830–890 nm), SWIR (1500–1750 nm), red (660–690 nm), green (550–570 nm), and red edge (690–750 nm) spectral regions. They showed the importance of considering narrow-band vegetation indices and absorption band parameters associated with vegetation structure, canopy biochemistry, and plant physiology in the discrimination procedure. After feature selection to avoid the Hughes effect and reduce data dimensionality, the combination of different metrics (reflectance, reflectance ratios, NVIs and absorption band parameters) increased crop classification accuracy. However, such results should be analyzed with care. For instance, a single Hyperion scene with multiple crops at variable phenological or growth stages was not considered in the analysis due to the narrow swath width of the sensor (7.7 km) to register such situation. In addition, bands placed between 2000 and 2400 nm were generally less efficient for discriminating crops than bands positioned in other spectral intervals. This result was affected to some extent by instrumental SNR because the SWIR-2 interval had noisier Hyperion reflectance data than the VNIR and SWIR-1 spectral regions.

Overall, the numerous opportunities offered by hyperspectral data are of great advantage when compared with the possibilities offered by broadband data. However, narrowband selection in these broad spectral intervals depends on each study area or on the set of crops/cultivars tested for hyperspectral classification. Considering the great data variability introduced by the factors previously discussed, the generation of a universal set of narrow bands for crop type discrimination requires large number of studies conducted in wide array of locations, crop types, and crop species in distinct agroecosystems around the world.

Finally, a new era of optical remote sensing science is emerging with forthcoming space-borne imaging spectrometer missions (Verrelst et al., 2016). For instance, the Environmental Mapping and Analysis Program (EnMAP) (planned for 2019; Guanter et al., 2015) and the Hyperspectral Infrared Imager (HyspIRI) (Hochberg et al., 2015) have been proposed to acquire images with much better SNR and larger swath width than those of Hyperion (Jacon et al., 2017; Toniol et al., 2017). Hyperspectral data coupled with Lidar observations, collected from airplanes and unmanned aerial vehicles (UAV), have been used recently to classify vegetation and to measure its structural characteristics (Sankey et al., 2017). These technological advances represent new opportunities for studies of crop type discrimination. Because these hyperspectral sensors have been designed to have much higher SNR than the Hyperion, the discriminatory power of the metrics will probably be higher than that observed in previous sections of this chapter. New approaches combining hyperspectral metrics and classifiers will be proposed using these data. The phenological response of different crops could also be added into the classification approach if hyperspectral data were acquired regularly over time and large areas.

ACKNOWLEDGMENTS

The authors are grateful to Drs. Prasad Thenkabail, John Lyon, and Alfredo Huete for the invitation to contribute a book chapter and for revising the manuscript. Thanks are also due to Conselho Nacional de Desenvolvimento Científico e Tecnológico (CNPq) (grant number 301486/2017-4) and to Fundação de Amparo à Pesquisa do Estado de São Paulo (FAPESP).

REFERENCES

Apan, A., Held, A., Phinn, S., and Markley, J. 2004. Detecting sugarcane "orange rust" disease using EO-1 Hyperion hyperspectral imagery. *International Journal of Remote Sensing*, 25, 489–498.

Bajcsy, P., and Groves, P. 2004. Methodology for hyperspectral band selection. *Photogrammetric Engineering and Remote Sensing*, 70, 793–802.

Bandos, T.V., Bruzzone, L., and Camps-Valls, G. 2009. Classification of hyperspectral images with regularized linear discriminant analysis. *IEEE Transactions on Geoscience and Remote Sensing*, 47, 862–873.

Bannari, A., Khurshid, K.S., Staenz, K., and Schwarz, J. 2008. Potential of Hyperion EO-1 hyperspectral data for wheat crop chlorophyll content estimation. *Canadian Journal of Remote Sensing*, 34, 139–157.

Bauer, M.E. 1975. The role of remote sensing in determining the distribution and yield of crops. *Advances in Agronomy*, 27, 271–304.

Camps-Valls, G., Gomez-Chova, L., Calpe-Maravilla, J., Soria-Olivas, E., Martin-Guerrero, J.D., and Moreno, J. 2003. Support vector machines for crop classification using hyperspectral data. *Lecture Notes in Computer Science*, 2652, 134–141.

Clark, R.N., and Roush, T.L. 1984. Reflectance spectroscopy: Quantitative analysis techniques for remote sensing applications. *Journal of Geophysical Research*, 89, 6329–6340.

Chen, J.K., Xia, J.S., Du, P.J., Chanussot, J., Xue, Z.H., and Xie, X.J. 2016. Kernel supervised ensemble classifier for the classification of hyperspectral data using few labelled samples. *Remote Sensing*, 8, article number 601.

Curran, P.J., Windham, W.R., and Gholz, H.L. 1995. Exploring the relationship between reflectance red edge and chlorophyll concentration in Slash Pine leaves. *Tree Physiology*, 15, 203–206.

Dalponte, M., Orka, H.O., Gobakken, T., Gianelle, D., and Naesset, E. 2013. Tree species classification in boreal forests with hyperspectral data. *IEEE Transactions on Geoscience and Remote Sensing*, 51, 2632–2645.

Damodaran, B.B., Courty, N., and Lefèvre, S. 2017. Sparse Hilbert Schmidt independence criterion and surrogate-kernel-based feature selection for hyperspectral image classification. *IEEE Transactions on Geoscience and Remote Sensing*, 55, 2385–2397.

D'Arco, E., Alvarenga, B.S., Rizzi, R., Rudorff, B.F.T., Moreira, M.A., and Adami, M. 2006. Geotechnologies to estimate flooded rice crop area. *Revista Brasileira de Cartografia*, 58, 247–253.

Duca, R., and Del Frate, F. 2009. Hyperspectral and multiangle CHRIS-PROBA images for the generation of land cover maps. *IEEE Transactions on Geoscience and Remote Sensing*, 46, 2857–2866.

Eddy, P.R., Smith, A.M., Hill, B.D., Peddle, D.R., Coburn, C.A., and Blackshaw, R.E. 2008. Hybrid segmentation: Artificial neural network classification of high resolution hyperspectral imagery for site-specific herbicide management in agriculture. *Photogrammetric Engineering and Remote Sensing*, 74, 1249–1257.

Everingham, Y., Lowe, K.H., Donald, D., Coomans, D., and Markley, J. 2007. Advanced satellite imagery to classify sugarcane crop characteristics. *Agronomy for Sustainable Development*, 27, 111–117.

Falqueto, A.R., Cassol, D., Magalhães Júnior, A.M., Oliveira, A.C., and Bacarin, M.A. 2009. Physiological analysis of leaf senescence of two rice cultivars with different yield potential. *Pesquisa Agropecuária Brasileira*, 44, 695–700.

Filippi, A.M., Archibald, R., Bhaduri, B.L., and Bright, E.A. 2009. Hyperspectral agricultural mapping using support vector machine-based endmember extraction (SVM-BEE). *Optics Express*, 17, 23823–23842.

Galvão, L.S., Vitorello, I., and Formaggio, A.R. 1997. Relationships of spectral reflectance and color among surface and subsurface horizons of tropical soil profiles. *Remote Sensing of Environment*, 61, 24–33.

Galvão, L.S., Formaggio, A.R., and Tisot, D.A. 2005. Discrimination of sugarcane varieties in southeastern Brazil with EO-1 Hyperion data. *Remote Sensing of Environment*, 94, 523–534.

Galvão, L.S., Formaggio, A.R., and Tisot, D.A. 2006. The influence of spectral resolution on discriminating Brazilian sugarcane varieties. *International Journal of Remote Sensing*, 27, 769–777.

Galvão, L.S., Roberts, D.A., Formaggio, A.R., Numata, I., and Breunig, F.M. 2009. View angle effects on the discrimination of soybean varieties and on the relationships between vegetation indices and yield using off-nadir Hyperion data. *Remote Sensing of Environment*, 113, 846–856.

Gamon, J.A., Serrano, L., and Surfus, J.S. 1997. The photochemical reflectance index: An optical indicator of photosynthetic radiation use efficiency across species, functional types and nutrient levels. *Oecologia*, 112, 492–501.

Gao, B.C. 1996. NDWI: A normalized difference water index for remote sensing of vegetation liquid water from space. *Remote Sensing of Environment*, 58, 257–266.

Gausman, H.W. 1985. *Plant Leaf Optical Properties in Visible and Near-Infrared Light*. Graduate Studies Texas No. 29, Texas Tech Press, Lubbock, TX, 78 p.

Ghamisi, P., Plaza, J., Chen, Y., Li, J., and Plaza, A. 2017. Advanced spectral classifiers for hyperspectral images: A review. *IEEE Geoscience and Remote Sensing Magazine*, 5, 8–32.

Ghimire, B., Rogan, J., Galiano, V.R., Panday, P., and Neeti, N. 2012. An evaluation of bagging, boosting, and random forests for land-cover classification in Cape Cod, Massachusetts, USA. *GIScience and Remote Sensing*, 49, 623–643.

Ghosh, A., Fassnacht, F.E., Joshi, P.K., and Koch, B. 2014. A framework for mapping tree species combining hyperspectral and LiDAR data: Role of selected classifiers and sensor across three spatial scales. *International Journal of Applied Earth Observation and Geoinformation*, 26, 49–63.

Gilabert, M.A., Segarra, D., and Melia, J. 1994. Simulation of citrus orchard reflectance by means of a geometrical canopy model. *International Journal of Remote Sensing*, 15, 2559–2582.

Gitelson, A.A., Merzlyak, M.N., and Lichtenthaler, H.K. 1996. Detection of red edge position and chlorophyll content by reflectance measurements near 700 nm. *Journal of Plant Physiology*, 148, 501–508.

Gitelson, A.A., Kaufman, Y.J., Stark, R., and Rundquist, D. 2002a. Novel algorithms for remote estimation of vegetation fraction. *Remote Sensing of Environment*, 80, 76–87.

Gitelson, A.A., Merzlyak, M.N., and Chivkunova, O.B. 2001. Optical properties and nondestructive estimation of anthocyanin content in plant leaves. *Photochemistry and Photobiology*, 71, 38–45.

Gitelson, A.A., Zur, A., Chivkunova, O.B., and Merzlyak, M.N. 2002b. Assessing carotenoid content in plant leaves with reflectance spectroscopy. *Photochemistry and Photobiology*, 75, 272–281.

Goetz, A.F.H., Vane, G., Solomon, J., and Rock, B.N. 1985. Imaging spectrometry for Earth remote sensing. *Science*, 228, 1147–1153.

Gonçalves, B., Correia, C.M., Silva, A.P., Bacelar, E.A., Santos, A., and Moutinho-Pereira, J.M. 2008. Leaf structure and function of sweet cherry tree (*Prunus avium* L.) cultivars with open and dense canopies. *Scientia Horticulturae*, 116, 381–387.

Guanter, L., Kaufmann, H., Segl, K., Foerster, S., Rogass, C., Chabrillat, S., Küster, T. et al. 2015. The EnMAP spaceborne imaging spectroscopy mission for Earth observation. *Remote Sensing*, 7, 8830–8857.

Haboudane, D., Tremblay, N., Miller, J.R., and Vigneault, P. 2008. Remote estimation of crop chlorophyll content using spectral indices derived from hyperspectral data. *IEEE Transactions on Geoscience and Remote Sensing*, 46, 423–437.

Heilman, J.L. and Boyd, W.E. 1986. Soil background effects on the spectral response of a three-component rangeland scene. *Remote Sensing of Environment*, 19, 127–137.

Hochberg, E.J., Roberts, D.A., Dennison, P.E., and Hulley, G.C. 2015. Special issue on the Hyperspectral Infrared Imager (HyspIRI): Emerging science in terrestrial and aquatic ecology, radiation balance and hazards. *Remote Sensing of Environment*, 167, 1–5.

Huete, A.R., Didan, K., Miura, T., Rodriguez, E.P., Gao, X., and Ferreira, L.G. 2002. Overview of the radiometric and biophysical performance of the MODIS vegetation indices. *Remote Sensing of Environment*, 83, 195–213.

Hughes, G.F. 1968. On the mean accuracy of statistical pattern recognizers. *IEEE Transactions on Information Theory*, 14, 55–63.

Hunt, E.R. Jr., and Rock, B.N. 1989. Detection of changes in leaf water content using near- and middle-infrared reflectances. *Remote Sensing of Environment*, 30, 43–54.

Jacon, A.L., Galvão, L.S., Santos, J.R., and Sano, E.E. 2017. Seasonal characterization and discrimination of savannah physiognomies in Brazil using hyperspectral metrics from Hyperion/EO-1. *International Journal of Remote Sensing*, 38, 4494–4516.

Jackson, R.D., and Pinter, P.J. 1986. Spectral response of architecturally different wheat canopies. *Remote Sensing of Environment*, 20, 43–56.

Jacquemoud, S., Verhoef, W., Baret, F., Bacour, C., Zarco-Tejada, P.J., Asner, G.P., François, C., and Ustin, S.L. 2009. PROSPECT+SAIL models: A review of use for vegetation characterization. *Remote Sensing of Environment*, 113, 56–66.

Kaufman, Y.J., and Tanré, D. 1992. Atmospherically resistant vegetation index (ARVI) for EOS-MODIS. *IEEE Transactions on Geoscience and Remote Sensing*, 30, 261–270.

Kaufman, Y.J., Tanré, D., Remer, L.A., Vermote, E.F., Chu, A., and Holben, B.N. 1997. Operational remote sensing of tropospheric aerosol over land from EOS Moderate Resolution Imaging Spectroradiometer. *Journal of Geophysical Research*, 102, 17051–17067.

Kokaly, R.F., and Skidmore, A.K. 2015. Plant phenolics and absorption features in vegetation reflectance spectra near 1.66 μm. *International Journal of Applied Earth Observation and Geoinformation*, 43, 1–29.

Li, Q.M., Hu, B.X., and Pattey, E. 2008. A scale-wise model inversion method to retrieve canopy biophysical parameters from hyperspectral remote sensing data. *Canadian Journal of Remote Sensing*, 34, 311–319.

Lobell, D.B., and Asner, G.P. 2003. Comparison of Earth Observing-1 ALI and Landsat ETM+ for crop identification and yield prediction in Mexico. *IEEE Transactions on Geoscience and Remote Sensing*, 41, 1277–1282.

Lu, S., Oki, K., Shimizu, Y., and Omasa, K. 2007. Comparison between several feature extraction/classification methods for mapping complicated agricultural land use patches using airborne hyperspectral data. *International Journal of Remote Sensing*, 28, 963–984.

Ma, L., Fu, T., Blaschke, T., Li, M., Tiede, D., Zhou, Z., Ma, X., Chen, D. 2017. Evaluation of feature selection methods for object-based land cover mapping of unmanned aerial vehicle imagery using random forest and support vector machine classifiers. *ISPRS International Journal of Geo-Information*, 6, article number 51, 1–21.

Merzlyak, M.N., Gitelson, A.A., Chivkunova, O.B., and Rakitin, V.Y. 1999. Non-destructive optical detection of pigment changes during leaf senescence and fruit ripening. *Physiologia Plantarum*, 106, 135–141.

Moreira, M.A., Adami, M., and Rudorff, B.F.T. 2004. Spectral and temporal behavior analysis of coffee crop in Landsat images. *Pesquisa Agropecuária Brasileira*, 39, 223–231.

Nidamanuri, R.R., and Zbell, B. 2011. Use of field reflectance data for crop mapping using airborne hyperspectral image. *ISPRS Journal of Photogrammetry and Remote Sensing*, 66, 683–691.

Numata, I., Roberts, D.A., Chadwick, O.A., Schimel, J.P., Galvão, L.S., and Soares, J.V. 2008. Evaluation of hyperspectral data for pasture estimate in the Brazilian Amazon using field and imaging spectrometers. *Remote Sensing of Environment*, 112, 1569–1583.

Pal, M. 2009. Margin-based feature selection for hyperspectral data. *International Journal of Applied Earth Observation and Geoinformation*, 11, 212–220.

Pal, M., and Foody, G.M. 2010. Feature selection for classification of hyperspectral data for SVM. *IEEE Transactions on Geoscience and Remote Sensing*, 48, 2297–2307.

Pearlman, J.S., Barry, P.S., Segal, C.C., Shepanski, J., Beiso, D., and Carman, S.L. 2003. Hyperion, a space-based imaging spectrometer. *IEEE Transactions on Geoscience and Remote Sensing*, 41, 1160–1173.

Penuelas, J., Baret, F., and Filella, I. 1995. Semi-empirical indices to assess carotenoids/chlorophyll-a ratio from leaf spectral reflectance. *Photosynthetica*, 31, 221–230.

Penuelas, J., Pinol, J., Ogaya, R., and Filella, I. 1997. Estimation of plant water concentration by the reflectance water index WI (R900/R970). *International Journal of Remote Sensing*, 18, 2869–2875.

Piiroinen, R., Heiskanen, J., Mottus, M., and Pellikka, P. 2015. Classification of crops across heterogeneous agricultural landscape in Kenya using AisaEAGLE imaging spectroscopy data. *International Journal of Applied Earth Observation and Geoinformation*, 39, 1–8.

Rao, N.R. 2008. Development of a crop-specific spectral library and discrimination of various agricultural crop varieties using hyperspectral imagery. *International Journal of Remote Sensing*, 29, 131–144.

Roberts, D.A., Roth, K.L., and Perroy, R.L. 2012. Hyperspectral vegetation indices. In: Thenkabail, P.S, Lyon, J.G., and Huete, A. (Eds.), *Hyperspectral Remote Sensing of Vegetation*. CRC Press, Boca Raton, FL, pp. 309–327.

Rouse, J.W., Haas, R.H., Schell, J.A., and Deering, D.W. 1973. Monitoring vegetation systems in the Great Plains with ERTS. *Proceedings of Third ERTS-1 Symposium*, Washington, DC, December 10–14, NASA, SP-351, Vol. 1, pp. 309–317.

Rudorff, B.F.T., Aguiar, D.A., Silva, W.F., Sugawara, L.M., Adami, M., and Moreira, M.A. 2010. Studies on the rapid expansion of sugarcane for ethanol production in São Paulo state (Brazil) using Landsat data. *Remote Sensing*, 2, 1057–1076.

Sankey, T., Donager, J., McVay, J., and Sankey, J.B. 2017. UAV lidar and hyperspectral fusion for forest monitoring in the southwestern USA. *Remote Sensing of Environment*, 195, 30–43.

Smith, A.M., Bourgeois, G., Teillet, P.M., Freemantle, J., and Nadeau, C. 2008. A comparison of NDVI and MTVI2 for estimating LAI using CHRIS imagery: A case study in wheat. *Canadian Journal of Remote Sensing*, 34, 539–548.

Sugawara, L.M., Rudorff, B.F.T., and Adami, M. 2008. Feasibility of the use of Landsat imagery to map soybean crop areas in Parana, Brazil. *Pesquisa Agropecuária Brasileira*, 43, 1777–1783. (In Portuguese.)

Taskın, G., Kaya, H., and Bruzzone, L. 2017. Feature selection based on high dimensional model representation for hyperspectral images. *IEEE Transactions on Image Processing*, 26, 2918–2928.

Thenkabail, P.S., Smith, R.B., and De Pauw, E. 2002. Evaluation of narrowband and broadband vegetation indices for determining optimal hyperspectral wavebands for agricultural crop characterization. *Photogrammetric Engineering and Remote Sensing*, 68, 607–621.

Thenkabail, P.S., Enclona, E.A., Ashton, M.S., and van der Meer, B. 2004. Accuracy assessments of hyperspectral waveband performance for vegetation analysis applications. *Remote Sensing of Environment*, 91, 354–376.

Thenkabail, P.S., Mariotto, I., Gumma, M.K., Middleton, E.M., Landis, D.R., and Huemmrich, K.F. 2013. Selection of hyperspectral narrowbands (HNBs) and composition of hyperspectral two-band vegetation indices (HVIs) for biophysical characterization and discrimination of crop types using field reflectance and Hyperion/EO-1 data. *IEEE Journal of Selected Topics in Applied Earth Observations and Remote Sensing*, 6, 427–439.

Thenkabail, P.S., Gumma, M.K., Teluguntla, P., and Mohammed, I.A. 2014. Hyperspectral remote sensing of vegetation and agricultural crops. *Photogrammetric Engineering and Remote Sensing*, 80, 697–709.

Toniol, A.C., Galvão, L.S., Ponzoni, F.J., Sano, E.E., and Amore, D.J. 2017. Potential of hyperspectral metrics and classifiers for mapping Brazilian savannas in the rainy and dry seasons. *Remote Sensing Applications: Society and Environment*, 8, 20–29.

Van Niel, T.G., and McVicar, T.R. 2004. Current and potential uses of optical remote sensing in rice-based irrigation systems: A review. *Australian Journal of Agricultural Research*, 55, 155–185.

Verrelst, J., Rivera, J.P., Gitelson, A., Delegido, J., Moreno, J., and Camps-Valls, G. 2016. Spectral band selection for vegetation properties retrieval using Gaussian processes regression. *International Journal of Applied Earth Observation and Geoinformation*, 52, 554–567.

Vogelmann, J.E., Rock, B.N., and Moss, D.M. 1993. Red edge spectral measurements from sugar maple leaves. *International Journal of Remote Sensing*, 14, 1563–1575.

Yang, C., Everitt, J.H., Bradford, J.M., and Murden, D. 2009. Comparison of airborne multispectral and hyperspectral imagery for estimating grain sorghum yield. *Transactions of the ASABE*, 52, 641–649.

Yanqun, Z., Yuan, L., Haiyan, C., and Jianjun, C. 2003. Intraspecific differences in physiological response of 20 soybean cultivars to enhanced ultraviolet-B radiation under field conditions. *Environmental and Experimental Botany*, 50, 87–97.

Xia, J.S., Falco, N., Benediktsson, J.A., Du, P.J., and Chanussot, J., 2017. Hyperspectral image classification with rotation random forest via KPCA. *IEEE Journal of Selected Topics in Applied Earth Observations and Remote Sensing*, 10, 1601–1609.

Xue, Z.H., Du, P.J., Li, J., and Su, H.J. 2017. Sparse graph regularization for robust crop mapping using hyperspectral remotely sensed imagery with very few in situ data. *ISPRS Journal of Photogrammetry and Remote Sensing*, 124, 1–15.

Zarco-Tejada, P.J., Ustin, S.L., and Whiting, M.L. 2005. Temporal and spatial relationships between within-field yield variability in cotton and high-spatial hyperspectral remote sensing imagery. *Agronomy Journal*, 97, 641–653.

Zhang, S.Z., Li, S.T., Fu, W., and Fang, L.Y. 2017. Multiscale superpixel-based sparse representation for hyperspectral image classification. *Remote Sensing*, 9, article number 139.

7 Identification of Canopy Species in Tropical Forests Using Hyperspectral Data

Matthew L. Clark

CONTENTS

7.1 INTRODUCTION

Tropical forests are globally important due to their extremely high species diversity and their major role in biogeochemical cycles. Spatial and temporal sampling of these forests in the field is greatly restricted by prohibitive costs and inaccessibility at the ground and canopy levels, and extrapolating plot data to broader spatial scales is problematic as species distributions and ecosystem processes vary at different spatial scales due to overlapping factors such as climate, soil, and past disturbance. Remote sensing plays an important role in mapping tropical forests over larger areas than available from the ground, especially with use of new sensors, processing and modeling techniques, and integration with field data [1]. In particular, a new generation of hyperspectral, hyperspatial, and hypertemporal sensors is making rapid and innovative advances in the understanding of tropical forest composition, structure, and function over broad spatial and temporal scales [1]. These advances include: canopy chemistry by hyperspectral sensors [2–6]; analysis of individual tree crowns [7,15] and canopy components, such as lianas [16] by hyperspatial (<4 m) sensors; and, new insights into canopy chemistry, physiology, and phenology by hypertemporal sensors [4,17,18].

This chapter reviews hyperspectral and hyperspatial remote sensing of tropical forest (TF) canopy species, with an emphasis on trees in wetter forests but excluding mangroves. The fundamental biochemical, structural (biophysical), phenological and site-specific factors that control plant spectral properties across the visible (VIS: 400–700 nm), near-infrared (NIR: 700–1300 nm), and short-wave-infrared (SWIR: 1500–2500 nm) regions of the electromagnetic spectrum are reviewed, beginning with fine-scale canopy photosynthetic and nonphotosynthetic components, and then broadening the scope to the three-dimensional (3D) scale of crowns in the canopy. Applications with hyperspectral and hyperspatial imagery from TFs have taken a pixel-based view of analysis, leading to wall-to-wall maps of canopy floristic diversity or fractional species abundance [19,20], or an object-based analysis of individual tree crowns (ITCs) [7–9]. This chapter concentrates on ITC species mapping with hyperspectral data, although advances from other research will be discussed in order to provide a holistic understanding of the factors influencing species discrimination using hyperspectral technology. For reference, Table 7.1 provides a summary of hyperspectral research from tropical forests discussed in this chapter and Table 7.2 lists the important properties of sensors used in existing, and potentially future, studies

7.2 DRIVERS OF SPECTRAL VARIATION IN TROPICAL FOREST CANOPIES

7.2.1 LEAVES, BARK, AND OTHER FINE-SCALE CANOPY COMPONENTS

Our ability to remotely detect TF canopy species using hyperspectral imagery hinges upon species having interspecies (among/between species) spectral differences that are detectable and consistently measured, despite intraspecific (within species) spectral variation. Humid TF canopies are characterized by hemispherical, broadleaf, multilayered crowns with high leaf area index (LAI) and perforated by gaps covered by understory vegetation (Figure 7.1). Spectral variation over a tropical forest is thus primarily determined by the biochemistry (e.g., plant pigments, water, structural carbohydrates) and structure (e.g., leaf thickness, air spaces) of leaves, and the scaling of these spectral properties due to volumetric scattering of photons in the canopy. However, nonphotosynthetic tissues (e.g., bark, flowers, seeds) and other photosynthetic canopy organisms (e.g., vines, epiphytes, epiphylls) can mix into the photon signal, and vary depending on a complex interplay of species, structure, phenology and site differences, none of which are well understood. Research on the spectral connection to biochemical and structural properties of fine-scale canopy tissues—mainly leaves—has accelerated and now includes several sites [6,16,21–29], thereby permitting a foundation for understanding how these properties scale to the crown or canopy level.

Leaf reflectance in the visible spectrum is dominated by absorption features created by plant pigments, such as chlorophyll *a* and chlorophyll *b*, carotenoids (e.g., β-carotene, lutein), and anthocyanins [30]. Photosynthetic pigment chlorophyll *a* (Chl-*a*) absorbs at 410–430 and 600–690 nm and chlorophyll *b* (Chl-*b*) absorbs at 450–470 nm [30,31]. Other accessory pigments increase absorption of light and act in plant defense functions, such as avoiding damage from ultraviolet light. Carotenoids have peak absorption in lower wavelengths <500 nm that overlaps Chl-*a* and Chl-*b* absorption features [30,31]. Reflectance in the NIR is relatively high, with absorption features created mainly by water around 970 and 1200 nm, and to a lesser extent by broad absorption from structural carbohydrates such as lignin and cellulose [31,32]. In photosynthetic leaves, the SWIR region is characterized by relatively low reflectance and strong absorption by water that masks other absorption features [31,33]. However, dry leaves do not have strong water absorption and reveal overlapping absorptions by carbon compounds, such as lignin and cellulose, and other plant biochemicals, including protein nitrogen, starch, and sugars [31,33].

Leaf structure also plays a role in a species reflectance signature. Surface reflectance can be influenced by surface topography caused by features such as leaf hairs and waxes [34]. Reflectance in NIR is high due to multiple scattering of photons inside the leaf caused by internal structures, air spaces, and air–cell interfaces, such as in spongy mesophyll [34]. An important summary variable of leaf structure is specific leaf area (SLA—projected leaf area per unit leaf dry mass), which can be thought of the amount of leaf biomass spread over an area. Leaf design scales as a stoichiometric balance of photosynthetic (e.g., Chl-*a*, Chl-*b*) and other plant biochemicals (e.g, water, nitrogen, cellulose), and different chemical constituents are correlated with structural metrics, such as SLA [5,6,25].

Do species have unique biochemical-structural properties that translate into hyperspectral signatures that can discriminate canopy species? We are only in the beginning stages of answering this question. A pioneering study in Pará, Brazil [21] acquired laboratory leaf spectra (450–950 nm) for 11 TF tree species, and averaged these spectra to analyze differences in species discrimination at simulated branch and crown scales. Using a novel shape filter technique, species discrimination was possible at crown scales and declined at branch and leaf scales. The distributions of red edge wavelength and position, calculated from derivatives of reflectance, were not normally distributed

TABLE 7.1

Tropical Forest Research Using Hyperspectral Data Reviewed in This Chapter

Reference (Year)	Life Forms (Forest)	Objective	Location	Spectral Region (Sensor)
		Laboratory Spectrometer		
[21] (2000)	Trees (wet forest)	Explore spectral separability of 11 species	Amazon, Brazil	VIS, NIR (FieldSpec, ASD)
[27] (2004)	Trees, lianas (dry & wet)	Discrimination of lianas and trees with 9 classifiers; leaf chlorophyll	PNM, FS Panama	VIS, NIR (UniSpec)
[7,8] (2005)	Trees (wet forest)	Discriminate 7 tree species based on LDA, DT and 2 other classifiers	LSBS, Costa Rica	VIS, NIR, SWIR (FieldSpec, ASD)
[22] (2006)	Trees (dry & wet)	Spectral separability and spectral-Biochemical links across sites	Mexico, Panama, CR	VIS, NIR (UniSpec, FieldSpec, ASD)
[16] (2007)	Trees, lianas (dry forest)	Discrimination of lianas and trees with data reduction and nine classifiers	PNM, Panama	VIS, NIR, SWIR (FieldSpec)
[6] (2008)	Trees (wet forest)	Link between spectral, biochemical and structural properties	Amazon, Brazil	VIS, NIR, SWIR (FieldSpec, ASD)
[23] (2008)	Trees (wide range)	Discriminate 20 tree species based on LDA classifier	Costa Rica	VIS, NIR, SWIR (FieldSpec, ASD)
[29] (2008)	Trees (wet forest)	Spectral and biochemical differences between sex of two species	LSBS, Costa Rica	VIS, NIR, SWIR (FieldSpec, ASD)
[25] (2008)	Trees (wet/ elev. Grad.)	Canopy reflectance leaf properties with RT variation in crown structure	Queensland Australia	VIS, NIR, SWIR (FieldSpec, ASD)
[24] (2009)	Trees (wet/ elev. Grad.)	Link between spectral, biochemical and structural properties	Queensland Australia	VIS, NIR, SWIR (FieldSpec, ASD)
[26] (2009)	Trees, lianas (dry & wet)	Differences in leaf reflectance, biochemistry, structure, trees and lianas	PNM, FS Panama	VIS, NIR (UniSpec)
[28] (2009)	Trees (wet forest)	Influence of epiphylls on canopy reflectance from RT; link to indices	Amazon, Brazil	VIS, NIR (PS2, ASD)
		Airborne Hyperspatial and Hyperspectral		
[7–9] (2005)	Trees (wet forest)	Discriminate emergent tree species based on LDA, DT and three other classifiers (DT and LDA with lidar)	LSBS, Costa Rica	VIS, NIR, SWIR (HYDICE, FLIMAP lidar)
[3] (2005)	Trees (wet forest)	How biological invasion alters forest canopy chemistry	Hawai'i	VIS, NIR, SWIR (AVIRIS)
[40] (2006)	Trees (wet forest)	Explore spectral separability of five species; data reduction	LSBS, Costa Rica	VIS, NIR (HYDICE)
[16] (2007)	Trees, lianas (dry forest)	Discrimination of lianas and trees with data reduction and nine classifiers	PNM, Panama	VIS, NIR, SWIR (HYDICE)
[19] (2007)	Trees (wet forest)	Link biochemistry to spectral diversity; map species richness	Hawai'i	VIS, NIR, SWIR (AVIRIS)
[5] (2008)	Trees (wet forest)	Canopy spectral, biochemical and structural differences between native and invasive species	Hawai'i	VIS, NIR, SWIR (AVIRIS)
[20] (2008)	Trees (wet forest)	Mapping the fractional abundance of invasive species	Hawai'i	VIS, NIR, SWIR (AVIRIS - CAO-beta)
[39] (2008)	Trees (dry forest)	Quantify nonphotosynthetic vegetation with spectral mixture analysis	PNM, Panama	VIS, NIR, SWIR (HYDICE)

(Continued)

TABLE 7.1 (*Continued*)

Tropical Forest Research Using Hyperspectral Data Reviewed in This Chapter

Reference (Year)	Life Forms (Forest)	Objective	Location	Spectral Region (Sensor)
		**Spaceborne Medium Resolution (30 m) Hyperspectral**		
[2] (2004)	Trees (wet forest)	Quantify differences in canopy water and productivity from drought	Amazon, Brazil	VIS, NIR, SWIR (Hyperion)
[4] (2006)	Trees (wet forest)	Track canopy physiology and biochemistry through time	Hawai'i	VIS, NIR, SWIR (Hyperion)
[18] (2008)	Trees (wet forest)	Spectral seasonality related to canopy phenology	Amazon, Brazil	VIS, NIR, SWIR (Hyperion)
[41] (2010)	Trees (wet forest)	Discriminate five tree species based on LDA classifier	Amazon, Peru	VIS, NIR, SWIR (Hyperion)

Abbreviations: LDA, linear discriminant analysis; DT, decision trees; RT, radiative transfer model; LSBS, La Selva Biological Station; PNM, Parque Natural Metropolitano; FS, Fort Sherman; VIS, visible; NIR, near-infrared; SWIR, short-wave infrared; ASD, Analytical Spectral Devices.

TABLE 7.2

Summary of Hyperspectral Sensors Discussed in This Chapter

Name	Wavelength Range (nm)	No. of Bands	Signal-to-Noise Ratio[a]	Ground Resolution (m)	Developer
		**Laboratory Spectrometer**			
FieldSpec3	350–2500	2151 (interpolated)	160,000 VIS, 1200 SWIR	<1	ASD
FieldSpec FR	350–2500	2151 (interpolated)	87,000 VIS, 805 SWIR	<1	ASD
UniSpec	350–1100	256	$\sim$250 VIS	<1	PP Systems
		**Airborne Hyperspatial and Hyperspectral**			
HYDICE	400–2500	210	$\sim$300 VIS, $\sim$300 SWIR @ 1.5 km (5000 ft)	$\geq$1	U.S. Naval Research Laboratory
AVIRIS[b]	400–2500	224	$\sim$1000 VIS, $\sim$500 SWIR	$\geq$3	NASA [42]
CAO-alpha (CASI-1500)	369–1052	$\leq$288	$\sim$400 VIS	0.45–1.35	CAO [43]; ITRES
HyMap	450–2480	128	>500	3–10	HyVista
		**Spaceborne Medium Resolution (30 m) Hyperspectral**			
Hyperion	400–2500	220	$\sim$60 VIS, $\sim$30 SWIR	30	NASA
HyspIRI[c]	380–2500	212	$\sim$900 VIS, $\sim$400 SWIR	60	NASA
Enmap[d]	420–2450	249	$\sim$500 VIS, $\sim$150 SWIR	30	German consortium

[a] VIS = 550 nm, SWIR = 2100 nm.

[b] Also used in Carnegie Airborne Observatory beta (CAO-beta) system with lidar sensor [43].

[c] Estimated launch date $\geq$2022; hyspiri.jpl.nasa.gov [accessed: September, 21, 2018].

[d] Estimated launch date $\geq$2020; www.enmap.org [accessed: September, 2018].

FIGURE 7.1 Humid TF canopies are characterized by hemispherical, broadleaf, multilayered crowns with high leaf area index (LAI) and perforated by gaps covered by understory vegetation.

within a species, indicating intraspecific spectral variability, yet there was potential to separate some species with these metrics. Clark et al. [7] measured laboratory reflectance for leaves from seven tree species in a wet TF at the La Selva Biological Station (LSBS), Costa Rica. Interspecific variability was included in the analysis by measuring bidirectional reflectance (i.e., not hemispherical reflectance in an integrating sphere) and leaves with epiphylls, herbivory and galls (effects discussed later). A linear discriminant analysis (LDA) classifier with a stepwise band selection procedure showed that just 10 reflectance bands could classify species with 90% overall accuracy, while accuracy increased to 100% when using 40 bands. The 10-band classifier included mostly bands in NIR and SWIR, where variability was largely controlled by leaf structure, water content, and possibly other biochemicals (e.g., lignin and cellulose), while VIS bands were in blue, green peak, and red edge wavelengths. Castro et al [22] collected leaf reflectance data, chlorophyll content and structural (mesophyll attributes, thickness) data from a range of species across seven sites within wet and dry TF in Mexico and Costa Rica, and Panama. Species could be accurately classified with leaf spectra within sites, with overall accuracy declining with species richness (85% for 20 species and 80% for 40 species). The best classifier and wavelengths used varied among sites, and spectral signatures from one site were not useful for classifying the same species at another site. The most frequently selected bands were from blue, blue-green edge, and red to red edge, while NIR bands were not as important in leaf-scale species discrimination (SWIR not analyzed). Rivard et al. [23] focused on discriminating species of 20 tropical tree species from two sites in Costa Rica using LDA and stepwise selection of reflectance bands and narrow-band indices. Accuracies ranged from 70% to 97%. Bands and indices that correlated with leaf water content, followed by pigment properties, were most important in discriminating these species, with indices having an advantage over individual bands. In both [7] and [23] it was found that the highest overall accuracy was achieved when information from across the visible to SWIR spectrum (i.e., full range) was used. Taken as whole, these studies represent the initial steps toward statistically demonstrating that for tree leaves, intraspecific variability can be less than interspecific variability, thereby permitting tree species discrimination.

Ground-breaking research from humid TF of Australia [24] provides a framework for linking leaf spectral, biochemical, and structural properties from many species together in a unified analysis, and thereby helps understand the success of these earlier leaf-scale studies. Asner et al. [24] measured the sunlit leaf chemistry (nitrogen [N], phosphorus [P], Chl-a, Chl-b, carotenoids, anthocynanins, water), structure (SLA), and full-range laboratory spectrometer data for 162 canopy species from sites spanning a lowland to montane climate gradient. A partial least-squares (PLS) regression technique was used to estimate leaf chemicals and SLA, as well as assess the relative contribution of wavelengths (i.e., weightings) from the entire spectrum to the model. A cluster analysis was used to sort related species based on their spectral or chemical signatures. Species were found to have high variation in leaf properties at site and regional scales. Leaf chemicals and SLA were weakly to moderately correlated, except for highly correlated Chl-a and Chl-b that are functionally linked. In particular, lowland forests had the highest variation in chemicals and SLA among species, attributed to a wider range of chemical and physiological adaptive strategies among species. It was also found that certain families could dominate chemical diversity at any site. The PLS weightings revealed that multiple parts of the spectrum are related to the various leaf biochemical and structural properties, often due to scaling between leaf SLA and a stoichiometric balance of chemical constituents. For example, carotenoids and Chl-a and b were important contributors to visible reflectance, which is not surprising given their absorption properties, but also in SWIR where SLA was best predicted, indicating correlations between SLA and pigments. There was weak clustering of species based on their reflectance spectra as species tended to have unique signatures. In particular, clustering was lowest in the lowland site, which matched the trends in chemical diversity. These results indicate that species spectral diversity is linked to chemical diversity in the canopy, a finding that the authors then used to ask if spectral and chemical variation tracks species diversity. With a Monte Carlo simulation, as more species were added to a model "virtual" forest (i.e., more richness), both chemical and spectral diversity increased without saturation across all sites. Spectral-chemical diversity increased most rapidly with inclusion of more species in the lowland forest. Similar patterns of increasing spectral and chemical diversity with species richness were also seen in a similar simulation with leaf properties from 150 species in the Brazilian Amazon [6]. It is important to note that these results are based on leaf-level spectral and chemical/structural analyses, but they indicate that these properties can be linked to species richness. The scaling of these results to actual humid TF has also made progress [5,19,25] and will be discussed in the next section.

All of these studies discussed so far have involved controlled laboratory conditions, and do not include the multitude of factors (i.e., variable view angle, poor atmospheric conditions, radiometric scattering) that are confronted when using airborne or spaceborne hyperspectral data over tropical forest canopies. In addition, these studies included a limited subset of tree species from TF canopies, which may include hundreds of tree species, and they generally ignore other canopy components in TFs that ultimately mix in the radiance received by a hyperspectral sensor.

One characteristic life form in tropical forests are lianas, or vines, which are woody climbers that use trees for support and form a monolayer of leaves that adds to canopy-level species diversity (Figure 7.2a). In neotropical forests, reports of trees covered with lianas have ranged from 43% to 86% [cited in 26]. Our knowledge of how the spectral-biochemical properties of liana and tree leaves contrast is still limited, but several studies from Central American wet and dry TF have focused intensively on this topic [16,26,27]. One study focused on biochemicals, structure and VIS–NIR optical properties for liana and tree leaves sampled from canopy cranes at tropical dry (Parque Natural Metropolitano—PNM) and wet forest sites (Fort Sherman—FS) in Panama [26]. Lianas in the dry forest had significantly lower chlorophyll and carotenoid concentrations relative to their host trees, and these differences were detected by spectral indices and the red edge. Lianas had higher water content than trees at both sites, although these differences were not detected by water indices. Leaf thickness was low and SLA was high for lianas relative to trees, but there was no difference in mesophyll air spaces due to high variability between life forms, which led to no differences in NIR reflectance. Supervised parametric and nonparametric classifiers applied to VIS–NIR reflectance

FIGURE 7.2   A. liana (darkcr green) on a tree crown. B. Flowering *D. panamensis* C. Deciduous tree with epiphytes.

spectra [27] could separate tree and liana leaves only at the dry forest site, which was attributed to differences in drought stress and phenological strategies between the two groups under dry season conditions. A follow-up study [16] at PNM measured full-range laboratory reflectance, and, using the reduction techniques and classifiers in [27], there was a >96% accuracy in distinguishing tree and liana spectra. Important bands were mostly in the VIS and SWIR regions, attributed to greater differences of chlorophyll and other pigments between lianas and tree leaves, rather than differences in internal leaf structure.

Life-cycle changes in trees and liana tissues, or phenology, will also add to intraspecific spectral variation. Important phenological factors to consider are leaf drop, period without leaves, leaf flush, flowering and fruit maturation. In wctter TF, a relatively constant growing season fosters a range of phenological traits, and leaf turnover and reproductive events may follow annual to irregular cycles, with synchronous to asynchronous timing of events among individuals of a species (cited in [7]). Drier tropical forests typically have a more characteristic community-wide syndrome, with more synchronized flowering events and many trees in a leaf-off (deciduous) stage in the dry season [35]. As leaves age, they can bccome more susceptible to epiphyll infestations (e.g., lichens, liverworts, fungi, algae, and bacteria), internal galls, herbivory and necrosis. Epiphyll infestation can be high. For example, it has been reported to occur on 16% of overstory plants at Barro Colorado Island, a dry forest in Panama, with cover exceeding 50% (cited in [28]). A study in Amazonian *caatinga* and *terra firme* forests indicates that as epiphyll cover increases, leaves tend to have lower green-peak reflectance and a red edge shifted to longer wavelengths, while leaf NIR transmittance is reduced [28].

Examples of factors that can increase intraspecific spcctral variation in tree crowns are shown in Figure 7.3 (photographs) and Figures 7.4 through 7.6 (reflectance graphs), with data from my research at LSBS, Costa Rica [8]. Hyperspectral metrics have been computed from these reflectance spectra to illustrate the power of having contiguous, narrowband data (Table 7.3). These metrics, calculated following methods in [8] (adapted from [36] and [37]), describe absorption features and scattering properties across the VIS to SWIR, including the center wavelength, depth and width of absorptions in blue, red, NIR and SWIR, and the wavelength and magnitude of derivative inflection points in the blue, yellow, and red edges. The narrow-band equivalents of the normalized difference vegetation index (NDVI) and normalized difference water index (NDWI) are also shown in Table 7.3.

Thc cffect of epiphylls on leaf reflectance can be seen for *Hymenolobium mesoamericanum* leaves (Figure 7.4; Table 7.3—HYME), where relative to an epiphyll-free leaf (Figure 7.3a), there is lower green-peak reflectance (GP-Refl) and less steep yellow and red edges (YE-Mag, RE-Mag) for a leaf completely covered with a lichen (Figure 7.3b). Herbivory affected 4% of leaves from canopy-emergent trees sampled at LSBS. As seen for *Lecythis ampla* leaves (Figure 7.3e and f; Table 7.3—LEAM), a leaf miner insect removes photosynthetic pigments, thereby reducing blue and red absorption depth and width (Blue-D,W and Red-D,W) and increasing green-peak reflectance (GP-Refl); SWIR reflectance increases due to lower water content and more exposed, dry leaf structure, possibly allowing deeper expression of SWIR features at near 1762 and 2301 nm

FIGURE 7.3 Examples of factors that can increase intraspecific spectral variation in tree crowns are shown in photographs here and in reflectance graphs in Figures 7.4 through 7.6.

(Figure 7.4; Table 7.3—SWIR1,3-D). Leaves from *Dipteryx panamensis* reveal that as a leaf senesces (Figure 7.3i and j; Figure 7.5; Table 7.3—DIPA), lower concentrations of chlorophyll reduce blue and red absorption depth (Blue, Red-D), increase the green-peak reflectance (GP-Refl), shifts the blue and red edge inflection points to lower wavelengths (BE, RE-λ) while shifting the yellow edge to longer wavelengths (YE- λ). Canopy tree flowers can be very conspicuous and possibly help discriminate species or detect richness at canopy scales. For example, canopy-emergent *Dipteryx panamensis* trees in Costa Rica have conspicuous purple-pink flowers (Figures 7.2b and 7.3c) that can be accurately identified with automated processing of color aerial photographs [38], and have distinct hyperspectral reflectance properties (Figure 7.5; Table 7.3—DIPA), such as absence of chlorophyll red absorption, low NDVI, high NIR reflectance and deeply expressed NIR water absorption (NDWI, NIR1,2-D).

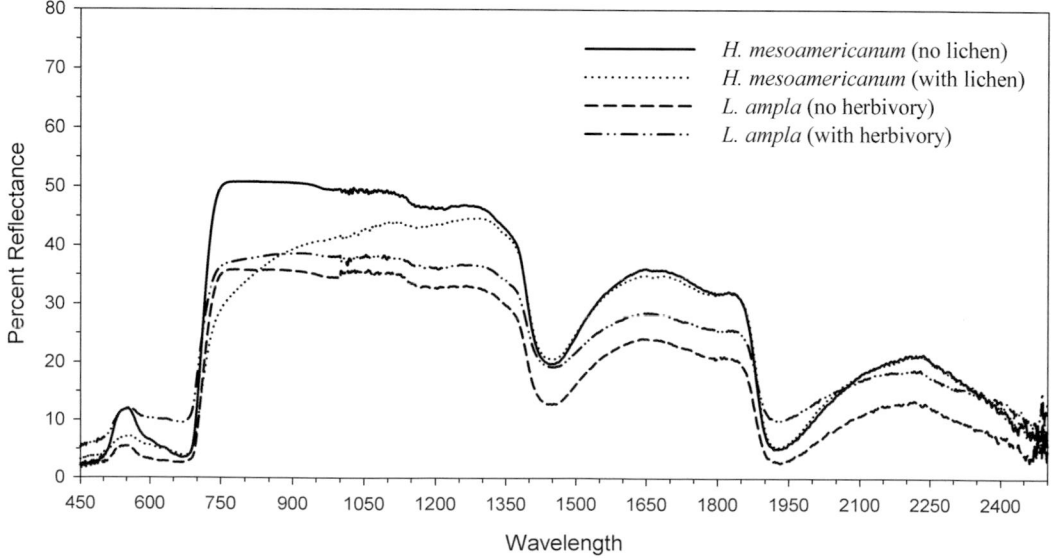

FIGURE 7.4 The effect of epiphylls on leaf reflectance for *Hymenolobium mesoamericanum* leaves.

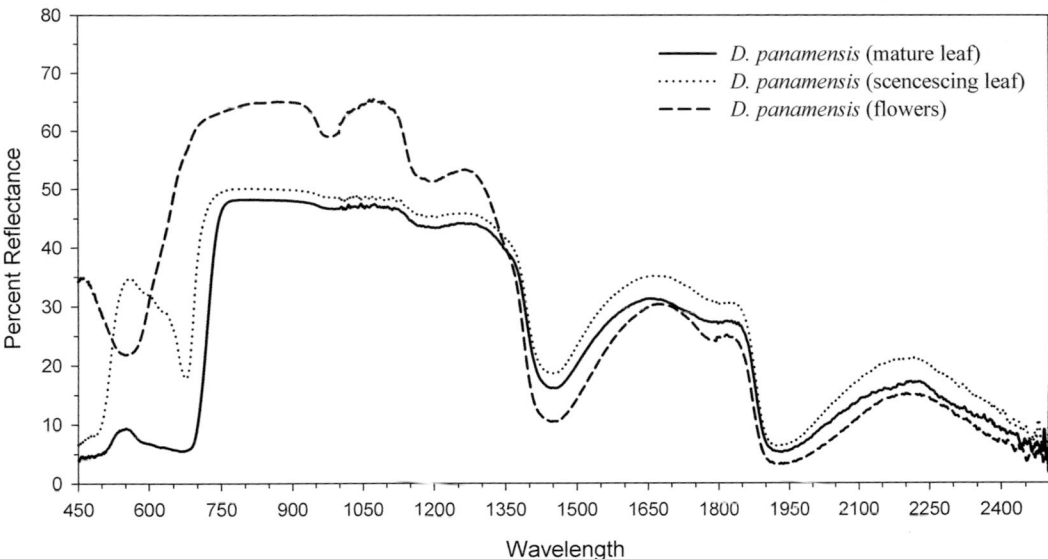

FIGURE 7.5 *Dipteryx panamensis* trees in Costa Rica have conspicuous purple-pink flowers (Figures 7.2b and 7.3c) that can be accurately identified with automated processing of color aerial photographs [38], and have distinct hyperspectral reflectance properties.

Other canopy components such as understory vegetation, bark and canopy epiphytes may have considerable influence on canopy spectral variation when trees have lower to no LAI due to seasonal phenological cycles (e.g., deciduousness, leaf exchange) or stress [7,8,39]. An example of a tree at LSBS without leaves is shown in Figure 7.2c. Note that most of the exposed bark is on the trunk and outer branches, while larger lateral branches are covered with epiphytes. This is not atypical for wet TF. Reflectance data from LSBS show that tree bark on branches and trunks can have considerable

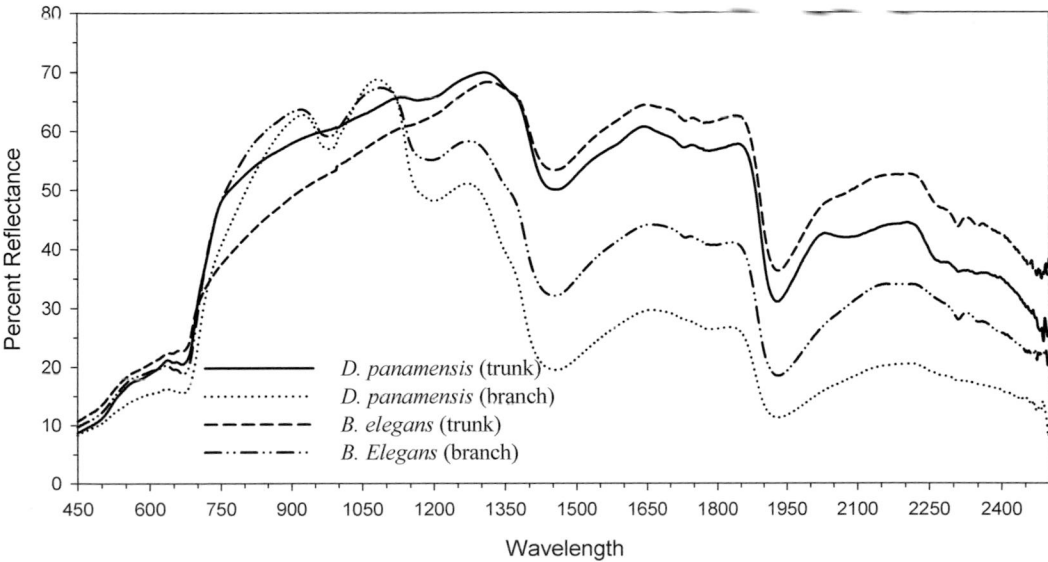

FIGURE 7.6 Branches taken from the upper canopy of younger trees (Figure 7.3h and l) show photosynthetic pigment absorption, seen as relatively deep red (Red-D) and NIR (NIR1,2-D) absorption features, related to chlorophyll and water, respectively.

TABLE 7.3

Hyperspectral Metrics for Components in Figures 7.4 through 7.9

Metric[a]	HYME No Lichen	HYME Lichen	LEAM No Herbivory	LEAM Herbivory	DIPA- Green Leaf	DIPA Senesced Leaf	DIPA Flowers	DIPA Trunk	DIPA Branch	Field Epiphyte	Field PEMA
NDVI	0.86	0.79	0.85	0.58	0.79	0.47	0.06	0.42	0.51	0.86	0.93
NDWI	0.04	−0.08	0.04	0.02	0.05	0.04	0.10	−0.09	0.08	0.09	0.03
Blue-λ	492	496	502	503	505	496	528	499	494	493	498
Blue-D	54.1	23.9	34.4	22.5	27.0	50.0	3.0	9.5	5.1	55.7	59.5
Blue-W	51.4	36.3	33.3	46.2	46.9	46.3	23.6	45.0	42.5	48.5	50.0
BE-λ	521	520	523	522	524	520	n/a	520	519	521	523
BE-Mag	0.3	0.1	0.1	0.2	0.2	0.8	n/a	0.1	0.1	0.3	0.1
GP-λ	549	553	548	554	550	560	553	n/a	n/a	555	552
GP-Refl	11.9	7.1	5.4	12.0	9.3	34.7	21.8	7.1	7.1	11.2	5.1
YE-λ	569	570	568	571	570	576	n/a	574	573	572	570
YE-Mag	−0.20	−0.06	−0.09	−0.08	−0.10	−0.12	n/a	0.03	0.03	−0.13	−0.09
Red-Wvl	678	679	674	679	682	678	n/a	676	677	677	676
Red-D	88.2	78.9	86.7	60.3	80.6	57.3	n/a	39.3	45.5	88.4	93.9
Red-W	104.6	98.8	111.2	90.1	112.4	49.8	n/a	72.5	81.5	88.1	114.6
RE-λ	710	708	711	704	724	688	n/a	694	702	702	720
RE-Mag	1.108	0.578	0.781	0.649	0.954	0.982	n/a	0.483	0.536	0.726	0.621
NIR1-λ	1018	1037	996	1015	990	1007	980	996	989	977	965
NIR1-D	2.1	1.2	2.8	4.3	2.1	2.7	9.2	0.9	12.3	13.7	5.5
NIR1-W	50.5	3.9	48.0	14.1	101.9	41.1	59.4	40.9	70.7	70.3	75.4
NIR2-λ	1158	1158	1168	1178	1167	1170	1162	1193	1167	1155	1159
NIR2-D	3.3	2.3	4.3	3.2	4.3	3.4	12.3	2.2	17.3	18.3	10.2
NIR2-W	82.3	72.5	70.1	78.4	77.0	70.7	78.3	73.2	83.1	80.2	78.9
SWIR1-λ	1770	1768	1776	1762	1770	1776	n/a	1725	1771	1777	1776
SWIR1-D	0.2	0.4	0.1	0.5	0.5	0.2	n/a	1.8	0.4	0.0	0.5
SWIR1-W	5.5	14.3	1.5	9.6	19.5	1.9	n/a	25.4	5.3	0.7	3.8
SWIR2-λ	n/a	2215	2202	n/a	2217	2047	n/a	2093	n/a	2049	n/a
SWIR2-D	n/a	0.7	0.0	n/a	0.4	0.1	n/a	2.0	n/a	2.3	n/a
SWIR2-W	n/a	3.7	0.0	n/a	1.8	3.1	n/a	62.9	n/a	3.9	n/a
SWIR3-λ	2354	2303	2321	2301	2344	2338	2350	2266	2309	2257	2300
SWIR3-D	5.2	2.7	5.8	6.7	8.1	3.4	0.7	6.3	3.4	2.4	6.4
SWIR3-W	7.1	47.4	29.4	14.6	33.9	7.8	15.3	86.3	22.5	109.1	60.6

Abbreviations:　HYME, *Hymenolobium mesoamericanum*; LEAM, *Lecythis ampla*; DIPA, *Dipteryx panamensis*; NDVI, normalized difference vegetation index; NDWI, normalized difference water index; BE, blue edge; GP, green peak; RE, red edge; Mag, magnitude of derivative; Ref, % reflectance; λ, wavelength (nm); D, depth of absorption feature (% reflectance); W, width of absorption feature (nm); n/a, absorption feature faint or nonexistent.

[a] Notation in this column is explained in the text.

intraspecific reflectance variation [8]. Branches taken from the upper canopy of younger trees (Figure 7.3h and l) show photosynthetic pigment absorption, seen as relatively deep red (Red-D) and NIR (NIR1,2-D) absorption features, related to chlorophyll and water, respectively (Figure 7.6; Table 7.3—DIPA). In contrast, bark from old-growth tree trunks (Figure 7.3g and k), taken lower in the canopy where branches split from the main bole, have shallower red and NIR absorption features, red edge shifted toward longer wavelengths (RE- λ), and deeper expression of SWIR features (SWIR1,2,3-D). Examples of field reflectance spectra taken from a suspension bridge at LSBS are

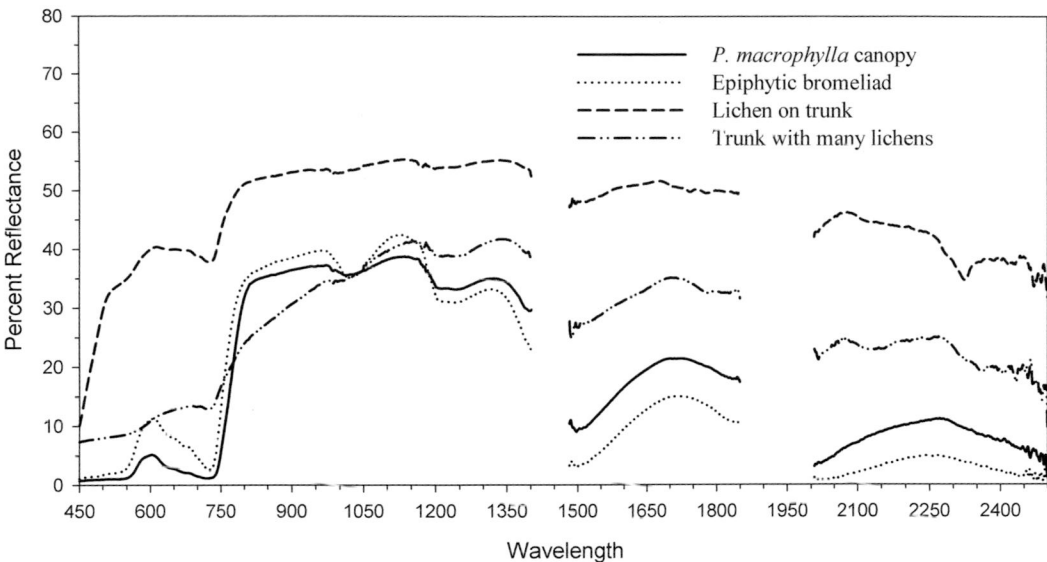

FIGURE 7.7 Field reflectance spectra taken from a suspension bridge at La Selva Biological Station. An example canopy spectrum for *Pentaclethera macrophylla* is shown as a contrast with a spectrum of an epiphytic tank bromeliad.

shown in Figure 7.7. Lichens tend to have high reflectance throughout the spectrum, with a deep absorption at 2300 nm possibly related to protein nitrogen. These surface organisms can add to tree bark intraspecific spectral variability. An example canopy spectrum for *Pentaclethera macrophylla* is shown as a contrast with a spectrum of an epiphytic tank bromeliad (Figure 7.3d—note contains water), which has higher green-peak reflectance (GP-Refl), steeper yellow edge (YE-Mag), and deeper NIR water absorption and higher NDWI (Figure 7.7; Table 7.3—Epiphyte, PEMA).

7.2.2 PIXEL TO CANOPY SCALES

Operational mapping of individual tree species or overall species diversity in tropical forest canopies using airborne or spaceborne hyperspectral sensors is a grand challenge for remote sensing. As in all remote sensing, moving from the controlled laboratory environment to a canopy scale introduces additional noise to the radiance signal, such as atmospheric moisture absorption and poorer radiometric calibration. Beyond the linkages between leaf reflectance and biochemical-structural properties, canopy reflectance is heavily influenced by species-level differences in crown biophysical structure, which determines the level of three-dimensional scattering of photons among the various tissues within the canopy, including leaves but also bark, flowers, fruits, lianas, and epiphytes.

Much understanding of how tissue-scale biochemical/structure is expressed in canopy-scale reflectance is through radiative transfer (RT) models [3,6,16,19,21]. Canopy-scale reflectance is best understood with the concept of "effective photon penetration depth" (EPPD, Figure 7.8), which is the canopy depth to which a downward (nadir) viewing hyperspectral sensor is most sensitive—conceptualized as the number of leaf layers (reported in LAI units) from which the sensor can detect biochemicals due to wavelength-specific absorption and scattering among canopy components [6]. In VIS, strong absorption by leaf pigments translates into a very short path length of photons scattering within the canopy, especially in blue and red (Figure 7.8, EPPD LAI < 2). In contrast, photons are heavily scattered in NIR, creating increasing reflectance as canopy LAI increases until the "NIR plateau" is saturated, yielding an EPPD LAI < 4 (Figure 7.8); however, water absorption features at

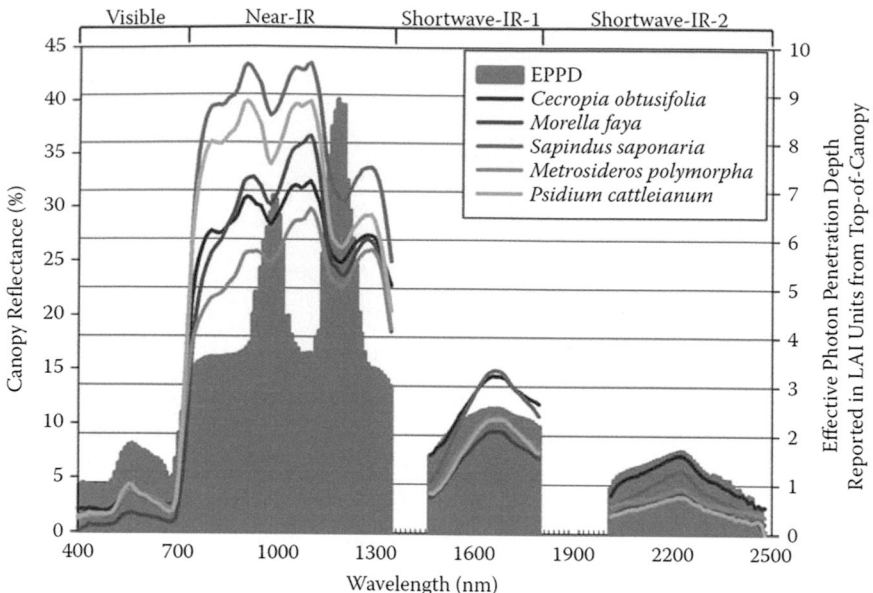

FIGURE 7.8 Effective photon penetration depth (EPPD) of a tropical forest canopy in Hawaii. This EPPD estimate was derived by combining field measurements of leaf hyperspectral optical properties, canopy shape, and architectural data with LAI in an 3D canopy radiative transfer model. EPPD was calculated as the depth from the top of the canopy to which photons no longer contribute to the top-of-canopy reflectance. (Reprinted from Asner, G.P., Hyperspectral remote sensing of canopy chemistry, physiology and biodiversity in tropical rainforests, in *Hyperspectral Remote Sensing of Tropical and Sub-tropical Forests*, CRC Press, Boca Raton, 2008, pp. 261–269.)

980 and 1190 nm will continue to broaden and deepen with increasing canopy LAI, making these features highly sensitive to canopy water content deep into the canopy (Figure 7.8, EPPD LAI < 8). In SWIR, the photon path length is relatively shallow due to strong water absorption, and so this region returns mostly information from the upper canopy (Figure 7.8, EPPD LAI < 3).

Asner and Martin [25] modeled canopy reflectance using their leaf reflectance and transmission spectra from Queensland, Australia [24], with a goal of estimating biochemicals and SLA at leaf and canopy scales with PLS regression. As reviewed in the last section, analysis of this leaf-scale dataset demonstrated a link of species-level spectral variation to inherent biochemical-structural diversity [24]. The RT model considered an area at the pixel scale (<3 m), not an entire crown, and included several structural parameters, the most important of which was leaf area index (LAI). The model also considered variable sensor and view zenith and azimuth angles, but not intracrown gaps and shadows, crown density, or spectral properties from other canopy tissues (e.g., bark, lianas). Foliar chlorophyll, carotenoids, and SLA were highly correlated with leaf reflectance measured in the laboratory ($r = 0.89$–0.91), with higher PLS regression weights for VIS wavelengths for pigments and NIR to SWIR wavelengths for SLA. Foliar N, P, and water were moderately correlated ($r = 0.79$–0.85). For N, high weights were for VIS and SWIR wavelengths, associated with chlorophyll and proteins, respectively, while for water, high weights were for SWIR wavelengths. The correlation of P with reflectance was likely due to a stoichiometric link with N, rather than expression of absorption features. At the canopy scale, with the RT model held constant at a crown LAI of 5.0 (an average level for humid TF), correlations of reflectance and pigments, N, P, and SLA were 3%–4% higher, and water was 18% higher than when using leaf-scale reflectance, attesting to the power of the canopy to amplify leaf biochemical and structural properties. In other simulations, pigments and SLA remained highly correlated to canopy reflectance with variable LAI conditions (LAI 1.5–8.0), which can be attributed to low EPPD in VIS (Figure 7.8), and subsequent insensitivity to variation in upper-canopy leaf layering [24,25]. In contrast, foliar water, and particularly N and P, had less predictive capacity with increasing variation in modeled LAI. These chemicals (especially water) are strongly expressed in the NIR, the region with the highest EPPD due to volumetric scattering (Figure 7.8); and thus, variations in crown LAI at greater depths will lead to more variable expression of chemical absorptions, thereby weakening predictive models.

Actual airborne hyperspectral data were used by Asner et al. [5] to explore the spectral, biochemical and structural differences of native, introduced, and invasive, N-fixing and nonfixing tree species on Hawai'i island. The sensor used was the Airborne Visible/Infrared Imaging Spectrometer (AVIRIS), which has very high fidelity (i.e., high signal-to-noise ratio, SNR; Table 7.2). Groups of trees were found to be spectrally unique for natives vs. introduced, N-fixers vs. nonfixers, and introduced vs. invasive. These differences between groups in reflectance and derivative spectra were related to variation in leaf pigments, water, N, P, and SLA and canopy LAI, which caused reflectance and absorption features at both specific and multiple spectral regions. For example, PLS regression weightings showed N and SLA were well predicted by SWIR wavelengths. Pigments heavily influenced VIS reflectance (e.g., 670 nm of chlorophyll absorption), but also NIR and SWIR reflectance due to an indirect relationship with SLA as all are expressed on a per-area basis. The main message of this paper was that no single band or spectral region, nor any underlying biochemical or structural property, could distinguish the groups of species, and so full-range measurements are needed to separate groups; in this case, PLS regression was used to analyze the full spectrum simultaneously to target the important predictive features.

Can linkages between canopy reflectance and leaf biochemistry and structural properties help us achieve species mapping with an airborne or spaceborne hyperspectral sensor? There are two important studies from Hawai'i that show that this may be possible. In a pioneering study, Asner et al. [20] took the evidence that invasive and native trees have unique hyperspectral properties [5], and sought to map the fractional abundance of representative species from these groups using AVIRIS imagery (spatial resolution: 3 m) acquired in tandem with small-footprint lidar. Lidar was used to mask forest gaps, intracrown and intercrown shadows, and low vegetation in the hyperspectral imagery, while spectral mixture analysis (SMA) based on SWIR was used to mask nonphotosynthetic vegetation. The remaining pixels in the imagery thus had the highest information content. Full-range spectra from two native and three invasive species were used as end-members in another SMA to map the per-pixel fractional abundance of each species, with a <7% error rate in detection of invasive species for pixels with 75% cover.

In another Hawai'i study, Carlson et al. [19] sought to link leaf biochemistry to woody species richness through AVIRIS hyperspectral data (spatial resolution: 3.3–3.6 m). Woody species richness was measured from plots in 17 lowland tropical forest sites across the Hawaiian islands. Leaf biochemicals (total chlorophyll, water, N) sampled from species in plots were used in a Monte Carlo simulation to show that as species were progressively added to a model forest community, biochemical diversity also had a nonlinear increase without saturation—a result later confirmed with a much larger dataset from Australia, reviewed in Section 7.2.1 [24]. Carlson et al. [19] next set out to map per-pixel woody species richness from AVIRIS imagery. The range of reflectance and reflectance derivative were highly correlated with plot-based species richness across many wavelengths, but the derivative had fewer intercorrelated bands and so was selected for modeling. Linear regression was used to find the best wavelength in each of four regions determined to add to biochemical diversity: green edge (500–550 nm) and red edge (700–800 nm) affected by total chlorophyll; the 1140–1250 nm feature affected by water absorption; and the 1500–1700 nm feature affected by water and N absorption. The final regression model with four optimal derivative bands had a strong linear relationship to species richness ($r = 0.85, p < 0.01$). The greatest predictor of species richness was a SWIR band (1525 nm), although the bands from other regions helped strengthen the model. It should be noted that, with 44 species sampled across all sites, these Hawaiian forests are relatively species poor compared to other humid TF, and so we do not yet know how the findings and techniques in this study will scale to other sites.

Tree phenology also has an important role in determining canopy reflectance. For one, structural properties of upper-canopy tree or liana tissues will vary through time—older leaves may accumulate epiphylls, and at certain times of the year, there will be senesced leaves, flowers, and fruits (see Section 7.2.1). Little is known about how these fine-scale tissues will change the crown photon scattering environment and alter the biochemical properties that contribute to overall radiance measured by the sensor.

Toomey et al. [19] used a GeoSAIL RT model to investigate the effect of epiphylls on canopy reflectance. Amazonian *caatinga* forests had a 23%–35% decrease in canopy NIR reflectance and 11%–20% decrease in green reflectance with infestations at 50%–100% leaf area cover, while *terra firme* forests had lower infestation rates and exhibited a 6%–11% decrease in NIR and 5%–10% decrease in green reflectance, respectively. Although only seven species were investigated in their simulations, there were differences in epiphyll effects among species related to factors such as leaf longevity and differences in transmittance [19].

If crown LAI is low due to crown leaf-drop (e.g., deciduousness) or general crown architecture (e.g., sparse leaves and branches), photons can penetrate deeper into the canopy, and the biochemical-structural properties of bark, canopy components (e.g., epiphytes, lianas) and understory vegetation, can contribute to canopy-scale reflectance [7,8,16,37,38]. Clark [8] and Bolhman [39], working in tropical wet (LSBS, Costa Rica) and dry (PNM, Panama) forests, respectively, explored SMA applied to hyperspectral imagery (<1.6 m) to estimate the relative fractional contribution of nonphotosynthetic tissues to canopy reflectance. Both studies used data from the Hyperspectral Digital Imagery Collection (HYDICE;Table 7.2) sensor acquired in March, 1998 during the dry season. These data are some of the earliest hyperspatial, hyperspectral imagery available from humid tropical forests for research use. Pixels were unmixed to provide fractional abundance of green vegetation (GV: leaves from trees, lianas, epiphytes, understory vegetation), nonphotosynthetic vegetation (NPV: bark) and shade. At LSBS, Clark [8] investigated the same seven species analyzed at tissue scales [7] using pixel- and crown-scale reflectance spectra from HYDICE. Mean values of SMA fractions from pixels within individual tree crowns across the seven species were roughly 40% GV, 15% NPV, and 44% shade. Canopies of deciduous tree species (*D. panamensis* and *L. ampla*) had relatively high fractions of NPV and low fractions of GV (Figures 7.9 and 7.10). In contrast, leaf-on, broadleaf species (*Ceiba pentandra*, *Hyeronima alchorneoides*, and *Terminalia oblonga*) had relatively high fractions of GV and low fractions of NPV (Figures 7.9 and 7.10). There was variability in the crown-level proportions of GV, NPV, and shade among individuals of the same species (i.e., intraspecific variability). For example, the

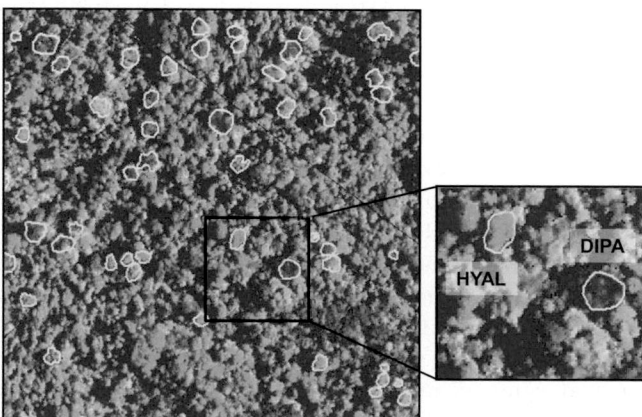

FIGURE 7.9 Spectral mixture analysis (SMA) fractions from HYDICE hyperspectral imagery over the La Selva Biological Station, Costa Rica. The SMA fractions are Red = NPV, Green = GV, and Blue = Shade. Individual tree crowns are yellow polygons, with species labeled for a *Dipteryx panamensis* (DIPA) and a *Hyeronima alchorneoides* (HYAL).

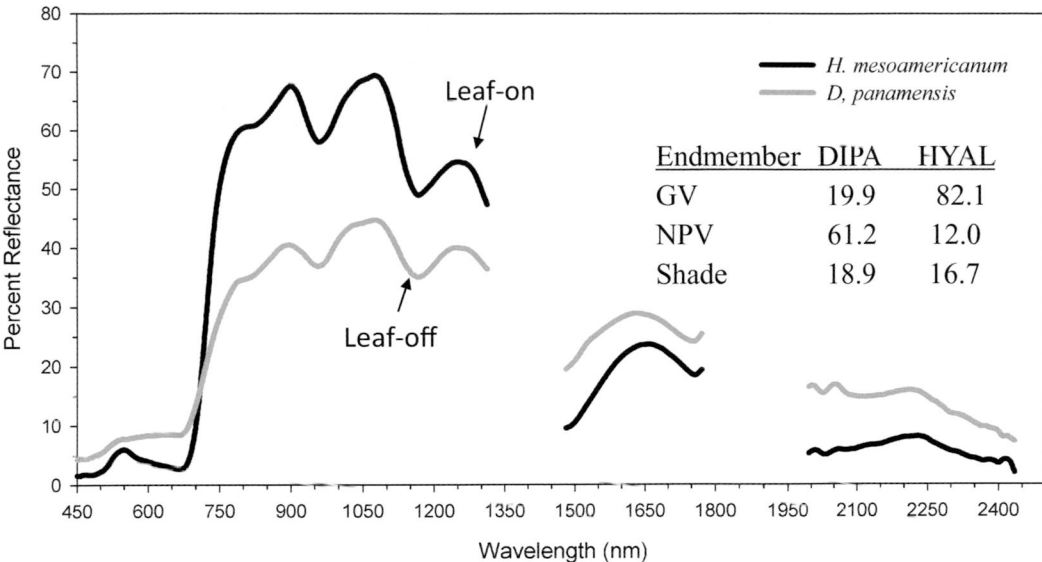

FIGURE 7.10 Crown-level reflectance spectra for *Dipteryx panamensis and Hyeronima alchorneoides* taken from HYDICE hyperspectral imagery at the La Selva Biological Station, Costa Rica. Individuals from crowns shown in Figure 7.9.

tree species *B. elegans* and *H. mesoamericanum* were fully-leaved, yet they had fine compound leaves and sparse branching architecture; as a consequence, some individuals of these species had NPV and GV fractions similar to deciduous crowns, while other individuals had fractions that resembled the broadleaf, leaf-on species. At the drier PNM forest site, which had more distinct tree leaf phenology and fewer species, Bohlman [39] used SMA at pixel (1 m) to canopy scales (30 m) using image GV, NPV, and shade end-members. For individual tree crowns, the percentage of GV increased and percentage of NPV decreased with higher levels of deciduousness, as measured in the field. The model also accurately determined the percent cover of nondeciduous (GV, 53%), deciduous (NPV, 32%), and intercrown shade/shadow (shade, 15%) over a 5-ha test area at these three scales. Deciduous species had a dominant NPV fraction, yet there was still GV cover (<40%), attributed to lianas and exposed understory vegetation. There was also variation in GV and NPV fractions for leaf-on species. For example, *Luehea seemanii* had relatively high NPV despite being fully-leaved, possibly due to lower LAI and presence of senescing leaves and high load of upper-canopy woody seeds that had spectral properties of NPV. Taken together, these studies at LSBS and PNM show that tropical forest canopies should not be viewed as a continuous blanket of green leaves with randomly mixed structure. Tree phenology and site characteristics will lead to important species-level variation in the relative exposure of differently aged leaves, bark, and other canopy components that form the crown's reflectance signature; these factors vary through time and, even within an individual crown, and may provide an important signal for species discrimination that goes beyond just leaf-scale spectral-biochemical-structural properties.

As discussed in the preceding section, tropical forests are particularly influenced by lianas and epiphytes, which can contribute to both canopy species richness and overall spectral diversity. Zhang et al. [40], working with HYDICE data at the LSBS, found that pixel-scale reflectance from a single *T. oblonga* crown with heavy epiphyte load had distinct spectral properties than other individuals of the species, thereby contributing to higher intraspecific spectral variation. Kalaskca et al. [16] used HYDICE imagery at the PNM dry forest to scale their hyperspectral discrimination of liana and trees from the leaves to crowns. Tree crowns with <40% liana cover could be separated from those with ≥40% liana cover with 93% overall accuracy, indicating that lianas have a profound impact on ITC spectral response, which may be independent of the tree's species. Top wavelengths for discriminating crowns from lianas were in SWIR (>2000 nm), which possibly indicates that lianas affect the water absorption characteristics of the upper-canopy. Leaf-level analysis revealed that lianas tend to have more water than trees [26], especially in the dry season when the image was acquired; and in this season, reflectance from deciduous crowns may come mostly from lianas, which tend to lose leaves after host leaves or retain them through the year.

7.3 MAPPING CANOPY SPECIES OVER BROAD SPATIAL SCALES

The operational mapping of individual species or species diversity of TF canopy plants is only in the initial stages of development. Hyperspatial, hyperspectral data with high fidelity are now available from commercial (e.g., HyMap [www.hyvista.com], CASI [www.itres.com], SpecTIR [www.spectir.com]) and experimental (e.g., AVIRIS [42], Carnegie Airborne Observatory, CAO [43]) airborne sensors (Table 7.2), yet to date there have been relatively few published studies using these datasets for mapping canopy species of humid TF, with most work in Hawai'i [3,5,19,20]—a site with relatively low species diversity.

Hyperion was the only spaceborne sensor offering publicly-accessible hyperspectral data over the large spatial extents needed for regional species mapping; however, the sensor has relatively low fidelity (Table 7.2), and the 30 m resolution of the imagery made it difficult to identify pure spectra from smaller tree crowns. To date, there has been one attempt to use Hyperion for species-level mapping in humid TF. Working in the Peruvian Amazon, Papeş et al. [41] used hyperspatial imagery from the Quickbird satellite to identify Hyperion pixels that were >50% covered by the crown of a single study species (called the "clean" dataset) relative to those pixels that included more spectral mixtures from multiple crowns. Five genera were then classified with an LDA classifier with 5–25 optimally selected bands, with wet and dry season images. The "clean" dataset achieved 100% overall accuracy with 25 bands for both seasons, while the raw dataset never reached more than 50% overall accuracy. This result indicates that when the crown from a single species covers a large portion of a moderate-resolution pixel, its spectral response may be dominant enough for accurate pixel-scale classification. However, in practice this study's landscape-scale maps of study genera were difficult to assess, accuracy was not quantified, and some distributions of taxa were obviously related to sensor noise. Given the high species diversity and range of crown sizes in humid tropical forests, it is inevitable that pixels will be spectrally mixed with medium-resolution pixels, even when screened with hyperspatial imagery. Although not yet tested, a sub-pixel multiple-end-member spectral mixture analysis (MESMA)/Auto-MCU-S [8,20,45] approach that maps fractional abundance of species within a pixel may be more fruitful.

As seen in [41], ITCs in hyperspatial imagery can be resolved as groups of image pixels, or objects, especially for larger crowns (e.g., canopy emergents) [7–15]. This object-based perspective may permit the mapping of species at a finer spatial scale than offered by medium resolution sensors, especially when hyperspectral data are available from the objects [7,8]. Operational mapping of ITCs to species could allow broad-scale delineation of communities types [11,13], measures of species richness, and monitoring of important keystone, endemic, rare or commercial tree species [1,7,38]. As found in [41], panchromatic or multispectral hyperspatial imagery (e.g., Quickbird, IKONOS) are also useful for masking pixels from medium-resolution spaceborne hyperspectral imagery (e.g., Hyperion) that have many small crowns, and thus more spectrally mixed—a step that can greatly improve classification accuracy for the remaining pixels.

There are two important steps in the automated mapping ITC species with hyperspectral data and an object-based perspective: (1) location and delineation of canopy objects, that is, image segmentation, and (2) classification of species based on predictor variables, that is, spectral reflectance, metrics, texture. Both of these steps are areas of active research, and few techniques have been applied to hyperspectral imagery from tropical forests.

7.3.1 AUTOMATED DELINEATION OF INDIVIDUAL TREE CROWNS OR CROWN CLUSTERS

Image segmentation can be accomplished with manual digitization [7,10]; however, for consistency and reduction of production costs, object-based species inventory over large spatial extents will require automated algorithms for crown detection and delineation. Algorithms are reviewed more extensively in [11,12,14]. Techniques generally involve using panchromatic bands or collapsing spectral data into one illumination/albedo band to form a "hillscape," where radiometric maxima are from upper crown areas, or "hills," and radiometric minima are assumed to be intercrown shadows, or "valleys." Algorithms specifically tailored to tree crown delineation include valley following, bright point expansion, and template matching [11,12,14]. High spatial resolution, mutispectral to panchromatic imagery is generally used to develop techniques, and most algorithms have been designed for forests with fewer species and more regular crown structure (e.g., conifers) than found in subtropical to tropical forests. A novel study with multilayered, mixed-species subtropical woodland in Australia had more success delineating crowns with the popular eCognition image segmentation and classification software (www.ecognition.com) than with existing crown-delineation techniques [12]. The technique used multispectral CASI imagery (14 bands, 446–840 nm) to first segment the images into forest/no forest, then worked on forest types for iterative delineation and splitting of crown clusters using both spectral and shape properties. Characteristics of the red edge were found to highlight illuminated foliage and help separate crowns from the open understory [11,12]. Accuracies in delineation were >70% within scenes with sparse trees, but accuracy dropped below 50% for forests with dense, multilayered trees. All automated crown delineation studies indicate that tropical forests are the ultimate challenge, as trees may intertwine and overlap, intercrown shadows are narrow, intracrown shadows are prevalent, and lianas connect multiple crowns, causing segmentation schemes to identify clusters of trees rather than ITCs. Working in wet tropical forests, Palace et al. [14] designed a hybrid algorithm that performs an iterative local maximum filtering (e.g., finding the bright apex of tree) and local minima-finding (e.g., crown edges) radiating in transects from the apex. When applied to IKONOS panchromatic data in forest stands in the Brazilian Amazon, crown widths estimated from the technique were within 3% of field measurements and estimates were better than manual crown delineation [10,14]. The algorithm was applied to Quickbird imagery over a moist tropical semideciduous forest of lowland Bolivia and analyzed at the level of individual trees rather than stands [15]. Emergent tree crowns were generally segmented into smaller crowns, thus appearing as multiple subcrowns. There is still much work needed to refine crown delineation algorithms for tropical forests, and this will require a greater understanding of tree and liana diversity, incorporation of spectral data to differentiate adjacent tree crowns, and possibly inclusion of biophysical information from small-footprint lidar sensors [9,15,20,43].

7.3.2 Classification Schemes

An object-based approach to mapping individual canopy tree species allows considerable flexibility in choosing a species-level classification scheme. There are many (sometimes hundreds) of pixels within crowns, allowing selection of well-lit pixels with potentially more biochemical-structural signal [7,11,13,20]. Reflectance spectra can be analyzed at the pixel scale or averaged to form crown-scale spectra, and bands can be smoothed with filters to remove noise [7,27,40], transformed to derivatives to reduce the effect of illumination and highlight spectral shape [19,21,27,40], and processed with various data reduction techniques that seek to isolate spectral features that maximize species discrimination and reduce error. In tropical species and life-form classification applications using field spectrometers or hyperspectral sensors, data reduction or "feature selection" techniques have included PCA [16,27], wavelet transforms [16,27,40], stepwise band selection [7,22,23], narrow-band indices [8,13,23,29], and hyperspectral metrics based on derivatives, absorption fitting and SMA [8]. In all of these applications, predictor variables are used in supervised classifiers, including spectral angle mapper (SAM) [7,21], MESMA [8], maximum likelihood ML) [7], LDA [7–9,13,23,41], decision trees (DT) [8,9,16,22,27,29], and neural networks, log linear, quadratic, and k-nearest-neighbor [16,22,27,29]. These studies have focused on discriminating different sets of tree and liana species, with hyperspectral data from leaf to crown scales acquired at different seasons and forest types, in concert with multiple data reduction/feature selection techniques—making intercomparison and generalization of findings difficult; however, an in-depth review of these applications with a focus on classifiers and feature selection is available in [44].

My research in the species-rich, tropical rain forests of LSBS, Costa Rica using airborne HYDICE reflectance data (437–2434 nm, 161 bands, 1.6 m) provides a unique exploration of different techniques for automated, object-based ITC species discrimination, including ML, SAM, LDA, DT, and MESMA classifiers [7,8]. The hyperspectral data were smoothed with a filter and included the full suite of factors that compound operational species mapping, such as variable illumination and viewing geometry, noise from atmospheric moisture, and spectral mixing with nontarget canopy (e.g., lianas, epiphytes) and understory components. As the focus was on species classification with hyperspectral techniques, I bypassed experimenting with automated crown delineation by manually digitizing 214 canopy-emergent ITCs from seven species.

In a first analysis [7], all reflectance bands were used to classify species using SAM, ML, and LDA, in combination with a stepwise band selection procedure. Classifiers were applied to all and sunlit pixel spectra within ITCs, and to crown-scale spectra from averaged pixel spectra. ITCs were also classified using a pixel-majority approach in which a species label was assigned according to the majority class of classified pixels within the crown. Overall classification accuracy decreased from leaf scales measured in the laboratory (reviewed in Section 7.2.1) to pixel and crown scales measured from the airborne sensor, but reasonable accuracy was still achieved. The highest ITC classification accuracies for crown-scale spectra and pixel-majority techniques were 92% and 86%, respectively, with LDA and 30 bands. The most optimal bands for classifying species at pixel and crown scales were concentrated in the yellow edge to red edge, around NIR water absorption features and throughout the SWIR, indicating that the full-range data are important for identifying species at these scales. Furthermore, classifications were significantly more accurate with 10 narrow bands from hyperspectral data relative to using simulated multispectral, broadband data.

In a second analysis [8], classification variables included hyperspectral metrics that responded to crown structure and absorption features from photosynthetic pigments, water, and other biochemicals. The metrics included narrow-band indices, derivative-based metrics, absorption-fitting metrics, and SMA fractions (GV, NPV, shade). Differences in spectral metrics among species at pixel and crown scales were largely dependent on tree leaf phenology and structure, which controlled the relative amounts of leaf and bark tissues within a crown. For example, leaf-off trees had fewer canopy leaves and less overall canopy water, and so those species had lower values of water indices and higher values of metrics responding to overlapping SWIR absorption features. A DT classifier was used to discriminate tree species using crown-scale spectra and pixel spectra with a majority vote, an approach similar to [7]. Some of the most heavily used metrics were related to water absorption in NIR and other nonpigment features in SWIR. The best classification scheme was with crown-scale metrics and had 70.1% overall accuracy, which was lower than with LDA and reflectance bands from the first analysis [7]. However, hyperspectral metrics and the DT classifier were instructive for identifying key spectral reflectance properties for tropical tree species discrimination, which helped establish linkages to underlying biochemical properties in tissues that warrant analysis with chemical assays. A third analysis [8] with MESMA [45] as a classifier only reached accuracies of LDA and reflectance bands [7] when hundreds of end-members per species were included, similar to the MESMA approach in [20] (i.e., Auto-MCU-S); however, I considered my result to be too optimistic given that model end-members came from the same crowns being classified.

Crown structure metrics derived from small-footprint lidar were also analyzed in concert with reflectance bands and hyperspectral metrics in classifying ITC species, with six target and 18 "other" species grouped together [8,9]. There were significant differences in the majority of lidar-derived metrics among the six target species, indicating that species have unique crown structural properties. Crown leaf cover, especially in deciduous leaf-off trees, was the primary factor controlling variation in lidar metrics. Both DT and LDA classifiers were compared for classifying the species of ITCs with lidar and hyperspectral metrics and reflectance bands. As with [7], the best classifier was stepwise-selection LDA applied to reflectance bands, with an overall accuracy of 88.9% and 81.5%

when classifying six target species alone or with the "other" species class, respectively. The addition of lidar-derived structure information to the classifier did not improve overall species classification accuracy, highlighting the importance of hyperspectral data for tropical tree species discrimination.

In summary, important findings from these series of analyses at LSBS showed that: (1) elaborate methods and labor involved in implementing the DT classifier with hyperspectral metrics or the MESMA classifier with optimal endmembers did not translate into improved species classification accuracy relative to using reflectance bands with a more traditional LDA classifier; (2) crown-scale spectra provided more accurate ITC species classification, with both LDA and DT classifiers, than the pixel-majority approach; (3) there were no major improvements in accuracy by isolating sunlit pixels in crowns, suggesting that image spatial resolution can be coarse as long as it avoids many mixed pixels and can isolate crowns [see 41]; (4) full-spectrum data provided optimal species discrimination; (5) an advantage of hyperspectral data is that they are over-sampled, and information that does not optimize separation among species can be discarded; and (6) crown-scale LAI and its changes with leaf phenology are important considerations in operational mapping of tree species, as they altered the volume-scattering and spectral mixing properties at pixel to crown scales. For further details of "feature" extraction methods, refer to the other chapters in the four volumes, especially to Chapter 10 of Volume 1.

7.4 CONCLUSIONS AND FUTURE CHALLENGES

The operational mapping of individual species and canopy diversity in tropical forests of the world from remote sensing is still a future goal as we are in the initial stages of this research frontier. This chapter has reviewed a growing body of evidence on the advances made and promise shown by hyperspectral sensors in mapping species in tropical forests.

Half of the research reviewed here (Table 7.1) worked with leaf spectral reflectance from tree and liana species in neotropical and Australian forests, with the goal of establishing statistical separation among species with data reduction, selection, and classification techniques [7,8,16,21–23,27,29] and connecting spectral differences to their underlying generating mechanisms—leaf biochemical and structural properties [6,22,26,29]. Although limited in breadth of sites and species, these analyses have laid the foundation of basic principles, showing that species have unique biochemical and structural properties at the tissue scale that translate into species-level spectral diversity across the entire VIS to SWIR range (i.e., full range); and these spectral properties, when expressed at the canopy scale, have potential for discriminating individual species or species diversity despite the myriad of factors that cause intraspecific variation. Some of the most important advances in understanding the link of spectral response to biochemistry and structure have been accomplished with high-fidelity, full-range hyperspectral data from laboratory spectrometers, taken in conjunction with foliar chemical assays and physical measurements [6,22,24,25]. These types of analyses are critical to replicate at other field sites in order to cover the extremely broad range of site conditions and species diversity that exist in the tropical forests. Our understanding of how leaf spectral properties scale to canopies, where the time-varying, volumetric scattering of photons is critically important, has been advanced with radiative transfer modeling [6,25,28,31]. Future RT modeling of tropical forest canopies should incorporate additional spectral (e.g., bark, epiphytes, lianas) and temporal diversity (e.g., older leaves with epiphylls and herbivory, deciduousness) to better understand variability in canopy-scale reflectance. It is difficult to undertake these types of research with limited funding, and field sites must have adequate access (e.g., roads, rivers, trails, and sampling and equipment permits), logistical support (e.g., transport, supplies, botanists, climbers/shooters, field assistants), and infrastructure (e.g., shelter, electricity, canopy access). There are few tropical forest sites that meet these requirements, thus partially explaining the limited sampling and analysis that has been achieved to date.

In order to map canopy plant species across soil, climate, and diversity gradients, and to track changes through time, we will need more hyperspectral datasets from airborne or spaceborne sensors. The applications using hyperspatial, hyperspectral data reviewed in this chapter indicate some general conclusions: (1) full-range data are best for discriminating species and detecting diversity; (2) high-fidelity spectral data are most useful, especially when trying to understand spectral connections to subtle biochemical absorption properties, such as those found in SWIR; (3) an advantage of narrow-band, hyperspectral data is that the most important spectral features can be utilized through analytical techniques, without a priori information, and other redundant or noisy data can be discarded or minimized; (4) hyperspatial imagery allows crown delineation, screening of pixels with high information content (e.g., sunlit), and analysis of intracrown spectral variability; (5) there is a wide range of data processing and analytical techniques being used, with most research teams settling on a unique approach—there is no existing synthetic analysis that compares the relative advantages/disadvantages of these methods across site conditions and species assemblages with a common dataset; (6) there is a paucity of detailed field and hyperspectral datasets from tropical forests, greatly limiting synthesis and advances in the field; and, (7) the potential of the temporal domain, which contains valuable information on plant phenological cycles, has not been exploited for tropical species identification and diversity mapping using hyperspectral imagery. The hyperspectral AVIRIS [43] and CAO [44] sensors offer cutting-edge performance, spectral range and spatial resolution for researchers (Table 7.2). In terms of producing results with greater ecological understanding, these airborne hyperspectral sensors, and coupled small-footprint waveform lidar, have made the greatest strides [3,5,6,19,20]. High-fidelity hyperspectral data can detect detailed biochemical information, while lidar peers deeper into the canopy and helps in analyzing the hyperspectral data by interpreting its structural signal and isolating information-rich content. Commercial hyperspectral and lidar sensors are already being combined for joint flights, and improvements in performance and capabilities are also advancing rapidly. These types of sensors need to fly over more sites where detailed field data can be collected, and just as important, these datasets need to reach a broader research community to be fully exploited.

Spaceborne sensors allow mapping at broader spatial and temporal scales, with much less monetary cost and restrictions than airborne sensors, albeit with lower spatial resolution and signal to noise. Hyperion provided multitemporal, full-range hyperspectral datasets at 30 m for almost two decades (Table 7.2). Although not the focus of this chapter, there has been some research on species-level mapping [41] and tracking landscape-scale biochemical patterns and shifts due to aggregated canopy phenology and climate [2,4,18]. A newer generation of future spaceborne hyperspectral sensors, HyspIRI (hyspiri.jpl.nasa.gov, 30 m) and EnMap (www.enmap.org, 30 m), will have higher fidelity (Table 7.2) and undoubtedly increase our understanding of variability in biochemical, structural, and phenological properties of tropical forest canopies across a range of sites; this, in turn, will both inform and help scale results from species-mapping applications conducted at finer spatial scales.

Our lack of knowledge of tropical species distributions has important implications for understanding changes in biodiversity due to climate and land change, locating and designing representative protected areas, planning extractive timber operations, and implementing global carbon agreements that require species information (e.g., United Nations REDD+, www.un-redd.org [46]). The important goal of having species-level or species diversity maps of tropical forests over broad spatial and temporal scales is not yet a reality, but hyperspectral remote sensing technology has a bright and exciting future in this endeavor!

REFERENCES

1. Chambers, J.Q. et al., Regional ecosystem structure and function: ecological insights from remote sensing of tropical forests, *Trends in Ecology and Evolution*, 22, 414–423, 2007.
2. Asner, G.P. et al., Drought stress and carbon uptake in an Amazon forest measured with spaceborne imaging spectroscopy, *Proceedings of the National Academy of Sciences*, 101, 6039–6044, 2004.
3. Asner, G.P. and Vitousek, P.M., Remote analysis of biological invasion and biogeochemical change, *Proceedings of the National Academy of Sciences*, 102, 4383–4386, 2005.

4. Asner, G.P. et al., Vegetation-climate interactions among native and invasive species in Hawaiian rainforest, *Ecosystems*, 9, 1106–1117, 2006.
5. Asner, G.P. et al., Remote sensing of native and invasive species in Hawaiian forests, *Remote Sensing of Environment*, 112, 1912–1926, 2008.
6. Asner, G.P., Hyperspectral remote sensing of canopy chemistry, physiology and biodiversity in tropical rainforests, in *Hyperspectral Remote Sensing of Tropical and Sub-tropical Forests*, Kalaskca, M. and Sánchez-Azofeifa, G.A. (Eds.), CRC Press, Boca Raton, FL, 261–296, 2008.
7. Clark, M.L. et al., Hyperspectral discrimination of tropical rain forest tree species at leaf to crown scales, *Remote Sensing of Environment*, 96, 375–398, 2005.
8. Clark, M.L., An Assessment of Hyperspectral and Lidar Remote Sensing for the Monitoring of Tropical Rain Forest Trees, *Doctoral Dissertation*, University of California, Santa Barbara, USA, 2005.
9. Clark, M.L., Relative advantages of airborne lidar and hyperspectral data for individual tropical tree classification, *Proceedings of the 32nd International Symposium on Remote Sensing of Environment*, San Jose, Costa Rica, June 25–29, 2007.
10. Asner, G.P., Estimation canopy structure in an Amazon forest from laser range finder and IKONOS satellite observations, *Biotropica*, 34, 483–492, 2002.
11. Lucas, R. et al., Hyperspectral Data for Assessing Carbon Dynamics and Biodiversity of Forests, in *Hyperspectral Remote Sensing of Tropical and Sub-tropical Forests*, Kalaskca, M. and Sánchez-Azofeifa, G.A. (Eds.), CRC Press, Boca Raton, FL, 47–86, 2008.
12. Bunting, P. and Lucas, R., The delineation of tree crowns in Australian mixed species forests using hyperspectral Compact Airborne Spectrographic Imager (CASI) data, *Remote Sensing of Environment*, 101, 230–248, 2006.
13. Lucas, R. et al., Classification of Australian forest communities using aerial photography, *CASI and HyMap data, Remote Sensing of Environment*, 112, 2088–2103, 2008.
14. Palace, M. et al., Amazon forest structure from IKONOS satellite data and the automated characterization of forest canopy properties, *Biotropica*, 40, 141–150, 2008.
15. Broadbent, E.B. et al., Spatial partitioning of biomass and diversity in a lowland Bolivian forest: linking field and remote sensing measurements, *Forest Ecology and Management*, 255, 2602–2616, 2008.
16. Kalaska, M. et al., Hyperspectral discrimination of tropical dry forest lianas and trees: Comparative data reduction approaches at the leaf and canopy levels, *Remote Sensing of Environment*, 109, 406–415, 2007.
17. Huete, A.R. et al., Amazon rainforests green-up with sunlight in dry season, *Geophysical Research Letters*, 33, L06405, 2006.
18. Huete, A.R. et al., Assessment of Phenological Variability in Amazon Tropical Rainforests Using Hyperspectral Hyperion and MODIS Data, in *Hyperspectral Remote Sensing of Tropical and Sub-tropical Forests*, Kalaskca, M. and Sánchez-Azofeifa, G.A. (Eds.), CRC Press, Boca Raton, FL, 233–259, 2008.
19. Carlson, K.M. et al., Hyperspectral remote sensing of canopy biodiversity in Hawaiian lowland rainforests, *Ecosystems*, 10, 536–549, 2007.
20. Asner, G.P. et al., Invasive species detection in Hawaiian rainforests using airborne imaging spectroscopy and LiDAR, *Remote Sensing of Environment*, 112, 1942–1955, 2008.
21. Cochrane, M.A., Using vegetation reflectance variability for species level classification of hyperspectral data, *International Journal of Remote Sensing*, 21, 2075–2087, 2000.
22. Castro-Esau, K.L. et al., Variability in leaf optical properties of Mesoamerican trees and the potential for species classification, *American Journal of Botany*, 93, 517–530, 2006.
23. Rivard, B. et al., Species classification of tropical tree leaf reflectance and dependence on selection of spectral bands, in *Hyperspectral Remote Sensing of Tropical and Sub-tropical Forests*, Kalaskca, M. and Sánchez-Azofeifa, G.A. (Eds.), CRC Press, Boca Raton, FL, 261–296, 2008.
24. Asner, G.P. et al., Leaf chemical and spectral diversity in Australian tropical forests, *Ecological Applications*, 19, 236–253, 2009.
25. Asner, G.P. and Martin, R.E., Spectral and chemical analysis of tropical forests: Scaling from leaf to canopy levels, *Remote Sensing of Environment*, 112, 3958–3970, 2008.
26. Sánchez-Azofeifa, G.A. et al., Differences in leaf traits, leaf internal structure, and spectral reflectance between two communities of lianas and trees: Implications for remote sensing in tropical environments, *Remote Sensing of Environment*, 113, 2076–2088, 2009.
27. Castro-Esau, K.L. et al., Discrimination of lianas and trees with leaf-level hyperspectral data, *Remote Sensing of Environment*, 90, 353–372, 2004.
28. Toomey, M. et al., The influence of epiphylls on remote sensing of humid forests, *Remote Sensing of Environment*, 113, 1787–1798, 2009.

29. Arroyo-Mora, J.P. et al., Spectral expression of gender: A pilot study with two dioecious neotropical tree species, in *Hyperspectral Remote Sensing of Tropical and Sub-tropical Forests*, Kalaskca, M. and Sánchez-Azofeifa, G.A. (Eds.), CRC Press, Boca Raton, FL, 125–140, 2008.

30. Ustin, S.L. et al., Retrieval of foliar information about plant pigment systems from high resolution spectroscopy, *Remote Sensing of Environment*, 113, S67–S77, 2009.

31. Asner, G.P., Biophysical and biochemical sources of variability in canopy reflectance, *Remote Sensing of Environment*, 64, 234–253, 1998.

32. Ustin, S.L. et al., Using imaging spectroscopy to study ecosystem processes and properties, *Bioscience*, 54, 523–534, 2004.

33. Kokaly, R.F. et al., Characterizing canopy biochemistry from imaging spectroscopy and its application to ecosystem studies, *Remote Sensing of Environment*, 113, S78–S91, 2009.

34. Grant, L., Diffuse and specular characteristics of leaf reflectance, *Remote Sensing of Environment*, 22, 309–322, 1987.

35. Castro-Esau, K.L. and Kalacka, M., Tropical dry forest phenology and discrimination of tropical tree species using hyperspectral data, in *Hyperspectral Remote Sensing of Tropical and Sub-tropical Forests*, Kalaskca, M. and Sánchez-Azofeifa, G.A. (Eds.), CRC Press, Boca Raton, FL, 1–25, 2008.

36. Pu, R., Ge, S., Kelly, N.M., & Gong, P., Spectral absorption features as indicators of water status in coast live oak (Quercus agrifolia) leaves. *International Journal of Remote Sensing*, 24, 1799–1810, 2003.

37. Pu, R.L. et al., Extraction of red edge optical parameters from Hyperion data for estimation of forest leaf area index, *IEEE Transactions on Geoscience and Remote Sensing*, 41, 916–921, 2003.

38. Chun, S., The Utility of Digital Aerial Surveys in Censusing *Dipteryx panamensis*, the Key Food and Nesting Tree of the Endangered Great Green Macaw (*Ara ambigua*) in Costa Rica, *Doctoral Dissertation*, Duke University, Durham, USA, 2008.

39. Bohlman, S., Hyperspectral remote sensing of exposed wood and deciduous trees in seasonal tropical forests, in *Hyperspectral Remote Sensing of Tropical and Sub-tropical Forests*, Kalaskca, M. and Sánchez-Azofeifa, G.A. (Eds.), CRC Press, Boca Raton, FL, 177–192, 2008.

40. Zhang, J. et al., Intra- and inter-class spectral variability of tropical tree species at La Selva, Costa Rica: Implications for species identification using HYDICE imagery, *Remote Sensing of Enviornment*, 105, 129–141, 2006.

41. Papeş, M. et al., Using hyperspectral satellite imagery for regional inventories: A test with tropical emergent trees in the Amazon Basin, *Journal of Vegetation Science*, 21, 342–354, 2010.

42. Green, R.O. et al., Imaging spectroscopy and the Airborne Visible Infrared Imaging Spectrometer (AVIRIS), *Remote Sensing of Environment*, 65, 227–248, 1998.

43. Asner, G.P. et al., Carnegie Airborne Observatory: In-flight fusion of hyperspectral imaging and waveform light detection and ranging (LiDAR) for three-dimensional studies of ecosystems, *Journal of Applied Remote Sensing*, 1, 013536, 2007.

44. Ghiyamat, A. and Shafri, H.Z.M., A review on hyperspectral remote sensing for homogeneous and heterogeneous forest biodiversity assessment, *International Journal of Remote Sensing*, 31, 1837–1856, 2010.

45. Dennison, P.E. and Roberts, D.A., Endmember selection for multiple endmember spectral mixture analysis using endmember average RMSE, *Remote Sensing of Environment*, 87, 123–135, 2003.

46. Stickler, C.M. et al., The potential ecological costs and cobenefits of REDD: A critical review and case study from the Amazon region, *Global Change Biology*, 15, 2803–2824, 2009.

8 Characteristics of Tropical Tree Species in Hyperspectral and Multispectral Data

Matheus Pinheiro Ferreira, Cibele Hummel do Amaral, Gaia Vaglio Laurin, Raymond Kokaly, Carlos Roberto de Souza Filho, and Yosio Edemir Shimabukuro

CONTENTS

8.1 INTRODUCTION

Tropical forests provide essential ecosystem services for humanity such as the storage of 55% of the global forest carbon stock (Pan et al., 2011). They also harbor at least two-thirds of the world's terrestrial biodiversity (Gardner et al., 2009) and have a critical role on biogeochemical cycles, fixing, for example, 70% of the terrestrial nitrogen (Wang and Houlton, 2009). The provisioning of these ecosystem services relies on the maintenance and conservation of forested areas. However, deforestation and forest depletion have reached alarming levels, with the tropics alone accounting for 32% of the global forest loss from 2000 to 2012 (Hansen et al., 2013). Conservation strategies of tropical forests, which include the establishment of new protected areas, restoration efforts, and management of forest resources, depend on high-quality information about its remnants.

To date, most of our knowledge on these important ecosystems is limited to field studies performed at the plot level (~1 ha). However, these studies sampled a small fraction of the total forest area. Saatchi et al. (2015) showed that only 0.0001% of the total area of the Amazon Forest biome has been surveyed since the 1950s. Another study performed in the Brazilian Atlantic Forest revealed that just 0.01% of the total number of forest remnants were sampled (de Lima et al., 2015). These examples highlight the need for data encompassing broad spatial extents to better understand tropical forest structure and function. Remote sensing holds excellent promise to generalize and extrapolate insights emerging from plot-based studies to whole landscapes. Specifically, images featuring metric and submetric spatial resolution (pixel size ≤1 m) enable the detection of individual tree crowns (ITCs). These images also provide structural and spectral information that can be used to discriminate and map tree species.

Remotely sensed data are usually acquired with hyperspectral and multispectral sensors onboard airborne or spaceborne platforms. While hyperspectral imaging sensors measure reflected light from the forest canopy over many narrow spectral bands, multispectral sensors acquire data over broad noncontiguous channels. For tree species discrimination, the use of narrow-band data usually results in an increased classification accuracy when compared to broad-band multispectral data (Fassnacht et al., 2016). The increased accuracy probably stems from the fact that hyperspectral sensors enable the detection of subtle variations in the spectral response of tree species (Amaral et al., 2015; Baldeck et al., 2015; Féret and Asner, 2013; Ferreira et al., 2016; Vaglio Laurin et al., 2016).

Tropical tree species classification using hyperspectral data, also known as imaging spectroscopy, has been carried out for more than a decade, since the pioneering work of Clark et al. (2005). However, as pointed out by Fassnacht et al. (2016), most studies focused on the improvement of the classification accuracy without examining the vegetation traits that cause variations in the remote sensing signal and, thereby, enable or hamper species detection.

Understanding the factors affecting tree species discrimination can improve and optimize the use of hyperspectral data. For example, airborne acquisitions can be scheduled to a given period in which the target species are more prone to be detected (e.g., the presence or absence of leaves). However, such examinations are difficult to perform because they usually rely upon field information about tree species traits that can be costly to obtain, particularly in tropical environments. Thus, it is important to identify spectral characteristics derived directly from the remote sensing data that can be used as proxies to assess vegetation traits. Moreover, it is of great value to understand how the spectral resolution impacts the spectral characteristics of ITCs and consequently the discrimination among species.

Here, our main objective is to show how spectral characteristics of tropical tree species derived from hyperspectral data are related to structural and chemical properties of individual tree crowns. For this purpose, we used individual tree crown datasets and airborne hyperspectral data acquired over tropical forest sites in South America and in Africa. We present two methods used for the extraction of spectral characteristics of ITCs and discuss how they are related to chemical and structural vegetation traits. Finally, we compare the use of hyperspectral and multispectral data for tree species classification and assessment of spectral separability.

8.2   STUDY AREAS

In this section, we briefly describe the tropical forest sites from which hyperspectral data were acquired.

8.2.1   BRAZILIAN SITE

In Brazilian, the study area is the Santa Genebra Forest Reserve located in the municipality of Campinas, São Paulo state, southeast of Brazilian (22°49′13.4″S 47°06′43.6″W) (Figure 8.1). The reserve comprises 251.8 ha of a seasonal semideciduous forest formation (Oliveira-Filho and Ratter, 1995) and features high species richness (>100 tree species per hectare, according to Farah et al., 2014). The area is subjected to a dry season of up to five months (May to September), resulting in a heterogeneous canopy cover composed of evergreen and deciduous trees. The annual precipitation is about 1500 mm and temperature ranges from 11°C to 28.5°C.

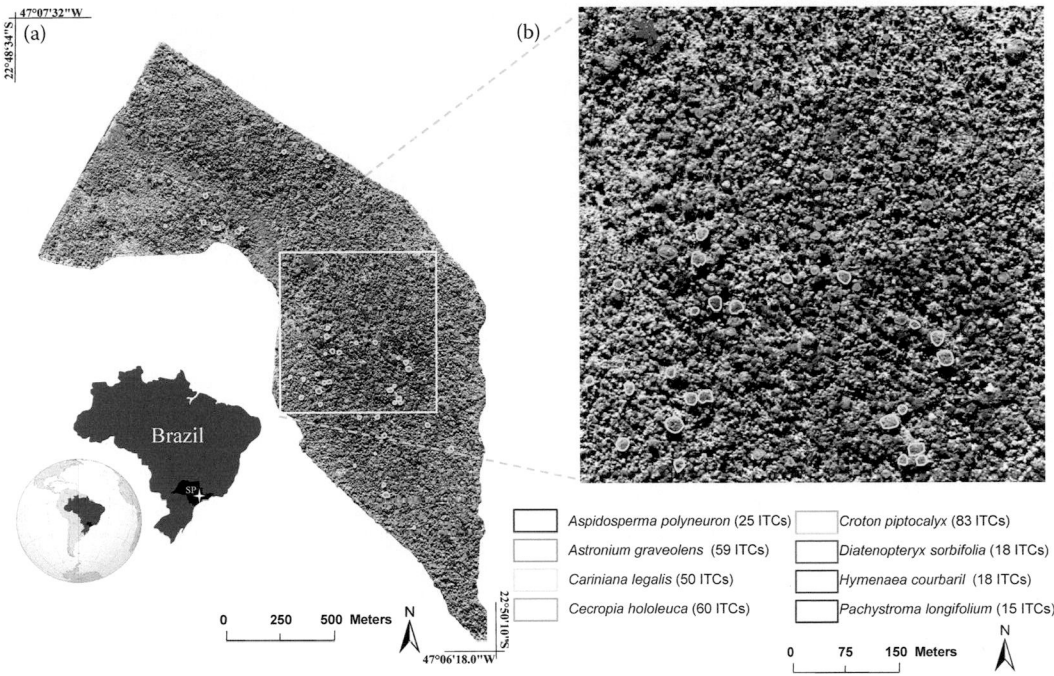

FIGURE 8.1 (a) Location of the study area in Brazilian, detailing the St. Genebra Forest Reserve with a true color composite (R = 639 nm, G = 548 nm, B = 460 nm) from hyperspectral data. (b) Individual tree crowns (ITCs) manually delineated and identified to the species level in the field. (Adapted from Ferreira, M. P. et al. 2016. *Remote Sensing of Environment*, 179, 66–78.)

8.2.2 AFRICAN SITES

The African sites include two areas in the southwestern portion of Ghana, the Ankasa and Bia conservation areas, covering, respectively, 509 and 306 km² (Vaglio Laurin et al., 2016). The mean monthly temperature in both areas ranges from 24°C to 28°C and the climate is characterized by a bimodal rainfall pattern. The Ankasa area receives between 2000 and 2200 mm of precipitation and its vegetation is characterized as wet evergreen forest. The Bia area covers a transition zone between moist evergreen and moist semideciduous forests, in which the mean annual precipitation ranges from 1500 to 1800 mm.

8.3 HYPERSPECTRAL ACQUISITIONS AND PREPROCESSING

8.3.1 BRAZILIAN SITE

In Brazilian, hyperspectral data were acquired with the AisaEAGLE and AisaHAWK (Spectral Imaging, Inc., Oulu, Finland) sensors that cover the visible/near-infrared (VNIR, 450–970 nm) and short-wave infrared (SWIR, 970–2500 nm) wavelength ranges, respectively (Table 8.1). These sensors were integrated into the ProSpecTIR-VS system (SpecTIR, Inc., Reno, NV, USA) to acquire VNIR and SWIR images synchronously. Geocoding information was collected with an inertial navigation system (INS) to create a geographic look-up table (GLT) that contained the map locations of each pixel. ProSpecTIR-VS was mounted on an aircraft that flew 1350 m above ground level at a speed of 241 km/h (130 knots), resulting in a spatial resolution of 1 m. A total of 357 spectral bands spaced by 5 nm were acquired with a radiometric resolution of 12 bits for AisaEAGLE and 14 bits for AisaHAWK (Table 8.1). Ten flightlines were acquired in 28 minutes on June 7, 2010 (at the beginning of the dry season) under clear sky conditions, starting at 1:27 p.m.

TABLE 8.1

Specifications of the Hyperspectral Instrument

ProSpecTIR-VS Dual Sensor

Sensor name	AisaEAGLE	AisaHAWK
Detector	Progressive scan CCD	MCT with maintenance-free cooler
Spectral range (nm)	400–970	970–2500
Spectral resolution (nm)	4	6
No. of spectral bands	122	235
Spatial resolution (m)	1	1
Radiometric resolution (bits)	12	14

Source: Adapted from Ferreira, M. P. et al. 2016. *Remote Sensing of Environment*, 179, 66–78.

Geometric correction of the radiance images of each flight line was performed with their respective GLTs. Then, the images were mosaicked and converted to surface reflectance by using the fast line-of-sight atmospheric analysis of spectral hypercubes (FLAASH) algorithm (Felde et al., 2003; ITT Visual Information Solutions, 2009). Channels around 1400 and 1900 nm were removed because of strong atmospheric water vapor absorption. Bands located in the transition zone between sensors (970–1045 nm) showed low signal-to-noise ratios (SNR) and were discarded. Noisy wavelengths below 400 nm and above 2400 nm were also discarded. A total of 260 spectral bands that covered the 450–2400 nm range were retained.

8.3.2 AFRICAN SITES

In Ghana, airborne hyperspectral data were collected in flight lines covering parts of the study areas, with the AisaEAGLE sensor (Spectral Imaging, Inc., Oulu, Finland) set to record 244 bands with a 2.3 nm sampling interval in the 400–1000 nm range. After orthorectification the nominal pixel size was 1 m. FLAASH (Felde et al., 2003) was used to perform the atmospheric correction of the strips. The resulting dataset had 186 bands, after removing 58 noisy bands. Additionally, the minimum noise fraction (MNF, Green et al., 1988) was applied to reduce noise. Only MNF components 9–15 were retained, in which the crown shapes were not confused by noise.

From the 186-band dataset the following vegetation indices were computed (Table 8.2): Normalized difference vegetation index (NDVI); simple ratio index (SRI); atmospherically resistant vegetation index (ARVI); red edge normalized difference vegetation index (RENDVI); Vogelmann red edge index (VREI); photochemical reflectance index (PRI); red-green ratio index (GRI); carotenoid reflectance indices 1 and 2 (CRI1, CRI2), and anthocyanin reflectance indices 1 and 2 (ARI1, ARI2) The gray levels co-occurrence matrix (GLCM), mean, variance, homogeneity, contrast, dissimilarity, entropy, second moment, and correlation textural features (Haralick, 1979) were also computed using a 5×5 window size, consistent with crown dimensions (generally comprising between 5 and 15 m radius).

8.4 INDIVIDUAL TREE CROWN DATASETS

8.4.1 BRAZILIAN SITE

Airborne hyperspectral data were acquired with high spatial resolution (1 m pixel size, Table 8.1), which allowed visualization of the tree crowns. Using a true color composite (R = 639 nm, G = 548 nm, B = 460 nm; Figure 8.1) we outlined crowns that were clearly seen by visual interpretation. These crowns were then visited with the aid of a GPS device and identified to the

TABLE 8.2
Vegetation Indices That Were Used in the Study with Their Respective Equations and References

Index[a]	Equation[b]	Reference
ARI1	$(1/\rho_{550}) - (1/\rho_{700})$	Gitelson et al. (2001)
ARI2	$\rho_{800} - [1/\rho_{550} - 1/\rho_{700}]$	
ARVI	$\rho_{800} - [\rho_{680} - (\rho_{450} - \rho_{680})]/\rho_{800} + [\rho_{680} - (\rho_{450} - \rho_{680})]$	Kaufman and Tanré (1996)
CRI-1	$(1/\rho_{510}) - (1/\rho_{550})$	Gitelson et al. (2002)
CRI-2	$(1/\rho_{510}) - (1/\rho_{700})$	Gitelson et al. (2002)
NDVI	$(\rho_{865} - \rho_{670})/(\rho_{865} + \rho_{670})$	Rouse et al. (1973)
PRI	$(\rho_{534} - \rho_{572})/(\rho_{534} + \rho_{572})$	Gamon et al. (1992)
RENDVI	$(\rho_{750} - \rho_{700})/(\rho_{750} + \rho_{700})$	Sims and Gamon (2002)
RGRI	$\sum_{i=600}^{699} \rho_i / \sum_{j=600}^{699} \rho_j$	Gamon and Surfus (1999)
SRI	$(\rho_{865} - \rho_{670})$	Sellers (1985)
VREI	$(\rho_{740} - \rho_{720})$	Vogelmann et al. (1993)

[a] Names of the vegetation indices are as given in the text.

[b] ρ is the closest AISAEagle band (the index number indicates the wavelength center in nanometers) to the original index formulation.

species level in the field. The presence of lianas was carefully inspected, avoiding liana-dominated crowns because these plants significantly influence the spectral response of tropical tree species at the canopy level (Kalacska et al., 2007). A total of 328 ITCs that belonged to eight species were identified (Table 8.3), 55 more trees than used by Ferreira et al. (2016). For each species, the within-crown pixel spectra were extracted from the hyperspectral image to establish a library of their spectral signatures, which is shown in Figure 8.2. Photographs of some tree species were taken during the field work and are presented in Figure 8.3.

TABLE 8.3
Description of the Trees from the Brazilian Site with Crowns Manually Delineated in the Hyperspectral Image and Identified to the Species Level in the Field: Species List, Species Code, Number of Individual Tree Crowns (ITC), Mean ITC Size (Pixels), Minimum ITC Size (Pixels), Maximum ITC Size (Pixels), and Total Number of Pixels Per Species

Species	Code	No. of ITCs	Mean Crown Size (pixels)	Min Crown Size (pixels)	Max Crown Size (pixels)	No. of Pixels
Aspidosperma polyneuron	AP	25	112	41	248	2631
Astronium graveolens	AG	59	104	21	400	5929
Cariniana legalis	CL	50	268	54	594	13,280
Cecropia hololeuca	CH	60	41	8	93	2277
Croton piptocalyx	CP	83	80	30	166	6341
Diatenopteryx sorbifolia	DS	18	21	8	35	376
Hymenaea courbaril	HC	18	210	70	467	3650
Pachystroma longifolium	PL	15	59	19	113	898

Source: Adapted from Ferreira, M. P. et al. 2016. *Remote Sensing of Environment*, 179, 66–78.

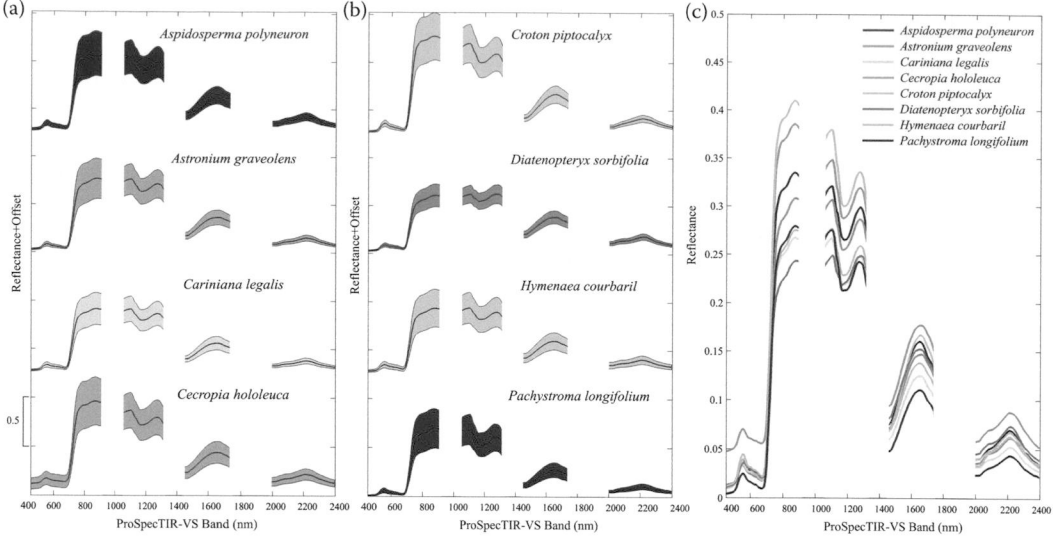

FIGURE 8.2 (a) and (b) Mean (bold lines) ± 1 standard deviation envelopes of the spectral responses of the tree species from the Brazilian site. An offset of 0.5 was applied to the reflectance spectra of each species for clarity. (c) Mean spectral responses of the studied species. Wavelength regions strongly affected by the atmosphere have been removed from the plots. (Adapted from Ferreira, M. P. et al. 2016. *Remote Sensing of Environment*, 179, 66–78.)

FIGURE 8.3 Individual tree crowns from species considered in the study. Photographs were taken during the dry season in the Brazilian site (Photos credit: M.P. Ferreira). (Adapted from Ferreira, M. P. et al. 2018. *Remote Sensing of Environment*, 211, 276–291.)

8.4.2 AFRICAN SITES

Crowns of 16 species with clear phenological and morphological characteristics were manually delineated in the two study sites using 10 cm spatial resolution RGB orthophotos. The method of species identification was similar to that employed in the Brazilian site, specifically, exploiting the ground information collected during various field surveys. At each site, three dominant canopy species featuring crowns that can be clearly seen in the hyperspectral image (Figure 8.4) were identified. A spectral library of each species was built by extracting their within crown pixels (Table 8.4).

The 16 identified species were divided into functional groups according to the plant response to illumination conditions. Following this approach, Hawthorne (1995) classified West African

ANKASA   **BIA**

0   50 100 150 200 m

FIGURE 8.4 Hyperspectral data collected over Ankasa and Bia conservation areas in a false color composite of 829 nm (R), 604 (G), and 465 (B) bands. The yellow polygons represent the delineated crown species. (Adapted from Vaglio Laurin, G. et al. 2016. *Remote Sensing of Environment*, 176, 163–176.)

TABLE 8.4

Selected Dominant Species for Ankasa and Bia Areas, Pixels in Hyperspectral Data, and Guild Type

Ankasa Conservation Area			Bia Conservation Area		
Species Name	No. of Pixels	Guild	Species Name	No. of Pixels	Guild
Cynometra ananta	4607	SB	*Pycnanthus angolensis*	4670	NPLD
Heritiera utilis	4373	NPLD	*Terminalia superba*	4578	PION
Protomegabaria stapfiana	4452	NPLD	*Triplochiton scleroxylon*	4739	PION

Source: Adapted from Vaglio Laurin, G. et al. 2016. *Remote Sensing of Environment*, 176, 163–176.
Abbreviations: NPLD, non-pioneer light demanding; PION, pioneer; SB, shade-bearer.

TABLE 8.5

Number of Pixels from Several Trees Species from the African Sites Belonging to Different Guilds per Area

Ankasa Conservation Area			Bia Conservation Area		
Guild	No. of Species	No. of Pixels	Species Name	No. of Species	No of Pixels
NPLD	6	2576	NPLD	3	2354
PION	2	2487	PION	4	2369
SB	5	2632	SB	3	2346

Source: Adapted from Vaglio Laurin, G. et al. 2016. *Remote Sensing of Environment*, 176, 163–176.
Abbreviations: NPLD, non-pioneer light demanding; PION, pioneer; SB, shade-bearer.

forest tree species into pioneer (PION), non-pioneer light demanding (NPLD), and shade-bearer (SB) guilds. PION species included *Alstonia boonei, Elaeis guineensis, Lophira alata, Myrianthus arboreus, Terminalia superba,* and *Triplochiton scleroxylon;* SB species included *Berlinia* spp., *Celtis mildbraedii, Cola gigantea,* and *Cynometra ananta;* and the NPLD species included *Albizia* spp., *Heritiera utilis, Piptadeniastrum africanum, Protomegabaria stapfiana, Pycnanthus angolensis,* and *Uapaca guineensis.* Again, the within-crown pixel spectra were extracted from the hyperspectral image to establish a library of the spectral signatures of the three guilds at each of the two African sites (Table 8.5).

8.5 EXTRACTING SPECTRAL CHARACTERISTICS OF TREE SPECIES

Here, we report on the extraction of spectral characteristics of ITCs in airborne hyperspectral data acquired in the visible to short-wave infrared (VSWIR, 450–2500 nm) wavelength range from the Brazilian site. For this purpose, we used spectral feature analysis and spectral mixture analysis that have been proven to be powerful tools to assess vegetation chemical and structural traits using hyperspectral data.

8.5.1 SPECTRAL FEATURE ANALYSIS: FEATURE DEPTH

Continuum removal was used to isolate and analyze features in reflectance spectra (see Clark and Roush, 1984; Kokaly and Clark, 1999). The continuum is an estimate of the other absorptions present in the spectrum, not including the one of interest (Clark, 1999; Clark and Roush, 1984). Continuum removal can be applied to dips in reflectance spectra (absorption features) and peaks in reflectance spectra (described as positive or emission features), although it is most often performed on absorption features. The continuum is defined by the line connecting the continuum end-points, that is, the channels at the high points on the left and right sides of an absorption feature. Continuum-removed reflectance values are calculated by dividing the original reflectance values by the corresponding values of the continuum line at the wavelength positions of each channel between the end-points of the absorption feature (Clark and Roush, 1984).

A module of the U.S. Geological Survey Processing Routines in IDL for Spectroscopic Measurements (PRISM) software, publicly available at http://pubs.usgs.gov/of/2011/1155/(Kokaly, 2011; last accessed January 27, 2018), was used to perform automated continuum removal of the commonly observed spectral features in the vegetation spectra. We analyzed the absorption features centered near 680, 1200, and 2100 nm. In addition to the three absorption features, the green peak feature (centered near 557 nm) was also analyzed. Table 8.6 lists the spectral features, associated biochemical components of plants related to each feature, the end-points used for linear continuum removal, and the number of imaging spectrometer channels between the continuum end-points that were used to compute spectral feature parameters (depth). For the absorption features, the feature depth values were calculated for the channel with the minimum value in continuum-removed reflectance between the continuum end-points. The depth of the feature was simply computed as 1 minus the continuum-removed reflectance value (Kokaly and Clark, 1999). For the green peak feature, the feature depth was calculated at the channel with the maximum in continuum-removed feature, resulting in negative values for green peak feature depth. Pixels within the manually delineated ITCs for the Brazilian site (Figure 8.1) were extracted from images corresponding to each spectral feature. Then, the pixel values of each ITC were averaged to assess differences on the spectral features among species. To facilitate comparison of feature depth values among spectral features, data were normalized to the [0–1] range based on the minimum and maximum value of each feature.

8.5.2 Spectral Mixture Analysis

Multiple end-member spectral mixture analysis (MESMA; Roberts et al., 1998) was performed on the hyperspectral images mosaic, to verify the within-crown differences in subpixel fractions of green vegetation (GV), nonphotosynthetic vegetation (NPV), and shade between the target species. Spectral mixture analysis (SMA; Adams et al., 1993) generates linear models using sets of end-members ("pure" spectra) from spectral libraries, and selects the model with the lowest root mean

---

TABLE 8.6
Spectral Features Computed in the Feature Depth Analysis of the Hyperspectral Data

Spectral Feature	Chemistry Related to the Absorption Feature	Left End-Point Wavelength (nm)	Right End-Point Wavelength (nm)	Number of Channels between Continuum End-Points
Chlorophyll (680 nm)	Chlorophyll and other plant pigments	516	753	51
Leaf water (1200 nm)	Leaf water	1103	1279	27
Nonpigment biochemical constituents (2100 nm)	Lignin, cellulose, and protein	2037	2200	27
Green peak (557 nm)	Chlorophyll and other plant pigments	498	677	39

square error (RMSE) for each pixel. MESMA is a SMA approach that allows the type and number of a single class end-members to vary per pixel (Roberts et al., 1998).

First, candidate end-members of GV and NPV were collected in the images, and an SMA (with GV, NPV, and photometric shade) was performed with no constraints. The resulting fraction images were analyzed in 2D scatterplots, for selecting three different GV and NPV end-members. These end-members present the highest values of their own fraction, and the lowest values of the other two fractions. The final fraction images were obtained using a MESMA with the following constraints: fraction thresholds between -0.05 and 1.05 and RMSE ≤ 0.025. SMA and MESMA were run in ViperTools (Roberts et al., 2007), and used the sum of the reflectances of a fixed set of end-members multiplied by their fractional cover:

$$\rho'_\lambda = \sum_{i=1}^{N} f_i * \rho_{i\lambda} + \varepsilon_\lambda \tag{8.1}$$

where ρ'_λ is the reflectance at a given wavelength (λ), $\rho_{i\lambda}$ is the reflectance of end-member i at λ, f_i is the cover fraction of end-member i, N is the total number of end-members in the model, and ε_λ is the residual error of the model. The end-member fractions are constrained by a sum equal to 1.0 (i.e., spectrum 100% modeled by the end-members):

$$\sum_{i=1}^{N} f_i = 1 \tag{8.2}$$

The evaluation of the best-fit model used the RMSE and the residual error of the model (ε_λ):

$$\text{RMSE} = \sqrt{\frac{\sum_{\lambda=1}^{M} (\varepsilon_\lambda)^2}{M}} \tag{8.3}$$

where M was the total number of bands. Pixels within ITCs were extracted from subpixel fraction images, averaged with respect to ITC and the values were scaled between 0 and 1 based on the minimum and maximum value of each feature to facilitate comparison among subpixel fractions.

8.6 RELATING SPECTRAL CHARACTERISTICS TO CHEMICAL AND STRUCTURAL PROPERTIES OF TREE SPECIES

In this section, we relate the depth of absorption features and subpixel fractions to biochemical and structural properties of tree species. Results of the feature depth analysis for the Brazilian species are shown in Figure 8.5. The nonpigment biochemical constituents feature depth (2100 nm) provided a high separation among species (Figure 8.5d). At this spectral feature, three species (*Aspidosperma polyneuron*, *Cecropia hololeuca*, and *Pachystroma longifolium*) were not statistically different from the others. It is worth noting that the ITCs of *Diatenopteryx sorbifolia* showed the highest feature depth at 2100 nm (mean of 0.5), suggesting a higher amount of nonpigment biochemical constituents. This arises from the low leaf density of *Diatenopteryx sorbifolia* in the dry season that allows sunlight to reach the forest floor composed mainly of nonphotosynthetic material (litter) (Nave, 1999). *Cariniana legalis* and *Hymenaea courbaril* also showed high depths at 2100 nm (0.46 and 0.44, respectively). Field observations revealed that they usually feature large crowns in which thick branches emerge from the main tree stem (Figure 8.3). Such branches are usually exposed due to the low leaf cover, which makes its spectral response more prone to be detected by the hyperspectral sensor.

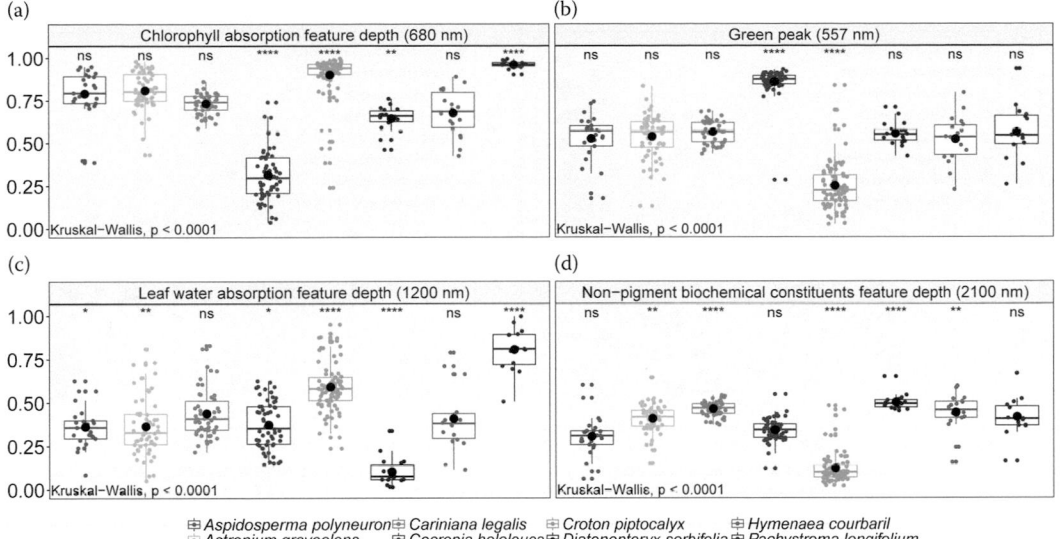

FIGURE 8.5   Boxplots showing the variability of absorption band features among individual tree crowns of the species under investigation from the Brazilian site (Table 8.3). (a) Chlorophyll absorption feature depth (680 nm); (b) green peak (557 nm); (c) leaf water absorption feature depth (1200 nm); and (d) nonpigment biochemical constituents feature depth (2100 nm). The central lines within each box are the medians and the central black dots are the means. The upper and lower quartiles are represented by the edges of the boxes. The Kruskal–Wallis test (Theodorsson-Norheim, 1986) was performed to assess differences among species. The global *p*-value of the Kruskal–Wallis test is shown in the lower left corner of each panel, while the *p*-values resulting from multiple comparisons are represented above each boxplot. ns, nonsignificant (*p*-value > 0.05); * *p*-value ≤ 0.05; ** *p*-value ≤ 0.01; *** *p*-value ≤ 0.001; **** *p*-value ≤ 0.0001.

Conversely, species with high amounts of leaves within the crown such as *Croton piptocalyx* showed low depths at 2100 nm (mean of 0.12, Figure 8.5d). *Croton piptocalyx* is of semi-deciduous type and leaf flushing during the dry season is high (Morellato, 1991). It is not uncommon to find trees with fully foliated crowns, as shown in Figure 8.3. This may cause a low exposure of branches and nonphotosynthetic material, decreasing the absorption feature depth near 2100 nm (Figure 8.5d) and, at the same time, increasing depths of the chlorophyll absorption band (Figure 8.5a). This is corroborated by the inverse relationship observed between the chlorophyll and the nonpigment biochemical constituents feature depth (Figure 8.6b).

A positive relationship between the chlorophyll and the leaf water absorption feature depth was verified for most of the species (Figure 8.6a). At these absorption features *Pachystroma longifolium* reached the highest values, which can be explained by its leaf biochemical properties and canopy structure. *Pachystroma longifolium* exhibits dark green leathery leaves with high chlorophyll content (Alcalá, 2010) that might increase the depth of the absorption feature near 680 nm. Moreover, it is an evergreen species that forms dense monospecific patches (Gandolfi et al., 2007). Álvarez-Yépiz (2017) showed that the spatial segregation between evergreen and deciduous species confers on the former a higher water use efficiency. Evergreen species usually maintain higher leaf water content than deciduous species during the dry season (Sobrado, 1986), which probably influenced the absorption feature around 1200 nm. An inverse relationship is observed between the chlorophyll and the nonpigment biochemical absorption feature depths (Figure 8.6b), enabling us to discriminate evergreen species (e.g., *Pachystroma longifolium*) from deciduous ones (e.g., *Diatenopteryx sorbifolia* or *Croton piptocalyx*).

After performing the spectral mixture analysis to obtain subpixel fractions for each ITC in the Brazilian site, we found that the proportion of NPV, GV, and shade varies among species (Figure 8.7).

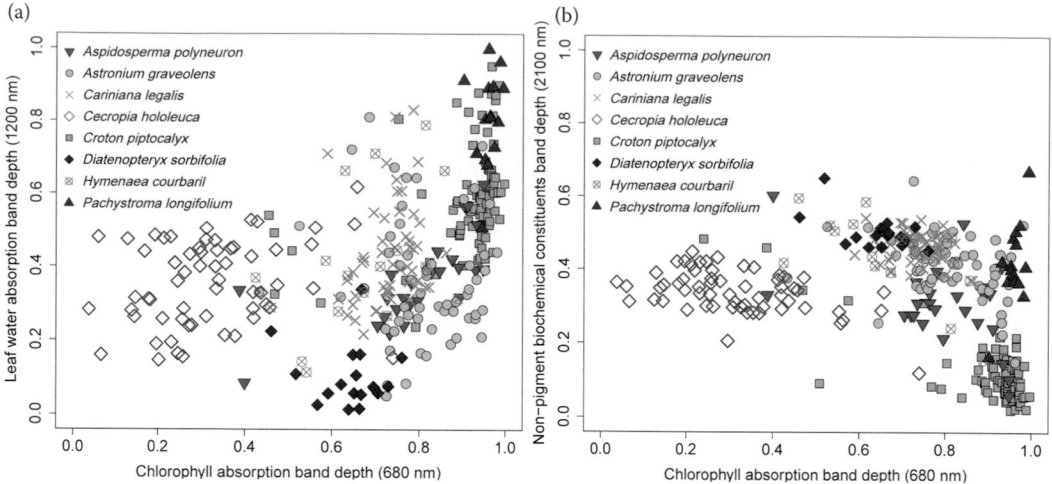

FIGURE 8.6 (a) Relationship between chlorophyll and leaf water absorption feature depth for individual tree crowns from the Brazilian site (Table 8.3). (b) Relationship between chlorophyll and nonpigment biochemical constituents feature depth for individual tree crowns from the Brazilian site (Table 8.3).

This agrees with field observations that revealed different crown structural characteristics defined by size, shape, branching, leaf area, and leaf insertion angle among the species under investigation. The highest values of the shade fraction were observed in *Diatenopteryx sorbifolia*, *Cariniana legalis,* and *Pachystroma longifolium*, reaching mean values of 0.71, 0.66, and 0.65, respectively. For *Cecropia hololeuca*, *Croton piptocalyx*, *Aspidosperma polyneuron*, *Astronium graveolens,* and *Hymenaea courbaril* the shade fraction showed lower values (0.22, 0.39, 0.55, 0.56, and 0.62, respectively). The tree species showed the highest separability on the green vegetation and on the shade fraction, in which only one showed nonsignificant values of the Kruskal–Wallis test (Figure 8.7b and c).

It is interesting to note the strong inverse relationship between the GV and the shade fraction (Figure 8.8a). Conversely, the shade fraction is positively related to NPV for most of the species,

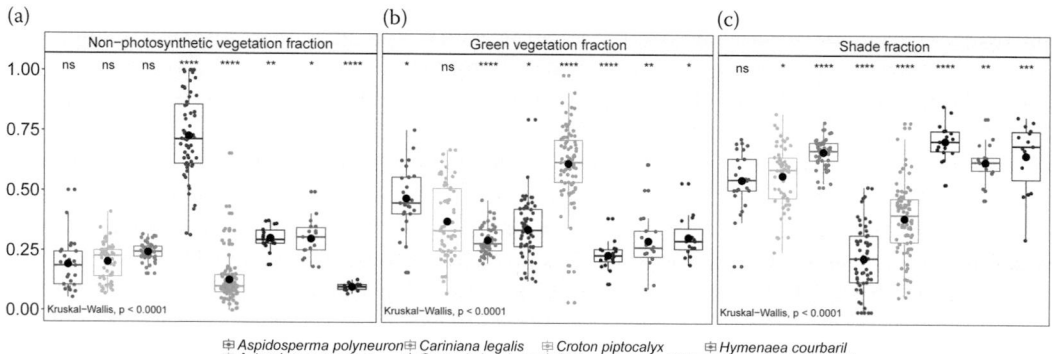

FIGURE 8.7 Boxplots showing the variability of subpixel fractions among individual tree crowns of the species under investigation from the Brazilian site (Table 8.3). (a) Nonphotosynthetic vegetation fraction; (b) green vegetation fraction; and (c) shade fraction. The central lines within each box are the medians and the black dot are the means. The upper and lower quartiles are represented by the edges of the boxes. The Kruskal–Wallis test (Theodorsson-Norheim, 1986) was performed to assess differences among species. The global p-value of the Kruskal–Wallis test is shown in the lower left corner of each panel, while the p-values resulting from multiple comparisons are represented above each boxplot. ns, nonsignificant (p-value > 0.05); * p-value ≤ 0.05; ** p-value ≤ 0.01; *** p-value ≤ 0.001; **** p-value ≤ 0.0001.

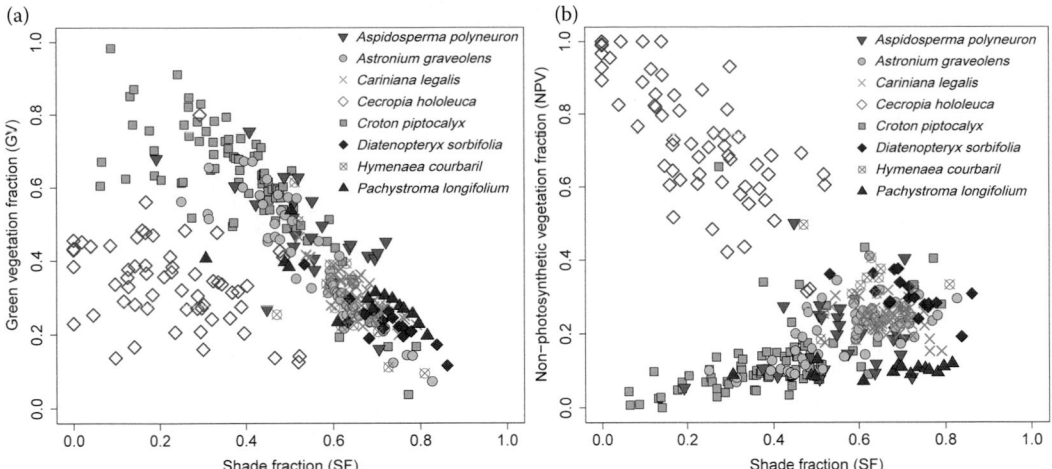

FIGURE 8.8 Relationships between the shade fraction and (a) the green vegetation fraction and (b) the nonphotosynthetic vegetation fraction for each tree crown from the Brazilian site. Subpixel fractions were obtained by spectral mixture analysis of the hyperspectral data.

except for *Pachystroma longifolium,* which showed low NPV, despite the high shade fraction values (Figures 8.6c and 8.8b). NPV and shade are positively related (Figure 8.8b). Although this relationship is not as strong as that of Figure 8.8a, some species showed clustered patterns with low variations of NPV and shade among ITCs. This is the case of *Cariniana legalis,* which in general has high values of NPV and shade (Figure 8.7a and c).

The proportion of shade, green, and nonphotosynthetic vegetation within ITCs can be explained by the architecture of the crown. For example, field observations showed that *Cariniana legalis* and *Hymenaea courbaril* are emergent trees featuring large crowns (most >15 m radius) with internal gaps caused by thick branches emerging from the main tree stem (Figure 8.3). Such branches can divide the tree crown into one or more parts, exposing nonphotosynthetic material and increasing the amount of internal-crown shadows, which produced high values of the shade fraction (Figure 8.7c). *Pachystroma longifolium* and *Diatenopteryx sorbifolia* are found in monospecific patches over the study area (Gandolfi et al., 2007; Nave, 1999) and also showed a high proportion of shade within the crown (Figure 8.7c). *Pachystroma longifolium* is an evergreen species with fully foliated crowns that present internal gaps caused by the heterogeneous crown structure (Alcalá, 2010). *Diatenopteryx sorbifolia* forms a continuous well-defined canopy, 25–30 m high, without the presence of emergent trees and a short understory layer (less than 4 m tall) (Nave, 1999). This species is deciduous and leaf fall during the dry season exposes the branches, producing shade within the crown. Moreover, the low leaf cover of *Diatenopteryx sorbifolia* crowns at the time of the overflight and the sparse understory layer beneath its crowns allowed detection of reflected sunlight from the forest floor, which explains its high NPV fraction (Figure 8.7a).

8.7 COMPARISON BETWEEN HYPERSPECTRAL AND SIMULATED MULTISPECTRAL DATA

When moving from airborne hyperspectral to spaceborne multispectral data, a consistent loss of spectral and spatial resolution occurs, with negative impacts on discrimination capabilities. However, considering the free and growing availability of enhanced multispectral datasets, it is worth exploring how much information can still be obtained using multispectral sensors. Here, we compare hyperspectral and simulated multispectral data in terms of spectral separability and classification accuracy of tropical tree species. As discussed, chemical and structural properties of individual tree

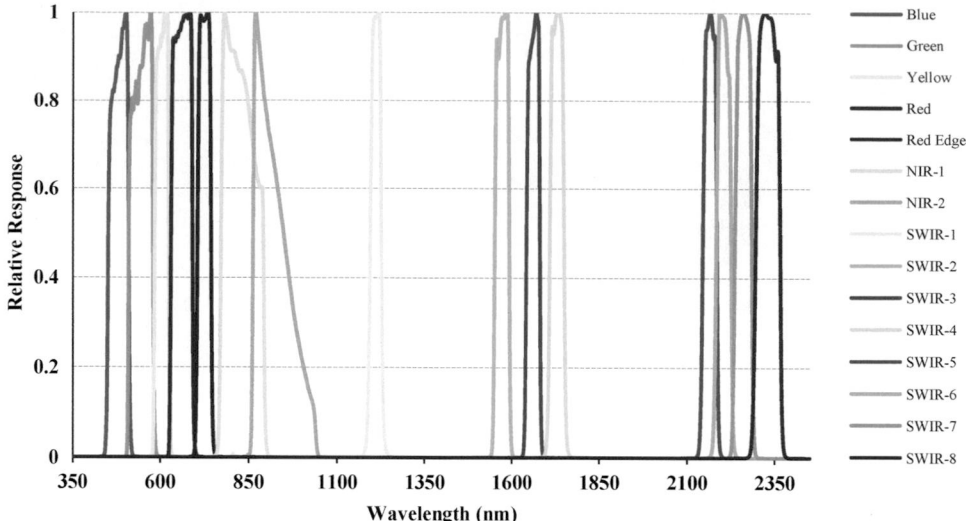

FIGURE 8.9 Spectral filter functions of the WorldView-3 satellite sensor. (Source: DigitalGlobe. Adapted from Ferreira, M. P. et al. 2016. *Remote Sensing of Environment,* 179, 66–78.)

crowns produce variations in the remote sensing signal that can be detected through the extraction of spectral characteristics. It is interesting to understand how the spectral resolution impacts the discrimination among species and the spectral separability between them. For this purpose we simulated, for the Brazilian site, high spatial resolution multispectral data of the WorldView-3 (WV-3) satellite sensor and, for the African sites, moderate spatial resolution data of the Sentinel-2 (S2) satellite mission.

WV-3 is currently the only high-resolution orbital platform with SWIR sensing capabilities. Moreover, it is equipped with a VNIR imaging system capable of acquiring data in eight spectral bands spanning 400–1040 nm. Aiming at performing an experimental comparison where the atmospheric, radiometric and geometric errors would be very similar for the hyperspectral and multispectral images, we used a simplified approach to mimic the spectral characteristics of the WV-3 satellite using the airborne hyperspectral data. Narrow bands were aggregated and the spatial resolution of the airborne hyperspectral imagery (1 m) was maintained. Figure 8.9 shows the spectral filter functions used to simulate 15 WV-3 bands.

The S2 sensor is equipped with 13 bands from 443 to 2190 nm, and a 10-day repeat cycle; 3 red edge bands are especially valuable for detection of fine differences in chlorophyll pigments, leaf area, and fractional cover (Verrelst et al., 2012). Using hyperspectral imagery from the African sites, we simulated the S2 data with the respective spectral response functions (SRFs) and the approach developed by D'Odorico et al. (2013). Due to the limited spectral range of the available hyperspectral data (450–900 nm) only eight bands were simulated, and centered at 490, 560, 665, 705, 740, 783, 842, and 865 nm. The spatial resolution of the airborne hyperspectral imagery was degraded to 10 m with nearest-neighbor interpolation to match the S2 pixel size. With the simulated bands, four vegetation indices (NDVI, SRI, RENDVI, ARI1) and GLCM textural features for each band were calculated using the smallest possible window size (3 × 3).

8.7.1　Classification Accuracy

Tree species classification was performed using hyperspectral and simulated WV-3 data to assess the impact of the spectral resolution in discriminating among species of the Brazilian site. We used the RUSBoost (Random Under Sampling) algorithm (Seiffert et al., 2010) to perform species classification at the pixel level. RUSBoost was designed to deal with the class imbalance problem for which one class outnumbers other classes by a large proportion. For example, the number of pixels of *Cariniana legalis* is 35 times greater than the number of pixels of *Diatenopteryx sorbifolia* (Table 8.3). The class imbalance problem significantly hinders the performance of machine learning methods (Japkowicz and Stephen, 2002).

Pixels within the 328 ITCs belonging to eight species (Table 8.3) were extracted from the hyperspectral and from the simulated WV-3 image to compose two datasets with 35,382 samples, 260 and 15 VSWIR (450–2500 nm) features, respectively. It is worth highlighting that the simulated WV-3 image was obtained by aggregating the spectral bands from the hyperspectral image; thus both hyperspectral and WV-3 simulated images feature a spatial resolution of 1 m. These datasets were partitioned into 60% for training and 40% for testing in such a way that training and testing pixels came from different ITCs. This procedure was repeated 100 times, changing the ITCs used to train and test the classifier at each realization, which allowed us to evaluate how variable the classification accuracy is depending on the samples used in the process. Classification was performed using all VNIR and VSWIR bands of the WV-3 sensor and set of 30 hyperspectral bands selected by stepwise regression. The RUSBoost algorithm provided by the *fitcensemble* function in MATLAB® (MathWorks, Natick, Massachusetts, USA) was used.

Table 8.7 shows the classification accuracy achieved with hyperspectral and simulated WV-3 multispectral data. The average accuracy was 17.6% and 15.4% higher when using VNIR and VSWIR hyperspectral bands, respectively. For both hyperspectral and simulated WV-3 datasets, discrimination accuracy increased for all the species after incorporating SWIR features.

The highest increase was observed in *Cariniana legalis* (13.8% with hyperspectral and 21.7% with WV-3 data), suggesting a distinctive reflectance pattern in the SWIR that was able to separate it from the other species. This is corroborated by the high values of the nonpigment absorption feature depth observed for this species (Figure 8.5d). Conversely, the smallest variation in classification accuracy after the incorporation of SWIR features was observed for *Croton piptocalyx* and *Cecropia hololeuca*. For the former, the classification accuracy increased 2% when using hyperspectral data and decreased 2.1% with simulated WV-3 data; for the latter it increased 0.4% with hyperspectral and 0.1% with simulated WV-3 data (Table 8.7). *Cecropia hololeuca* has a peculiar reflectance pattern in the visible range (450–650 nm, Figure 8.2c) that

---

TABLE 8.7

Accuracy of Species Classification Using Hyperspectral and Simulated WorldView-3 Data in the VNIR (450–970 nm) and VSWIR (450–2500 nm) Ranges

Species	Simulated WorldView-3		Hyperspectral	
	VNIR (7 Bands) (Mean±1SD,%)	VSWIR (15 Bands) (Mean±1SD,%)	VNIR (30 Bands) (Mean±1SD,%)	VSWIR (30 Bands) (Mean±1SD,%)
Aspidosperma polyneuron	38.8(±4.9)	44.3(±6.2)	63.5(±15.1)	65.3(+16.6)
Astronium graveolens	56.3(±4.6)	63.0(±4.3)	76.7(±9.0)	85.6(±8.9)
Cariniana legalis	50.9(±3.9)	72.6(±3.3)	83.8(±9.1)	97.6(±4.5)
Cecropia hololeuca	94.3(±4.0)	94.4(±3.8)	96.9(±5.6)	97.3(±5.3)
Croton piptocalyx	72.1(±3.9)	74.3(±3.7)	87.0(±5.8)	89.0(±5.1)
Diatenopteryx sorbifolia	77.1(±10.6)	86.0(±6.7)	88.3(±15.2)	91.2(±9.2)
Hymenaea courbaril	29.6(±5.3)	38.2(+6.6)	57.6(±23.4)	63.0(±24.5)
Pachystroma longifolium	85.3(±8.4)	85.1(±9.1)	92.0(±14.4)	92.0(±14.4)
Average accuracy	63.1(±5.7)	69.7(±5.5)	80.7(±12.2)	85.1(±11.1)

Abbreviations: VNIR, visible/near-infrared; VSWIR, visible to short-wave infrared.

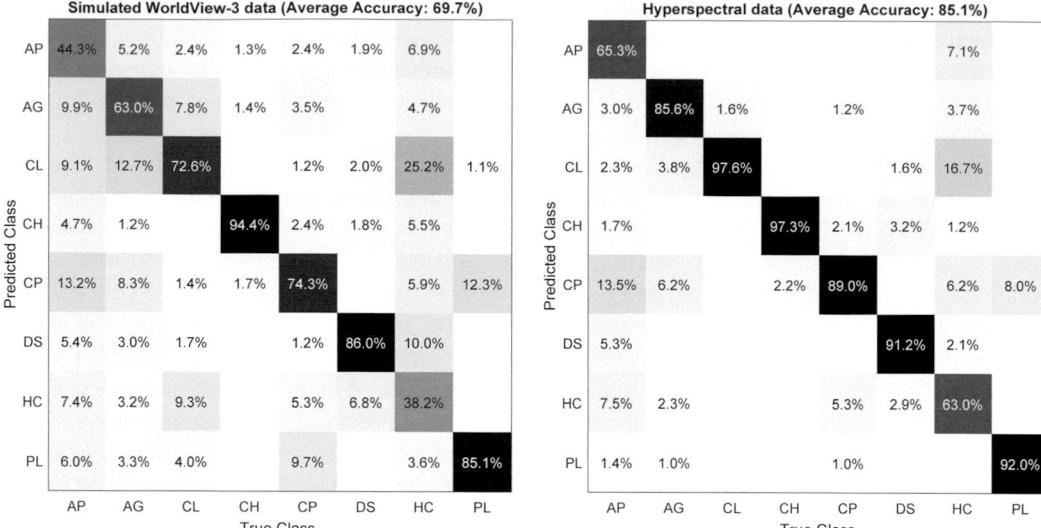

FIGURE 8.10 Confusion matrices showing the accuracy of tree species classification using simulated WorldView-3 (left panel) and hyperspectral data (right panel). The diagonal cells show the classification accuracy of each species. The off-diagonal cells show the percentage of classification error; empty white cells refer to less than 1% misclassification. AP, *Aspidosperma polyneuron*; AG, *Astronium graveolens*; CL, *Cariniana legalis*; CH, *Cecropia hololeuca*; CP, *Croton piptocalyx*; DS, *Diatenopteryx sorbifolia*; HC, *Hymenaea courbaril*; PL, *Pachystroma longifolium*.

clearly distinguishes it from the other species, which explains its high classification accuracies (>95% accuracy with VSWIR hyperspectral data, Table 8.7). The same applies for *Pachystroma longifolium* that was also well classified (>90% accuracy with VSWIR hyperspectral data, Table 8.7) and showed the lowest reflectance values in the visible (Figure 8.2c). The highest classification errors were obtained between *Cariniana legalis* and *Hymenaea courbaril*, reaching 25.2% for simulated WV-3 data and 16.7% for hyperspectral data (Figure 8.10). These species are deciduous (Lorenzi, 2014) and feature similar crown characteristics (Figure 8.3) that may lead to similar spectral responses (Figure 8.2).

8.7.2 SPECTRAL SEPARABILITY

Tropical tree species discrimination using hyperspectral data is intrinsically related to the spectral separability. For example, Ferreira et al. (2016) showed that classification accuracy depends on the amount of intraclass variation and a significant difference between intra- and interclass separability. Here, we evaluated how a decrease in the spectral resolution impacts the spectral separability.

Previous studies have shown that the Bhattacharyya distance (Kailath, 1967) is a suitable measure of the separability within and among tropical tree species (Féret and Asner, 2013). The Jeffries–Matusita (J-M) distance is an extension of the Bhattacharyya distance (Richards and Jia, 1999), ranging from 0 to 2.0, and was used here to analyze how well crowns belonging to different light response functional groups (guilds) could be separated by hyperspectral and simulated S2 data in the African sites. A J-M value >1.8 indicates separability. The analysis has been performed at two levels: analyzing the separability of crowns from few dominant species of different guilds types, and analyzing the separability after aggregating all the species crowns per group.

The two tropical forested areas feature different moisture levels (wet and semideciduous forests). Results indicate that dominant species crowns from wet tropical forests (Ankasa site) are more difficult to distinguish than those from semideciduous forest (Bia site) (Tables 8.8 and 8.9). The

TABLE 8.8

Jeffries–Matusita (J-M) Values at Crown Level, for a Few Dominant Species Belonging to Different Functional Guilds

	Hyper Bands	Hyper VIs	Hyper Bands + Textures	S2 Bands	S2 VIs	S2 Bands + Textures
			Ankasa (Wet Forest)			
Cynometra (SB)-*Heritiera* (NPLD)	1.49	1.46	1.97	1.02	0.83	1.8
Cynometra (SB)-*Protomegabaria* (NPLD)	1.37	1.45	1.81	1.01	0.36	1.97
Heritiera (NPLD)- *Protomegabaria* (NPLD)	1.29	1.14	1.8	0.99	0.72	1.98
			Bia (Semideciduous Forest)			
Pycnanthus (PION)-*Terminalia* (NPLD)	1.67	1.75	1.84	1.22	0.57	1.93
Pycnanthus (PION)-*Triplochiton* (NPLD)	1.61	1.79	1.87	1.29	0.57	1.91
Terminalia (NPLD)-*Triplochiton* (NPLD)	1.9	1.98	1.99	1.55	1.29	1.95

Source: Adapted from Vaglio Laurin, G. et al. 2016. *Remote Sensing of Environment*, 176, 163–176.
Abbreviations: NPLD, non-pioneer light demanding; PION, Pioneer; SB, shade-bearer; VIs, vegetation indices.

highest separability at the Bia site, particularly the NPLD guild, can be explained by phenological features, such as deciduousness or flowering, observed in orthophotos.

At crown level (Table 8.8) or after grouping species into functional guilds (Table 8.9) the hyperspectral bands yielded higher separability values. This demonstrates that even 13 systematically sampled hyperspectral bands are more informative for species discrimination than eight broad multispectral bands. However, it is worth noting that separability was not reached for most of the species based only on the hyperspectral data; the use of texture descriptors

TABLE 8.9

Jeffries–Matusita (J-M) Separability Values at Functional Guild Level, for Crowns of 16 Different Tree Species Aggregated according to Guild Type

	Hyper Bands	Hyper VIs	Hyper Bands + Textures	S2 Bands	S2 VIs	S2 Bands + Textures
			Ankasa (Wet Forest)			
SB—NPLD	1.53	1.77	1.82	1.23	0.77	1.96
PION—NPLD	1.78	1.9	1.92	1.41	0.45	1.97
SB—PION	1.76	1.86	1.91	1.62	1.11	1.97
			Bia (Semi-Deciduous Forest)			
SB—NPLD	1.31	1.53	1.83	1.21	1.87	1.8
PION—NPLD	1.74	1.9	1.99	1.47	1.94	1.97
SB—PION	1.71	1.86	1.98	1.54	1.17	1.97

Source: Adapted from Vaglio Laurin, G. et al. 2016. *Remote Sensing of Environment*, 176, 163–176.
Abbreviations: NPLD, non-pioneer light demanding; PION, pioneer; SB, shade-bearer.

was necessary to produce J-M scores >1.8. The same applies for S2 bands, in which the use of texture features became essential to reach species discrimination (Tables 8.8 and 8.9). The use of GLCM features increased separability because some features were able to enhance crown edges in the imagery (Figure 8.11). Moreover, brightness variations within the crowns are related to crown-internal shadows, branching, and foliage properties and vary among species (Fassnacht et al., 2016).

Species-level discrimination cannot be usually performed with multispectral data, unless the vegetation is composed of a reduced number of species with clear phenological differences that can, for example, detected by multitemporal analysis, or species show large differences in leaf biochemical composition. Differences among species that show a distinct functional behavior can still be evidenced by multispectral data, as in the case of the response to light availability as shown here. Functional type information is ecologically important because it highlights the common response of certain group of plants to environmental influences. Shadow-tolerant species, pioneer species, and nonpioneer light demanding species are examples of groups with different leaf-level strategies in response to light availability, which result in a varied biochemical composition possibly detectable by certain remote sensing sensors. The comparison highlights the added value of hyperspectral data in tropical species discrimination, even when using data with a reduced spectral range that limits the separability. The results obtained with S2 data are encouraging, especially considering that they were based on a single image; the availability of temporal series certainly allows an increase in the ecological information that can be extracted from these data, thanks to species-level differences in phenological behavior.

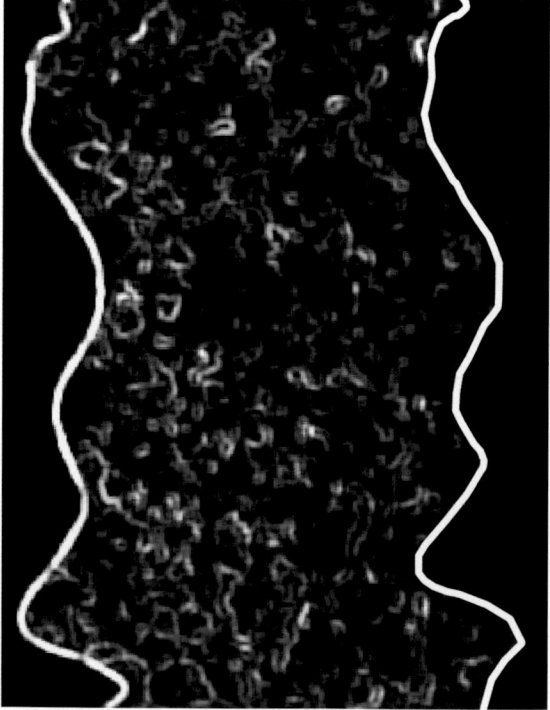

FIGURE 8.11 Falsecolor composite of hyperspectral variance textures from bands 1 (R), 5 (G), and 6 (B), which highlight crowns edges. (Adapted from Vaglio Laurin, G. et al. 2016. *Remote Sensing of Environment*, 176, 163–176.)

8.8 CONCLUSIONS

In this study we showed that feature depth analysis and spectral mixture analysis provided valuable insights on chemical and structural properties of individual tree crowns of tropical forest species. Specifically, the NPV, GV, and shade fractions were related to exposed branches, foliage, and gaps within the tree crown. The depth of absorption features over the 450–2500 nm spectral interval revealed chemical differences among the species. Interestingly, the nonpigment biochemical feature depth at 2100 nm was high for deciduous species such as *Diatenopteryx sorbifolia* in which the branches are exposed and the understory layer was sparse, allowing sunlight to reach the forest floor that is composed mainly of nonphotosynthetic material.

Hyperspectral data provided significantly higher classification results and separability among species than multispectral data. The incorporation of SWIR features systematically increased tree species classification accuracies, particularly simulated WV-3 bands, highlighting the potential of the SWIR sensing capabilities of this sensor for terrestrial plant species discrimination. The use of texture descriptors increased separability among species and its use in conjunction with hyperspectral and multispectral data is encouraged.

ACKNOWLEDGMENTS

This work was supported by the São Paulo Research Foundation (FAPESP) (grants n°. 2013/11.589-5 and n°. 2016/24977-1 to MPF) and by the Brazilian National Council for Scientific and Technological Development (CNPq) (grant n°. 303563/2008-7 to CRSF and grant n°. 301190/2013-5 to YES). We thank FotoTerra Atividades de Aerolevantamentos Ltda. for kindly providing the hyperspectral images used in this research. Any use of trade, firm, or product names is for descriptive purposes only and does not imply endorsement by the US government. The authors are grateful to Dr. Prasad Thenkabail for the invitation to contribute to the second edition of this seminal book and for revising the manuscript.

REFERENCES

Adams, J. B., Smith, M. O., and Gillespie, A. R. 1993. Imaging spectroscopy: Interpretation based on spectral mixture analysis. In C. M. Pieters, and P. A. J. Englert (Eds.), *Remote Geochemical Analysis: Elemental and Mineralogical Composition*. New York: Press Syndicate of the University of Cambridge.

Alcalá, M. 2010. *Ecologia da população de Pachystroma longifolium (NESS). I.M. JOHNST. em área fragmentada de um remanescente de Mata Atlântica* (Master dissertation (Ecology and Natural Resources)) São Carlos, Brasil: UFSCAR.

Álvarez-Yépiz, J. C., Búrquez, A., Martínez-Yrízar, A., Teece, M., Yépez, E. A., and Dovciak, M. 2017. Resource partitioning by evergreen and deciduous species in a tropical dry forest. *Oecologia*, 183(2), 607–618.

Amaral, C. H., Roberts, D. A., Almeida, T. I., and Souza Filho, C. R. 2015. Mapping invasive species and spectral mixture relationships with neotropical woody formations in southeastern Brazil. *ISPRS Journal of Photogrammetry and Remote Sensing*, 108, 80–93.

Baldeck, C. A., Asner, G. P., Martin, R. E., Anderson, C. B., Knapp, D. E., Kellner, J. R., and Wright, S. J. 2015. Operational tree species mapping in a diverse tropical forest with airborne imaging spectroscopy. *PloS one*, 10(7), e0118403.

Clark, M. L., Roberts, D. A., and Clark, D. B. 2005. Hyperspectral discrimination of tropical rain forest tree species at leaf to crown scales. *Remote Sensing of Environment*, 96, 375–398.

Clark, R. N. 1999. Spectroscopy of rocks and minerals, and principles of spectroscopy. *Manual of Remote Sensing*, 3, 3–58.

Clark, R. N., and Roush, T. L. 1984. Reflectance spectroscopy: Quantitative analysis techniques for remote sensing applications. *Journal of Geophysical Research: Solid Earth*, 89(B7), 6329–6340.

de Lima, R. A., Mori, D. P., Pitta, G., Melito, M. O., Bello, C. et al. 2015. How much do we know about the endangered Atlantic Forest? Reviewing nearly 70 years of information on tree community surveys. *Biodiversity and Conservation*, 24, 2135–2148.

D'Odorico, P., Gonsamo, A., Damm, A., and Schaepman, M. E. 2013. Experimental evaluation of Sentinel-2 spectral response functions for NDVI time-series continuity. *IEEE Transactions on Geoscience and Remote Sensing*, 51(3), 1336–1348.

Farah, F. T., Rodrigues, R. R., Santos, F. A. M., Tamashiro, J. Y., Shepherd, G. J., Siqueira, T., and Manly, B. J. F. 2014. Forest destructuring as revealed by the temporal dynamics of fundamental species–Case study of Santa Genebra Forest in Brazil. *Ecological Indicators*, 37, 40–44.

Fassnacht, F. E., Latifi, H., Stereńczak, K., Modzelewska, A., Lefsky, M., Waser, L. T., Straub, C., and Ghosh, A. 2016. Review of studies on tree species classification from remotely sensed data. *Remote Sensing of Environment*, 186, 64–87.

Felde, G. W., Anderson, G. P., Cooley, T. W., Matthew, M. W., Adler-Golden, S. M. et al. 2003. Analysis of Hyperion data with the FLAASH Atmospheric Correction Algorithm. *Geoscience and Remote Sensing Symposium 2003 IGARSS'03. Proceedings 2003 IEEE International*, 1. (pp. 90–92).

Féret, J. B., and Asner, G. P. 2013. Tree species discrimination in tropical forests using airborne imaging spectroscopy. *IEEE Transactions on Geoscience and Remote Sensing*, 51(1), 73–84.

Ferreira, M. P., Zortea, M., Zanotta, D. C., Shimabukuro, Y. E., and de Souza Filho, C. R. 2016. Mapping tree species in tropical seasonal semi-deciduous forests with hyperspectral and multispectral data. *Remote Sensing of Environment*, 179, 66–78.

Gamon, J. A., and Surfus, J. S. 1999. Assessing leaf pigment content and activity with a reflectometer. *The New Phytologist*, 143, 105–117.

Gamon, J. A., Penuelas, J., and Field, C. B. 1992. A narrow-waveband spectral index that tracks diurnal changes in photosynthetic efficiency. *Remote Sensing of Environment*, 41, 35–44.

Gandolfi, S., Joly, C. A., and Rodrigues, R. R. 2007. Permeability-impermeability: Canopy trees as biodiversity filters. *Scientia Agricola*, 64, 433–438.

Gardner, T. A., Barlow, J., Chazdon, R., Ewers, R. M., Harvey, C. A., Peres, C. A., and Sodhi, N. S. 2009. Prospects for tropical forest biodiversity in a human-modified world. *Ecology Letters*, 12, 561–582.

Gitelson, A. A., Merzlyak, M. N., and Chivkunova, O. B. 2001. Optical properties and nondestructive estimation of anthocyanin content in plant leaves. *Photochemistry and Photobiology*, 71, 38–45.

Gitelson, A. A., Zur, Y., Chivkunova, O. B., and Merzlyak, M. N. 2002. Assessing carotenoid content in plant leaves with reflectance spectroscopy. *Photochemistry and Photobiology*, 75, 272–281.

Green, A. A., Berman, M., Switzer, P., and Craig, M. D. 1988. A transformation for ordering multispectral data in terms of image quality with implications for noise removal. *IEEE Transactions on Geoscience and Remote Sensing*, 26, 65–74.

Hansen, M. C., Potapov, P. V., Moore, R., Hancher, M., Turubanova, S. A. et al. 2013. High-resolution global maps of 21st-century forest cover change. *Science*, 342(6160), 850–853.

Haralick, R. M. 1979. Statistical and structural approaches to texture. *Proceedings of the IEEE*, 67(5), 786–804.

Hawthorne, W. D. 1995. *Ecological Profiles of Ghanaian Forest Trees*. ODA tropical forestry papers 29. Oxford Forestry Institute 345 pp.

ITT Visual Information Solutions 2009. Atmospheric correction module: QUAC and FLAASH user's guide. Version 4.7. Boulder, CO: ITT Visual Information Solutions.

Japkowicz, N., and Stephen, S. 2002. The class imbalance problem: A systematic study. *Intelligent Data Analysis*, 6(5), 429–449.

Kailath, T. 1967. The divergence and Bhattacharyya distance measures in signal selection. *IEEE Transactions on Communication Technology*, 15, 52–60.

Kalacska, M., Bohlman, S., Sanchez-Azofeifa, G. A., Castro-Esau, K., and Caelli, T. 2007. Hyperspectral discrimination of tropical dry forest lianas and trees: Comparative data reduction approaches at the leaf and canopy levels. *Remote Sensing of Environment*, 109, 406–415.

Kaufman, Y. J., and Tanré, D. 1996. Strategy for direct and indirect methods for correcting the aerosol effect on remote sensing: From AVHRR to EOS-MODIS. *Remote Sensing of Environment*, 55(1), 65–79.

Kokaly, R.F. 2011. *PRISM:* Processing routines in IDL for spectroscopic measurements (installation manual and user's guide, version 1.0): U.S. Geological Survey Open-File Report 2011–1155, 432 p., available at https://pubs.usgs.gov/of/2011/1155/, last accessed February 2, 2018.

Kokaly, R. F., and Clark, R. N. 1999. Spectroscopic determination of leaf biochemistry using band-depth analysis of absorption features and stepwise multiple linear regression. *Remote Sensing of Environment*, 67(3), 267–287.

Lorenzi, H. 2014. *Árvores brasileiras: Manual de identificação e cultivo de plantas arbóreas nativas do Brasil*, vol. 1, Instituto Plantarum de Estudos da Flora, São Paulo.

Morellato, L. P. C. 1991. Estudo da fenologia de árvores, arbustos e lianas de uma floresta semidecídua no sudeste do Brasil. (*PhD. dissertation (Ecology)*) Campinas, Brasil: UNICAMP.

Nave, A. G. 1999. Determinação de unidades ecológicas num fragmento de floresta nativa, com auxílio de sensoriamento remoto (*Master dissertation (Forest science)*) Piracicaba, Brasil: ESALQ/USP.

Oliveira-Filho, A. T., and Ratter, J. A. 1995. A study of the origin of central Brazilian forests by the analysis of plant species distribution patterns. *Edinburgh Journal of Botany*, 52, 141–194.

Pan, Y., Birdsey, R. A., Fang, J., Houghton, R., Kauppi, P. E. et al. 2011. A large and persistent carbon sink in the world's forests. *Science*, 333, 988–993.

Richards, J. A., and Jia, X. 1999. *Remote Sensing Digital Imaging Analysis: An Introduction* (3rd ed.). Berlin: Springer.

Roberts, D. A., Gardner, M., Church, R., Ustin, S., Scheer, G., and Green, R. O. 1998. Mapping Chaparral in the Santa Monica Mountains using multiple endmember spectral mixture models. *Remote Sensing of Environment*, 65, 267–279.

Roberts, D. A., Halligan, K. Q., and Dennison, P. E. 2007. ViperTools. Retrieved from http://www.vipertools. org/, last Aaccessed 17 February 2, 2018)

Rouse, J.W., Haas, R.H., Schell, J.A., and Deering, D.W. 1973. Monitoring vegetation systems in the Great Plains with ERTS. In: *Proceedings of Third ERTS-1 Symposium*, vol. 1, Washington, DC, 10–14 December, NASA, SP-351, pp. 309–317.

Saatchi, S., Mascaro, J., Xu, L., Keller, M., Yang, Y., Duffy, P., Espírito-Santo, F., Baccini, A., Chambers, J., and Schimel, D. 2015. Seeing the forest beyond the trees. *Global Ecology and Biogeography*, 24(5), 606–610.

Seiffert, C., Khoshgoftaar, T. M., Van Hulse, J., and Napolitano, A. 2010. RUSBoost: A hybrid approach to alleviating class imbalance. *IEEE Transactions on Systems, Man, and Cybernetics-Part A: Systems and Humans*, 40(1), 185–197.

Sellers, P. J. 1985. Canopy reflectance, photosynthesis and transpiration. *International Journal of Remote Sensing*, 6(8), 1335–1372.

Sims, D. A., and Gamon, J. A. 2002. Relationships between leaf pigment content and spectral reflectance across a wide range of species, leaf structures and developmental stages. *Remote Sensing of Environment*, 81(2), 337–354.

Sobrado, M. A. 1986. Aspects of tissue water relations and seasonal changes of leaf water potential components of evergreen and deciduous species coexisting in tropical dry forests. *Oecologia*, 68(3), 413–416.

Theodorsson-Norheim, E. 1986. Kruskal-Wallis test: BASIC computer program to perform nonparametric one-way analysis of variance and multiple comparisons on ranks of several independent samples. *Computer Methods and Programs in Biomedicine*, 23(1), 57–62.

Vaglio Laurin, G., Puletti, N., Hawthorne, W., Liesenberg, V., Corona, P., Papale, D. and Valentini, R. 2016. Discrimination of tropical forest types, dominant species, and mapping of functional guilds by hyperspectral and simulated multispectral Sentinel 2 data. *Remote Sensing of Environment*, 176, 163–176.

Verrelst, J., Muñoz, J., Alonso, L., Delegido, J., Rivera, J. P., Camps-Valls, G., and Moreno, J. 2012. Machine learning regression algorithms for biophysical parameter retrieval: Opportunities for Sentinel-2 and-3. *Remote Sensing of Environment*, 118, 127–139.

Vogelmann, J. E., Rock, B. N., and Moss, D. M. 1993. Red edge spectral measurements from sugar maple leaves. *International Journal of Remote Sensing*, 14(8), 1563–1575.

Wang, Y. P., and Houlton, B. Z. 2009. Nitrogen constraints on terrestrial carbon uptake: Implications for the global carbon-climate feedback. *Geophysical Research Letters*, 36, L24403.

9 Detecting and Mapping Invasive Plant Species Using Hyperspectral Data

Ruiliang Pu

CONTENTS

9.1 INTRODUCTION

Invasions of plant species have caused significant changes in structure and function of ecosystems. Biological invasions have been identified as a major nonclimatic driver of global change [1,2]. Such changes usually have negative impacts on ecological functions of natural ecosystems at various scales and become a serious problem to environment-friendly sustainable ecosystems all over the world [3–6]. Invasive plant species (IPS) have caused costs associated with environment, economy, and culture. For example, in the United States, the estimated cost of environmental damage and associated management and control of IPS is about $137 billion per year; the total amount could be several times more if one considers native species extinctions, biodiversity reduction, ecosystem services, and aesthetics [7]. Therefore, monitoring the invasion extent and speed of the invasion by IPS and eradicating IPS across their invaded areas is an important task. Traditionally, this task relies heavily on field-based investigations and on methods that are usually expensive and time-consuming [6].

However, remote sensing techniques, especially hyperspectral remote sensing, offer an effective and economical alternative for gathering spatially distributed data over a large area and have potential for detecting and mapping IPS [8–13]. With remote sensing techniques, a large area of coverage can be acquired in a short period of time. Hyperspectral remote sensing, with its spectral information associated with phenological and structural characteristics of specific invasive species, allows for the species-level detection necessary to map invasive species [10,14,15]. During the last two decades, researchers have had an increasing interest in studies on the hyperspectral detection and mapping of invasions of aquatic species (e.g., [16,17]), grasses and weeds (e.g., [9,18–21]), scrubs and shrubs (e.g., [10,12,13,15,22,23]), and trees (e.g., [6,24–28]).

Based on a review of the existing literature, this chapter provides an overview on hyperspectral remote sensing techniques for detecting and mapping IPS. The main objectives of this chapter are:

- To review suitable techniques/methods and their applications in detecting and mapping IPS with hyperspectral data
- To address relevant considerations for detecting and mapping IPS
- To discuss challenges and indicate future directions for identifying and mapping invasive species using hyperspectral remote sensing techniques.

9.2 POTENTIAL FOR DETECTING AND MAPPING INVASIVE PLANT SPECIES

9.2.1 Physiological and Phenological Characteristics of IPS

Plant infestation can make several impacts on various ecosystems, such as (1) enhancing fire frequency and severity [29]; (2) increasing soil salinity, which reduces productivity of native plants and results in the loss of natural habitat [7]; (3) consuming soil water to such an extent that it can dry up streams and reduce water levels of rivers and lakes [30]; (4) depleting nutrients and making changes in microclimate and alterations in vegetation succession [31]; and (5) increasing nitrogen (N) deposition [32]. These impacts caused by IPS can directly (e.g., fire frequency and intensity) or indirectly (e.g., depleting nutrients and altering vegetation succession) mirror spectral differences compared with native species.

Compared with native species, most IPS usually have different phenological characteristics. Remote sensing methods have potential to detect such phenological differences, such as peak bloom, bud break, or senescence. Numerous researchers have investigated relationships between plant phenology and environmental factors, which could be exploited for species identification. For example, Ramsey et al. [25,26] applied the five characteristic spectra (senescing foliage, canopy shadow, green vegetation, yellow foliage, and red tallow) to the corrected and normalized Hyperion image data to identify an invasive species, Chinese tallow. Phenological timing information extracted from multitemporal hyperspectral data (two-seasonal image data for each of the CASI and AVIRIS sensors) helps detect tamarisk [12] via classification and cheatgrass [18] via color changes and altered seasonality relative to native species.

9.2.2 Canopy Structure and Biochemistry of IPS

Recent work has shown that invasive tree species often express biochemical and physiological properties distinct from those of native trees [27,33,34]. Resolving these particular leaf and canopy characteristics in remotely sensed imagery may provide a way to map and monitor invaders at the regional scale. For example, Asner et al. [27,28] found that invasive tree species have hyperspectral reflectance signatures distinct from those of native tree species. In their work, canopy reflectance properties in the 400–2500 nm wavelength range, collected from an airborne hyperspectral sensor, Airborne Visible and Infrared Imaging Spectrometer (AVIRIS), demonstrated spectral separability of native, introduced, and highly invasive species. Their further analyses showed systematic, wavelength-dependent spectral reflectance differences between plant functional types, such as nitrogen-fixing (*Morella faya*) and nonfixing trees (native species). Most importantly, they showed that the spectral separability of species was tightly linked to their biochemical composition associated with relative differences in measured leaf pigment (chlorophyll, carotenoids), nutrient (N, P), and structural (specific leaf area) properties, as well as to canopy leaf area index [27].

Others have found diagnostic characteristics of invasive plant species. With AVIRIS hyperspectral data, Underwood et al. [10] found that quantification of the water absorption feature (via the continuum removal approach) was sufficient to map the invasive succulent iceplant. Hunt et al. [35,36] detected the invasive species leafy spurge through estimating the pigment content of leaves and floral bracts from simulated, *in situ*, and hyperspectral image (AVIRIS) data.

9.3 TECHNIQUES AND METHODS

The goal of detecting and mapping IPS using hyperspectral data is to employ suitable analysis techniques and methods to compare spectral signatures characterized in spatial, spectral, and temporal domains, derived from hyperspectral data, between invasive species and native (background) species so as to separate them spatially. Focusing more on application, this section will review the techniques/methods that currently appear in the existing literature and comprising seven techniques/methods: spectral derivative analysis, spectral matching, vegetation index analysis, absorption analysis, hyperspectral transformation, spectral mixture analysis, and classification. For convenience, each technique/method will be reviewed by describing the characteristics of the technique/method itself, briefly summarizing the advantages and/or disadvantages of the technique/method when applied in detecting and mapping IPS with hyperspectral data with a few typical application cases, and listing several major factors to be considered when it is applied in practice. For the seven techniques/methods, the main characteristics, advantages and disadvantages, major factors affecting detecting and mapping results, and some application examples are provided in Table 9.1. The hyperspectral data used for implementing the techniques/methods include *in situ* spectral measurements taken with various spectrometers and image data acquired with airborne and spaceborne systems/sensors.

9.3.1 DERIVATIVE ANALYSIS

In situ or imaging hyperspectral data obtained in the field are rarely from a single object. They are contaminated by illumination variations caused by terrain relief, cloud, and viewing geometry [37]. For the first- and second-order derivative spectra, a finite approximation [38] can be applied to calculate them from hyperspectral data. The derivative spectrum is the normalized spectral difference of two continuous or neighboring narrow bands with their wavelength interval. Spectral derivative analysis has been considered a desirable tool for removing or compressing the effect of illumination variations of low frequency on target spectra, but it is sensitive to the signal-to-noise ratio (SNR) of hyperspectral data and higher-order spectral derivative processing is susceptible to the noise [39]. In other words, lower-order derivatives (e.g., the first-order derivative) are less sensitive to noise and hence more effective in operational remote sensing. When implementing spectral derivative analysis, the spectral resolution is required to be narrower than 10 nm and spectral bands to be continuous.

Derivative spectra have been successfully employed in hyperspectral data analysis for estimating biophysical and biochemical parameters, thus aiding in detecting and mapping IPS. For example, Asner et al. [27] conducted the spectral separability analysis between Hawaiian native and introduced (invasive) tree species with AVIRIS hyperspectral image data for detecting and assessing invasive species. They observed that the spectral differences (measured in reflectance, first- and second-derivative spectra, see Figure 9.1) in canopy spectral signatures are linked to relative differences in leaf pigment (chlorophyll, carotenoids), nutrient (N, P), and structural (specific leaf area) properties, as well as to canopy leaf area index (LAI). These relative differences associated with leaf and canopy properties of trees are helpful in separating invasive species from the background (native) species. Figure 9.1 shows that there is a greater derivative spectral difference between the invasive and introduced species than that of their reflectance spectra in the regions of the gray bars. Combining *in situ* spectral measurements with Airborne Imaging Spectroradiometer for Application (AISA) data, Wang et al. [19] mapped an invasive weed (*Sericea lespedeza*) in a public grass field in mid-Missouri, USA. The maximal first-order derivative in the red–near-infrared region (650–800 nm)

TABLE 9.1

Summary of Techniques Suitable for Detecting and Mapping Invasive Plant Species by Using Hyperspectral Data

Technique	Characteristics and Description	Advantages and Disadvantages	Major Factor	Typical Example [Ref.]
1. Derivative analysis	Normalized spectral difference of two continuous/neighbor narrow bands with their wavelength interval.	Removes or compresses the effect of illumination variations with low frequency on target spectra but sensitive to the SNR of hyperspectral data and higher-order spectral derivative processing is susceptible to the noise.	Spectral resolution <10 nm and also continuous, right threshold.	[19,27]
2. Spectral matching	With n-dimensional angles (distances, correlations) to match pixels to reference spectra with smaller angles (shorter distances, higher correlations) representing closer matches to the reference spectrum, otherwise not matching the reference spectrum.	It is a physically based spectral classification and is less sensitive to differences in curve magnitude caused by variation in lighting across a scene, but it is sensitive to noise in any particular band.	Determination of threshold of angle (distance, correlation).	[9,41,23,43]
3. Vegetation index analysis	Calculates the ratio of images of two bands or normalized difference of two or more than two bands.	Easy to use and reduce impact of sun angle, atmosphere, shadow, topography. However, VI image usually is not normal.	Identify suitable bands to construct VIs.	See Table 9.2
4. Absorption feature analysis	Caused by a combination of factors inside and outside the material surface including electronic processes, molecular vibrations, abundance of chemicals, granular size and physical structure, and surface roughness relative to electromagnetic wavelength.	Absorption features directly linked to material structure, constituents, concentration/content. Requires images with higher SNR.	Spectral resolution <10 nm and also continuous.	[10,17,22]
5. Hyperspectral transformation	A linear or nonlinear combination of raw data to reduce dimensionality and preserve variance contained in raw data as much as possible in the first several components images	Dimension reduction and usefully informative feature extraction; not easy to identify which are more signal components.	Identify informative features/components	[6,10,12,18]
6. Spectral mixture analysis	Spectral reflectances from different materials (>1 material) within a pixel are recorded as one spectral response (mixed spectrum). Use of linear or nonlinear spectral mixture model to derive fraction (end-member) images from mixed pixels.	The fraction representing the areal proportion of each end-member, but one does not know where the proportioned areas locate within a mixed pixel and sometime it is difficult to obtain end-member spectra and know all end-members in a scene.	Identify suitable/pure end-members and extract their individual spectra	[20,26,28,69]
7. Classification	Using supervised/unsupervised methods, parametric/nonparametric algorithms to assign one pixel or image-object into one of the classes (species). Usually for classifying hyperspectral data, dimension reduction and feature extraction are first considered.	There is a basis in statistics/probability or a rule for classification; usually it is difficult to obtain adequate training samples for supervised methods and labeling of unsupervised spectral clusters.	Identify suitable method/classifier for a specific task and gather adequate training samples.	[6,17,22,55]

Note: SNR, signal-to-noise ratio; VI, vegetation index.

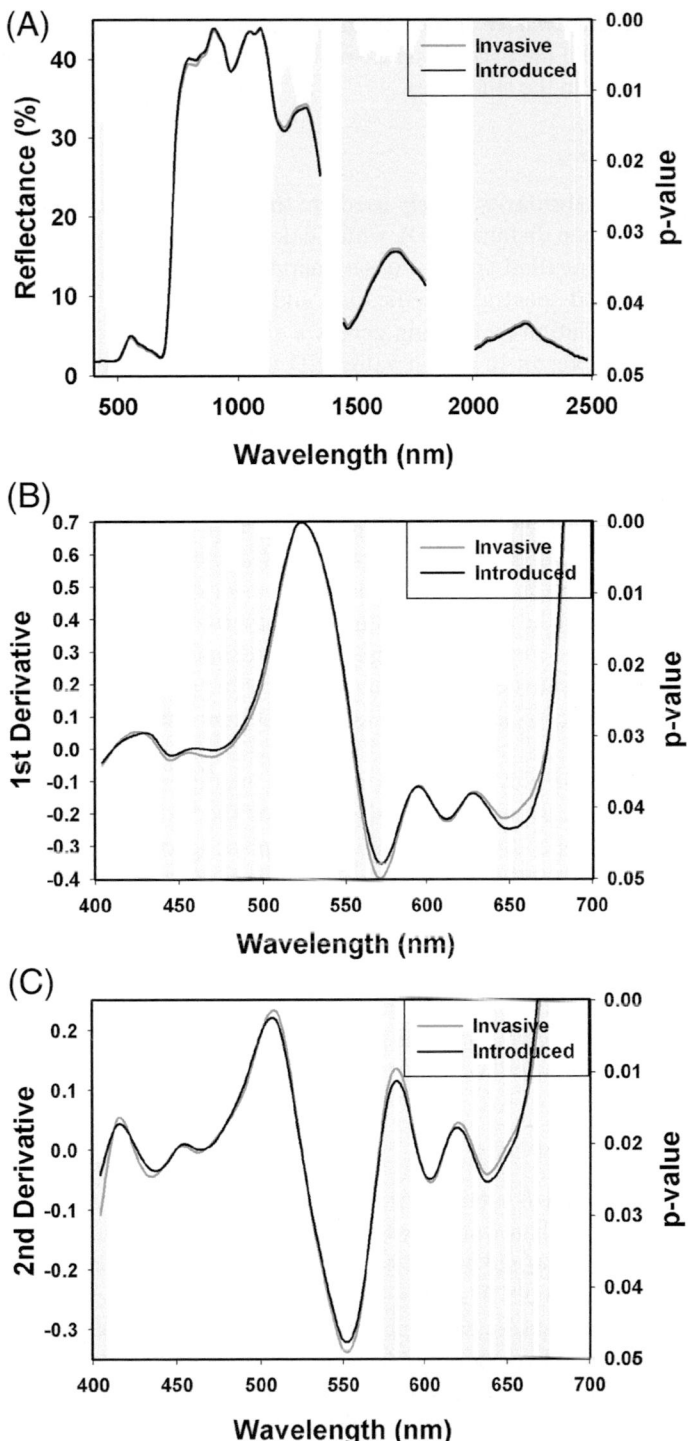

FIGURE 9.1 (A) Reflectance, (B) first-derivative, and (C) second-derivative spectra of highly invasive vs. other introduced tree species of Hawaiian tropical and subtropical forests, with band-by-band *t*-tests showing significant differences in the gray bars (*p*-values <0.05). (Reprinted from *Remote Sensing of Environment*, 112, Asner, G.P. et al., Remote sensing of native and invasive species in Hawaiian forests, 1912–1926, Copyright 2008, with permission from Elsevier [27].)

was derived to separate the invasive species from the target grass in pastures in Missouri. With a simple threshold approach for the maximum first-order derivative spectrum, *Sericea* of various sizes was successfully identified in the study area.

9.3.2 Spectral Matching

Two measures of spectral similarity widely used are the spectral angle, calculated between two curve vectors, and Euclidean distance (ED), which calculates the vector distance. The algorithm that uses spectral angle is called spectral angle mapper (SAM). It is commonly believed that SAM is a physically based spectral classification and is less sensitive to differences in curve magnitude caused by variation in lighting across a scene. As long as the shapes of spectral curves are similar, it will result in a high value. ED is a widely used measure that detects an absolute difference between two spectral curves. It is sensitive to the noise in any individual band spectrum.

A third measure of spectral similarity, named spectral correlation measure (SCM; also termed a cross correlogram spectral matching [CCSM]), as a modified version of SMA and ED, is calculated with the vector cross correlation between a known reference spectrum and an unknown target spectrum [40,41]. Spectral matching uses *n*-dimensional angles (distances, cross correlations) to match pixels to reference spectra, with smaller angles (shorter distances, higher cross correlations) representing closer matches to the reference spectrum, or otherwise representing non-matching to the reference spectrum. Usually, a key factor in applying the technique for detecting and mapping IPS with hyperspectral data is how to determine a suitable threshold of angle (or distance or correlation). Spectral matching techniques to identify and label vegetation categories have been discussed in detail elsewhere [93,94].

Many researchers have employed spectral matching methods to identify and map IPS. For example, with AISA airborne hyperspectral image data and the SAM analysis method, Narumalani et al. [23] quantified and mapped four dominant invasive plant species, including saltcedar, Russian olive, Canada thistle, and musk thistle, along the flood plain of the North Platte River, Nebraska, USA. Validation procedures confirmed an overall map accuracy of 74%. Also with the same analysis method (SAM) and airborne hyperspectral image data, Lass and Prather [42] detected the location of Brazilian pepper trees in the Everglades, and Hirano et al. [43] mapped wetland vegetation with the invasive species lather leaf (*Colubrina asiatica*). In addition, Pengra et al. [41] used the SCM to map an invasive plant, *Phragmites australis*, in coastal wetland using the Hyperion hyprspectral sensor's data to show a good overall accuracy of 81.4%.

9.3.3 Vegetation Index Analysis

When using hyperspectral data to conduct spectral vegetation index (VI) analysis, we can make use of the advantage of increasing chance and flexibility to choose spectral bands. With multispectral data one may have only one choice of red and NIR bands to use. However, with hyperspectral data, one can choose many of such red and NIR narrow-band combinations [44,95–98]. Accordingly, spectral VIs applied to hyperspectral data are called narrow-band VIs [45–47]. Table 9.2 lists a set of 21 VIs that have been developed from hyperspectral data and suitable for detecting and mapping IPS. These VIs have appeared in existing literature. For the reader's convenience in locating a VI or group of VIs, the 21 VIs are organized into three categories based on the characteristics and functions of the VIs: multiple bioparameters, pigments, and foliar chemistry. Within these individual categories, the VIs are arranged in alphabetical order. The explicit advantages of VIs are that they are easy to use and can also reduce the impact of sun angle, atmosphere, shadow, and topography on target spectra. However, the probability distribution of VI image for different land cover types usually is not normal. A key factor in determining the usefulness of a VI depends on identifying suitable wavebands.

Developing various VIs (Table 9.2) from hyperspectral data helps detect and map invasive species. For example, Hestir et al. [17] used mSR_{705} VI to map invasive species (three weeds: perennial pepperweed, water hyacinth, and Brazilian waterweed) with airborne hyperspectral data (HyMap). They achieved a moderate to high mapping accuracy. Andrew and Ustin [20,21] used 13 VIs, derived from HyMap data, together with a minimum noise fraction (MNF) transformation, mixture tuned matched filtering (MTMF), and a decision tree classification method (CART: classification and regression tree) to identify and map the perennial pepperweed from three sites of California's San Francisco Bay/Sacramento–San Joaquin Delta Estuary. Their approach was sufficiently flexible and robust to detect the invasive species with similar accuracies (~90%) at both Rush Ranch and Jepson Prairie sites, but was unsuccessful at Cosumnes River Preserve due to the different environmental context. Also using VIs derived from airborne hyperspectral image data (AVIRIS and CASI), Underwood et al. [10] detected and mapped iceplant and jubata grass in California coastal habitat, and Pu et al. [13] mapped change of saltcedar after a biocontrol measure was taken by USDA. Both studies show high mapping accuracy. Compiling a time series of Hyperion image data, Asner et al. [32] studied the variations in upper-canopy leaf chlorophyll and carotenoid content during a climatological transition with the remotely sensed photochemical and carotenoid reflectance indices (PRI, CRI). They found that the PRI and CRI were related to differences in light-use efficiency between invasive and native tree species and thus helped in separating invasive species (*Myrica faya*) from native species (*Metrosideros polymorpha*).

9.3.4 ABSORPTION FEATURES ANALYSIS

Quantitative characterization of absorption features allows for abundance estimation of materials from hyperspectral data. Spectral absorption features mirror diagnostic spectral features of materials [48] and are caused by a combination of factors inside and outside the material surface, including electronic processes, molecular vibrations, abundance of chemical constituents, granular size and physical structure, and surface roughness relative to electromagnetic wavelength. Therefore, the absorption features are directly linked to material structure, constituents, concentration, and content. For example, the central wavelengths of *in situ* chlorophyll-*a* absorption are at 0.45 and 0.67 µm; the central wavelengths of water absorption are near 0.97, 1.20, 1.40, and 1.94 µm; the central wavelengths of N absorption are near 1.51, 2.06, 2.18, 2.30, and 2.35 µm; the central wavelengths of lignin absorption features are near 1.12, 1.42, 1.69, and 1.94 µm; and the central wavelengths of cellulose absorption features are near 1.20, 1.49, 1.78, 1.82, 2.27, 2.34, and 2.35 µm [49]. Extraction of these absorption features requires hyperspectral data with higher SNR and spectral resolution <10 nm and continuous neighbor bands.

In order to analyze the absorption features of a spectral reflectance curve, one needs to normalize the spectral curve so that the spectral values inside the absorption features will be less than 1 (100%). This can be done using a continuum removal technique proposed by Clark and Roush [50]. Quantitative measures can be determined from each absorption peak after the normalization of the raw spectral reflectance curve. The quantitative measures can be used to determine abundances of certain compounds in a pixel thus linked to characteristics of plant species. For example, Hamada et al. [22] explored the effectiveness of the depth of the chlorophyll absorption feature of tamarisk species from airborne hyperspectral image data with the continuum removal technique. They combined the absorption feature and other spectral variables extracted from the hyperspectral data with a parallelepiped classifier to yield the most accurate and reliable tamarisk classification products. Underwood et al. [10] successfully used the continuum removal technique to extract water absorption feature from AVIRIS hyperspectral image data to map nonnative species—iceplant and jubata grass—in California's coastal habitat. With the same technique, Hestir et al. [17] also estimated foliar water absorption from hyperspectral image data HyMap to successfully indentify and map invasive species in the California Delta ecosystem.

TABLE 9.2

Summary of 21 Spectral Indices Extracted from Hyperspectral Data Used in Detecting and Mapping Invasive Plant Species

Spectral Index	Characteristics and Functions	Definition	Reference
	Multiple Bioparameters		
LI, lepidium index	Sensitive to the uniformly bright reflectance displayed by *Lepidium* in the visible range.	R_{630}/R_{586}	[20]
NDVI, normalized difference vegetation index	Responds to change in the amount of green biomass and more efficiently in vegetation with low to moderate density.	$(R_{NIR} - R_R)/(R_{NIR} + R_R)$	[74]
PSND, pigment-specific normalized difference	Estimate LAI and carotenoids (Cars) at leaf or canopy level	$(R_{800} - R_{470})/(R_{800} + R_{470})$	[74,75]
SR, simple ratio	Same as NDVI	R_{NIR}/R_R	[76,77]
	Pigments		
Chl$_{green}$, chlorophyll index using green reflectance	Estimate chlorophyll (Chl) content in anthocyanin-free leaves if NIR is set beyond 760 nm	$(R_{760-800}/R_{540-560}) - 1$	[78]
Chl$_{red edge}$, chlorophyll index using red edge reflectance	Estimate Chl content in anthocyanin-free leaves if NIR is set	$(R_{760-800}/R_{690-720}) - 1$	[78]
LCI, leaf chlorophyll index	Estimate Chl content in higher plants, sensitive to variation in reflectance caused by Chl absorption	$(R_{850} - R_{710})/(R_{850} + R_{680})$	[79]
mND$_{680}$, modified normalized difference	Quantify Chl content and sensitive to low content at leaf level.	$(R_{800} - R_{680})/(R_{800} + R_{680} - 2R_{445})$	[80]
mND$_{705}$, modified normalized difference	Quantify Chl content and sensitive to low content at leaf level. mND$_{705}$ performance better than mND$_{680}$	$(R_{750} - R_{705})/(R_{750} + R_{705} - 2R_{445})$	[80,81]
mSR$_{705}$, modified simple ratio	Quantify Chl content and sensitive to low content at leaf level.	$(R_{750} - R_{445})/(R_{705} - R_{445})$	[80]
NPCI, normalized pigment chlorophyll ratio index	Assess Cars/Chl ratio at leaf level	$(R_{680} - R_{430})/(R_{680} + R_{430})$	[82]
PBI, plant biochemical index	Retrieve leaf total Chl and nitrogen concentrations from satellite hyperspectral data	R_{810}/R_{560}	[83]
PRI, photochemical/ physiological reflectance index	Estimate Car pigment contents in foliage	$(R_{531} - R_{570})/(R_{531} + R_{570})$	[84]
PI2, pigment index 2	Estimate pigment content in foliage	R_{695}/R_{760}	[85]
RGR, red:green ratio	Estimate anthocyanin content with a green and a red band	R_{683}/R_{510}	[80,86]
SGR, summed green reflectance	Quantify Chl content	Sum of reflectances from 500 to 599 nm	[81]
	Foliar Chemistry		
CAI, cellulose absorption index	Cellulose and lignin absorption features, discriminates plant litter from soils	$0.5(R_{2020} + R_{2220}) - R_{2100}$	[87]
NDLI, normalized difference lignin index	Quantify variation of canopy lignin concentration in native shrub vegetation	$[\log(1/R_{1754}) - \log(1/R_{1680})]/[\log(1/R_{1754}) + \log(1/R_{1680})]$	[88]
NDWI, normalized difference water index	Improving the accuracy in retrieving the vegetation water content at both leaf and canopy levels	$(R_{860} - R_{1240})/(R_{860} + R_{1240})$	[89,90]
RVI$_{hyp}$, hyperspectral ratio vegetation index	Quantify LAI and water content at canopy level.	R_{1088}/R_{1148}	[91]
WI, water index	Quantify relative water content at leaf level	R_{900}/R_{970}	[92]

9.3.5 HYPERSPECTRAL TRANSFORMATION

Hyperspectral transformation is a linear or nonlinear combination of raw data to reduce dimensionality and preserve variance as much as possible in the first several component images. In detecting and mapping IPS with hyperspectral data, the most popular transformation techniques are principal component analysis (PCA) and its modified version termed maximum noise fraction (MNF). The PCA technique has been applied to reduce the data dimension and feature extraction from hyperspectral data for assessing leaf or canopy parameters (e.g., [51,52]). With a covariance (or correlation) matrix calculated from vegetated pixels only, it is commonly believed that the eigenvalues and corresponding eigenvectors of the first several principal component images, computed from the covariance (or correlation) matrix, are expected to be able to enhance vegetation variation. Green et al. [53] developed the transform method called MNF transform to maximize the SNR when choosing principal components with increasing component number. Then several MNFs with maximum SNR are selected for further analysis of hyperspectral data, such as for determining end-member spectra for spectral mixture analysis [54,55]. Although canonical discriminant analysis (CDA) and the wavelet transform (WT) have been more attractive for dimension reduction and feature extraction for classification recently (e.g., [52,56]), they are not seen to be used currently in detecting and mapping IPS with hyperspectral data. Accordingly, their applications as transformation techniques will not be reviewed here. In general, for all the transformation methods, identification of informative features/components is still a difficult task because some subtle and useful features may not be included in first several component images [17,20].

During the last two decades, PCA and MNF techniques have been extensively applied for extracting spectral features for detecting and mapping IPS with either spectral mixture analysis (SMA) or other supervised classifiers such as maximum likelihood classifier (MLC) and artificial neural networks (ANN). For example, Pu et al. [12] and Tsai et al. [6] used PCA (segmented PCA) transformation method for extracting several important component images from Compact Airborne Spectrographic Imager (CASI) and Hyperion hyperspectral image data to successfully detect and map the invasive species saltcedar and *Leucaena leucocephala* in Lovelock, Nevada, USA, and in southern Taiwan, respectively. For detecting and mapping four invasive species—saltcedar, Russian olive, Canada thistle, and musk thistle—along the flood plain of the North Platte River, Nebraska, USA, Narumalani et al. [23] transformed AISA airborne hyperspectral image data with the MNF technique to extract first several minimum noise fraction images that were used as input for running SAM to separate the four invasive species. The overall accuracy of mapping the four species was 74%. Underwood et al. [10,15] employed several MNFs transformed from AVIRIS hyperspectral image data in coastal California to map nonnative plant species (iceplant, jubata grass, and blue gum). They used a standard supervised classifier, MLC with the several MNFs as an input to map the three invasive species, which resulted in a higher accuracy. For mapping some invasive species in the California Delta ecosystem, Hestir et al. [17] also transformed HyMap hyperspectral images to extract several MNFs, together with other extracted spectral variables, as inputs to run a few supervised classifiers to map the invasive species. They achieved moderate to high success in the task.

9.3.6 SPECTRAL MIXTURE ANALYSIS

Spectral reflectances measured from different materials within a pixel are recorded as one spectral response or as a mixed spectrum. A large portion of remotely sensed data is spectrally mixed because the spatial resolution (pixel size) of image data cannot resolve individual materials. In order to identify various "pure materials" and to determine their spatial proportions from the remotely sensed data, the spectral mixing process has to be properly modeled. Then the model can be inverted to derive the spatial proportions and spectral properties of those "pure materials."

There are two types of spectral mixing: linear spectral mixing and nonlinear spectral mixing. Linear spectral mixing modeling and its inversion have been widely used since the late 1980s. A linear spectral mixture model analysis (SMA) has been extensively applied to extract the abundance of various components within mixed pixels. Nonlinear spectral mixture model can be found in Sasaki et al. [57] and Zhang et al. [58].

In a linear spectral mixture model analysis, there are two solutions: a linear least-square solution and a nonlinear solution (e.g., a neural network-based, nonlinear, subpixel classifier by Walsh et al. [55]). At present, since the SMA method is easy to use, it has been widely and successfully applied for mapping the abundance of a certain number of invasive species with hyperspectral image data. For instance, with airborne and spaceborne hyperspectral image data (AVIRIS, HyMap, and Hyperion), the invasive species that were detected and mapped include Brazilian pepper [42], perennial pepperweed, water hyacinth, and Brazilian waterweed [16,17], Chinese tallow [25,26], and *Psidium guajava* [55].

In addition, mixture tuned matched filtering (MTMF) is an advanced spectral unmixing algorithm that does not require that all materials within a scene are known and have been identified as end-members [59]. This method represents an improved alternative to SMA analysis for cases where the number of similar spectra are large or where it is problematic to collect spectra of all potential end-member components within the scene. MTMF treats each end-member independently and, at each pixel for each end-member, models the pixel as a mixture of the end-member and an undefined background material. It outputs a matched filter (MF) score and an infeasibility value for each end-member. The MF score is analogous to the fraction value from simple SMA, and the infeasibility is a measure of how likely a pixel is to contain the material of interest. Pixels are likely to contain materials for which they receive high MF scores and low infeasibilities [20]. MTMF has proven to be a very powerful tool to detect specific materials that differ slightly from the background. For example, with airborne hyperspectral image data (e.g., AVIRIS and HyMap), MTMF has successfully mapped a variety of invasive species, including tamarisk [22], perennial pepperweed [20,21], and cheatgrass [18]. Figure 9.2 presents a map of the *Lepidium* infestation detected at Rush Ranch, California, USA, overlaid on a true-color image of the site, created with MTMF outputs. With the classification algorithm, CART, the MTMF outputs were the most important variables for *Lepidium* detection [20].

As a result of spectral unmixing, the fractions represent the areal proportions of end-members, but we do not know where the proportioned areas of the then end-members locate within a mixed pixel. A key factor to unmix mixed spectra is to identify suitable/pure end-members and extract their individual spectra for training and test purposes.

9.3.7 Hyperspectral Image Classification

To overcome difficulties faced by traditional classifiers, caused by high dimensionality of hyperspectral data and high correlation of adjacent bands with limited number of training samples, it is necessary to reduce dimension and extract features for classifying IPS. Based on review of existing literature, there are currently several hyperspectral transformation techniques/methods (see Section 9.3.5) that have been successfully applied in dimension reduction and feature extraction for classifying plant species. These transformation techniques/methods include PCA, MNF, linear discriminant analysis (LDA), CDA, and WT, among others. Theoretically, after features/component images in a relatively lower dimension are produced by running appropriate transformation algorithms, some supervised algorithms, such as MLC, LDA, ANN, and classification and regression tree (CART), and unsupervised methods, such as ISODATA, are utilized for detecting and mapping a variety of invasive species. For example, after extracting several MNFs from the AVIRIS hyperspectral image

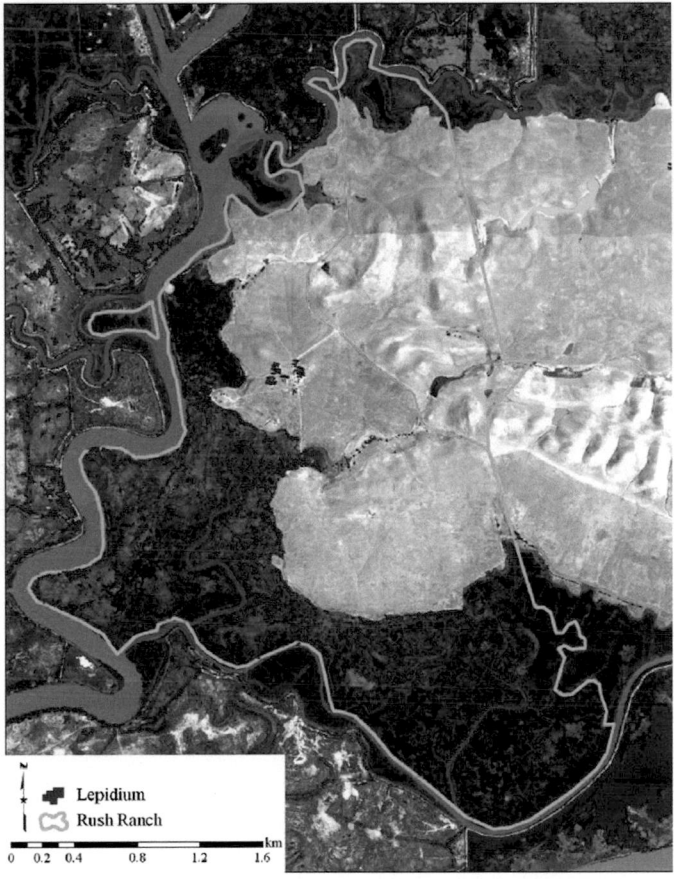

FIGURE 9.2 Map of the *Lepidium* infestation detected at Rush Ranch, California, overlaid on a true-color image of the site, created with MTMF outputs. (Reprinted from *Remote Sensing of Environment*, 112, Andrew, M.E., and Ustin, S.L., The role of environmental context in mapping invasive plants with hyperspectral image data, 4301–4317, Copyright 2008, with permission from Elsevier [20].)

data, Underwood et al. [10,15] used the MLC to classify three nonnative plant species (iceplant, jubata grass, and blue gum) in California's coastal areas. With several PC images transformed from CASI and Hyperion hyperspectral image data, Pu et al. [12, Figure 9.3] utilized ANN and LDA algorithms to map saltcedar invasive species, and Tsai et al. [6] used MLC to map *Leucaena leucocephala*. Figure 9.3 presents change maps of saltcedar invasive species from July, 2002 (JUL02) to August, 2002 (AUG02), from AUG02 to September, 2003 (SEP03), and from JUL02 to SEP03 with multitemporal CASI hyperspectral image data. In addition, some researchers applied a rule-based algorithm, CART, to detecting and mapping a number of IPS, including some aquatic species (water hyacinth and Brazilian waterweed) with inputs of several MNFs and other spectral variables, derived from HyMap hyperspectral image data, and achieved a varying degree of success [17,20,21,60].

For hyperspectral image classification it is in practice sometimes difficult to obtain adequate training samples for supervised method and labeling unsupervised spectral clusters due to the requirement of a large amount of field work. The major factors for detecting and classifying invasive species with the various hyperspectral data are identifying suitable classification methods and gathering adequate training samples for a specific task.

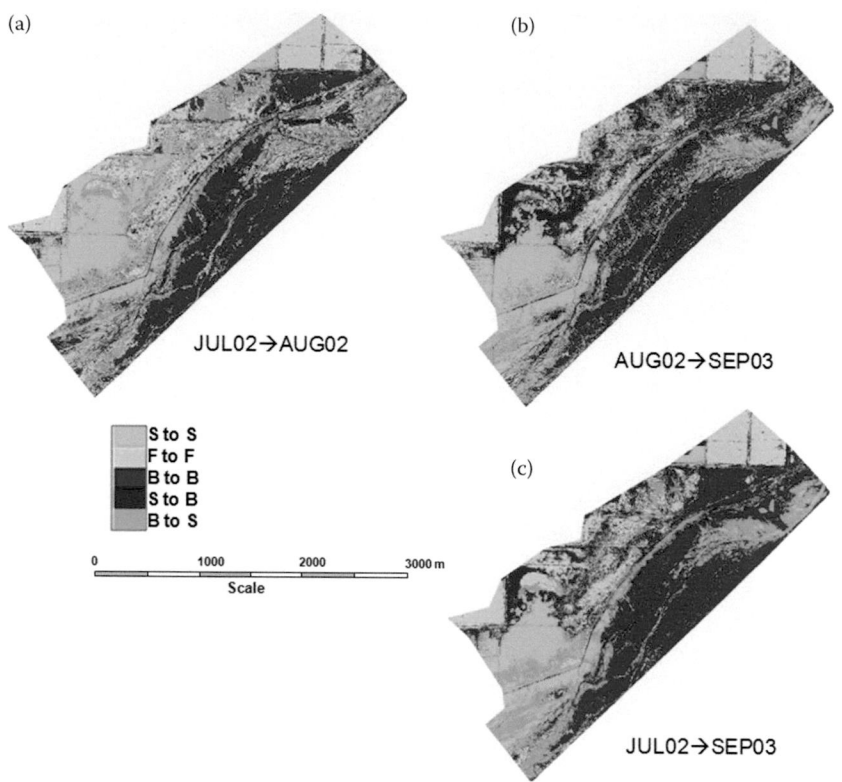

FIGURE 9.3   Change detection resultant maps produced by ANN with principal components extracted from the CASI data from JUL02 to AUG02 (a), from AUG02 to SEP03 (b), and from JUL02 to SEP03 (c). In the key, "S to S," "F to F," and "B to B" represent no changes of saltcedar, farmland, and bare/wildland, respectively; "S to B" means saltcedar changed to bare/wildland, while "B to S" means bare/wildland changed to saltcedar. (With kind permission from Springer Science+Business Media: *Environmental Monitoring and Assessment*, 140, 2008, 15–32, Pu et al., [12].)

9.4   CONSIDERATIONS

Successfully implementing detecting and mapping IPS using hyperspectral data requires careful consideration of hyperspectral sensors/systems, image preprocessing methods, vegetation types and characteristics, phenological change, and environmental context before starting a project. This is because the temporal, spatial, spectral, and radiometric resolutions of hyperspectral data have a significant impact on the success of remote sensing detection and mapping of IPS.

In general, atmospheric correction for hyperspectral data is optimal for conversion of radiance to reflectance [61]. Of the requirements of image preprocessing for detecting and mapping IPS, conversion of digital numbers to radiance or surface reflectance is a most important task for quantitative analysis of plant spectra. This is because the radiometrically corrected and calibrated image data can enhance spectral distinguishability between different plant species due to removal or compression of atmospheric effects on target spectra. A variety of methods [62] including Atmospheric Correction Now program (ACORN), Atmospheric Correction program (ATCOR), Atmospheric Removal program (ATREM), ENVI's fast line-of-sight atmospheric analysis of spectral hypercubes (FLAASH) [63], Second Simulation of the Satellite Signal in the Solar Spectrum (6S), and High-accuracy Atmosphere Correction for Hyperspectral data (HATCH), have been developed [27,61,64,65]. In rugged or mountainous areas, correcting for slope and aspect effects may be necessary. More detailed information about topographic correction can be found in Meyer et al. [66], Allen [67], and Adler-Golden et al. [68].

Different vegetation types comprising individual invasive species should be considered first. The vegetation type information may help in selecting appropriate hyperspectral sensor data and image analysis techniques. For the aquatic vegetation type, its water background may significantly absorb and scatter a sufficient portion of the radiometric energy to make reflected signal very weak compared to most terrestrial vegetation types. This requires hyperspectral data with a high SNR, such as HyMap and AVIRIS. Due to the relatively small size of individual water plants, suitable analysis techniques may include MNF transformation to extract several features/component images, and SMA/CART for mapping abundance of IPS (e.g., [16,17,43]). For terrestrial vegetation types comprising grass, weed, shrub, and scrub species, due to their individual sizes being relatively small and several species frequently growing together, this requires hyperspectral data with high spectral (<5 nm)/spatial (<5 m) resolutions besides a high SNR (e.g., CASI, HyMap and AVIRIS). The corresponding analysis techniques/methods may involve using data transformation methods (PCA, MNF, CDA) and mapping and classification methods (SAM, SMA, MTMF, CART, MLC, ANN) (e.g., [9,12,15,18–23,55,69]). For trees and forest types, due to their individual sizes being relatively large and spectral difference between different species being relatively distinct, we may consider using high-altitude AVIRIS data and Hyperion satellite image data. Suitable analysis techniques/ methods for data transformation and invasive species detecting and mapping may use those for terrestrial vegetation types for identifying and mapping grass, weed, shrub, and scrub species (e.g., [6,25–28,70]). Finally, the concept of plant functional types may also be cited here in consideration of a some invasive species that distribute in the same areas and may have similar structural, functional, and/or phenological properties. After considering both ecology and remote sensing characteristics and the capabilities of new remote sensing instruments (e.g., hyperspectral sensor/system), Ustin and Gamon [71] proposed a new concept of optically distinguishable plant functional types ("optical types") as a unique way to address ecological and remote sensing properties for plant functional types. The proposed concept of "optical type" is based on the assessment of vegetation structure, physiology, and phenology. All the three ecological variables affect vegetation optical properties and contribute to the definition of the concept of "optical type." This would ensure more direct relationships between ecological information and remote sensing observations [71].

Species life history stages or phenological change have different spectral characteristics. When collecting hyperspectral data (either *in situ* measurements or image data), we should consider the

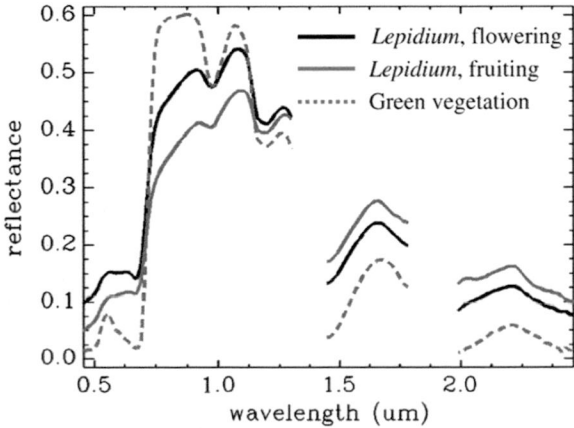

FIGURE 9.4 Field spectra of flowering and fruiting phenologies of *Lepidium*, along with a typical reflectance spectrum of green vegetation for reference. (Reprinted from *Remote Sensing of Environment*, 112, Andrew, M.E., and Ustin, S.L., The role of environmental context in mapping invasive plants with hyperspectral image data, 4301–4317, Copyright 2008, with permission from Elsevier [20].)

life history of the target (invasive) species and find the appropriate phenological stage when spectral difference between invasive species and native species is maximum. For example, the three species (Brazilian waterweed, perennial pepperweed, and water hyacinth) in Hestir et al. [17]'s case studies all differ at different phenological stages: flowering, growing peaks, fruiting, and senescing. The flowering and fruiting phenologies of *Lepidium* can be spectrally distinct between its phenologies and spectrally different from those of co-occurring species [20,99] and other general green plants (Figure 9.4). The figure presents the field spectral difference of flowering and fruiting phenologies of *Lepidium*, also showing a typical reflectance spectrum of green vegetation for reference. When mapping downy brome using multidate AVIRIS image data, Noujdina and Ustin [18] found that the temporal offset in phenology of the downy brome relative to native vegetation provided a basis for spectral differences.

9.5 CHALLENGES AND FUTURE DIRECTIONS

The largest challenge to successful detection of invasive species from their background species is perhaps that all plants are spectrally similar because they are composed of the same spectrally active materials: pigments, water, cellulose, etc. [72]. The spectral distinctiveness requisite for hyperspectral detection most often occurs when invaders possess physiological traits or phenological characteristics that are novel to the invaded ecosystems [20]. To obtain adequate spectral information unique for IPS, we should consider the suitability of spatial and temporal resolutions of existing hyperspectral sensors/systems. Currently, most of the spatial and temporal resolutions are insufficient to decipher the complexity of natural environment and further delineate the distribution of invasive species. Although the high spatial resolution (0.5–5 m) of most airborne hyperspectral sensors/systems (e.g., AVIRIS, HyMap, CASI) has been proven to be a valid method for mapping these invasive plants, there are still several drawbacks to these approaches. One of them involves aircraft scheduling. This would limit the flexibility of data collection that catches the chance of maximizing spectral difference between invasive and native species across their phenology states. As a result, detecting and mapping IPS from native plants cannot effectively utilize the maximum difference information of the phenology of the different plants. Although this issue (i.e., low temporal resolution) can be resolved by utilizing spaceborne hyperspectral sensors/systems, such as Hyperion (30 m resolution) on board the EO-1 satellite, CHRIS (Compact High Resolution Imaging Spectrometer) (18 m resolution) on board the ESA PROBA satellite, and the HyspIRI (60 m resolution) [73] planned mission, their spatial resolutions are relatively low compared to the individual sizes of many IPS. Therefore, many nonnative species are still not discernible, especially grasses, weeds, and shrubs. Such challenges from the spatial and temporal resolutions of hyperspectral sensors/systems are not easy to solve because they are related to technical and economic issues as well as huge data volume storage and processing requirements.

Another challenge in detecting an invasive species is its biological heterogeneity. Such biological heterogeneity may lead to intraspecies variation, causing overlapping spectral features between co-occurring species in the same study area. Hyperspectral image data acquired for suitable phenological states may provide sufficient information to overcome these challenges, allowing the application of more complex spectral analyses and spectral unmixing techniques [17] to the detection and mapping of IPS. In other words, with the use of multitemporal hyperspectral image data, the accuracy of detecting and mapping such invasive species can be expected to increase.

The application of multisensor data, which can synergize the high spatial resolution (airborne hyperspectral data) and high temporal resolution (satellite hyperspectral data), provides the potential to more accurately detect IPS through integration of different features of sensor data. The disadvantage of using multisensor data for detecting IPS is the difficulty in image acquisition or processing and use of appropriate detecting and mapping techniques. In the future, when the satellite hyperspectral sensor HyspIRI is in operation, high-frequency hyperspectral data will be available globally. Application of multisensor data will become increasingly important in future studies of detection and mapping of IPS, and thus more advanced image processing and invasive species detection and mapping techniques are needed. Accurately detecting and mapping IPS with hyperspectral data remains an active research topic and new techniques continue to be developed. Any advanced invasive species mapping/detecting technique is required to be easy to use and able to provide accurate detection and mapping results.

ACKNOWLEDGMENTS

Three anonymous reviewers' comments and suggestions were greatly valuable in improving the chapter. The author sincerely appreciates their efforts.

REFERENCES

1. Beck, K.G., Zimmerman, K., Schardt, J.D., Stone, J., Lukens, R.R., Reichard, S., Randall, J., Cangelosi, A.A., Cooper, D., and Thompson, J.P., Invasive species defined in a policy context: Recommendations from the Federal Invasive Species Advisory Committee, *Invasive Plant Science and Management*, 1, 414–421, 2008.
2. Huang, C.-Y., and Asner, G.P., Applications of remote sensing to alien invasive plant studies, *Sensors*, 9, 4869–4889, 2009.
3. Mack, R.N., Simberloff, D., Lonsdale, W.M., Evans, H., Clout, M., and Bazzaz, F., Biotic invasions: Causes, epidemiology, global consequences and control, *Ecological Application*, 10, 689–710, 2000.
4. Simberloff, D., Biological invasions: How are they affecting us and what can we do about them? *Western North American Naturalist*, 61, 308–315, 2001.
5. Jackson, R.B., Banner, J.L., Jobbagy, E.G., Pockman, W.T., and Wall, D.H., Ecosystem carbon loss with woody plant invasive of grasslands, *Nature*, 418, 623–626, 2002.
6. Tsai, F., Lin, E.E., and Yoshino, K., Spectrally segmented principal component analysis of hyperspectral imagery for mapping invasive plant species, *International Journal of Remote Sensing*, 28(5), 1023–1039, 2007.
7. Pimentel, D., Lach, L., Zuniga, R., and Morrison, D., Environmental and economic costs associated with non-indigenous species in the United States, *BioScience*, 50, 53–65, 2000.
8. Ustin, S.L., Scheer, G., DiPietro, D., Underwood, E., and Olmstead, K., Hyperspectral remote sensing for invasive species detection and mapping, *Abstracts of Papers of the American Chemical Society*, 221, U50, 2001.
9. Lass, L.W., Thill, D.C., Shafii, B., and Prather, T.S., Detecting spotted knapweed (*Centaurea maculosa*) with hyperspectral remote sensing technology, *Weed Technology*, 16, 426–432, 2002.
10. Underwood, E.C., Ustin, S.L., and DiPietro, D., Mapping nonnative plants using hyperspectral imagery, *Remote Sensing of Environment*, 86, 150–161, 2003.
11. Miao, X., Gong, P., Swope, S., Pu, R., Carruthers, R., Anderson, G.L. et al., Estimation of yellow starthistle abundance through CASI-2 hyperspectral imagery using linear spectral mixture models, *Remote Sensing of Environment*, 101, 329–341, 2006.
12. Pu, R., Gong, P., Tian, Y., Miao X., and Carruthers, R., Invasive species change detection using artificial neural networks and CASI hyperspectral imagery, *Environmental Monitoring and Assessment*, 140, 15–32, 2008.
13. Pu, R., Gong, P., Tian, Y., Miao, X., Carruthers, R., and Anderson, G.L., Using classification and NDVI differencing methods for monitoring sparse vegetation coverage: A case study of saltcedar in Nevada, USA, *International Journal of Remote Sensing*, 29(14), 1987–4011, 2008.
14. Clark, M.L., Roberts, D.A., and Clark, D.B., Hyperspectral discrimination of tropical rain forest tree species at leaf to crown scales, *Remote Sensing of Environment*, 96, 375–398, 2005.
15. Underwood, E.C., Ustin, S.L., and Ramirez, C.M., A comparison of spatial and spectral image resolution for mapping invasive plants in coastal California, *Environmental Management*, 39, 63–83, 2007.
16. Underwood, E.C., Mulitsch, M.J., Greenberg, J.A., Whiting, M.L., Ustin, S.L., and Kefauver, S.C., Mapping invasive aquatic vegetation in the Sacramento-San Joaquin Delta using hyperspectral imagery, *Environmental Monitoring and Assessment*, 121, 47–64, 2006.
17. Hestir, E.L., Khanna, S., Andrew, M.E., Santos, M.J., Viers, J.H., Greenberg, J.A., Rajapakse, S.S., and Ustin, S.L., Identification of invasive vegetation using hyperspectral remote sensing in the California Delta ecosystem, *Remote Sensing of Environment*, 112, 4034–4047, 2008.

18. Noujdina, N.V., and Ustin, S.L., Mapping downy brome (*Bromus tectorum*) using multidate AVIRIS data, *Weed Science*, 56, 173–179, 2008.
19. Wang, C., Zhou, B., and Palm, H.L., Detecting invasive Sericea lespedeza (*Lespedeza cuneata*) in mid-Missouri pastureland using hyperspectral imagery, *Environmental Management*, 41, 853–862, 2008.
20. Andrew, M.E., and Ustin, S.L., The role of environmental context in mapping invasive plants with hyperspectral image data, *Remote Sensing of Environment*, 112, 4301–4317, 2008.
21. Andrew, M.E., and Ustin, S.L., The effects of temporally variable dispersal and landscape structure on invasive species spread, *Ecological Applications*, 20(3), 593–608, 2010.
22. Hamada, Y., Stow, D.A., Coulter, L.L., Jafolla, J.C., and Hendricks, L.W., Detecting Tamarisk species (*Tamarix* spp.) in riparian habitats of Southern California using high spatial resolution hyperspectral imagery, *Remote Sensing of Environment*, 109, 237–248, 2007.
23. Narumalani, S., Mishra, D.R., Wilson, R., Reece, P., and Kohler, A., Detecting and mapping four invasive species along the floodplain of North Platte River, Nebraska, *Weed Technology*, 23, 99–107, 2009.
24. Asner, G.P., and Vitousek, P.M., Remote analysis of biological invasion and biogeochemical change, *Proceedings of the National Academy of Sciences of the USA*, 102, 4383–4386, 2005.
25. Ramsey III, E., Rangoonwala, A., Nelson, G., and Ehrlich, R., Mapping the invasive species, Chinese tallow, with EO1 satellite Hyperion hyperspectral image data and relating tallow occurrences to a classified Landsat Thematic Mapper land cover map, *International Journal of Remote Sensing*, 26(8), 1637–1657, 2005.
26. Ramsey III, E., Rangoonwala, A., Nelson, G., Ehrlich, R., and Martella, K., Generation and validation of characteristic spectra from EO1 Hyperion image data for detecting the occurrence of the invasive species, Chinese tallow, *International Journal of Remote Sensing*, 26(8), 1611–1636, 2005.
27. Asner, G.P., Jones, M.O., Martin, R.E., Knapp, D.E., and Hughes, R.F., Remote sensing of native and invasive species in Hawaiian forests, *Remote Sensing of Environment*, 112, 1912–1926, 2008.
28. Asner, G.P., Knapp, D.E., Kennedy-Bowdoin, T., Jones, M.O., Martin, R.E., Boardman, J., and Hughes, R.F., Invasive species detection in Hawaiian rainforests using airborne imaging spectroscopy and LiDAR, *Remote Sensing of Environment*, 112, 1942–1955, 2008.
29. Brooks, M.L., D'Antonio, C.M., Richardson, D.M., Grace, J.B., and Keeley, J.E., Effects of invasive alien plants on fire regimes, *BioScience*, 54(7), 677–688, 2004.
30. Friederici, P., The alien saltcedar, *American Forests*, 101, 45–47, 1995.
31. D'Antonio, C.M., and Vitousek, P.M., Biological invasions by exotic grasses, the grass/fire cycle, and global change. *Annual Review of Ecology and Systematics*, 23, 63–87, 1992.
32. Asner, G.P., Martin, R., Carlson, K., Rascher, U., and Vitousek, P., Vegetation-climate interactions among native and invasive species in Hawaiian rainforest, *Ecosystems*, 9, 1106–1117, 2006.
33. Hughes, F.R., and Denslow, J.S., Invasion by a N2-fixing tree alters function and structure in wet lowland forests of Hawaii, *Ecological Applications*, 15, 1615–1628, 2005.
34. Funk, J.L., and Vitousek, P.M., Resource-use efficiency and plant invasion in low-resource systems, *Nature*, 446, 1079–1081, 2007.
35. Hunt, E.R., McMurtrey, J.E., Parker, A.E., and Corp, L.A., Spectral characteristics of leafy spurge (*Euphorbia esula*) leaves and flower bracts, *Weed Science*, 52, 492–497, 2004.
36. Hunt, E.R., Daughtry, C.S., Kim, M.S.S., and Williams, A.E.P., Using canopy reflectance models and spectral angles to assess potential of remote sensing to detect invasive weeds, *Journal of Applied Remote Sensing*, 1, 013506–013519, 2007.
37. Pu, R., and Gong, P., Hyperspectral remote sensing of vegetation bioparameters, in: *Advances in Environmental Remote Sensing: Sensors, Algorithms, and Applications*, Q. Weng (ed.), CRC Press/Taylor & Francis Group, Chapter 5, 2011, 101–142.
38. Tsai, F., and Philpot, W., Derivative analysis of hyperspectral data, *Remote Sensing of Environment*, 66(1), 41–51, 1998.
39. Cloutis, E.A., Hyperspectral geological remote sensing: Evaluation of analytical techniques, *International Journal of Remote Sensing*, 17(12), 2215–2242, 1996.
40. van der Meer, F., The effectiveness of spectral similarity measures for the analysis of hyperspectral imagery, *International Journal of Applied Earth Observation and Geoinformation*, 8(1), 3–17, 2006.
41. Pengra, B.W., Johnston, C.A., and Loveland, T.R., Mapping an invasive plant, *Phragmites australis*, in coastal wetlands using the EO-1 Hyperion hyperspectral sensor, *Remote Sensing Environment*, 108, 74–81, 2007.
42. Lass, L.W., and Prather, T.S., Detecting the locations of Brazilian pepper trees in the everglades with a hyperspectral sensor, *Weed Technology*, 18, 437–442, 2004.

43. Hirano, A., Madden, M., and Welch, R., Hyperspectral image data for mapping wetland vegetation, *Wetlands*, 23(2), 436–448, 2003.

44. Gong, P., Pu, R., Biging, G.S., and Larrieu, M., Estimation of forest leaf area index using vegetation indices derived from Hyperion hyperspectral data, *IEEE Transactions on Geoscience and Remote Sensing*, 41(6), 1355–1362, 2003.

45. Zarco-Tejada, P.J., Miller, J.R., Noland, T.L., Mohammed, G.H., and Sampson, P.H., Scaling-up and model inversion methods with narrowband optical indices for chlorophyll content estimation in closed forest canopies with hyperspectral data, *IEEE Transactions on Geoscience and Remote Sensing*, 39(7), 1491–1507, 2001.

46. Eitel, J.U.H., Gessler, P.E., Smith, A.M.S., and Robberecht, R., Suitability of existing and novel spectral indices to remotely detect water stress in Populus spp, *Forest Ecology and Management*, 229(1–3), 170–182, 2006.

47. He, Y., Guo, X., and Wilmshurst, J., Studying mixed grassland ecosystems I: Suitable hyperspectral vegetation indices, *Canadian Journal of Remote Sensing*, 32(2), 98–107, 2006.

48. Hunt, G.R., Electromagnetic radiation: The communication link in remote sensing, in *Remote Sensing in Geology*, B. Siegal and A. Gillespia (eds), Wiley, New York, p. 702, 1980.

49. Curran, P.J., Remote sensing of foliar chemistry, *Remote Sensing of Environment*, 30, 271–278, 1989.

50. Clark, R.N., and Roush, T.L., Reflectance spectroscopy: Quantitative analysis techniques for remote sensing applications, *Journal of Geophysical Research*, 89, 6329–6340, 1984.

51. Gong, P., Pu, R., and Heald, R.C., Analysis of *in situ* hyperspectral data for nutrient estimation of giant sequoia, *International Journal of Remote Sensing*, 23(9), 1827–1850, 2002.

52. Pu, R., and Gong, P., Wavelet transform applied to EO-1 hyperspectral data for forest LAI and crown closure mapping, *Remote Sensing of Environment*, 91, 212–224, 2004.

53. Green, A.A., Berman, M., Switzer, P., and Craig, M.D., A transformation for ordering multispectral data in terms of image quality with implications for noise removal, *IEEE Transactions on Geoscience and Remote Sensing*, 26, 65–74, 1988.

54. Pu, R., Gong, P., Michishita, R., and Sasagawa, T., Spectral mixture analysis for mapping abundance of urban surface components from the Terra/ASTER data, *Remote Sensing of Environment*, 112, 939–954, 2008.

55. Walsh, S.J., McCleary, A.L., Mena, C.F., Shao, Y., Tuttle, J.P., González, A., and Atkinson, R., QuickBird and Hyperion data analysis of an invasive plant species in the Galapagos Islands of Ecuador: Implications for control and land use management, *Remote Sensing of Environment*, 112, 1927–1941, 2008.

56. van Aardt, J.A.N., and Wynne, R.H., Examining pine spectral separability using hyperspectal data from an airborne sensor: An extension of field-based results, *International Journal of Remote Sensing*, 28(2), 431–436, 2007.

57. Sasaki, K., Kawata, S., and Minami, S., Estimation of component spectral curves from unknown mixture spectra, *Applied Optics*, 23, 1955–1959, 1984.

58. Zhang, L., Li, D., Tong, Q., and Zheng, L., Study of the spectral mixture model of soil and vegetation in Poyang Lake area, China, *International Journal of Remote Sensing*, 19, 2077–2084, 1998.

59. Boardman, J.W., Kruse, F.A., and Green, R.O., Mapping target signatures via partial unmixing of AVIRIS data, in *Summaries of the Fifth JPL Airborne Geoscience Workshop JPL Publication, 95-1* (pp. 23–26), NASA Jet Propulsion Laboratory, Pasadena, CA, 1995.

60. Andrew, M.E., and Ustin, S.L., Habitat suitability modelling of an invasive plant with advanced remote sensing data, *Diversity and Distributions*, 15, 627–640, 2009.

61. Goetz, A.F.H., Ferri, M., Kindel, B., and Qu, Z., Atmospheric correction of Hyperion data and techniques for dynamic scene correction, in *2002 IEEE International Geoscience and Remote Sensing Symposium and the 24th Canadian Symposium on Remote Sensing*, Toronto, Canada, June 24–28, 2002.

62. Jensen, J.R., *Introductory Digital Image Processing: A Remote Sensing Perspective*, 3rd edition, Pearson Prentice Hall, pp. 175–222, 2005.

63. FLAASH User's Guide, ENVI FLAASH Version 4.1, September, 2004 Edition, Research Systems, Inc. pp. 1–80, 2004.

64. Gao, B.-C., Heidebrecht, K.B., and Goetz, A.F.H., Derivation of scaled surface reflectance from AVIRIS data, *Remote Sensing of Environment*, 44, 165–178, 1993.

65. Qu, Z., Kindel, B.C., and Goetz., A.F.H., The high accuracy atmospheric correction for hyperspectral data (HATCH) model, *IEEE Transactions on Geoscience and Remote Sensing*, 41, 1223–1231, 2003.

66. Meyer, P., Itten, K.I., Kellenberger, T., Sandmeier, S., and Sandmeier, R., Radiometric corrections of topographically induced effects on Landsat TM data in alpine environment, *ISPRS Journal of Photogrammetry and Remote Sensing*, 48, 17–28, 1993.

67. Allen, T.R., Topographic normalization of Landsat thematic mapper data in three mountain environments, *Geocarto International*, 15(2), 13–19, 2000.

68. Adler-Golden, S.M., Matthew, M.W., Anderson, G.P., Felde, G.W., and Gardner, J.A., An algorithm for de-shadowing spectral imagery, in *Proceedings of the 11th JPL Airborne Earth Science Workshop*, March 5–8, 2002, JPL Publication 03-04, Pasadena, USA, 2002.

69. Judd, C., Steinberg, S., Shaughnessy, F., and Crawford, G., Mapping salt marsh vegetation using aerial hyperspectral imagery and linear unmixing in Humboldt Bay, *California, Wetlands*, 27(4), 1144–1152, 2007.

70. Asner, G.P., Martin, R.E., Knapp, D.E., and Kennedy-Bowdoin, T., Effects of Morella faya tree invasion on aboveground carbon storage in Hawaii, *Biological Invasions*, 12, 477–494, 2010.

71. Ustin, S.L., and Gamon, J.A., Remote sensing of plant functional types, *New Phytologist*, 186, 795–816, 2010.

72. Jacquemoud, S., and Baret, F., Prospect—A model of leaf optical properties spectra, *Remote Sensing of Environment*, 34, 75–91, 1990.

73. NRC's Decadal Survey report, Earth Science and Applications from Space: National Imperatives for the Next Decade and Beyond, http://www.nap.edu/catalog/11820.html, 2007.

74. Rouse, J.W., Haas, R.H., Schell, J.A., and Deering, D.W., Monitoring vegetation systems in the Great Plains with ERTS, in *Proceedings, Third ERTS Symposium*, 1, pp. 48–62, 1973.

75. Blackburn, G.A., Quantifying chlorophylls and caroteniods at leaf and canopy scales: An evaluation of some hyperspectral approaches, *Remote Sensing of Environment*, 66, 273–285, 1998.

76. Jordan, C.F., Derivation of leaf area index from quality of light on the forest floor, *Ecology*, 50, 663–666, 1969.

77. Tucker, C.J., Red and photographic infrared linear combinations for monitoring vegetation, *Remote Sensing of Environment*, 8, 127–150, 1979.

78. Gitelson, A.A., Keydan, G.P., and Merzlyak, M.M., Three-band model for non-invasive estimation of chlorophyll, carotenoids and anthocyanin contents in higher plant leaves, *Geophysical Research Letters*, 33, L11402, 2006.

79. Datt, B., A new reflectance index for remote sensing of chlorophyll content in higher plants: Tests using Eucalyptus leaves, *Journal of Plant Physiology*, 154, 30–36, 1999.

80. Sims, D.A., and Gamon, J.A., Relationships between leaf pigment content and spectral reflectance across a wide range of species, leaf structures and developmental stages, *Remote Sensing of Environment*, 81, 337–354, 2002.

81. Fuentes, D.A., Gamon, J.A., Qiu, H.-L., Sims, D.A., and Roberts, D.A., Mapping Canadian boreal forest vegetation using pigment and water absorption features derived from the AVIRIS sensor, *Journal of Geophysical Research*, 106, 33565–33577, 2001.

82. Peñuelas, J., Gamon, J.A., Fredeen, A.L., Merino, J., and Field, C.B., Reflectance indices associated with physiological changes in nitrogen- and water-limited sunflower leaves, *Remote Sensing of Environment*, 48, 135–146, 1994.

83. Rama Rao, N., Garg, P.K., Ghosh, S.K., and Dadhwal, V.K., Estimation of leaf total chlorophyll and nitrogen concentrations using hyperspectral satellite imagery, *Journal of Agricultural Science*, 146, 65–75, 2008.

84. Gamon, J.A., Peñuelas, J., and Field, C.B., A narrow waveband spectral index that tracks diurnal changes in photosynthetic efficiency, *Remote Sensing of Environment*, 41, 35–44, 1992.

85. Zarco-Tejada, P.J., *Optical indices as bioindicators of forest sustainability*, Report to the Graduate Programme in Earth and Space Science, York University, Toronto, 1998.

86. Gamon, J.A., and Surfus, J.S., Assessing leaf pigment content and activity with a reflectometer, *New Phytologist*, 143, 105–117, 1999.

87. Nagler, P.L., Daughtry, C.S.T., and Goward, S.N., Plant litter and soil reflectance, *Remote Sensing of Environment*, 71, 207–215, 2000.

88. Serrano, L., Peñuelas, J., and Ustin, S.L., Remote sensing of nitrogen and lignin in Mediterranean vegetation from AVIRIS data: Decomposing biochemical from structural signals, *Remote Sensing of Environment*, 81, 355–364, 2002.

89. Datt, B., McVicar, T.R., Van Niel, T.G., Jupp, D.L.B., and Pearlman, J.S., Preprocessing EO-1 Hyperion hyperpsectral data to support the application of agricultural indexes, *IEEE Transactions on Geoscience and Remote Sensing*, 41, 1246–1259, 2003.

90. Gao, B.C., NDWI—A normalized difference water index for remote sensing of vegetation liquid water from space, *Remote Sensing of Environment*, 58, 257–266, 1996.

91. Schlerf, M., Atzberger, C., and Hill, J., Remote sensing of forest biophysical variables using HyMap imaging spectrometer data, *Remote Sensing of Environment*, 95, 177–194, 2005.

92. Peñuelas, J., Piñol, J., Ogaya, R., and Filella, I., Estimation of plant water concentration by the reflectance water index WI (R900/R970), *International Journal of Remote Sensing*, 18, 2869–2875, 1997.

93. Thenkabail, P.S., GangadharaRao, P., Biggs, T., Krishna, M., and Turral, H., Spectral matching techniques to determine historical land use/land cover (LULC) and irrigated areas using time-series AVHRR pathfinder datasets in the Krishna River Basin, India, *Photogrammetric Engineering and Remote Sensing*, 73(9), 1029–1040, 2007.

94. Thenkabail, P.S., Lyon, G.J., Turral, H., and Biradar, C.M., *Remote Sensing of Global Croplands for Food Security*, CRC Press-Taylor and Francis group, Boca Raton, London, New York. p. 556 (48 pages in color). Published in June, 2009.

95. Thenkabail, P.S., Smith, R.B., and De-Pauw, E., Hyperspectral vegetation indices for determining agricultural crop characteristics, *Remote Sensing of Environment*, 71, 158–182, 2000.

96. Thenkabail, P.S., Smith, R.B., and De-Pauw, E., Evaluation of narrowband and broadband vegetation indices for determining optimal hyperspectral wavebands for agricultural crop characterization, *Photogrammetric Engineering and Remote Sensing*, 68(6), 607–621, 2002.

97. Thenkabail, P.S., Enclona, E.A., Ashton, M.S., and Van Der Meer, V., Accuracy assessments of hyperspectral waveband performance for vegetation analysis applications, *Remote Sensing of Environment*, 91(2–3), 354–376, 2004.

98. Thenkabail, P.S., Enclona, E.A., Ashton, M.S., Legg, C., and Jean De Dieu, M., Hyperion, IKONOS, ALI, and ETM+ sensors in the study of African rainforests, *Remote Sensing of Environment*, 90, 23–43, 2004.

99. Andrew, M.E., and Ustin, S.L., Spectral and physiological uniqueness of perennial pepperweed (*Lepidium latifolium*), *Weed Science*, 54, 1051–1062, 2006.

10 Visible, Near Infrared, and Thermal Spectral Radiance On-Board UAVs for High-Throughput Phenotyping of Plant Breeding Trials

Scott C. Chapman, Bangyou Zheng, Andries B. Potgieter, Wei Guo, Frederic Baret, Shouyang Liu, Simon Madec, Benoit Solan, Barbara George-Jaeggli, Graeme L. Hammer, and David R. Jordan

CONTENTS

10.1 INTRODUCTION

Together with increased scientific understanding of plant growth, improvements in instrument technology and survey platforms are now enabling the application of spectral techniques at higher resolutions (down to sub-centimeter) than have been previously possible in field-scale research. In field trials, we are interested in being able to measure the "phenotypes" of plants, which are any of the observable traits or derived calculations of traits that may vary across genotypes of the same species. Hence, compared to "remote sensing," the "proximal" sensing (in the range of millimeters to 50 meters or so) that we are concerned with in this chapter is frequently related to fine-scale identification of objects and detection of signals from noncontiguous pixels. We consider here how proximal sensing can replace manual measurements of these objects and signals and extend the number and types of features that can be monitored by plant scientists in field trials. Other chapters in this volume, and recent reviews (e.g., Adão et al., 2017), focus more on interpreting plant signals in direct relationships to hyperspectral sensing per se, and its relationship to management and disease challenges. This chapter attempts to provide a bridge between image analysis methods and their interpretation into crop characteristics via the science of crop physiology.

In field experiments, collected phenotypes are normally aggregated for plots (sometimes called microplots) of plants grown together in minicanopies of 1–20 m² depending on purpose and crop type. These "plot-level" phenotypes are often termed "traits," with typical examples being canopy height, crop cover percentage (COVER%), flowering date, biomass per unit area, and grain yield per unit area, aggregated at the plot level. The application of proximal sensing methods in plant breeding and agronomy research is frequently termed "high-throughput plant phenotyping" (HTPP) in the plant science literature, where HTPP includes a range of technologies and collection methods such as ground-based vehicles, and automated measurements of plant characteristics in the laboratory as well as the field. Field-based Phenotyping (FBP) is also used to refer more specifically to this type of research for field research applications (Chapman et al., 2014; Pauli et al., 2016), as opposed to many other applications of HTPP in glasshouses and controlled environments (e.g., Cabrera-Bosquet et al., 2016). In the field environment, high resolution may not be essential for application of these technologies on farms in production agriculture, but it becomes useful in applications related to the underlying research infrastructure of agriculture—the genetics, agronomy, and physiology of how crops grow and adapt to the abiotic and biotic challenges of environments (e.g., Liebisch et al., 2015; Zaman-Allah et al., 2015; Elazab et al., 2016; Kefauver et al., 2017).

10.1.1　Plant Breeding and Physiological Adaptation

Crop plants are exposed to a vast range of environmental conditions (abiotic effects) and challenges from other organisms—pests, weeds, diseases (biotic effects). Over time, public and private investments have developed complex networks of interacting breeding programs that identify the product (variety) profiles and geographic markets for varieties. Plant breeding is therefore one of the major engines of increased food production, and acceleration of this process is essential to the maintenance and the continued increase in on-farm yields of crops to support an increasing population. Within plant breeding, HTPP methods have a role in "conventional" field-based breeding whereby varieties are identified through a pipeline of genetic recombination (crossing of lines) and testing for adaptation to abiotic and biotic conditions and stresses. In this chapter, we focus on the characterization of abiotic traits and their physiology rather than biotic effects for which readers are directed to other chapters and recent papers (e.g., DeChant et al., 2017, who used machine learning/neural networks to develop a highly efficient detector of northern leaf blight in maize). Further, we focus mainly on field cereal crops, where most of our experience lies.

With the advent of high-throughput genomic (DNA) characterization of breeding lines, plant breeding is transiting to a new model. In these approaches, based on developments first achieved in animal breeding, "genomic selection" aims to utilize genetic markers (statistically associated with plant traits) and candidate genes (verified genes with known roles in specific gene to trait pathways) in the identification and selection of superior plant genotypes (see Potgieter et al., in press for extended description). The extent of application of genomics is such that plant breeders are now able to select new genotypes based on their DNA alone (genomic selection; Meuwissen et al., 2001), providing that they have a sufficiently precise prediction model of how the allelic combinations of genes in the DNA contribute to yield and/or quality via the many traits that are affected by these genes. Rutkoski et al. (2016) have shown that accuracy of selection for grain yield in wheat can be improved by the inclusion of secondary traits derived from UAV imagery and canopy temperature. A further elaboration of these methods is where HTPP data are used as input to ecophysiological crop models in order to compute "derived" traits, which are predicted at the level of genomic information, rather than genotype (see Technow et al., 2015; Messina et al., 2018). Hence, the next era of plant breeding is to develop plant phenomics (the science of phenotyping [quantifying] plant and crop traits) to achieve the throughput, precision, and integration that is required to support the continual update of genomic prediction models.

The combination of miniaturized proximal sensing instruments and unmanned aerial vehicles (UAVs) allows remote measurement of multiple attributes of crops that are specifically related to how efficiently the crops develop their leaf canopies to intercept photosynthetically active radiation, and convert this radiation to photosynthesis. With appropriate interpretation, visible and hyperspectral (especially near-infrared) imagery can contribute to the monitoring of the structure and biochemical composition of these canopies, and their health as affected by abiotic and biotic stresses. The mechanisms in dehydration tolerance and heat dissipation provide an example. As photosynthesis requires the direct exchange of CO_2 across membranes inside the stomata of leaves, a direct and necessary consequence of photosynthesis is the loss of water via evapotranspiration from those same membranes. Plants need to evaporate water to dissipate the energy of sunlight incident on canopies, and so there is a need, through the season, to balance the capacity of leaf tissues and biochemical processes for tolerance to heat versus the conservation of water that is yet to be extracted from the soil. In irrigated conditions, plants that maintain cooler canopies via higher stomatal conductance have been shown in several species to produce higher biomass and yield (e.g., sugarcane [Basnayake et al., 2012]; wheat [Rebetzke et al., 2016]; wheat and sorghum [Chenu et al., 2018]).

10.1.2   EXPLOITING HTPP BEYOND "CONVENTIONAL" TRAITS: A HIERARCHY OF INFORMATION FROM SENSORS TO TRAITS

Initial research in applications of HTPP was largely focused on how to accelerate existing methods of phenotyping plots; for example, measuring the canopy height using UAV imagery instead of requiring a technician with a ruler. However, HTTP methods allow the measurement of plants using higher-resolution information that provides the opportunity to create new types of traits, as well as to assess the within-plot variability—for example, analyzing a height trait using point cloud data from thousands of points per plot rather several ruler measurements. In order to match traits to what breeders currently do and to provide access to new traits, it helps to consider how traits are derived from proximal information.

To be utilized by plant breeders, raw sensor data needs to be transformed from a signal to a static or functional trait (Figure 10.1), for example, successive steps in image processing and computation allow the structural characterization of plant canopies and their components; for example, counting of plants or reproductive structures at various stages of the season, and visualization and analysis of the three-dimensional structure of canopies (including height, leaf display, position of different organs [Espana et al., 1999]) and the consequent microenvironment (light, temperature etc.) within canopies. Examples of this computation process are given in Table 10.1 and are discussed in more detail later in this chapter. Lelong et al. (2008) and Haghighattalab et al. (2016) outline some of the key corrections that are required to transform the L0 signal from a RGB or multispectral camera to a normalized quantity (L1) in UAV-type systems, and these corrections need to be determined for each type of camera system. Hence the types of traits being generated from these systems include both traits that breeders traditionally measure (L1 or L2 level traits such as plant density [Liu et al., 2017] and canopy height [Madec et al., 2017]) as well as more complex static L2 traits such as green area index (Verger et al., 2014; Potgieter et al., 2017) and leaf chlorophyll content (Jay et al., 2017). Applied to UAV imagery, machine learning methods are now being used to segment objects at the plant and leaf level (Chen et al., 2017b). To continue the example of canopy traits, it is possible to derive the fraction of light intercepted by the canopy over several weeks (using static L2 traits interpolated through time by statistical or ecophysiological models) and to multiple this vector by the vector of observed incident radiation at a nearby weather station in order to estimate the sum of radiation intercepted by a canopy on a per unit area basis, for each experimental plot. Finally, L3 dynamic traits like light use efficiency (LUE) can then be calculated as the ratio of change in

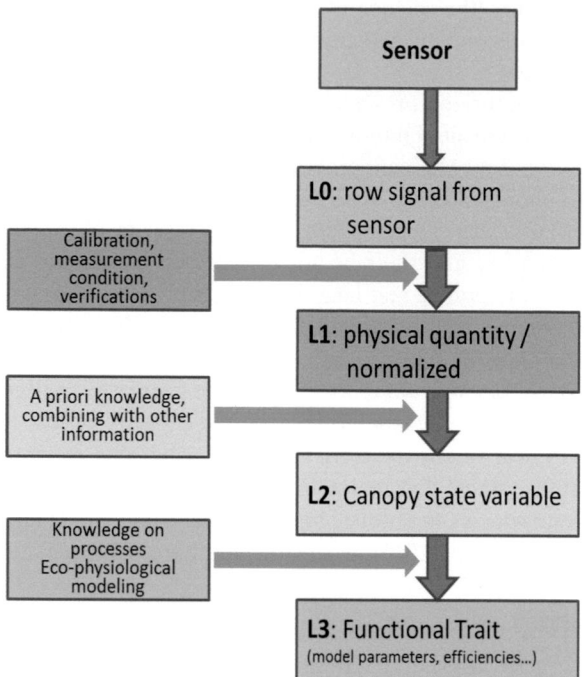

FIGURE 10.1 Hierarchy of data values between raw signal and "functional" trait (F. Baret, unpublished). This framework can be applied to explain the way in which traits are expressed (see Table 10.1).

biomass over time and the sum of radiation intercepted, where biomass may be directly measured or, ideally, directly estimated from multispectral imagery collected by UAV (e.g., Bendig et al., 2014).

The level of precision required in plant breeding is substantially more than that needed in production agriculture for several reasons: (1) plant breeding grows trials of hundreds or thousands of genotypes in small plots (typically 2–15 m^2) and so requires higher spatial resolutions compared to the information needed to program tractor management operations in a production field; (2) plant breeders are attempting to discern small differences among genotypes that are distinct from those differences due to variation in field soil condition; and (3) plant breeders are interested in deriving a range of traits that may not be of direct interest to growers but which have utility in the selection of genotypes as related to the potential differences in yield performance. These steps require substantial rigor in the methods being used, and at the same time, they offer new opportunities to use subplot data; for example, to use image analysis to quantify "gappiness" or heterogeneity within plots, rather than using only a human-generated "score" of plot quality. In this chapter, we use a pipeline workflow to outline the issues and current practices required to ensure good-quality data collection in HTPP using UAVs.

10.2 A WORKFLOW FOR IMAGE PROCESSING IN PLANT BREEDING TRIALS

Plant science researchers have been utilizing satellite and airborne imagery since the mid 1970s, around the time that film-based cameras could start to be used for the characterization of infrared and thermal signals from field trials (Sankaran et al., 2015; Martínez et al., 2016) that can be related to "traits" (the selectable phenotypes) screened by plant breeders. For this early technology, the main challenges were the expense of flights and films and timeliness in collecting this type of imagery. Field-based phenotyping methods using UAVs began to appear in the literature around 2005 (e.g., LeLong et al., 2008) when consumer-level UAVs started to become affordable for use in plant science research programs (military UAVs were either unavailable or prohibitively expensive). Since 2010,

TABLE 10.1

Example Traits and Hierarchy of Information and Processes Needed to Derive the Canopy-Related Traits from UAV-Borne Systems

Trait and Hierarchy	L0—Source	L0 to L1 Processing	L1	L1 to L2 Processing	L2	L2 to L3 Processing	L3	Aggregation to Plot Level
Area-related traits								
Plant height—RGB (L2 static)	RGB images	Normalise and georeferenced imagery and mosaic	Point cloud or raster digital surface model	Subtract ref ground, extract percentile for canopy top	Canopy height from ground	NA	NA	Mean or median of pixels in 99th percentile[a]
Plant height—LIDAR	LIDAR	Georeferenced	Point cloud	As above	As above	NA	NA	As above
COVER%	RGB imagery	As for plant height RGB	Image mosaic and/or calibrated flight images	Masking of background using thresholding or other methods (e.g., decision trees)	Proportion of ground area covered by target vegetation	NA	NA	Single value for plot
Green Area Index	Multispectral camera	As for plant height RGB	Multi-spec mosaic	Local calibration of mosaic data against ground imagery	Green area index	PROSAIL to reverse calculate		
FIPAR								
Light Use Efficiency of Canopy						Ratio of FIPAR*RADN to delta BIOMASS	LUE in g/MJ	Plot mean value
Leaf rolling due to stress								
Canopy temperature								
Lodging percentage								
Object counting								
Plant density—larger species	RGB imagery							
Plant density—small plants								
Head density per unit area					Head number per unit area	NA	NA	Single value for plot
Tiller number per plant						Ratio of plant density (earlier date) and head density	Tiller number per plant	

Note: For some traits, the first box to right indicates that the trait is a combination of other traits.

[a] May be computed as subplot slices.

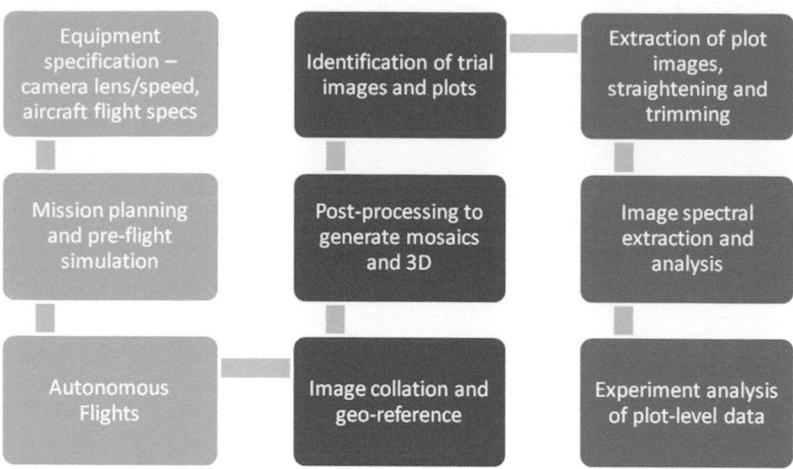

FIGURE 10.2   Aerial image analysis workflow showing preflight (left), postflight (centre), and image processing (right) tasks. (Reproduced from Chapman SC et al. 2014. *Agronomy* 42: 279–301.)

there have been numerous reviews published on the utility (usually the potential utility) of these methods for use in the acceleration of plant breeding. These reviews cover multiple aspects of the state-of-the-art for measurement of plant traits (Cairns et al., 2012; Pauli et al., 2016), comparisons of different types of phenotyping platforms (White et al., 2012), through to descriptions of how these methods might be integrated with other breeding technologies such as genome-wide association analysis in a breeding program (Rutkoski et al., 2016).

The workflow that was outlined by Chapman et al. (2014) (Figure 10.2) is generally applicable to the needs for FBP by UAVs. Other research groups are developing similar types of workflows related to plant breeding which incorporate these steps (e.g., Haghighattalab et al., 2016), with a variety of refinements that may relate to the crop species and specific challenges associated with measuring different traits. In the UQ/CSIRO (The University of Queensland/Commonwealth Scientific and Industrial Research Organisation) research group, part of this workflow is managed via a software database and viewer ("PhenoCopter"; see http://phenocopter.csiro.au, for example) that manages and summarizes flight data from multiple camera types, assists in the processing of images via commercial mosaicking software and stores the resulting files (orthophoto mosaic, point cloud, digital surface model). In Figure 10.2, the left-hand tasks are undertaken by a range of open-source or in-house software; the centre column of tasks are largely handled by PhenoCopter workflow; and the right-hand column of tasks where data is extracted and summarized for plots are either completed via customized R scripts or are undertaken by the integration pipeline described by Potgieter et al. (this series, Vol 1, Chapter 5). While we refer the reader to Chapman et al. (2014) for the original description of these steps, the remainder of this paper attempts to summarize issues affecting the quality and value of UAV collected data, to provide some guidance as to approximate current best practice in order to derive plot-level traits. We also aim to identify the opportunities to further refine methodologies and measure plant traits using UAV sensing. There are an increasing number of service providers, including most of the camera hardware providers, who have established cloud-based or local analysis software to provide most of the steps in the pipeline, from mission planning through to computation of indices, and in some cases, extraction of plot level data.

10.3   AN EXAMPLE DATA PROCESSING PLATFORM

Compared to ground vehicles, unmanned aerial vehicles (UAVs) are more convenient for capturing images over large scales in order to extract crop phenotypes. However, both types of vehicles face key challenges in how to efficiently manage all meta-information, process large numbers of images,

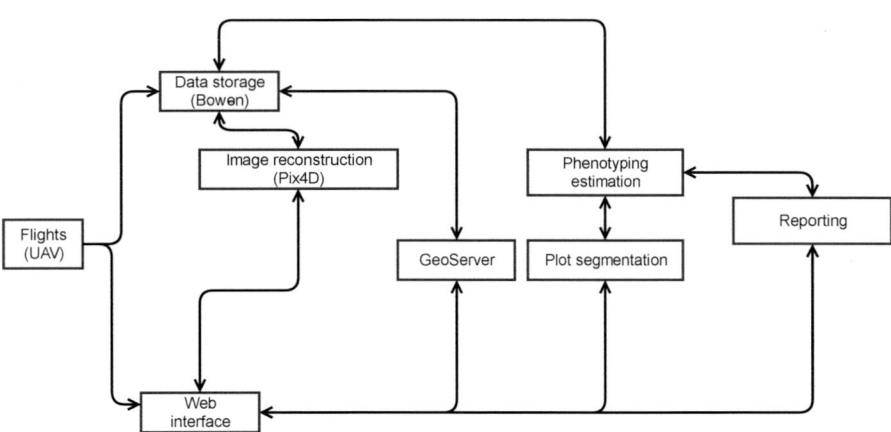

FIGURE 10.3 A pipeline of the PhenoCopter platform for UAV data processing. (Contact authors B.Z. or S.C.C. for information about accessing PhenoCopter software.)

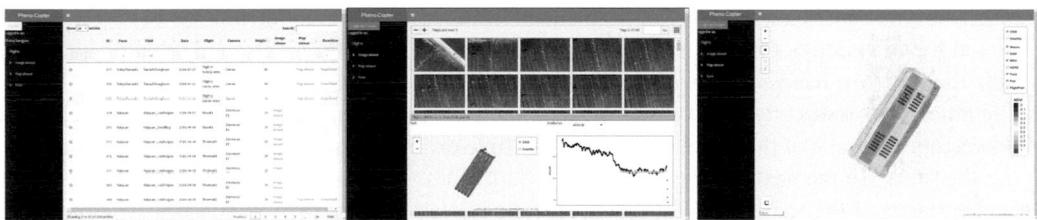

FIGURE 10.4 The major features of PhenoCopter platform including the flight management, image viewers, and mosaic viewers. The PhenoCopter website is developed using the R Shiny framework.

visualize intermediate outputs, and extract phenotypic values. A cloud based platform, PhenoCopter, is a tool for handling the challenge of data processing (Figure 10.3). The platform captures data processing steps of the whole workflow (Figure 10.2), and separates it into several steps: collecting meta-information, geotagging and checking raw images, reconstructing and tiling orthomosaic, segmenting mosaic into individual plots, and extracting plot data for conversion into different levels of phenotype traits, as discussed above.

Several servers are utilized, combining open source and commercial software. All meta-information and the status of data processing are stored in a MySQL database. The hierarchical structure is designed for separating flights and management of flight information including farm, field, date, and flight. Basic information on flights includes drone, camera, flight height, speed, capture interval, weather condition, etc. The raw mages, intermediate outputs, and final phenotypes are stored in the CSIRO cloud-based storage. The raw images are stitched together using commercial software*, which requires a Windows GPU (graphical processing unit) server and produces processed outputs for each flight, including a point cloud, orthomosaics, and digital surface model (DSM). At the time this is done, ground control points (GCPs) may be included as well as adjustments to calibrate spectral data using colored or gray-scale panels (see Haghighattalab et al., 2016, for appropriate methods). The orthomosaics, generated at same resolution as the original GSD (ground sample distance) are used to calculated vegetative indices and other phenotypic traits, and then are georetiled into GeoServer for web map service (Figure 10.4).

---

* Pix4DMapper, Pix4D S.A., Lausanne, Switzerland. pix4d.com

All raw images and outputs are visualized in a web interface, which is built using R Shiny framework, including three major features, that is, metaviewer for filter flights, image viewer for exploring raw images, and mosaic viewers for viewing mosaic on the zoomable maps (http://phenocopter.csiro.au, Figure 10.2). The plot segmentation is the key component for UAV applications in agriculture and breeding trials, which is implemented in a simple way using a four-cornered fishnet generated for a user-positioned rectangle defining the field. In the next step, all outputs are segmented into individual plots to extract phenotypic values. The final report is automatically generated using Rmarkdown language.

10.4   PROCESSING CORRECTED SIGNALS TO PLOTS AND DERIVING TRAIT DATA

There are multiple software solutions (desktop, web-based cloud providers, etc.) to process imagery to generate georectified results for a field or trial that usually comprise an orthomosaic photo of the original channels, a point cloud, and multiple raster files projected at the original or lower resolution. The raster outputs typically include a DSM) projection, a reflectance map, and layers of indices computed from the mosaic channels. Additional computations, particularly to identify "background" pixels from this nadir view, may be made to create layers of crop height (difference between ground level and top of canopy), COVER% (ratio of vegetation pixels to total pixels) and other object layers (pixels that contain heads or ears of grain, for example). From that point, researchers may aim to derive other traits indirectly, for example, by relating the time of maximum canopy height to the date of flowering. For most of these traits, there are multiple examples and refinements of methodologies in the literature. In the next sections, we mainly outline our own experiences in measuring crops at different stages of the season and for different traits related to plant breeding issues. The emphasis is on providing an overview of some of the different approaches being used for various traits and the remaining challenges.

10.4.1   EARLY-SEASON MEASUREMENTS: PLANT COUNTING AND GROUND COVER

Once processed from raw signal to L1 stage, the first derived L2 variables are often those related to segmentation of images due to differences in color, usually vegetation vs. background. Crop establishment is the most essential period in the growth of row crops. While some crops (tillering cereals like wheat, barley, and sorghum) are able to compensate for poor establishment, low plant number will still have impacts on early-season competitiveness against weeds, as well as slowing primary production per unit area due to lower leaf area indices (LAI). In crops where the planting density is relatively low (sorghum in Australia, ca. 50,000 plants per ha or fewer), direct segmentation of either RGB or multispectral imagery (Figure 10.5) at the right time after emergence can be sufficient to obtain accurate estimates (r^2 of 0.74–0.97) of plant counts in standard size plots (ca. 8 m^2). While multispectral cameras provide the opportunity to discriminate plants by using indices built from contrasting narrow-band wavelengths, there is still a role for higher pixel resolution RGB cameras, with their wider spectrum channels. Gnädinger and Schmidhalter (2017) found that a "decorrelation stretch" method could enhance the contrast sufficiently in RGB images to allow identification of individual maize plants, even when overlapping. In this process, the thresholding can be set to a value ("yellow and light green pixels") that identifies only the young leaves that are located in the center of the plants. These approaches become more difficult as the plants grow and leaves begin to overlap substantially. In these cases, advanced machine learning methods can be applied. For example Chen et al. (2017c) used a Multiple Instance Learning approach in order to classify pixels as being a plant center or a non–plant center. Even in relatively large overlapping sorghum crops, they achieved a precision of about 66%.

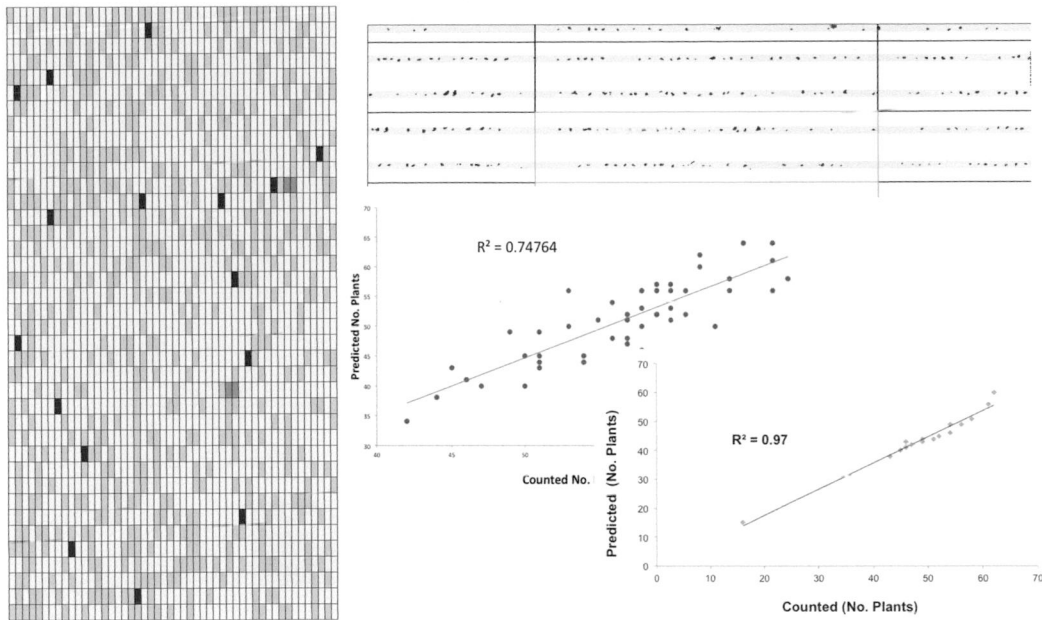

FIGURE 10.5 Counting sorghum plants as objects using a multispectral camera (A. Potgieter, unpublished data). From a segmentation of NDVI signal (top) for each trial plot in a mosaic, plants at about 3 emerged leaf stage (ca. 3 weeks after sowing) are able to be counted (plants in two rows × 4 m) in two separate trials (graphs) by thresholding of the foreground/background index to produce a heat map of plant count for a sorghum breeding trial in 2015 at Hermitage Research Station, Warwick, Australia.

Smaller plants, such as wheat, provide additional challenges to early-season counting due to their size relative to pixel resolutions of most UAV flights, and the overlapping of plants that occurs when sown at high within-row densities. Liu et al. (2017) have overcome this challenge by developing a method whereby they can fly low (ca. 3–5 m) across plots with a high-resolution camera angled at about 45°. The method then applies a series of steps (Figure 10.6) that allows the identification of wheat seedlings as "green pixel objects," the bases of which intersect with nongreen soil. The method is then trained to identify and classify objects as comprising one or more plants in order that they can be counted. These methods are currently being improved further through the development of databases of small plants and the generation of "artificial scenes" comprising collections of randomly sampled single plant images that can be orientated in structured ways. This allows the development of methods to better handle different plant sizes and arrangements, for example, row spacing and within-row density.

Plant counting is essential in most experimental situations in order to monitor variation in emergence due to differences in seed quality and/or responses of genotypes to different climatic, soil, and management conditions. Plant counts are often used by breeders as covariates to try to adjust final season yield data for establishment effects.

The next phenotype that is monitored by breeders is crop cover. In the early part of the season, this trait relates to ability of genotypes to grow quickly to cover the ground, which serves three purposes: (1) reducing evaporation of water from wet soil (which is lost to the plant); (2) competing against weed species so that they do not compete for light, water, and nutrients and are less likely to contaminate any crop product; and (3) efficiently intercepting the greatest amount of light per unit soil area in order to support higher crop growth rates. In lower-resolution situations (satellite data, etc.), the change in canopy cover (roughly estimated as the proportion of green to nongreen pixels) has typically been approached with the development of statistical models of reflectance indices as related to canopy cover (Wiesmair et al., 2016). However, these methods are less reliable

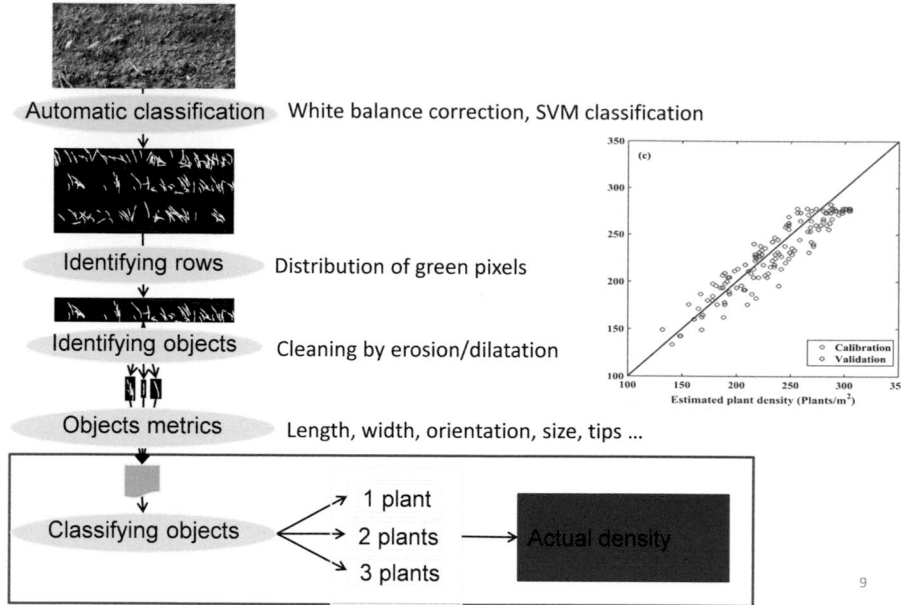

FIGURE 10.6 Identification of young wheat plants to count plant density per unit area. After objects are isolated and sized from rows, the data are processed to predict intersecting objects and estimate their number. (From Liu S et al. 2017. *Frontiers in Plant Science* 8.)

in low-resolution situations where there is substantial variation in color among genotypes, as lighter-colored genotypes (against a darker soil) would likely receive a lower index value even when their canopy cover was the same as that of a darker green genotype.

In UAV image processing, there are two major sources of error for estimation of canopy cover: error associated with resolution (as described above) and error associated with artifacts and issues of extracting data from image mosaics, which is the usual way in which large numbers of plots are analysed. Figure 10.7 shows a set of crop cover values on the x-axis extracted from about 90 plots of wheat (genotype and nitrogen treatments were applied) for hand-held images taken at about 1.8 m altitude with a 20 Mpx camera about every two weeks across the entire season. On the x-axis is the ground coverage, estimated by an efficient illumination-invariant method (Guo et al., 2013, 2017) for images where the ground sample distance (GSD) or pixel size, was about 0.2 mm (reference canopy cover). The pixels were then sequentially degraded using the cubic interpolation algorithm in the R package imager. Cubic interpolation is a standard method for image degradation, which involves fitting a series of cubic polynomials to the brightness values contained in 16 nearest pixels (4 × 4) surrounding the calculated pixel. Then the canopy cover was computed again, for different pixel sizes (equivalent flight heights, with the 20 Mpx camera). The canopy cover estimates were biased when pixel size was increased from 0.1 to 2 cm (Figure 10.8), which would be the pixel size for flights at about 50 m altitude, using the same camera. In general, the ground coverages were underestimated at larger pixel size when reference ground coverages were less than a certain value (about 0.7 for this resolution) and vice versa. The largest absolute errors occurred when reference ground coverage (0.1 cm pixels) was between about 0.2 and 0.5, while the lower resolution ground coverage (2 cm pixels) was up to 0.3 units underestimated. This effect is associated with mixed pixels (dark soil and wheat leaves) being more often classified as soil than as leaf. The expectation is that the magnitude of this error will change with the relative size of the leaves of the crop (we see smaller effects in larger-leafed crops like sorghum) and also with the color of the soil background. It is therefore possible to derive some kinds of approximate rules to plan the flights in order to achieve appropriate resolutions.

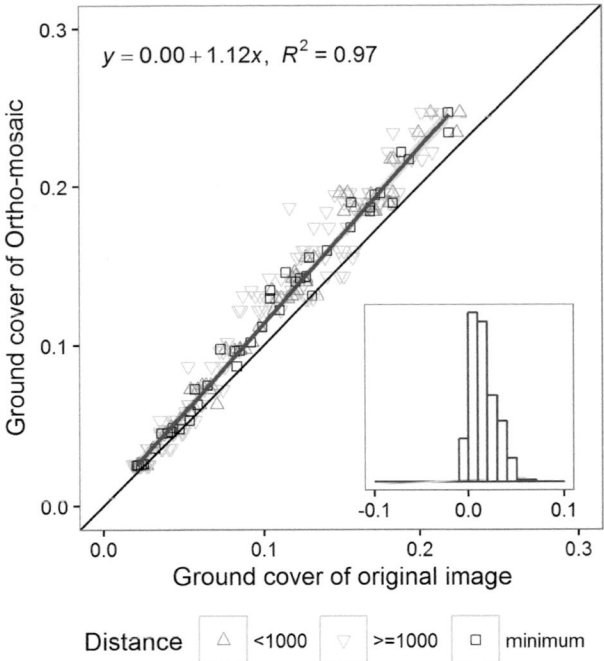

FIGURE 10.7   For a cotton density trial, overprediction of ground cover when extracted from the mosaic as compared to from original images. The symbols represent the plot position in the original images (i.e., the center of the plot either less than or greater than 1000 pixels). The "minimum" symbol represents the value of ground cover for a plot in the image in which it was closest to the center of the image. (Adapted from Duan T et al. 2017. *Functional Plant Biology* 44: 169–183.)

The orthophoto resulting from the mosaicking process typically contains artifacts that relate to combining data from pixels in multiple overlaying files (Duan et al., 2017). The artifacts include things like "ghost leaves" (where a leaf has moved between images and appears twice, usually once being semitransparent), and artifacts due to stitching algorithms that are sometimes only apparent as discontinuous lines through the vegetation. Typically the next step is to apply some kind of shape-file of the experimental layout of plots to the orthomosaic in order to extract subfiles of data that represent each plot. Alternatively, the projection relationship between the orthomosaic and the original images (which usually contain several plots) can be used to extract plot data from the original input images, that is, without the distortions and artifacts that occur in an orthomosaic (Duan et al., 2017). This procedure extracts plots from every image in which they appear, but some of these images should be discarded due to distorted views of the plot (Figure 10.7). It does have the advantage, of providing multiple samples of ground cover estimates per physical plot in the field, although data from images where the plots are close to the edge of the image are more likely to be slightly more variable in their estimates of ground cover.

Canopy cover and its dynamics are key L2 variables in quantifying plant growth, that is, response to environment (see examples for these in Potgieter et al. (2017) [sorghum] and Makanza et al. (2018) [maize]). Ground monitoring of NDVI using NDVI sensors has been a typical method for tracking the health of different treatments and Duan et al. (2017) showed that this type of dynamic monitoring throughout the season could be readily achieved using a multispectral camera that could be adjusted for canopy cover using high-resolution ground photos. These measurements can then estimate the crop condition. Together, canopy cover and canopy condition influence the light interception and radiation use efficiency, which are two L3 variables with direct associations with crop photosynthesis and growth.

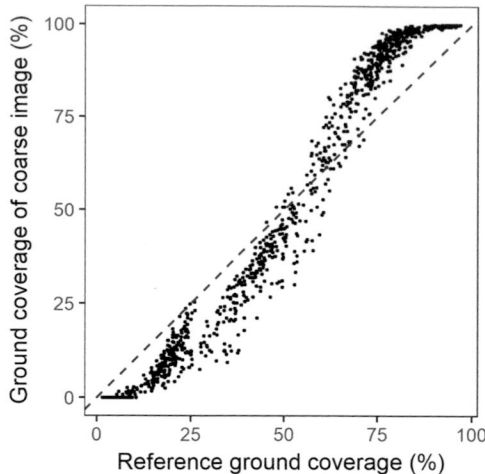

FIGURE 10.8   Influence of pixel-mixing as camera resolution is decreased from 0.1 cm per pixel to 2 cm per pixel for a season-long measurement of canopy cover in wheat (see text). (P. Hu, B. Zheng, S. Chapman, unpublished data.)

10.4.2   Midseason Measurements: Estimating Light Interception and Radiation Use Efficiency

Crop biomass can be described as the product of incident photosynthetically active radiation (PAR), the fraction of light intercepted (FIPAR) and the radiation use efficiency (Monteith, 1965). Given that canopies are comprised of layers of leaves, the canopy cover (as measured by a nadir image classification of vegetative vs. background pixels) is not quite the same as the fraction of light interception intercepted by the crop. Canopies vary in their transparency to PAR and also change with environmental conditions (e.g., in response to stress) and as they grow. Light is also reflected up from the soil and residue on the soil. Light interception by crop canopies is often approximated as a function of leaf area index and a light extinction coefficient (k) (FIPAR $= 1 - \exp^{(-k.LAI)}$), so it is helpful to be able to predict these two L3 variables. Potgieter et al. (2017) describe the seasonal monitoring of LAI indirectly using functions based on the normalized difference red edge (NDRE) index . The functions were not global and had to be re-fitted for early and midseason canopy states. However, in large-scale breeding trials of hundreds to thousands of plots, there are ways to deal with this issue by building "local" calibrations, that is, for any trait of interest (height, leaf area index, and possibly biomass), there can be a small number of plots (typically 20–50 are sufficient) that can be manually measured and sacrificed from the trial, either as whole plots or quadrat samples of 1–3 m^2 within plots. Across a wide range of LAI in sorghum, correlation coefficients between quadrat values and an NDRE index were greater than 0.8 for quadrats from 30 plots. In a similar way, Tewes and Schellberg (2018) showed that they could estimate both canopy cover and reflectance using modified RGB cameras (with NIR filter removed). By estimating the change in these parameters over time, as the canopies increased in LAI (increased canopy cover), and then senesced (reduced reflectance), they were able to compute the seasonal radiation use efficiency of the crop, after manually sampling the biomass at several points, that is, such that RUE, in the units of g biomass per MJ PAR intercepted, is computed as the total above-ground dry biomass (g m^{-2}) divided by the sum of the daily products of weather station PAR (MJ m^{-2}) and FIPAR (see equation in Figure 10.9). Local calibration, this time nondestructive, is again of assistance in estimating FIPAR from UAV. For a subset of microplots, the FIPAR phenotype can be calculated using a method based on digital hemispherical photography (DHP), taking photographs (ca. 10 per plot) upwards from the base of the canopy (Jonckheere et al., 2004). A flight was made using a six-band multispectral camera, and, averaged for each microplot,

$$\Delta_{t1}^{T2} Biomass = LUE \int_{t1}^{t2} FIPAR \cdot PAR_i \cdot dt$$

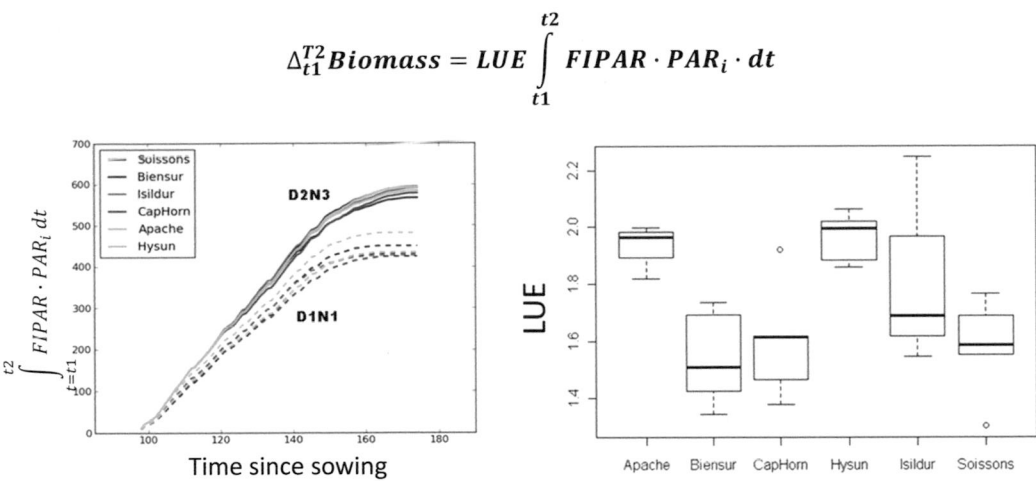

FIGURE 10.9 When FIPAR can be estimated over time, and biomass is measured at two points, then LUE can be computed. For six genotypes of wheat grown at two nitrogen fertilizer levels (left) Intercepted radiation was computed daily, with $FIPAR_{DHP}$ updated by weekly flights. The sum of intercepted radiation could be then divided into the change in biomass between two harvests (right) in order to estimate the seasonal LUE (S. Liu, F. Baret, INRA, unpublished data).

the set of corrected reflectances was used to develop a multiple linear regression of FIPAR across the plots in which the DHP method was used (Figure 10.10) (Baret et al., in press). In this paper, the authors also calibrated the relationship between $FIPAR_{DHP}$ for each flight during a hot day over a water-stressed crop. This allowed them to provide a quantitative score of "leaf-rolling" as the canopy changed during the day with stress condition. This score was well correlated with a visual score made by plant breeders. Hence, by measuring the $FIPAR_{DHP}$ for a subset of diverse plots, the derived relationship could be applied to hundreds or thousands of plots that had been captured during the same flight.

The method described above for maize has also been applied to wheat breeding trials in which biomass was assessed at two times in each plot. Using calibrated $FIPAR_{DHP}$ estimates through the season (approximately weekly), the light use efficiency (same as RUE above) was computed as differing substantially for six different wheat genotypes. This demonstrates that it is now possible to derive a complex L3 variable, such as LUE (or RUE) for a large number of genotypes, something it has previously not been possible to do at this scale.

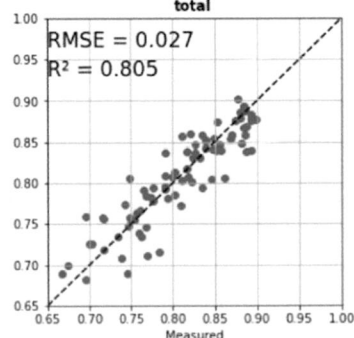

- Measurement of FIPAR over a set of microplots
  From DHP measurements
- Calibration of a regression for each flight:
  $$FIPAR = \alpha_0 \sum_{i=1}^{6} \alpha_i \cdot R_i / R_{ref}$$

FIGURE 10.10 A local ground calibration for a subset of plots allows estimation of FIPAR for the entire field (unpublished data, INRA https://www.biorxiv.org/content/biorxiv/early/2017/10/11/201665.full.pdf).

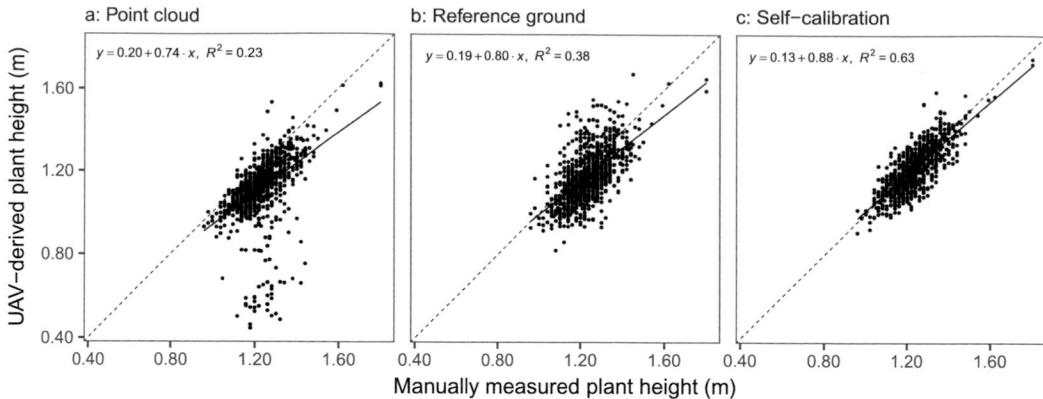

FIGURE 10.11   A local-calibration approach to estimate canopy height in sorghum. (From Hu P et al. 2018. *European Journal of Agronomy* 95: 24–32.)

10.4.3   Midseason Measurements: Canopy Height and Morphological Variation

Two measurements that are frequently made by breeders during the mid part of the season are scores or observations of flowering date, and scores or observations of canopy height. In large nurseries (thousands of plots), these scores can take a substantial amount of labor to collect. An additional trait of interest is to characterize the structure of the canopy as it relates to its photosynthetic function—the distribution of photosynthetic components and direct or indirect effects of crop lodging can both be measured through use of appropriate sensors and flight plans.

Estimation of crop height seems to be relatively straightforward, with multiple researchers approximating height to be about the 95th–99th percentile of data of a point cloud, for a given plot area (Holman et al., 2016; Watanabe et al., 2017). It is more challenging to estimate the actual value by identifying the absolute height of the largely obscured soil surface. A local (or self-) calibration approach was used to estimate canopy height of sorghum genotypes, within a relatively small range of about 0.5 m (Hu et al., 2018). This method required the manual measurement of about 30 plots (of more than 1000) which took about the same time as was needed to complete a UAV flight with an RGB camera on-board (Figure 10.11). This 'local-calibration' method was more precise than trying to use a reference ground height derived from an earlier flight, and also was better than a method of trying to measure the soil surface in a mature crop in which there were few gaps that could be used to estimate a surface.

In dense crops, LIDAR sensors provide a potential alternative, now that they are light enough to be carried on UAVs. In a dense crop of sugarcane, we were able to use a modified LIDAR* flown at 15 m altitude, to obtain sufficient returns that we could easily extract the ground surface using LAStools (https://rapidlasso.com/lastools/) (Figure 10.12).

The structure of a canopy is another phenotype measured by breeders. Chapman et al. (2014) demonstrated a method to estimate the proportion of lodging in a wheat crop based on analysis of a pixel height variance within a plot, and semiautomatic setting of a threshold to separate a bimodal distribution of heights (lodged/not-lodged). Taller crops like sugarcane exhibit variation in canopy structure (Figure 10.12) that, when derived from a multispectral camera, can provide a profile of NDVI versus height, for example. In addition to showing the degree of lodging, this type of analysis shows whether the lodged parts of the canopy remain photosynthetically active. Lodging is common in high-rainfall zones and breeders aim to select genotypes that resist lodging (unlike the genotype in the lower central panel) and that have canopies that continue to function while they recover from lodging events (Figure 10.13).

---

* HoverMap, Brisbane, Australia http://www.emesent.io/

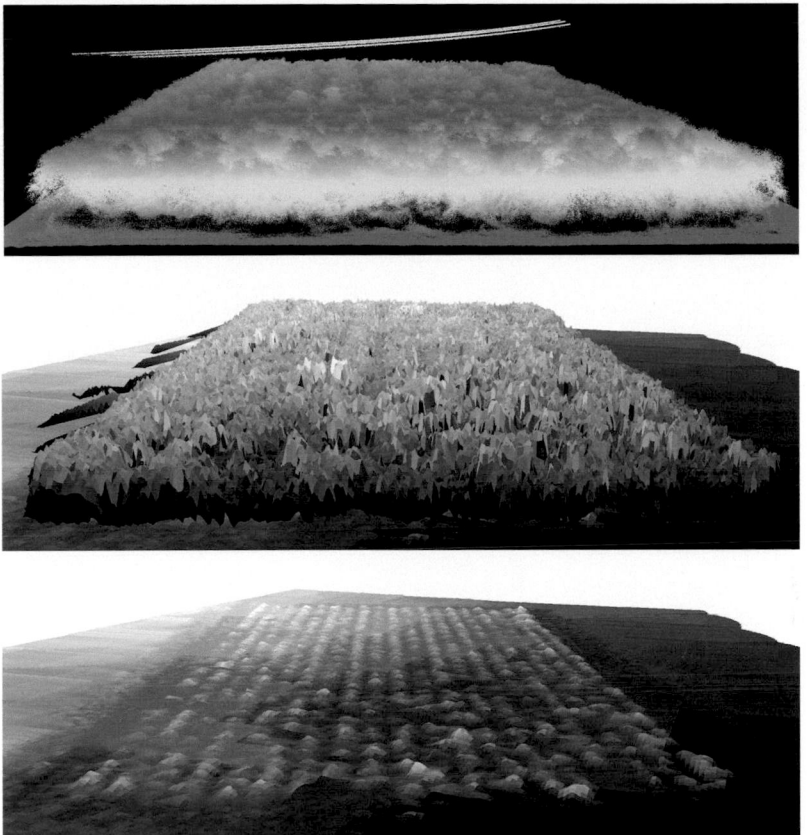

FIGURE 10.12   LIDAR mapping of sugarcane allows extraction of a mesh ground surface (bottom) from the canopy point cloud (top) and canopy mesh (middle). (J Sofonia, Y Shendryk, S Chapman, CSIRO, unpublished data.)

In cereal crops, the date of maximum canopy height can potentially be used as a way of estimating the date of flowering (Madec et al., 2017). This methodology was more reliable in a nonstressed treatment rather than one in which water stress was allowed to develop (Figure 10.14). These types of relationships become useful in conditions where it is difficult to obtain sufficiently high quality images to identify flowering heads, which may only flower for 1–3 days in any case.

10.4.4   MIDSEASON MEASUREMENTS: THERMAL CONDITION

Canopy or leaf temperature is an important physiological trait affecting the rates of key physiological processes such as biochemical reactions and cell growth and division, and is associated with heat and drought tolerance and yield. Furthermore, canopy temperature is recognized as a potential trait for the control of irrigation scheduling (Fischer et al., 1998; Pinto et al., 2010; Cossani and Reynolds, 2012) and for the selection of new varieties for high yield or drought tolerance. Compared with other thermal imagery techniques, UAV-based thermal imagery is more flexible in spatial and temporal resolution of data acquisition, since they allow coverage of specific areas at low flight altitude to obtain high-resolution data at key times of the day or stages of crop growth (Chapman et al., 2014).

Thermal imagery converts the energy of infrared radiation to imagery by camera/sensor, where the energy emitted from the object surface is known to be a function of surface temperature (Prashar and Jones, 2014; Sugiura et al., 2016). As a result, factors affecting energy emission, transmission, sensing, and conversion influence thermal imagery. These factors have been well reviewed by

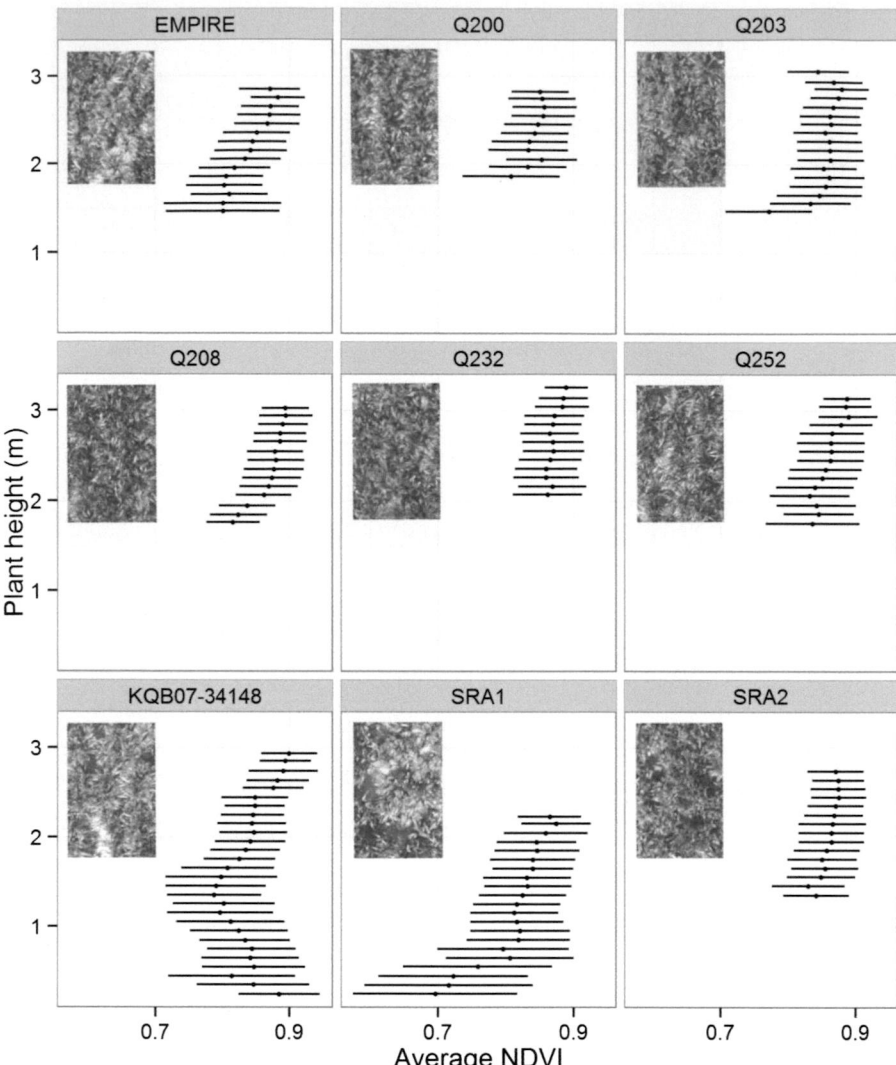

FIGURE 10.13 NDVI down the canopy in sugarcane. Distribution of plant height pixels for plots of partially–lodged sugarcane genotypes, as measured using a DSM derived from a multispectral camera. The NDVI for pixels in 5 cm height classes across each plot are shown. (S. Chapman, T. Duan, B. Zheng, CSIRO, unpublished data.)

(Prashar and Jones, 2014), including dynamic environmental conditions, such as air temperature, relative humidity, wind speed, cloud covering, and sun direction; camera or sensor settings, such viewing distance, viewing angle, field of view; and canopy characteristics, including leaf angle, canopy height, and albedo. One initial issue is simply trying to extract the data from vegetative tissue without including effects from the background. Figure 10.15 shows an example of doing this for higher-resolution thermal data that has been merged with RGB data, with both being collected from a ground vehicle (ca 3 m altitude) over sorghum. This approach should be able to be employed more easily with the availability of fused RGB/thermal cameras that can be carried on UAVs.

In order to improve usability of UAV-based thermal imagery, many methods were developed to acquire high-quality thermal images or correct thermal images, such as calibrating sensors in the laboratory before flight, setting ground thermal reference, image processing techniques to properly align thermal images, and atmospheric correction. These methods mainly corrected the factors that physically impact on canopy temperature. However, a major challenge with UAV sensors is that

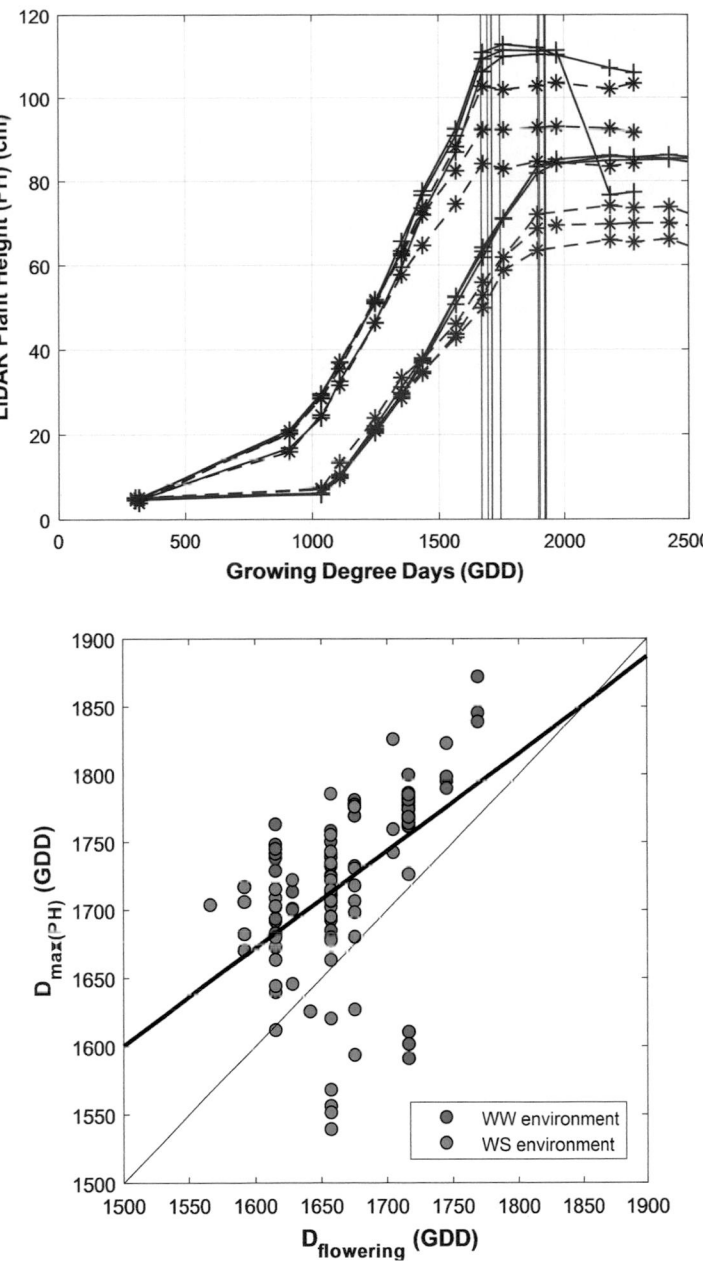

FIGURE 10.14 Estimation of date of maximum canopy height is correlated with flowering date in wheat. Data are shown for change in height of six different genotypes of wheat grown under two water stress environments (left) for comparison of the degree day date of flowering compared to the degree day date of maximum height. (From Madec et al., 2017 *Frontiers in Plant Science* 8) These data were derived from a ground vehicle but are being validated for data collected by UAV.

their size means that it is not possible to build in all of the correction (and cooling) of larger sensors. Figure 10.16 shows the preliminary effects of making adjustments for the drift in the temperature of the sensor as well as spatial artifacts associated with sensor angle (even while carried on a gimbal). This area is in need of further investment in order to develop reliable solutions that match those possible using piloted aircraft (Deery et al., 2016).

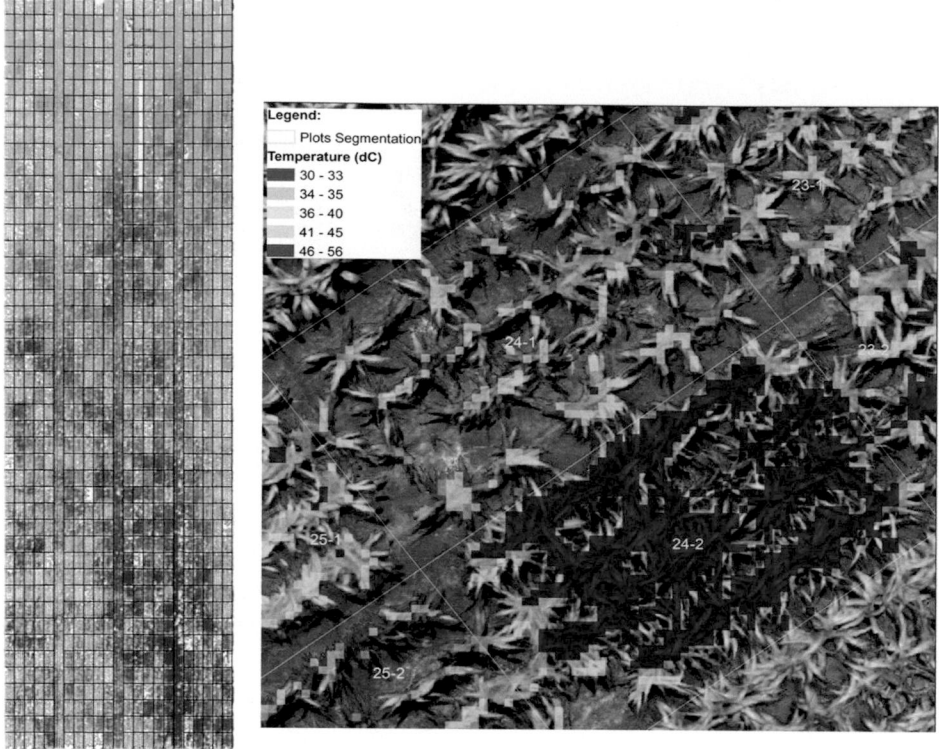

FIGURE 10.15   Subsetting thermal data from UAV using masks derived from RGB. The mask uses the pixels associated with vegetation and then trims the outermost pixel from the mask to reduce the error associated with mixing of pixels. This method is applied here using a thermal camera mounted at 3 m on a ground vehicle and is being tested for use on UAVs. (A. Potgieter, J. Watson, The University of Queensland, unpublished data.)

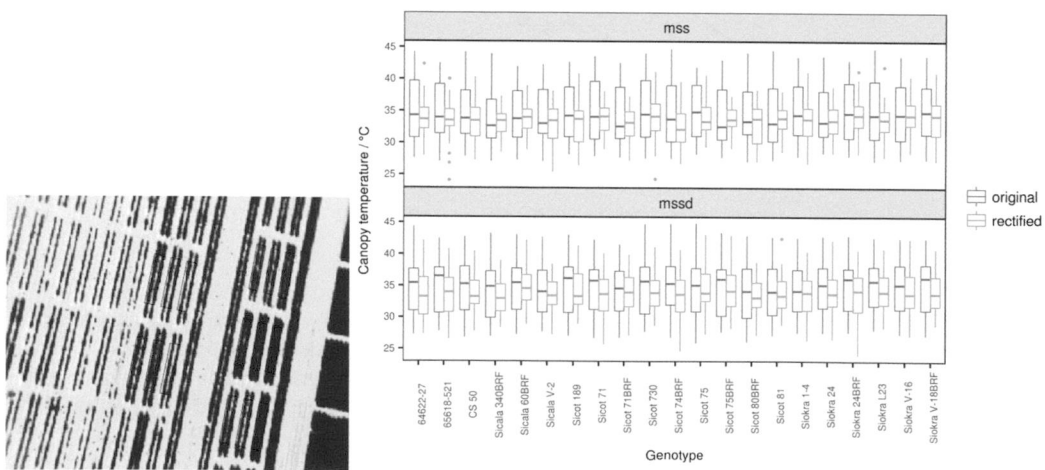

FIGURE 10.16   The genotypic variation of canopy temperature for a cotton population. The original temperature is captured by a thermal camera attached to a UAV (plots are illustrated in the false-color image). The rectified temperature accounts for the spatial and temporal correction of thermal images. These data are not yet sufficiently resolved for these two drought treatments (panels) in order to distinguish between genotypes (*x*-axis).

10.4.5 Late-Season Measurements: Ear or Head Counts and Canopy Senescence

The yield of cereal crops is distributed across multiple ears/heads/spikes in different branching arrangements. As the grain filling process occurs, the crop transfers resources from the canopy and stem into the grain, which results in crop canopy senescence. For crops like wheat and sorghum, each planted seed can give rise to multiple tillers, so that, per unit area, the number of heads can be greater than the number of plants. At harvest, plant breeders will often take subsamples of grain from the header in order to determine seed characteristics (mainly grain size or weight). However this does not provide information on the genotypic variation in the number or fertile heads produced per plant. Tillering through the season affects the development of leaf area, and also the competition among grains during grain filling, and, depending on the environment, it may be desirable to select for more or fewer tillers (heads) per plant. Counting of heads provides multiple challenges, from the need to distinguish between heads and background, or to count heads that are almost adjacent, and an issue that is often not anticipated—the difficulty of counting heads when the plants have leaned into adjacent plots, that is, the heads in the field can be counted, but it may be difficult to assign individual values accurately to the plots in the field.

In a collaboration between the University of Tokyo, UQ and CSIRO, we have prototyped a simple two-step machine learning–based image processing method to detect and count the number of heads from high-resolution images captured by UAVs in breeding trials. The first step is focusing on detecting possible sorghum head regions, based on the knowledge from a previous study (Potgieter et al., 2015; Guo et al., 2016), where we simply use color features to train a pixel-based segmentation model. First, seven classes (1, "Background soil"; 2, "Background shadow"; 3, "Background dead leaf"; 4, "Leaf"; 5, "Green heads"; 6, "Orange heads"; 7, "White heads") were defined and then for each class, a series of nine color features (r, g, b; H, S, V; L∗, a∗, b∗) from three color spaces (RGB, HSV, and L∗a∗b∗) was carefully collected from 17 images that had been chosen from the whole image dataset while considering light conditions and head colors. Then, using those features, a decision tree pixel classification model was trained and was applied to all the test images to classify their pixels into seven classes. Finally, the head-related pixels (class "Green heads," "Orange heads," "White heads") were selected and integrated together into "heads regions," as shown in (Figure 10.17).

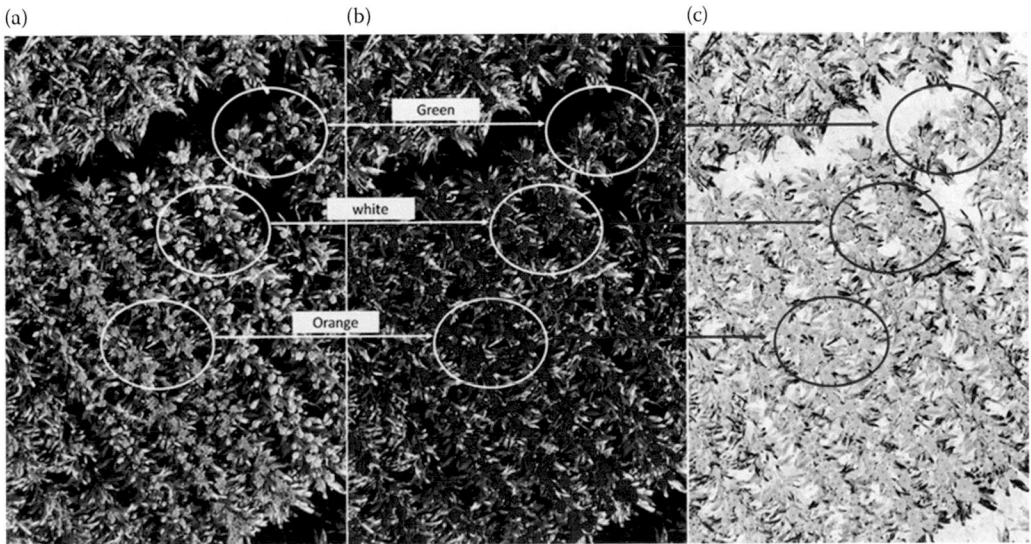

(a) (b) (c)

FIGURE 10.17 Head counting in sorghum using decision tree based on color and morphology. This method has about 83% accuracy across multiple types of heads in color and size. (W. Guo, S. Chapman, B. Zheng, A. Potgieter, University of Tokyo, CSIRO, UQ, unpublished data.)

The second step focused on counting the numbers of detected regions from the first step. First, all the training images were hand-labeled by point, then morphology features of all the candidate head regions and their corresponding numbers were extracted and used to traine a quadratic-SVM model. Then the model was applied to all the regions to count the numbers of heads in each region. To demonstrate the performance of the proposed method, fivefold cross-validation was applied to 52 manually labeled images. The precision and recall of head detection reached 0.87 and 0.98, respectively, and the coefficient of determination between manual and algorithm counting of heads number was 0.87. This methodology relied on appropriate selection of images to encompass a diversity of head types and colors, and now performs well for most grain sorghums that are grown in our breeding trials.

Recent advances in hardware, particularly regarding the GPU processing capacity, along with the availability of very large collection of labeled images from which computers can learn, have fostered enhanced methods in the field of computer vision and convolutional neural networks (CNNs) (Krizhevsky et al., 2012). Nowadays CNNs are achieving impressive results for natural image classification and object localization. In other crop species, studies have shown that CNNs outperform classic hand-crafted feature descriptors and offers an alternative approach (Chen et al., 2017a; Pawara et al., 2017).

The potential of convolutional neural networks to provide accurate ear count and density using RGB images in field conditions was investigated by the INRA team. The RGB images were taken from the nadir view direction at 2.9 m distance to the ground. This results in a ground sampling distance between 0.010 and 0.016 cm/pixel and a footprint area between 0.25 and 0.56 m^2 (depends on the focal length used and the height of the wheat). Two images per plots were acquired.

The problem of ear counting was formulated as a counting by detection problem. The state-of-the-art object detector Faster-RCNN (Ren et al., 2017) was employed along with its implementation within the tensor flow object detection API (Huang et al., 2017). This has the benefit of taking advantage of a fine-tuning approach as the network was already trained to classify millions of images within thousands of different categories (Krasin et al., 2016). The ears were interactively identified in all the images, resulting in 240 images (20 genotypes × 3 replicates × 2 modalities × 2 images). Between 80 and 170 ears were included in an image.

The training and validation datasets were populated with different genotypes: 14 of the 20 genotypes were randomly selected for training and the remaining six were used for the validation. This resulted in 168 images for training (70%) and 72 images for validation (30%).

The problem of counting of wheat ears in high-resolution RGB images was solved (Figure 10.18). The results were similar when resampling the original images by a factor of 5 using a bilinear aggregation function. This makes possible the high-throughput counting of wheat ears and density estimation with UAV.

However, two general questions remain important in solving the deployment in large scale phenotyping platform issues.

- Can the developed networks handle a sufficient diversity of genotype in different conditions and stages? In other words, what is the scalability of the CNNs?
- Do the acquired images contain enough information for estimation of the ear density within a microplot? Are all the ears visible in the images? What is footprint area in the images is needed for it to be representative of the microplot?

In their unpublished study, S. Madec and F. Baret made a start in answering these questions and commented on those issues. With the emergence of deep learning it is certainly important to merge the datasets because the diversity of the training dataset is the key to good performance.

Monitoring canopy senescence (e.g., Christopher et al., 2016; Makanza et al., 2018) in cereal crops has long been of interest to plant breeders, as slow senescence indicates varieties that are able to

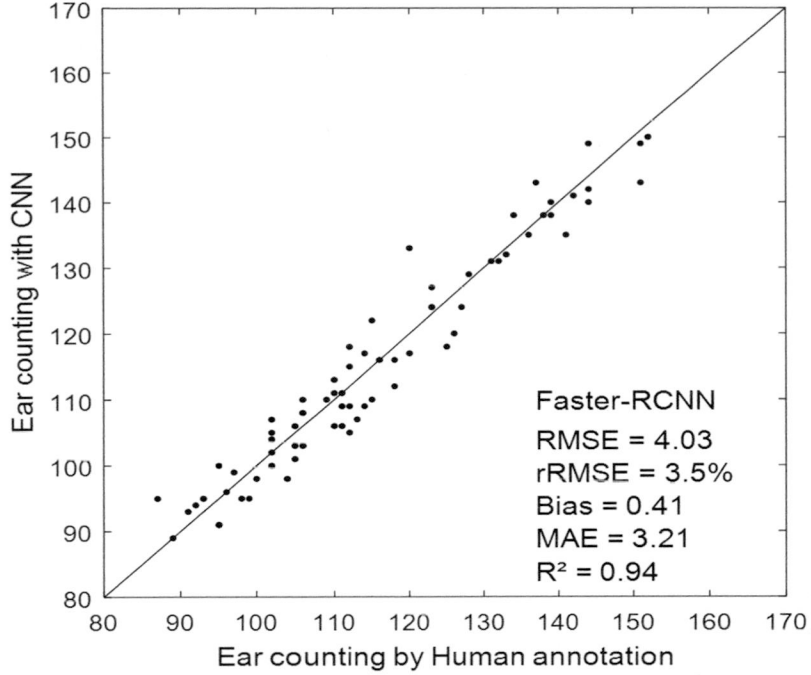

FIGURE 10.18   Comparison between the number of ears in each image by visual labeling and that estimated using the Faster-CNN and example of output from the original image.

maintain a functional canopy for a longer period, thereby having more time to accumulate biomass and increase yield. An example for sorghum (Figure 10.19) shows this type of association derived for a series of flights through the grain filling period (Potgieter et al., 2017).

Finally, it can be noted that phenotyping of breeding plots can continue even after crop harvest. For example, where stems are cut cleanly and stover is removed from the field, it is possible to use low UAV flights to count the stem density, as well as approximate stem diameter, by imaging the cut ends of the plants from a low altitude (F. Baret, unpublished data).

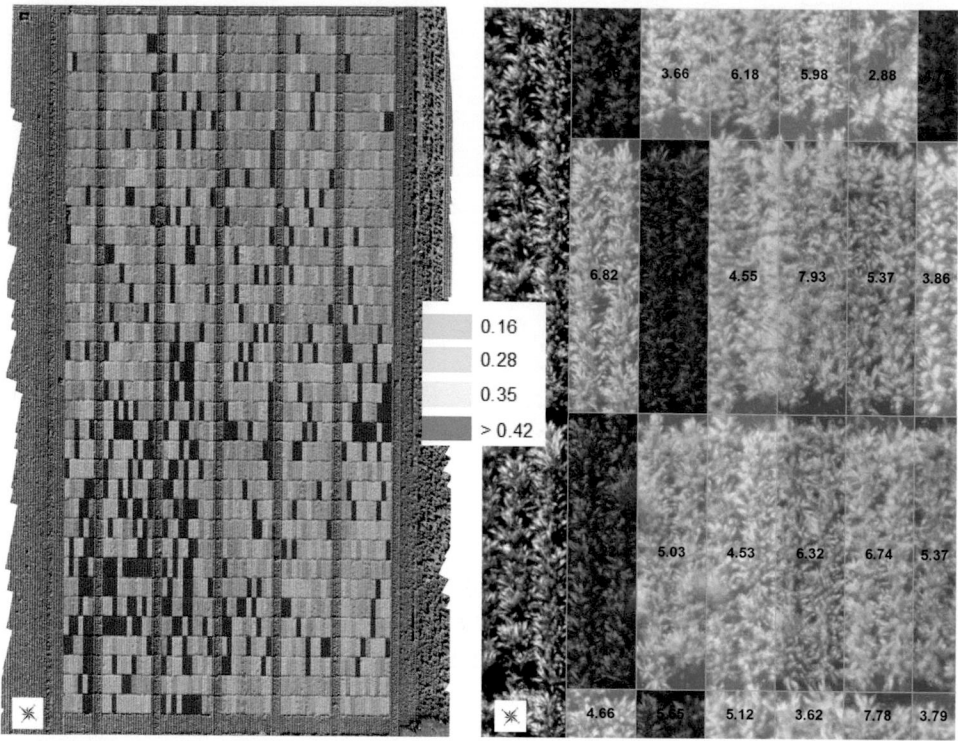

FIGURE 10.19 Change in NDVI as related to plot yield in a sorghum breeding trial. (*Left*) Difference in NDVI between maximum value and value at harvest. (*Right*) Colors show that shallow slope of (negative) change in NDVI from maximum to harvest are associated with higher grain yield (tonnes per ha) shown as text on graphic.

10.5 CONCLUSIONS

This area of application of proximal sensing to plant breeding and plant agronomy research is developing quickly. For breeding, these traits need to be shown to have sufficient precision to be able to contribute to improvement of adaptive growth traits. Using phenotypic traits to run simulation models is the current frontier in this research, that is, using models to generate new traits (e.g., light interception, water use) through the simulation of 3D canopies and/or the accumulation of biomass and utilization of resources (light, radiation, water) over the season. This chapter was arranged to provide a seasonal context for phenotyping, and to highlight where some of the opportunities and needs currently lie. From this we can suggest some guidelines around phenotyping for breeding applications where high resolutions are needed:

- The need for phenotyping and the development of new traits should be guided by physiological understanding of the controls (genetic and environmental) of plant growth. This provides the opportunity to select for more complex traits than can be directly measured—augmenting rather than simply replacing breeder-measured traits.
- Choice of equipment to obtain an appropriate resolution for the trait to be estimated at the stage of growth which it needs to be measured.
- Application of new methods to enhance image signals in order to better discriminate between vegetative and background pixels.
- Consideration of pixel resolution and mosaic artifacts in segmentation tasks.
- The use of "local" calibration of variables, by making direct measurements on small numbers of plots within large trials of hundreds to thousands of plots.

- Derivation of "hard-to-measure" traits like leaf-rolling and leaf senescence that have typically been "scored by eye."
- The need for development of public image databases that can be utilized to further train machine learning approaches to segmentation of vegetation characteristics.

ACKNOWLEDGMENTS

We would like to acknowledge other contributors to the work described here:

- Mark Elridge, Ken Law, James Watson (UQ/DAF)—contributions to sorghum trial analysis.
- Tao Duan, Pengcheng Hu (ex-CSIRO students)—contributions to wheat trial analysis.
- Yuri Shendryk and Jeremy Sofonia (CSIRO/UQ)—LIDAR data for sugarcane.
- Ayman Habib, Melba Crawford, Ed Delp, Javier Ribera, Yuhao Chen, Mitch Tuinstra (Purdue)—for various discussions about phenotyping methods.

REFERENCES

Adão T, Hruška J, Pádua L, Bessa J, Peres E, Morais R, Sousa JJ. 2017. Hyperspectral imaging: A review on UAV-based sensors, data processing and applications for agriculture and forestry. *Remote Sensing* 9: 1110.

Baret F, Madec S, Irfan K, Lopez J, Comar A, Hemmerlé M, Dutartre-Cohen D, Praud S, Tixier M. Leaf rolling in maize crops: from leaf scoring to canopy level measurements for phenotyping. *Journal of Experimental Botany*. In Press. doi: https://doi.org/10.1101/201665

Basnayake J, Jackson PA, Inman-Bamber NG, Lakshmanan P. 2012. Sugarcane for water-limited environments. Genetic variation in cane yield and sugar content in response to water stress. *Journal of Experimental Botany* 63: 6023 6033.

Bendig J, Bolten A, Bennertz S, Broscheit J, Eichfuss S, Bareth G. 2014. Estimating biomass of barley using crop surface models (CSMs) derived from UAV-based RGB imaging. *Remote Sensing* 6: 10395–10412.

Cabrera Bosquet L, Fournier C, Brichet N, Welcker C, Suard B, Tardieu F. 2016. High-throughput estimation of incident light, light interception and radiation-use efficiency of thousands of plants in a phenotyping platform. *New Phytologist* 212: 269–281. https://doi.org/10.1111/nph.14027.

Cairns JE, Sonder K, Zaidi PH, Verhulst N, Mahuku G, Babu R, Nair SK et al. 2012. Maize production in a changing climate: Impacts, adaptation, and mitigation strategies. In: *Advances in Agronomy*. Academic Press, 1–58.

Chapman SC, Merz T, Chan A, Jackway P, Stefan H, Dreccer MF, Holland E, Zheng BY, Ling J, Berni J. 2014. Pheno-Copter: A low-altitude, autonomous remote-sensing robotic helicopter for high-throughput field-based phenotyping. *Agronomy* 42: 279–301.

Chen SW, Shivakumar SS, Dcunha S, Das J, Okon E, Qu C, Taylor CJ, Kumar V. 2017a. Counting apples and oranges with deep learning: A data-driven approach. *IEEE Robotics and Automation Letters* 2: 781–788.

Chen Y, Ribera J, Boomsma C, Delp EJ. 2017b. Plant leaf segmentation for estimating phenotypic traits. In: *International Conference on Image Processing*. Beijing, China.

Chen Y, Ribera J, Boomsma C, Delp EJ. 2017c. Locating crop plant centers from UAV-based RGB imagery. In: *Proceedings of the IEEE Conference on Computer Vision and Pattern Recognition*, pp. 2030–2037.

Chenu K, Oosterom V, JE, McLean G, Deifel KS, Fletcher A, Geetika G, Tirfessa A et al. 2018. Integrating modelling and phenotyping approaches to identify and screen complex traits—Illustration for transpiration efficiency in cereals. *Journal of Experimental Botany*.

Christopher JT, Christopher MJ, Borrell AK, Fletcher S, Chenu K. 2016. Stay-green traits to improve wheat adaptation in well-watered and water-limited environments. *Journal of Experimental Botany* 67: 5159–5172.

Cossani CM, Reynolds MP. 2012. Physiological traits for improving heat tolerance in wheat. *Plant Physiology* 160: 1710–1718.

DeChant C, Wiesner-Hanks T, Chen S, Stewart EL, Yosinski J, Gore MA, Nelson R, Lipson H. 2017. Automated identification of northern leaf blight-infected maize plants from field imagery using deep learning. *Phytopathology*.

Deery DM, Rebetzke GJ, Jimenez-Berni JA, James RA, Condon AG, Bovill WD, Hutchinson P, Scarrow J, Davy R, Furbank RT. 2016. Methodology for high-throughput field phenotyping of canopy temperature using airborne thermography. *Frontiers in Plant Science* 7.

Duan T, Zheng B, Guo W, Ninomiya S, Guo Y, Chapman S. 2017. Comparison of ground cover estimates from experiment plots in cotton, sorghum and sugarcane based on images and ortho-mosaics captured by UAV. *Functional Plant Biology* 44: 169–183.

Elazab A, Ordóñez RA, Savin R, Slafer GA, Araus JL. 2016. Detecting interactive effects of N fertilization and heat stress on maize productivity by remote sensing techniques. *European Journal of Agronomy* 73: 11–24.

Espana ML, Baret F, Aries F, Chelle M, Andrieu B, Prevot L. 1999. Modeling maize canopy 3D architecture: Application to reflectance simulation. *Ecological Modelling* 122: 25–43.

Fischer RA, Rees D, Sayre KD, Lu Z-M, Condon AG, Saavedra AL. 1998. Wheat yield progress associated with higher stomatal conductance and photosynthetic rate, and cooler canopies. *Crop Science* 38: 1467–1475.

Gnädinger F, Schmidhalter U. 2017. Digital counts of maize plants by unmanned aerial vehicles (UAVs). *Remote Sensing* 9: 544.

Guo W, Rage UK, Ninomiya S. 2013. Illumination invariant segmentation of vegetation for time series wheat images based on decision tree model. *Computers and Electronics in Agriculture* 96: 58–66.

Guo W, Potgieter AB, Jordan D, Armstrong R, Lawn K, Kakeru W, Duan T et al. 2016. Automatic detecting and counting of sorghum heads in breeding field using RGB imagery from UAV. In: *CIGR-AgEng Conference*, Aarhus, Denmark.

Guo W, Zheng B, Duan T, Fukatsu T, Chapman S, Ninomiya S. 2017. EasyPCC: Benchmark datasets and tools for high-throughput measurement of the plant canopy coverage ratio under field conditions. *Sensors* 17: 798.

Haghighattalab A, González Pérez L, Mondal S, Singh D, Schinstock D, Rutkoski J, Ortiz-Monasterio I, Singh RP, Goodin D, Poland J. 2016. Application of unmanned aerial systems for high throughput phenotyping of large wheat breeding nurseries. *Plant Methods* 12: 35.

Holman FH, Riche AB, Michalski A, Castle M, Wooster MJ, Hawkesford MJ. 2016. High throughput field phenotyping of wheat plant height and growth rate in field plot trials using UAV based remote sensing. *Remote Sensing* 8: 1031.

Hu P, Chapman SC, Wang X, Potgieter A, Duan T, Jordan D, Guo Y, Zheng B. 2018. Estimation of plant height using a high throughput phenotyping platform based on unmanned aerial vehicle and self-calibration: Example for sorghum breeding. *European Journal of Agronomy* 95: 24–32.

Huang J, Rathod V, Sun C, Zhu M, Korattikara A, Fathi A, Fischer I, Wojna Z, Song Y, Guadarrama S. 2017. Speed/accuracy trade-offs for modern convolutional object detectors. In: *IEEE CVPR*.

Jay S, Maupas F, Bendoula R, Gorretta N. 2017. Retrieving LAI, chlorophyll and nitrogen contents in sugar beet crops from multi-angular optical remote sensing: Comparison of vegetation indices and PROSAIL inversion for field phenotyping. *Field Crops Research* 210: 33–46.

Jonckheere I, Fleck S, Nackaerts K, Muys B, Coppin P, Weiss M, Baret F. 2004. Review of methods for *in situ* leaf area index determination: Part I. Theories, sensors and hemispherical photography. *Agricultural and Forest Meteorology* 121: 19–35.

Kefauver SC, Vicente R, Vergara-Díaz O, Fernandez-Gallego JA, Kerfal S, Lopez A, Melichar JPE, Molins S, D M, Araus JL. 2017. Comparative UAV and field phenotyping to assess yield and nitrogen use efficiency in hybrid and conventional barley. *Frontiers in Plant Science* 8.

Krasin I, Duerig T, Alldrin N, Veit A, Abu-El-Haija S, Belongie S, Cai D, Feng Z, Ferrari V, Gomes V. 2016. Openimages: A public dataset for large-scale multi-label and multi-class image classification. Dataset available from https://github.com/openimages 2: 7.

Krizhevsky A, Sutskever I, Hinton GE. 2012. Imagenet classification with deep convolutional neural networks. In: *Advances in Neural Information Processing Systems*. pp. 1097–1105.

Lelong CCD, Burger P, Jubelin G, Roux B, Labbé S, Baret F. 2008. Assessment of unmanned aerial vehicles imagery for quantitative monitoring of wheat crop in small plots. *Sensors* 8: 3557–3585.

Liebisch F, Kirchgessner N, Schneider D, Walter A, Hund A. 2015. Remote, aerial phenotyping of maize traits with a mobile multi-sensor approach. *Plant Methods* 11: 9.

Liu S, Baret F, Andrieu B, Burger P, Hemmerlé M. 2017. Estimation of wheat plant density at early stages using high resolution imagery. *Frontiers in Plant Science* 8.

Madec S, Baret F, de Solan B, Thomas S, Dutartre D, Jezequel S, Hemmerlé M, Colombeau G, Comar A. 2017. High-throughput phenotyping of plant height: Comparing unmanned aerial vehicles and ground LiDAR estimates. *Frontiers in Plant Science* 8.

Makanza R, Zaman-Allah M, Cairns JE, Magorokosho C, Tarekegne A, Olsen M, Prasanna BM. 2018. High-throughput phenotyping of canopy cover and senescence in maize field trials using aerial digital canopy imaging. *Remote Sensing* 10: 330.

Martínez J, Egea G, Agüera J, Pérez-Ruiz M. 2016. A cost-effective canopy temperature measurement system for precision agriculture: A case study on sugar beet. *Precision Agriculture* 1–16.

Messina CD, Technow F, Tang T, Totir R, Gho C, Cooper M. 2018. Leveraging biological insight and environmental variation to improve phenotypic prediction: Integrating crop growth models (CGM) with whole genome prediction (WGP). *European Journal of Agronomy*.

Meuwissen THE, Hayes BJ, Goddard ME. 2001. Prediction of total genetic value using genome-wide dense marker maps. *Genetics* 157: 1819–1829.

Monteith JL. 1965. Light distribution and photosynthesis in field crops. *Annals of Botany* 29: 17–37.

Pauli D, Andrade-Sanchez P, Carmo-Silva AE, Gazave E, French AN, Heun J, Hunsaker DJ et al. 2016. Field-based high-throughput plant phenotyping reveals the temporal patterns of Quantitative Trait Loci associated with stress-responsive traits in cotton. *G3: Genes, Genomes, Genetics*. g3.115.023515.

Pawara P, Okafor E, Surinta O, Schomaker L, Wiering M. 2017. Comparing local descriptors and bags of visual words to deep convolutional neural networks for plant recognition. In: *ICPRAM*. pp. 479–486.

Pinto RS, Reynolds MP, Mathews KL, McIntyre CL, Olivares-Villegas J-J, Chapman SC. 2010. Heat and drought adaptive QTL in a wheat population designed to minimize confounding agronomic effects. *Theoretical and Applied Genetics* 121: 1001–1021.

Potgieter A, Jordan D, Hammer G, Armstrong R, Chapman S, Guo W. 2015. The use of *in situ* proximal sensing technologies to determine crop characteristics in sorghum crop breeding. In: *Tropical Agriculture Conference 2015*. Brisbane, Australia.

Potgieter AB, George-Jaeggli B, Chapman SC, Laws K, Cadavid SAL, Wixted J, Watson J et al. 2017. Multi-spectral imaging from an unmanned aerial vehicle enables the assessment of seasonal leaf area dynamics of sorghum breeding lines. *Frontiers in Plant Science* 8.

Potgieter AB, Watson J, George-Jaeggli B, Mclean G, Eldridge M, Chapman SC, Lawa K, Christopher J, Chenu K, Borrell A, Hammer GL, and Jordan DR. In Press. The use of hyperspectral proximal sensing for phenotyping of plant breeding trials. In: Hyperspectral Remote Sensing of Vegetation. Thenkabail, P.S., Lyon, G.J., and Huete, A. (Editors), pp. 125–145, Chapter 5. CRC Press/Taylor and Francis group, Boca Raton, London, New York.

Prashar A, Jones HG. 2014. Infra-Red thermography as a high-throughput tool for field phenotyping. *Agronomy* 4: 397–417.

Rebetzke GJ, Jimenez-Berni JA, Bovill WD, Deery DM, James RA. 2016. High-throughput phenotyping technologies allow accurate selection of stay-green. *Journal of Experimental Botany* 67: 4919–4924.

Ren S, He K, Girshick R, Sun J. 2017. Faster R-CNN: towards real-time object detection with region proposal networks. *IEEE Transactions on Pattern Analysis and Machine Intelligence* 39: 1137–1149.

Rutkoski J, Poland J, Mondal S, Autrique E, Pérez LG, Crossa J, Reynolds M, Singh R. 2016. Canopy temperature and vegetation indices from high-throughput phenotyping improve accuracy of pedigree and genomic selection for grain yield in wheat. *G3: Genes, Genomes, Genetics* 6: 2799–2808.

Sankaran S, Khot LR, Espinoza CZ, Jarolmasjed S, Sathuvalli VR, Vandemark GJ, Miklas PN et al. 2015. Low-altitude, high-resolution aerial imaging systems for row and field crop phenotyping: A review. *European Journal of Agronomy* 70: 112–123.

Sugiura R, Tsuda S, Tamiya S, Itoh A, Nishiwaki K, Murakami N, Shibuya Y, Hirafuji M, Nuske S. 2016. Field phenotyping system for the assessment of potato late blight resistance using RGB imagery from an unmanned aerial vehicle. *Biosystems Engineering* 148: 1–10.

Technow F, Messina CD, Totir LR, Cooper M. 2015. Integrating crop growth models with whole genome prediction through approximate Bayesian computation. *PLOS ONE* 10: e0130855.

Tewes A, Schellberg J. 2018. Towards remote estimation of radiation use efficiency in maize using UAV-based low-cost camera imagery. *Agronomy* 8: 16.

Verger A, Vigneau N, Chéron C, Gilliot J-M, Comar A, Baret F. 2014. Green area index from an unmanned aerial system over wheat and rapeseed crops. *Remote Sensing of Environment* 152: 654–664.

Watanabe K, Guo W, Arai K, Takanashi H, Kajiya-Kanegae H, Kobayashi M, Yano K et al. 2017. High-throughput phenotyping of sorghum plant height using an unmanned aerial vehicle and its application to genomic prediction modeling. *Frontiers in Plant Science* 8.

White JW, Andrade-Sanchez P, Gore MA, Bronson KF, Coffelt TA, Conley MM, Feldmann KA et al. 2012. Field-based phenomics for plant genetics research. *Field Crops Research* 133: 101–112.

Wiesmair M, Feilhauer H, Magiera A, Otte A, Waldhardt R. 2016. Estimating vegetation cover from high-resolution satellite data to assess grassland degradation in the georgian caucasus. *Mountain Research and Development* 36: 56–65.

Zaman-Allah M, Vergara O, Araus JL, Tarekegne A, Magorokosho C, Zarco-Tejada PJ, Hornero A et al. 2015. Unmanned aerial platform-based multi-spectral imaging for field phenotyping of maize. *Plant Methods* 11: 35.

Section III

Conclusions

11 Fifty Years of Advances in Hyperspectral Remote Sensing of Agriculture and Vegetation— Summary, Insights, and Highlights of Volume III
Biophysical and Biochemical Characterization and Plant Species Studies

Prasad S. Thenkabail, John G. Lyon, and Alfredo Huete

CONTENTS

The goal of this summary chapter is twofold. The first is to provide the reader an overview of the content of the preceding chapters. This they can read at the very beginning, before moving on to individual chapters in detail. Alternatively, they may read it at the very end to refresh their memory and to summarize the contents of the Volume. Second, this summary provides the editors' perspective, bringing in their rich collective experience and expertise to guide the reader. We have kept the summaries brief and illustrative, so that the reader can quickly gather essential initial knowledge and guidance. The in-depth details can be found in the individual chapters.

Hyperspectral remote sensing, also referred to as imaging spectroscopy, is the process of gathering data in narrow bands (10 nm or less) across the electromagnetic spectrum (e.g., 400–900 nm, or 400–2500 nm, or 400–14,000 nm). Hyperspectral data are typically gathered in 1–10 nm bandwidths per band, leading to hundreds or thousands of narrow bands over a wavelength range. However, even a few (e.g., 20) hyperspectral narrow bands (HNBs) selected continually along a wavelength (400–2500 nm) are also called hyperspectral data. It is not so much the number of bands that defines what is meant by hyperspectral data, but rather an adequate number (e.g., 20 or 30) of targeted nonredundant HNBs at specific portions of the spectrum. Nevertheless, in most cases, hyperspectral data are acquired continually in hundreds or thousands of HNBs, leading to spectral signatures of various earth features such as soils or vegetation of various types. Unlike the multispectral broad-band data, such as those collected from Landsat and the Advanced Land Imager, where data are acquired from only a selective broad-band wavelength range, hyperspectral data provides continuous signatures throughout the electromagnetic spectrum as illustrated using ground-based spectroradiometers and the spaceborne Earth Observing-1 (EO-1) Hyperion. Such data are rich spectrally, and also rich spatially when acquired with sufficient spatial resolution such as the centimeter resolution of spectroradiometers or the 30-m resolution of Hyperion. As a result, hyperspectral data bring opportunities to advance remote sensing studies of various kinds as well as create challenges due to the large data volume.

The chapters explore the challenges of overcoming data volumes through use of various data dimensionality reduction techniques to overcome redundant data and optimize nonredundant data for a given application; thereby unlocking the rich spectral signatures for analyses. Opportunities to advance remote sensing science in various vegetation and crop studies are presented and discussed in various chapters. Computation of hyperspectral vegetation indices (HVIs) and numerous other methods and techniques are central to most of the chapters. Recent advances in cloud computing, machine learning, and artificial intelligence have enabled a "paradigm shift" in how remote sensing science is approached and this greatly benefits hyperspectral remote sensing data analysis and applications.

11.1 REMOTE ESTIMATION OF CROP BIOPHYSICAL CHARACTERISTICS AT VARIOUS SCALES

Modeling and mapping of crop and vegetation characteristics are crucial for understanding their health and productivity as well as for their management. Their biophysical quantities are linked to climate, water, and energy (Bagley et al., 2014). The importance of biophysical characteristics such as land cover, leaf area index (LAI), and fraction of absorbed photosynthetically active radiation (FAPAR) as essential climate variables (ECVs) is recognized by the United Nations Framework Convention on Climate Change (UNFCCC) and the Intergovernmental Panel on Climate Change (IPCC) (GCOS, 2010; Frampton et al., 2013). A wide array of crop and vegetation characteristics has been quantified using broad-band remote sensing (Fernandes et al., 2003; Thenkabail, 2003; Ramoelo et al., 2015), and this study is advanced by hyperspectral remote sensing (Thenkabail et al., 2000a,b, 2004a,b; Galvão et al., 2003; Gitelson et al., 2003a,b; Pu et al., 2003, 2012; Roberts et al., 2004; Kokaly et al., 2009; Ustin et al., 2009; Heiskanen et al., 2013). Plant quantities such as biomass (wet or dry), LAI, chlorophyll content, FAPAR, plant height, crop or vegetation types, and plant moisture/water content have been characterized, modeled, and mapped using a wide array of hyperspectral vegetation indices (HVIs) such as the ones listed in Table 11.1 with far greater accuracies than using broad-band remote sensing (Thenkabail et al., 2013, 2015). HVIs are also known to advance our understanding on the impacts of the variables that affect vegetation characteristics such as soil arsenic and other heavy metals (Shi et al., 2016), various management practices, and inputs such as nitrogen. The limitations of the spectra obtained from broad-band remote sensors such as the Landsat Thematic Mapper (TM) and Advanced Land Imager (ALI) as opposed to narrow-band hyperspectral sensors such as the Analytical Spectral Devices' (ASD) ground-based spectroradiometer, and Earth

TABLE 11.1

Broad-Band and Hyperspectral Narrow-Band Vegetation Indices That Are Widely Used in the Study of Agricultural Crop and Vegetation Biophysical and Biochemical Characteristics

Vegetation Index	Reference
Broad-Band Indices	
$VI_{green} = (\rho_{green} - \rho_{red})/(\rho_{green} + \rho_{red})$	Gitelson et al. (2002)
$SR = \rho_{NIR}/\rho_{red}$	Tucker (1979)
$NDVI = (\rho_{NIR} - \rho_{red})/(\rho_{NIR} + \rho_{red})$	Tucker (1979)
$ISR = \rho_{NIR}/\rho_{SWIR}$	Fernandes et al. (2003)
$RSR = (\rho_{NIR}/\rho_{red}) \times [(\rho_{SWIR\_max} - \rho_{SWIR})/(\rho_{SWIR\_max} - \rho_{SWIR\_min})]$	Brown et al. (2000)
Narrow-Band Indices	
$CI_{green} = (\rho_{773-803}/\rho_{518-579}) - 1$	Gitelson et al. (2003a,b)
$CI_{rededge} = (\rho_{773-803}/\rho_{702-742}) - 1$	Gitelson et al. (2003a,b)
$MCARI = [(\rho_{700} - \rho_{670}) - 0.2(\rho_{700} - \rho_{550})] \times (\rho_{700}/\rho_{670})$	Daughtry et al. (2000)
$MCARI1 = 1.2[2.5(\rho_{800} - \rho_{670}) - 1.3(\rho_{800} - \rho_{550})]$	Haboudane et al. (2004)
$MCARI2 = \dfrac{1.5[2.5(\rho_{800} - \rho_{670}) - 1.3(\rho_{800} - \rho_{550})]}{\sqrt{(2\rho_{800} + 1)^2 - (5\rho_{800} - 5\sqrt{\rho_{670}})}0.5}$	Haboudane et al. (2004)
$MTCI = (\rho_{754} - \rho_{709})/(\rho_{709} - \rho_{681})$	Dash and Curran (2004)
$VOG1 = \rho_{740}/\rho_{720}$	Vogelmann et al. (1993)
$WI = \rho_{900}/\rho_{970}$	Peñuelas et al. (1993)
$NDWI = (\rho_{860} - \rho_{1240})/(\rho_{860} + \rho_{1240})$	Gao (1996)
$SRWI = \rho_{858}/\rho_{1240}$	Zarco-Tejada and Ustin (2001)
$SIWSI = (\rho_{859} - \rho_{1640})/(\rho_{859} + \rho_{1640})$	Fensholt and Sandholt (2003)
$DWSI\text{-}1 = \rho_{800}/\rho_{1660}$	Apan et al. (2004)
$LWVI\text{-}2 = (\rho_{1094} - \rho_{1205})/(\rho_{1094} + \rho_{1205})$	Galvão et al. (2005)
$NDWI_{2130} = (\rho_{858} - \rho_{2130})/(\rho_{858} + \rho_{2130})$	Chen et al. (2005)
$DLAI = \rho_{1725} - \rho_{970}$	le Maire et al. (2008)
$CAI = 0.5(\rho_{2021} + \rho_{2213}) - \rho_{2100}$	Daughtry et al (2005)
$PRI = (\rho_{531} - \rho_{570})/(\rho_{531} + \rho_{570})$	Gamon et al. (1992)
$PSRI = (\rho_{678} - \rho_{500})/\rho_{750}$	Merzlyak et al. (1999)
REIP (polynomial method)	Pu et al. (2003)
A (continuum removed area for the absorption feature)	Curran et al. (2001)
BNA (continuum removed band depth normalized by the area)	Curran et al. (2001)

Source: Heiskanen, J. et al. 2013. *ISPRS Journal of Photogrammetry and Remote Sensing*, 78:1–14, ISSN 0924-2716, https://doi.org/10.1016/j.isprsjprs.2013.01.001.

Observing-1 (EO-1) spaceborne Hyperion (HYP) sensors remain obvious, as illustrated in a study of submerged aquatic vegetation (SAV) (Figure 11.1, Pu et al., 2012). Broad-band sensors (TM, ALI) only gather data by averaging spectra across a broad range of the spectrum, whereas the narrow-band sensors (ASD, HYP) gather data by sampling in very narrow wavebands (e.g., 1-nm or 10-nm bandwidth intervals) across the spectrum (e.g., 400–2500 nm). While there is often a focus on the spatial resolution, there is also a need for equal importance assigned to issues of spectral resolution. Gathering data in fine spectral resolution (e.g., Figure 11.1) will lead to improved modeling of most crop and vegetation biophysical and biochemical quantities.

In Chapter 1, Dr. Gitelson presented and discussed five relationships between spectral indices and biophysical quantities (FAPAR, green LAI, vegetation fraction, chlorophyll content, and gross primary productivity). The spectral indices were based on ground-level spectroradiometer data acquired 6 m

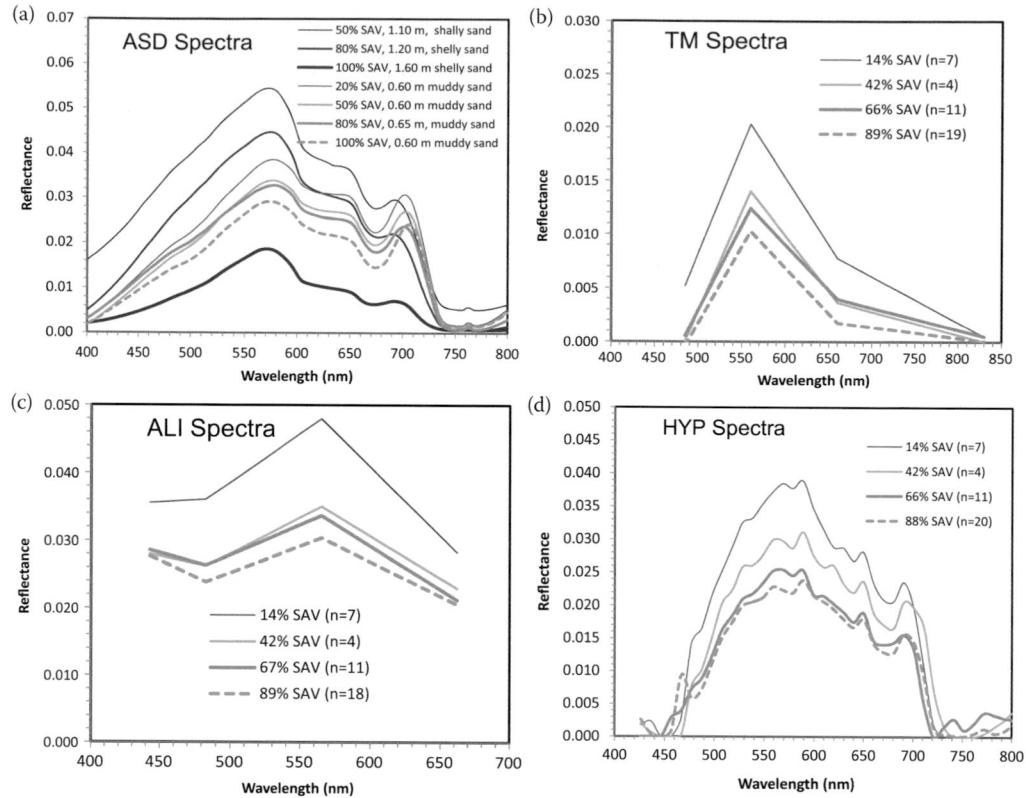

FIGURE 11.1 Presentation of sampling submerged aquatic vegetation (SAV) spectra, measured by ASD spectroradiometer (a), and three satellite sensors, Landsat TM (b, also including TM band 4), EO-1 ALI (c), and HYP (d), shows variable reflectance over different %SAV covers. (a) ASD spectral measurements from (1) deep water (1.1–1.6 m) and shelly sand with variable %SAV cover and (2) shallow water (0.60–0.65 m) and muddy sand with variable %SAV cover. (b–d) Mean spectra extracted from the sensors' images, corrected by running the FLAASH atmospheric correction code for four levels of %SAV cover (1%–24%, 25%–49%, 50%–74%, and ≥75%). The percentage of SAV cover in legends was averaged from number (*n* in parentheses) of transects/locations in each cover grouping. (From Pu, R. et al. 2012. *Estuarine, Coastal and Shelf Science*, 115:234–245.)

above the canopy. They also simulated Moderate Resolution Imaging Spectroradiometer (MODIS) and Medium Resolution Imaging Spectrometer (MERIS) bands from the spectroradiometer data, computed indices, and related them to the biophysical quantities. First, they found that various spectral vegetation indices such as the normalized difference vegetation index (NDVI), enhanced vegetation index (EVI), wide dynamic range vegetation index (WDRVI), atmospherically resistant vegetation indices (VARIs), modified chlorophyll absorption ratio index (MCARI), optimized soil-adjusted vegetation index (OSAVI), chlorophyll indices (CIs), and a few other indices were highly correlated with the biophysical quantities, typically, explaining 80%–90% of the variability in data. They reported results of studies conducted using corn and soybean crops in the mid-Western USA. They also determined that the relationships obtained using spectroradiometer data at 6 m above canopy were close to results obtained using simulated MODIS and MERIS data. They imply that these relationships hold with data acquired from satellites. Indeed, there are numerous researchers (Mariotto et al., 2013; Thenkabail et al., 2013; Marshall and Thenkabail, 2014; 2015a,b) who have conducted extensive studies relating crop biophysical quantities to broad bands as well as hyperspectral narrow-band indices to show strong relationships.

It is important to note three critical aspects to be kept in mind when studying crop and vegetation biophysical quantities using spectral indices. First, is the need to establish specific indices to study specific features. For example, Dr. Gitelson espoused the use of a chlorophyll index that uses a narrow band from the red edge portion (700–740 nm) of the spectrum. Second, there is a need to develop biophysical quantity relationships with spectral indices taking specific crops or vegetation types into account. For example, if multiple crops are pooled, relationships will be weak and/or may have significant uncertainties. In contrast, if quantities of specific crops are modeled using spectral wavebands or indices, one can expect strong relationships. Third, the best relationships are obtained when using specific wavebands from specific portions of the spectrum. For example, chlorophyll is best studied using a different set of wavebands and/or indices when compared with crop moisture and/or water content, as is apparent in Chapter 1 as well as numerous papers (Mariotto et al., 2013; Thenkabail et al., 2013, 2014a,b, 2015; Marshall and Thenkabail, 2014, 2015a,b).

11.2 HYPERSPECTRAL ASSESSMENT OF ECOPHYSIOLOGICAL FUNCTIONING FOR DIAGNOSTICS OF CROPS AND VEGETATION

Just like the biophysical properties discussed in Chapter 1, biochemical quantities (Gitelson and Solovchenko, 2018) are widely studied using remote sensing (Deel et al., 2012) and significant advances have been made in their study using hyperspectral remote sensing (Jacquemoud et al., 1996; Kokaly et al., 2009; Wang et al., 2015a,b; He et al., 2016). Biochemical properties of a leaf and plant include parameters such as chlorophyll, nitrogen, carotenoids, anthocyanins, tannin, cellulose, lignin, protein, flavonoids, and xanthophylls (Ustin et al., 2009; Gitelson and Solovchenko, 2018). Quantifying and mapping these plant biochemical variables helps us understand health and productivity of crops and vegetation and enables us to better manage crops and optimize their productivity in terms of biomass or yield. For example, Figure 11.2 shows reflectance and absorption spectra for various vegetation types in the 400–2500 nm spectral range that depict not only variations in vegetation type, but also variations in their biophysical and biochemical quantities. The causes of such variations are best studied by examples of specific crop or specific crop quantities. For example, Zarco-Tejada et al. (2013) showed that narrow-band hyperspectral vegetation indices (HVIs) like R_{515}/R_{570} were good indicators for modeling and mapping leaf carotenoid levels in vineyards. Wang

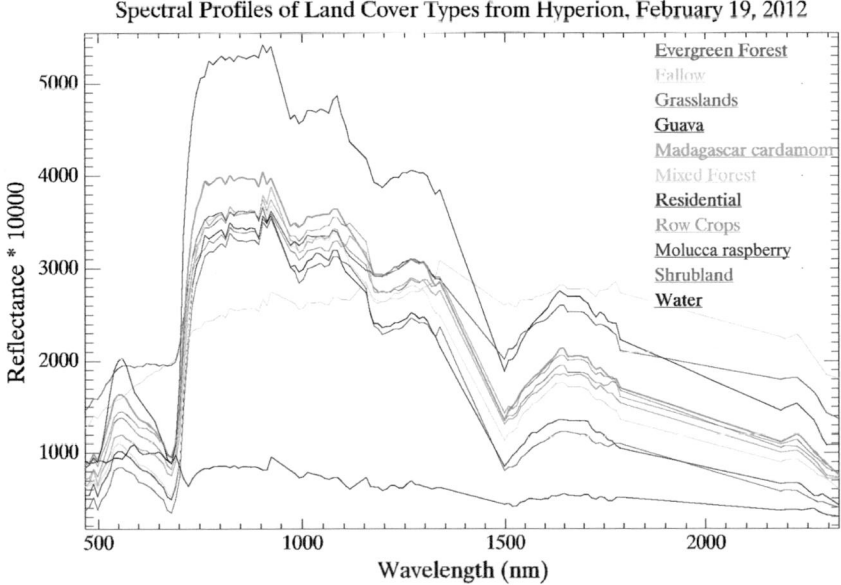

FIGURE 11.2 Mean reflectance spectra of invasive species along with dominant native trees and other background objects for reference. (From Ghulam, A. et al. 2014. *ISPRS Journal of Photogrammetry and Remote Sensing*, 88:174–192.)

et al. (2015a,b) performed a PROSPECT model (a radiative transfer model for studying leaf optical properties) inversion using hyperspectral narrow-band data to estimate cellulose, protein, and lignin. For this they established the spectral region of 2100–2300 nm as the best data for the 400–2500 nm range. Chen et al. (2015) showed that hyperspectral data in the range 900–1700 nm provided fast and accurate means of predicting anthocyanin content of wine grapes during ripening. Similarly, other biochemical quantities of crops and vegetation have been modeled using various hyperspectral narrow bands (HNBs) or hyperspectral vegetation indices (HVIs) by several authors (Gitelson et al., 2006; Merzlyak et al., 2008; Kokaly et al., 2009; Ustin et al., 2009; Kira et al., 2015; Wang et al., 2015a,b; Gitelson and Solovchenko, 2018).

In Chapter 2, Dr. Yoshio Inoue and coauthors used hyperspectral data to perform diagnostics of crops and vegetation in order to gain intimate and accurate knowledge, which in turn helped to better manage them. This process leads to advances into the domain of smart agriculture. To perform such diagnoses, they studied biophysical, biochemical, and structural properties of crops and vegetation using hyperspectral reflectivity in the 400–2500 nm range of crops in various growth stages or conditions. Various biophysical and biochemical quantities discussed above were plotted for their spectral behavior using both hyperspectral narrow bands (HNBs) and hyperspectral vegetation indices (HVIs). In order to analyze large volumes of data they recommended methods such as (1) HVIs, (2) multivariate regression, (3) powerful machine learning algorithms such as artificial neural networks (ANNs) and support vector machines (SVMs), and (4) physically based reflectance models like PROSAIL—PROSPECT and Scattering by Arbitrary Inclined Leaves (SAIL)—radiative transfer models. They provide comparisons between broad band vegetation indices derived from sensors like Landsat and narrow band–derived HVIs. Unique lambda vs. lambda plots that enabled evaluations of thousands of HVIs in modeling various crop or vegetation quantities are presented and discussed. Hyperspectral data have issues with data redundancy and the need for overcoming this has been presented and discussed in various chapters here. Further in Chapter 2, to address this issue, Dr. Inoue and coauthors recommended certain optimal HNBs for study of various crop biophysical and biochemical quantities. These optimal hyperspectral narrow bands (OHNBs) will help in selection of narrow bands, to optimize information content and reduce data redundancies. These OHNBs were also used to calculate HVIs.

11.3   SPECTRAL AND SPATIAL METHODS FOR HYPERSPECTRAL IMAGE ANALYSIS FOR ESTIMATION OF BIOPHYSICAL AND BIOCHEMICAL PROPERTIES OF AGRICULTURAL CROPS

To understand the pure characteristics of a feature or an object, we need adequate spectral resolution as well as adequate spatial resolution. What is this "adequate" resolution? That will depend on the object or feature we are studying. For example, in agricultural fields a 30-m pixel (0.09 hectares per pixel) from the spaceborne Earth Observing-1 (EO-1) Hyperion sensor gathers hyperspectral data that may include hundreds or thousands of plants (rice or wheat, for example). In such a case, the data are of adequate spectral purity because Hyperion samples a portion of an agricultural field that has similar plants throughout. In contrast, when Hyperion data are gathered from multiple objects (e.g., in a landscape with shrubs, grasses, barren land) within a 30-m pixel, the result is a hyperspectral dataset of all these objects or features in various combinations, resulting in mixed vegetation signatures. A ground-based spectroradiometer, on the other hand, often gathers pure hyperspectral signatures of an object or a feature because it not only has adequate spectral resolution (e.g., measurements taken in tens or hundreds or thousands of wavebands across the spectrum such as 400–2500 nm) but also gathers data in a very high spatial resolution of a few cm^2. Present-day drones can carry hyperspectral sensors and fly low to capture data that at spatial resolutions of a few cm^2 or at sub-meter level or often within 5 m. UAVs often carry only hyperspectral sensors that cover about half of the required spectrum, either the visible–near-infrared (VNIR; most common) sensor or short-wave infrared sensor (SWIR), due to the considerable weight of two sensors, which

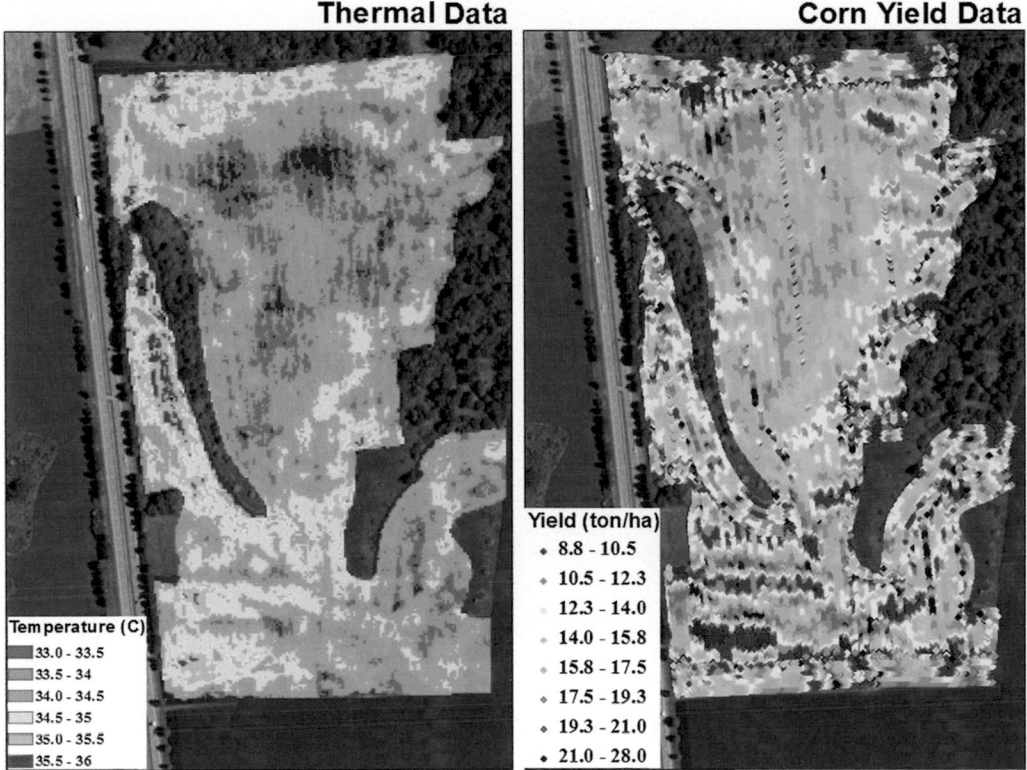

FIGURE 11.3 Temperature of a corn field acquired by aerial thermal sensors on May 8, 2015 (left) and corn yield data at the end of the growing season (right) at the Molly Caren Farm near London, OH, USA. (From Khanal, K. et al. 2017. *Computers and Electronics in Agriculture*, 139:22–32.)

cannot be supported by current UAVs. Low-weight hyperspectral sensors from 400 to 2500 nm, can be mounted on UAVs. However, 400–2500 nm sensors can be easily mounted in aircraft. Upcoming satellites such as HyspIRI and others (Panda et al., 2015) carry hyperspectral sensors that have spatial resolution of 60 m or less.

Precision farming and smart-agricultural applications require data gathered in both hyperspectral and hyperspatial modes. Vegetation species detection and species level classification with high accuracy are best achieved when data are gathered using sensors that have both hyperspectral and hyperspatial capabilities. When this is achieved, almost every plant biophysical and biochemical quantity can be quantified, modeled, and mapped using hyperspectral and/or hyperspatial data (Figures 11.3 through 11.6). Further, it is well established that these data need to be gathered not only in the visible, near-infrared (NIR), and short-wave infrared but also in the thermal (TIR; 3000–14,000 nm) range. For example, temperature quantified using thermal sensors is related to corn crop yield (Figure 11.3; Khanal et al., 2017) for precision farming studies. Satellite remote sensing–derived precision farming products need to be validated using ground-based measurements. Further, wheat yields measured using very fine-resolution combine-mounted sensor data were found to be related to Landsat Thematic Mapper (TM) 30-m wheat yields (Figure 11.4; Thenkabail et al., 2003). However, when hyperspatial 5-m IKONOS data were available, wheat yields could be modeled and mapped with much greater precision and accuracies (Figure 11.5; Enclona et al., 2004) relative to what is feasible using Landsat moderate resolution data (Figure 11.4; Thenkabail et al., 2003). Even better, when hyperspectral narrow-band vegetation indices are available, they can explain significantly greater variability in crop quantities such as biomass and leaf area index when compared with the same models using broad-band Landsat TM

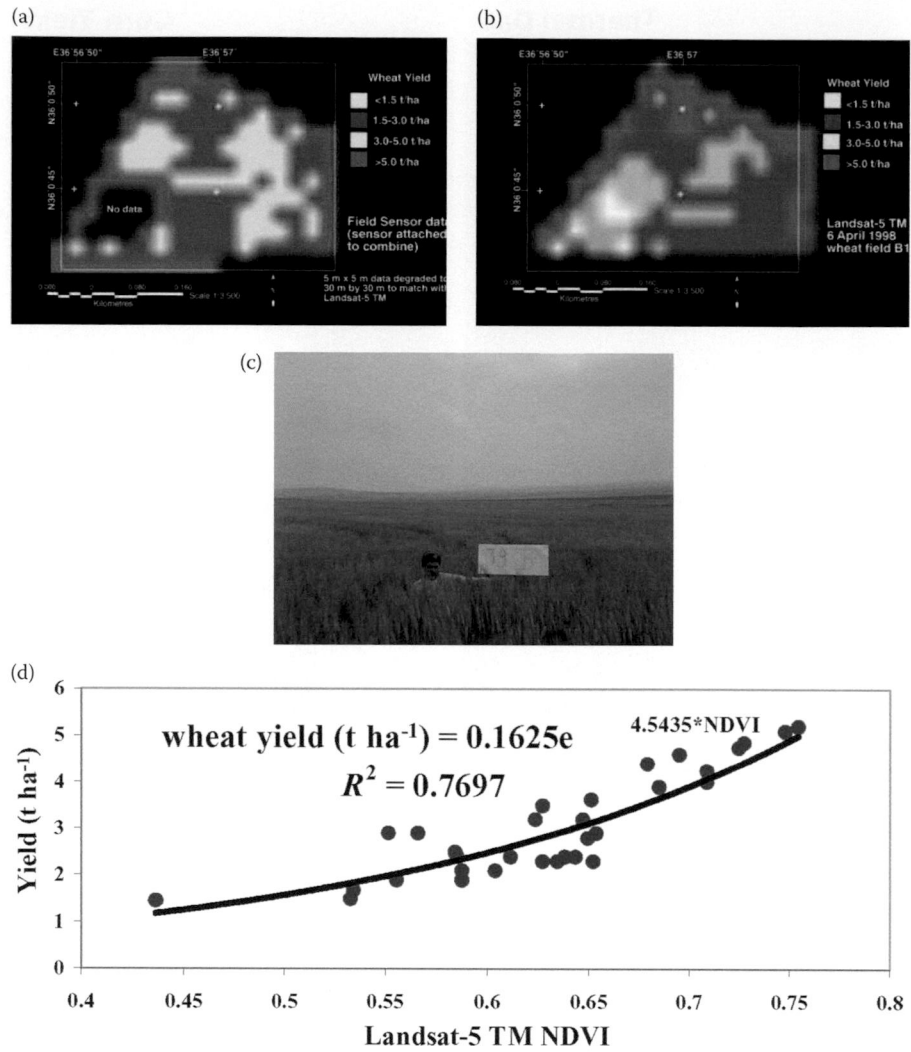

FIGURE 11.4 Yield variation in a wheat field measured using a combine-mounted sensor (a) and Landsat-5 TM sensor (b). The field as seen during ground data collection is shown in (c). Relationship between actual after harvest yields and TM NDVI is shown in (d). (From Thenkabail, P.S. 2003. *International Journal of Remote Sensing*, 24(14):2879–2904.)

data. For example, Figure 11.6 (Thenkabail et al., 2000a,b) shows 10%–14% greater variability explained by HVIs relative to Landsat TM broad-band indices for LAI and biomass modeling of potato and soybean crops.

In Chapter 3, Drs. Cohen and Alchanatis recommended that we use not only hyperspectral data in the visible, NIR, and SWIR portion of the spectrum (e.g., 400–2500 nm), but also that in the thermal portion (3000–14,000 nm; in their chapter they use the 8000–14,000 nm portion of the thermal data) to study biophysical quantities such as LAI, biomass, water status and biochemical quantities such as chlorophyll content, and nitrogen (N). In order to study these quantities, they explored spectral, spatial, and combination methods including: spectral band selection, two-band hyperspectral vegetation indices which they simply refer to as spectral indices, and multivariate methods that take into consideration *n*-band indices (e.g., best 1-band index, best 2-band index, to best *n*-band index; where *n* can be tens or hundreds or thousands of hyperspectral narrow bands or HNBs). Band selection methods employed were either unsupervised (principal component

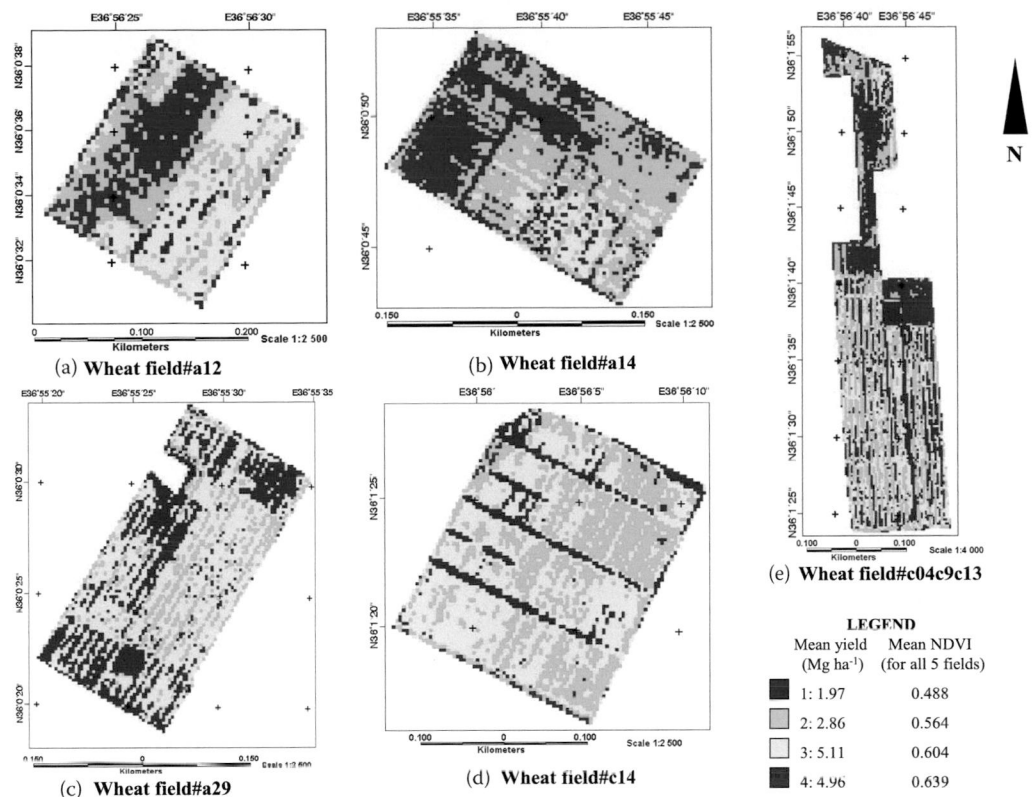

FIGURE 11.5 A matrix approach to mapping wheat yield using IKONOS data. Within-field wheat yield (Mg ha^{-1}) variability was mapped using IKONOS data by solving a system of simultaneous equations. (From Enclona, E.A. et al. 2004. *International Journal of Remote Sensing*, 25(2):377–388.)

analysis, lambda vs. lambda plots) or supervised (e.g., stepwise discriminant analysis). Two-band HVIs involved normalized indices that can be plotted as lambda vs. lambda plots. Multivariate methods included multiple linear regression involving tens or hundreds or thousands of wavebands as independent variables to model a dependent crop variable via partial least-squares regressions (PLSR) and ANNs. Not commonly attempted, but most necessary for crop water management studies, is thermal data analysis of crops. In this respect they used the crop water stress index (CWSI) which measures and models canopy temperature. They recommend use of the above methods to model and map several crop biophysical and biochemical quantities. They remind readers that hyperspectral data are, often, gathered as spectral signatures and analyzed as such, but without looking at the spatial aspect. They suggest that the data should be looked at as imaging spectroscopy and analyzed as such. This, hitherto, has been widely implemented already in the remote sensing community; but its geographic scope of use was limited by computing limitations. However, now with the power of cloud computing and machine learning, handling of hyperspectral data in its spatial context (e.g., hyperspectral image classification involving hundreds of bands) is becoming feasible and efficient over large geographic areas. In conclusion, the central theme is the study of crop biophysical and biochemical quantities as well as crop water management assessments through the combined use of (1) VIS/NIR/SWIR data along with TIR data, and (2) high–spectral and high–spatial resolution data. With availability of drones, they emphasize, these tools and methods should become affordable and more widely used for purposes of crop management and crop water management.

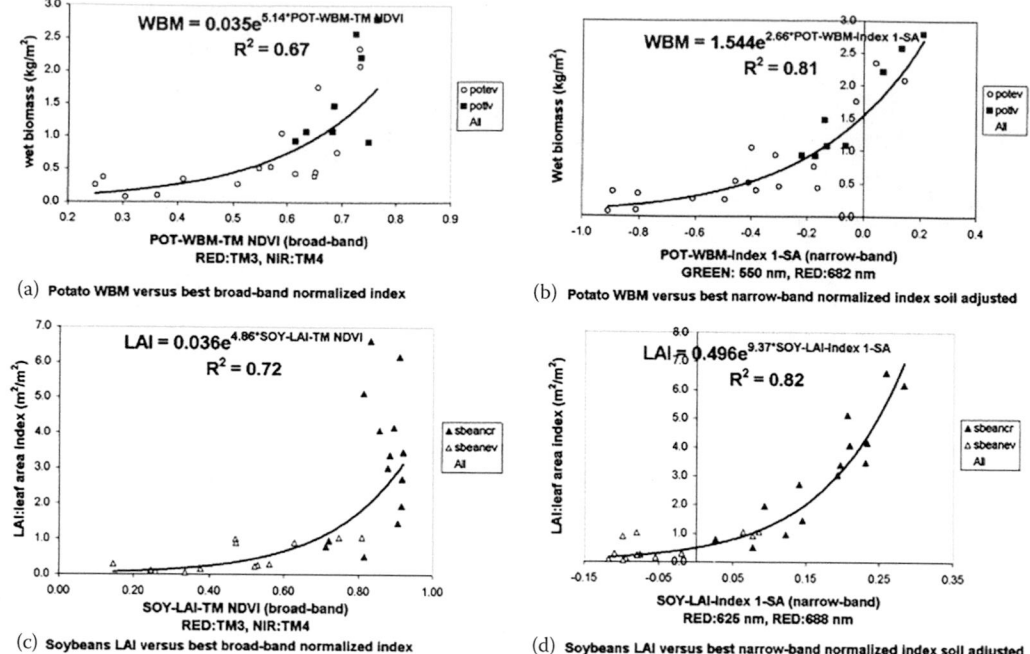

FIGURE 11.6 Comparison between the broad band and the narrow band soil-adjusted indices for potato wet biomass (WBM) (a and b), and soybeans leaf area index (LAI) (c and d). NIR and red-based indices of broad bands (a and c) vs. visible band based indices of narrow bands (b and d). (From Thenkabail et al. 2003. *International Journal of Remote Sensing*, 24(14):2879–2904.)

11.4 SPECTRAL AND 3D NONSPECTRAL APPROACHES TO CROP TRAIT ESTIMATION USING GROUND AND UAV SENSING

Crop and vegetation quantitative variables are widely studied using either multispectral broad-band vegetation indices (MBB VIs) derived from sensors such as Landsat Operational Land Imager (OLI) and Sentinel multispectral instrument (MSI) or hyperspectral narrow-band vegetation indices (HNB VIs) derived from sensors such as Earth Observing-1 (EO-1) Hyperion or Airborne Visible/ Infrared Imaging Spectrometer-Next Generation (AVIRIS-NG). Figure 11.7 shows crop variables such as LAI and biomass modeled using Landsat Thematic Mapper (TM). However, as we can see in Figure 11.8, Landsat TM and other broad-band instruments like IKONOS and Advanced Land Imager (ALI) have poor spectral resolutions that miss the fine sensitivity at various portions of the electromagnetic spectrum as gathered by the hyperspectral Hyperion sensor. It is well established that HNB VIs perform significantly better than MBB VIs (Mariotto et al., 2013; Thenkabail et al., 2013, 2014a,b, 2015; Marshall and Thenkabail, 2014, 2015a,b). One also should consider adopting multiband vegetation indices (MBVIs) such as multiple linear regression model–derived indices involving "*N*" number of bands rather than simple two-band vegetation indices (TBVIs) such as normalized difference vegetation index (NDVI) or other TBVIs—whether they are MBB VIs or HNB VIs (Thenkabail et al., 2000a,b; Bendig et al., 2015). Studies have shown nonlinear indices to perform better than linear indices, whether MBVIs or TBVIs (Thenkabail et al., 2004a,b). In a nutshell, it is best to use nonlinear models involving multiple hyperspectral narrow-band (HNBs) or hyperspectral vegetation indices (HVIs) to derive crop biophysical and biochemical variables such as biomass, leaf area index, chlorophyll, and nitrogen.

Nevertheless, further significant enhancement in crop variable models is feasible when height information is added. The best way to calculate height of crops and vegetation is through light

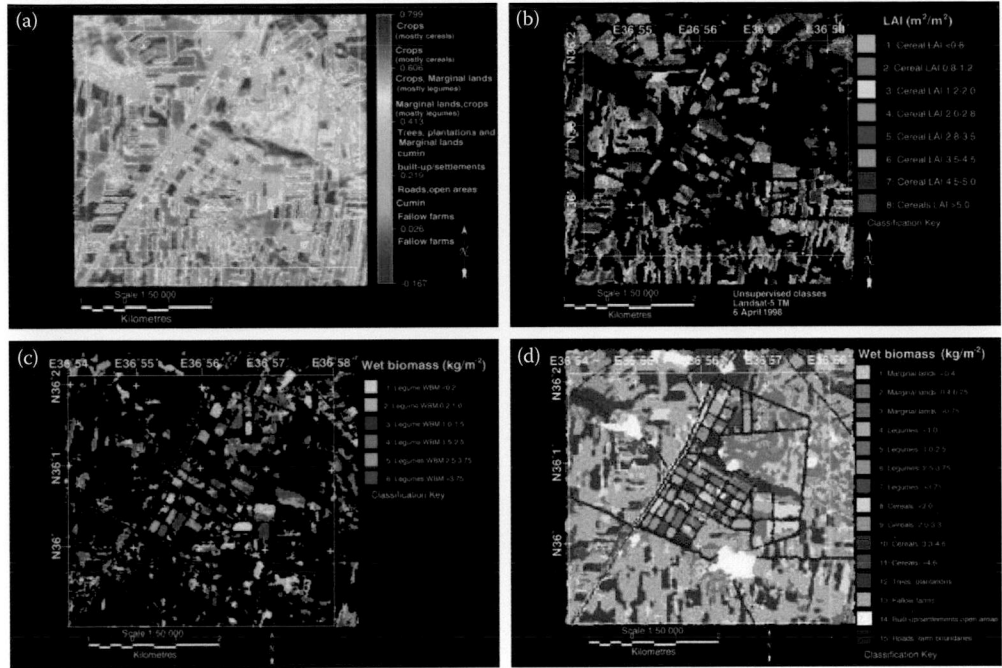

FIGURE 11.7   Steps in mapping LAI and WBM of agricultural crops, illustrated for the April 6, 1998 image: (a) NDVI vegetation gradient; (b) masking out nonagricultural areas to delineate LAI; (c) WBM; and (d) the resulting final biomass map. (From Thenkabail et al. 2003. *International Journal of Remote Sensing,* 24(14):2879–2904.)

detection and ranging (LiDAR) data (Eitel et al., 2014) or synthetic aperture radar (SAR) data such as polarimetric SAR interferometry (PolInSAR) (Lopez-Sanchez et al., 2017). Eitel et al. (2014) showed LiDAR capabilities in measuring crop variables such as dry weight and nitrogen. Lopez Sanchez et al. (2017) showed that when rice is taller than 25–40 cm, plant height can be accurately estimated without external reference data. Uncertainties are mainly in early growth stages (<40 cm) when background soil and water significantly influence estimates of crop variables. Plant height is also derived through crop surface models (Bendig et al., 2015). Unmanned aerial vehicles (UAVs)—or drones or unmanned aerial systems (UAS)—are widely used for gathering frequent and near–real time data of crops using hyperspectral and multispectral sensors. Series of images acquired that have significant overlap will allow for orthomosaics, from which height information is derived. Crop height models (CHMs) are derived by subtracting digital terrain models (DTMs) from digital surface models (DSMs) (Chang et al., 2017).

In Chapter 4, Drs. Aasen and Bareth presented and discussed tools, methods, and approaches to model and map agricultural crop traits (e.g., plant height, biomass, leaf area index) using spectral data and 3D DSMs. The chapter's distinct components are in modeling and mapping crop traits by (1) spectral data, (2) DSMs, and (3) combinations of the two. Both the spectral and the 3D DSM data can be acquired either through terrestrial/ground-based platforms or through UAVs. As shown, a number of plant traits such as plant height and dry biomass, and nitrogen uptake and dry biomass are highly correlated. Therefore, determining some of these accurately, will help us correlate the others accurately as well. The LiDAR data for deriving DSMs were obtained either through terrestrial laser scanning (TLS) data or through airborne laser scanning (ALS). The authors derived crop height (CH) by subtracting the ground-level (before crop is planted) digital elevation model (DEM)-derived height from the DSM-derived elevations obtained from either TLS or ALS. Such measurements are taken throughout the phenological growth phases as needed. The DSM-measured plant heights were

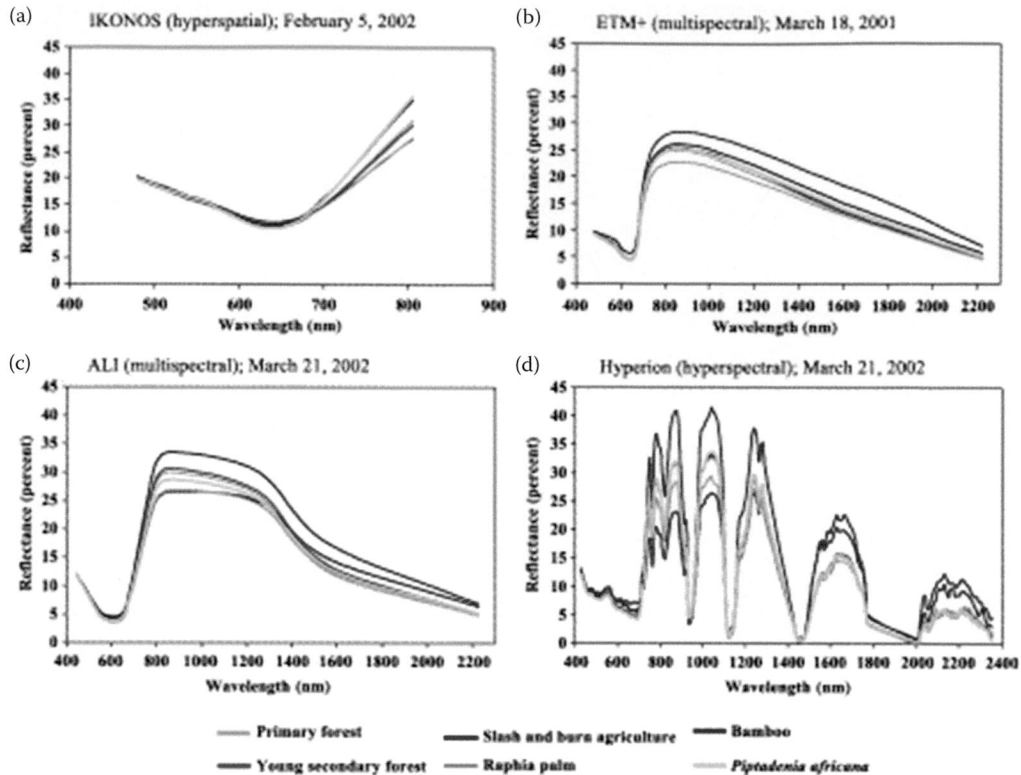

FIGURE 11.8 Mean spectral profile of rainforest vegetation using (a) hyperspatial, (b and c) advanced multispectral, and (d) hyperspectral sensors. (From Thenkabail, P.S. et al. 2004a. *Remote Sensing of Environment*, 90:23–43.)

within 1–2 cm of the actual height and, in general, outperformed spectral measured heights. Further, the DSM-based measurements were collected independently of clear-sky conditions. However, spectral measurements were important for quantifying biochemical characteristics that cannot be measured using DSMs. Combined spectral and 3D nonspectral instruments assess biophysical as well as biochemical traits. Indeed, the future of remote sensing involves the use of data from multiple sensors to study multiple plant traits. This is also becoming increasingly feasible from spaceborne data, covering the entire planet. Good examples are the upcoming NASA instruments like the Global Ecosystem Dynamics Investigation (GEDI), where high-resolution laser ranging of the Earth's forests and topography will be obtained from the International Space Station (ISS), as well as hyperspectral data from the Hyperspectral InfraRed Imager (HyspIRI).

11.5 SPECTRAL BIOINDICATORS OF PHOTOSYNTHETIC EFFICIENCY AND VEGETATION STRESS

Measuring light use efficiency (LUE) is crucial to the study of vegetation health, stress, and productivity. Hyperspectral remote sensing–derived photochemical reflectance indices (PRI) are increasingly used as a proxy for nondestructive measurement of LUE. Gamon et al. (1992) first proposed PRI as a normalized index using data from two hyperspectral narrow bands (HNBs) centered at 531 nm and another band centered typically at 570 nm (Table 11.2) but also at slightly different wavelengths (e.g., 670 or 680 nm). Subsequently, many others have proposed some variations of the original PRI, typically using HNBs from slightly different portions of the spectrum (Table 11.2). Table 11.2 also lists certain other hyperspectral vegetation indices (HVIs)

TABLE 11.2
Spectral Indices Used in This Study

VI	Formulation	References
	Greenness, LAI	
NDVI	$(R_{802.57} - R_{682.52})/(R_{802.57} + R_{682.52})$	Rouse et al. (1974)
SR	$R_{802.57}/R_{682.52}$	Jordan (1969)
MTVI2	$\dfrac{1.5 \times [1.2 \times (R_{802.57} - R_{549.79}) - 2.5 \times (R_{7668.27} - R_{549.79})]}{\sqrt{(2 \times R_{802.57} + 1)^2 - (6 \times R_{802.57} - 5 \times \sqrt{R_{668.27}}) - 0.5}}$	Haboudane et al. (2004)
NDVI$_g$	$(R_{802.57} - R_{549.79})/(R_{802.57} + R_{549.79})$	Gitelson et al. (1996)
	Chlorophyll Content	
CI$_{red-edge}$	$\left(\dfrac{R_{NIR}}{R_{red-edge}}\right) - 1$	Gitelson et al. (2003a,b)
TCARI/OSAVI	$\dfrac{3 \times [(R_{700} - R_{670}) - 0.2 \times (R_{700} - R_{550}) \times (R_{700}/R_{670})]}{(1 + 0.16) \times (R_{800} - R_{670})/(R_{800} + R_{670} + 0.16)}$	Haboudane et al. (2002)
	Leaf Water Content	
WI	$R_{972.11}/R_{899.41}$	Peñuelas et al. (1993)
WI/NDVI		Peñuelas et al. (1997)
	Light Use Efficiency	
PRI$_{570}$	$(R_{531} - R_{569})/(R_{531} + R_{569})$	Gamon et al. (1992)
PRI$_{512}$	$(R_{531} - R_{513})/(R_{531} + R_{513})$	Hernández-Clemente et al. (2011)
PRI$_{600}$	$(R_{531} - R_{602})/(R_{331} + R_{602})$	Hernández-Clemente et al. (2011)
PRI$_{670}$	$(R_{531} - R_{668})/(R_{531} + R_{668})$	Hernández-Clemente et al. (2011)

Source: Rossini, M. et al. 2013. *ISPRS Journal of Photogrammetry and Remote Sensing*, 86:168–177, ISSN 0924-2716, https://doi.org/10.1016/j.isprsjprs.2013.10.002.

Abbreviations: NDVI, normalized difference vegetation Index; SR, simple ratio; MTVI2, modified triangular vegetation index 2; NDVIg, green normalized difference vegetation index; CI$_{red-edge}$, red edge chlorophyll index; TCARI/OSAVI, transformed chlorophyll absorption in reflectance index normalized by the optimized soil-adjusted vegetation index; WI, water index; WI/NDVI, the ratio between WI and NDVI; PRI, photochemical reflectance index, computed with different reference bands. R is the reflectance at the specified AISA wavelength in nm. R_{NIR} is the reflectance in the near-infrared range from 770 to 800 nm and $R_{red-edge}$ is the reflectance in the red-edge range from 720 to 730 nm.

used to measure productivity and other vegetation characteristics such as the leaf water content and chlorophyll. Zhang et al. (2017) recommended a canopy-level PRI observation separated into sunlit and shaded PRI values to improve the performance of PRI. This is because of the variability in the internal factors (e.g., pigment concentrations) and external factors (e.g., environmental conditions and sun-target-view geometry) that affect PRI signals (Zhang et al., 2017). However, as Middleton et al. (2016) points out, the two HNBs centered at 531 and 570 nm are rarely available from most satellite sensors. In this context, Middleton et al. (2016) calculated a MODIS Terra/Aqua based PRI by using the MODIS ocean band centered at 531 nm and MODIS land band (in absence of 570 nm) centered at 620–670 nm and found good results for determining daily LUE of Canadian forests. However, PRI is heavily affected by canopy structural effects such as the presence of shadows in remote sensing pixels and directional effects associated with changes in illumination and viewing angles (Wu et al., 2015). In order to remove these structural influences, they suggest a structure-related signal in PRI (sPRI) as a function of LAI; consequently the residual PRI (rPRI = PRI − sPRI) would be independent of canopy structural characteristics (Wu et al., 2015). Magney et al. (2016) measured diurnal variations in wheat photosynthetic physiology using delta PRI (ΔPRI). They determined that ΔPRI was sensitive to abiotic environmental conditions across 16 wheat canopies, and that ΔPRI response to solar radiation was highly correlated with biomass accumulation. Wu et al. (2015) showed LAI explained 66% variability in PRI. Gitelson et al. (2017) have shown that corn and soybean stand chlorophyll content explained greater than 90% of the variability in PRI. Thus, it is clear that PRI is highly correlated with crop and vegetation biophysical and biochemical quantities such as biomass, LAI, and chlorophyll, and hence is ideal to quantify vegetation productivity parameters such as gross primary productivity (GPP) and net primary productivity (NPP).

In Chapter 5, Dr. Elizabeth Middleton and coauthors proposed a new scientific approach called "spectral bioindicators" to study vegetation health, physiology, and biodiversity by using HNBs, typically in the 400–2500 nm range of the spectrum. Each narrow band is defined as a band with bandwidth of 10 nm or less. These HNBs were acquired from different platforms (ground-based, airborne, spaceborne) and used in the study of various vegetation parameters at leaf to global scales. The authors put emphasis on the study of plant stress conditions. When plants are in stress, due to environmental and/or climatic and/or physiological factors, their productivity is less than optimal. The spectral bioindicators will help study vegetation in above-optimal, normal, and below-optimal conditions. They outline a host of parameters (e.g., absorbed photosynthetically active radiation, CO_2, GPP, LUE, PAR; see various Tables in Chapter 5) required to measure ecosystem photosynthesis and provide detailed methods and approaches to measure them using HNBs and HVIs in models. Planet-scale measurements of gross primary productivity (GPP, gC m^{-2} y^{-1}) can vary within and across seasons and years and can range from as low as <100 gC m^{-2} y^{-1} for deserts and snow covered areas to as high as 3500 gC m^{-2} y^{-1} for Amazonian and Southeast Asian tropical forests. Uncertainties will depend on various factors such as the spatial and spectral resolution of the imagery as well as uncertainties involved in model parameterization, and data gathering such as from eddy covariance towers used to measure canopy/ecosystem photosynthesis through net exchange of carbon, water, and energy. In this regard, they discuss the spread of global networks of eddy covariance towers from the early 1980s, now covering a wide spectrum of biomes. They present and discuss PRI used as a proxy to measure LUE. They calculate PRI using a band at 531 nm and one of the other bands centered at 551, or 488, to 670 nm. Further they discuss a pathway for plants to dissipate excess energy through chlorophyll fluorescence (ChlF) of leaves, which is emitted from chlorophyll-a molecules. They further discuss in detail the quantitative measurements of various spectral bioindicators using data from field-, aircraft-, and satellite-based platforms to derive PRI and solar induced chlorophyll fluorescence (SIF; nW^{-2} m μm^{-1} sr^{-1}), and then link PRI to LUE and ChlF through model relationships. HNBs and HVIs available from various two-band and multiband linear or nonlinear models will help advance our understanding of spectral bioindicators.

11.6 CROP TYPE DISCRIMINATION USING HYPERSPECTRAL DATA: ADVANCES AND PERSPECTIVES

One of the first applications of remote sensing since the launch of Landsat-1 way back in 1972 was to agriculture. The United States interagency projects or programs such as the Large Area Crop Inventory Experiment (LACIE) and the Agriculture and Resources Inventory Surveys through Aerospace Remote Sensing (AgRISTARS) demonstrated large scale application of remote sensing data for agricultural cropland studies such as mapping cropland extent, estimating cropland areas, mapping crop types, and modeling crop yields (Pinter et al., 2003). Those efforts were primarily dependent on Landsat-1 Multispectral Scanner (MSS) data, and aerial imagery. These early projects clearly demonstrated the power of remote sensing in agricultural applications with potential for mapping the entire world's croplands routinely, repetitively, and objectively. During the 1980s and 1990s, there was a steep increase in remote sensing data use for agricultural applications, primarily using data from Landsat-4/5 Thematic Mapper (TM), Système Pour l'Observation de la Terre High Resolution Visible (SPOT HRV), and the Indian Remote Sensing Satellite-1A, 1B (IRS 1A, 1B) by a broad

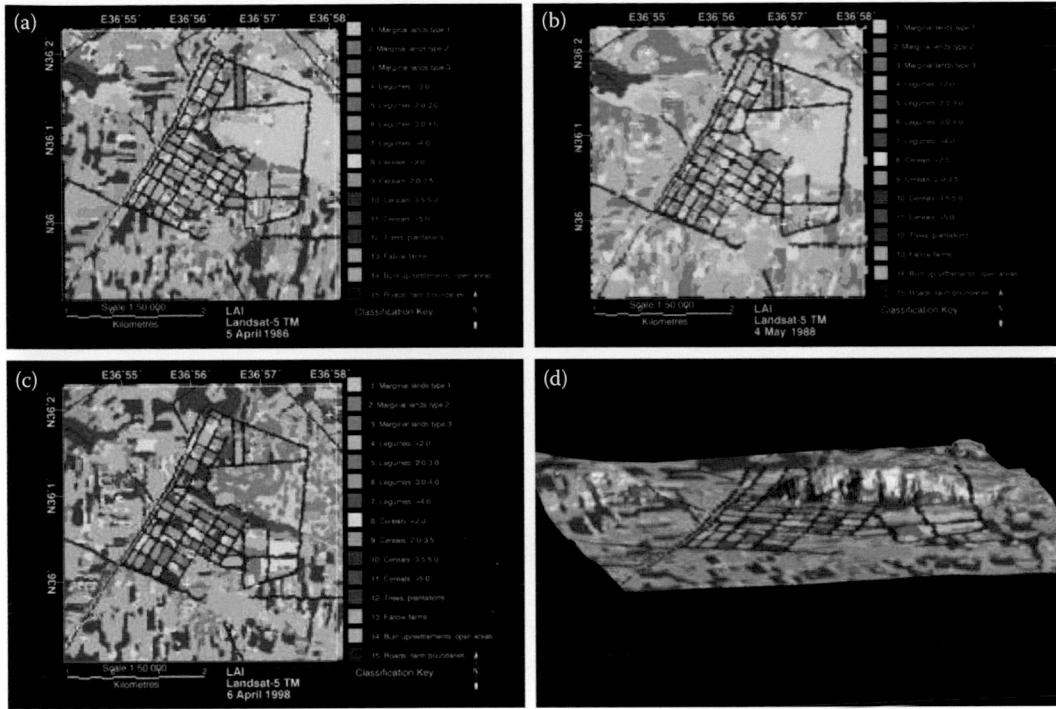

FIGURE 11.9   Mapping LAI using Landsat TM data. LAI maps of spring season crops at or near critical growth stages for (a) April 5, 1986; (b) May 4, 1988; and (c) April 6, 1998. (d) The LAI of 1998 draped over 10-m DEM data obtained from the Russian TK-350 camera system. Images were acquired for the ICARDA research farms South of Aleppo, Syria. (From Thenkabail, P.S. 2003. *International Journal of Remote Sensing,* 24(14):2879–2904.)

range of national and international organizations. These applications included agricultural drought monitoring, and mapping irrigated areas, forest type mapping, wetland mapping (Lopez et al. 2017), and assessing productivities. In the latter half of the 1990s and in the early 2000s the new millennium era of remote sensing started with the acquisition of hyperspatial data in submeter to 5-m very–high resolution imagery (VHRI) from satellites such as IKONOS, and QuickBird. This led to proliferation of remote sensing applications including mapping of many agricultural quantities (e.g., Figure 11.9) such as crop type, biomass, leaf area index (LAI), and many others (Atzberger, 2013; Mulla, 2013). Thus, precision farming applications from submeter to 5-m VHRI became common. In the last two decades, there has been a revolution in remote sensing data from a wide array of satellite sensors, from different countries, acquiring data across the electromagnetic spectrum (visible, near-infrared, short-wave infrared, mid-infrared, thermal, microwave, radar, and LiDAR). Data also started becoming freely available, which was matched by the advances in power of parallel processing in the cloud, and machine learning, leading to continental and global cropland applications (Thenkabail et al., 2012; Teluguntla et al., 2015; Teluguntla et al., 2017; Xiong et al., 2017a). This has led to mapping the entire Planet's agricultural extent (www.croplands.org; https://lpdaac.usgs.gov/about/news_archive/release_gfsad_30_meter_cropland_extent_products). Studies of these advances were documented by several researchers (Thenkabail, 2010; Thenkabail et al., 2012; Atzberger, 2013; Mulla, 2013; Brown, 2015; Teluguntla et al., 2015; Corgne et al., 2016; Teluguntla et al., 2017; Yang et al., 2017; Xiong et al., 2017a,b). However, all of these advances have been conducted using data from satellite sensors with a few broad bands. With modern capabilities to normalize and harmonize satellite data across multiple satellites and sensors, along with cloud computing and machine learning (Xiong et al., 2017a) there has been a paradigm shift in how remote sensing science is conducted.

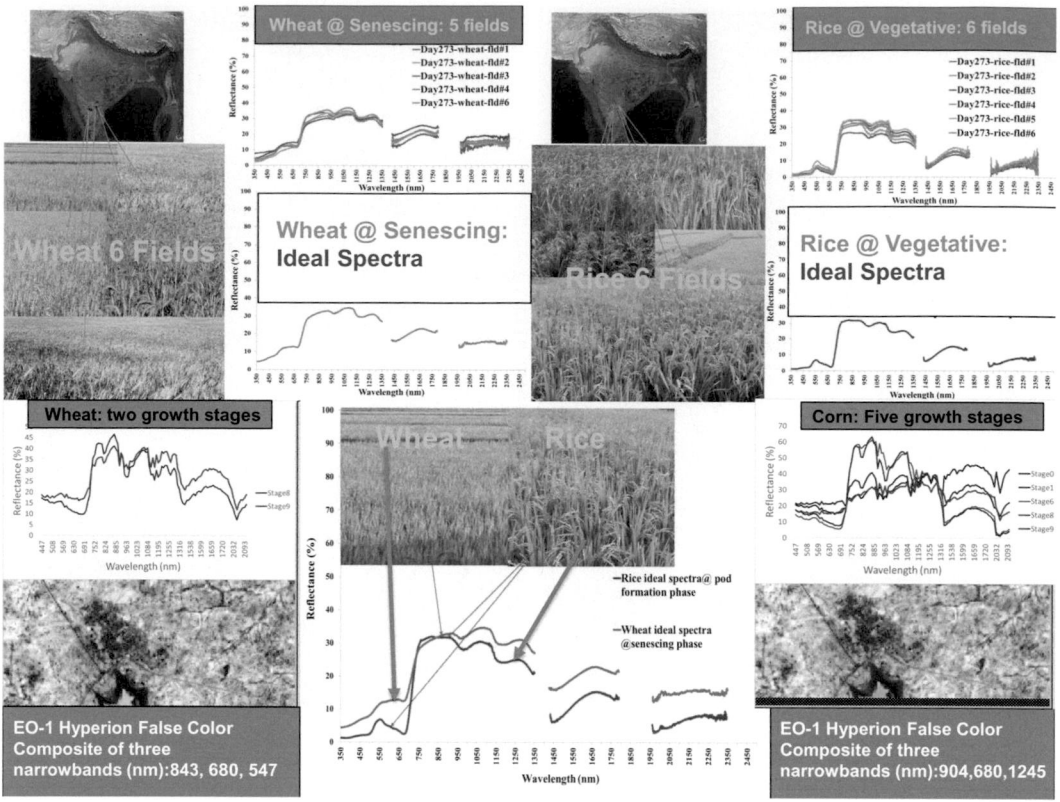

FIGURE 11.10 Hyperspectral data for separating crop types and their growth stages. Illustration of the ability of hyperspectral narrow-band data to separate crop types and their growth stages based on neat continuous spectra across the electromagnetic spectrum.

Nevertheless, substantially greater advances as well as new advances can be made in the study of crop types, plant species, as well as quantifying various crop and vegetation biophysical, biochemical, and structural quantities by using hyperspectral data (Roberts et al., 2012; Clark et al., 2018; Cleemput et al., 2018). These crop and vegetation characteristics have been demonstrated, modeled, and mapped with significantly greater precision and accuracies through hyperspectral remote sensing (Thenkabail et al., 2000a,b, 2002, 2004a,b; Mariotto et al., 2013; Marshall and Thenkabail, 2014, 2015a,b) relative to broad-band multispectral remote sensing. This is feasible because hyperspectral narrow-band (HNB) data are available in fine spectral wavebands ($\leq$10 nm) across the electromagnetic spectrum, helping separate crop characteristics such as crop types and growth stages (e.g., Figure 11.10).

In Chapter 6, Dr. Lênio Soares Galvão and coauthors provided an interesting study of crop types and crop species discrimination and classification using spaceborne Earth Observing-1 (EO-1) Hyperion hyperspectral data. For this they chose six crops: coffee (*Coffea arabica* L.), sugarcane (*Saccharum* spp.), flooded rice (*Oryza sativa* L.), common bean (*Phaseolus vulgaris* L.), corn (*Zea mays* L.), and soybean (*Glycine max* (L.) Merrill). In addition, they also selected pasture types for analyses. All of these are major crops in Brazil where the study was conducted. For crop species, they also selected seven sugarcane species to evaluate the performance of hyperspectral Hyperion narrow-band (HNB) data with multispectral broad-band (MBB) data from several sensors. Overall, they demonstrated the power of hyperspectral narrow bands (HNBs) in different ways. First, they plotted crop type or species types in two-dimensional plots using waveband reflectivity (e.g., a near-infrared 864 nm HNB vs. a red 660 nm HNB; or 1205 nm leaf water absorption band vs. 671 nm chlorophyll absorption band), or taking two selective indices (e.g., normalized difference vegetation index [NDVI] or NDVI vs. normalized difference water index [NDWI]), or two transformed bands (e.g., discriminant band 1 vs. discriminant band 2). Such plots provide useful visual, qualitative information on where and how crop types and species can be separated using HNBs and how they can perform better than multispectral broad bands (MBBs) when corresponding broad bands from MBBs are plotted to separate the same crops or species. Second, they highlight the importance of removing redundant bands and how using all the available HNBs can actually decrease accuracies when compared to using selective HNBs that can optimize classification accuracies. Third, they also demonstrated the power of HNBs such as 671 nm HNB to map chlorophyll absorption across an area. Similar specific information plots can be done taking other wavebands. For example, plant water content characteristics can be evaluated by using an HNB centered at 970 nm. Fourth, the chapter highlights the selection of 21 HNB-derived hyperspectral vegetation indices (HNBs) in the study of crops and crop species. The importance of HNBs is their ability to quantify characteristics. For example, plant water index, plant stress index, plant structural index, plant pigment indices, and several other such indices are nicely characterized. Fifth, they show that 36 HNBs from Hyperion classified crop species (they illustrate this taking sugarcane varieties) significantly better (about 10%–30% higher accuracy) than simulated multispectral data from several satellite sensors such as NOAA-17 AVHRR, CBERS-2 CCD, Terra ASTER, Landsat-7 ETM+ and Terra MODIS. They recognized that signal-to-noise ratio (SNR) of Hyperion was poorer than other sensors and if that can be improved as expected from new generations of hyperspectral sensors like HyspIRI, there will definitely be better performance of hyperspectral data in studies.

11.7   IDENTIFICATION OF CANOPY SPECIES IN TROPICAL FORESTS USING HYPERSPECTRAL DATA

Remote sensing is extensively used in mapping forest covers (Hansen et al., 2013) including change over time and space, as well as forest type or species mapping, selective logging, quantifying biomass, assessing carbon, and numerous other forest quantities such as diameter at breast height (DBH), stem volume, basal area, and crown diameter. Remote sensing methods are also known to reduce uncertainties in species-level canopy cover estimates compared to field-plot methods (Alonzo et al., 2016). Allometric equations relating above-ground biomass to spectral indices and the upscaling of these equations to larger areas to determine forest biomass and carbon are quite popular (Djomo and Chimi, 2017). Traditionally, multispectral remote sensing has been used widely and successfully in mapping, modeling, and assessments as was amply demonstrated in high-level studies conducted by the United Nations Collaborative Programme on Reducing Emissions from Deforestation and Forest Degradation for enhancing forest conservation, sustainable management of forests, and enhancement of forest carbon stocks (UN REDD+).

Since forests can be complex, especially tropical forests where tens or hundreds of species often occupy an area of 30 m × 30 m, reducing uncertainties in modeling and mapping of forests requires the use of advanced datasets, methods, and approaches. Multispectral broad-band data can map general forest covers quite well (Hansen et al., 2013), but can be limited when it comes to delineating species types or their dominance. Hyperspectral data have shown great promise in mapping and quantifying forest characteristics such as biomass (Thenkabail et al., 2004a,b) and species (Figure 11.11, Ferreira et al., 2016). However, most of the remote sensing literature on forests, calls for multidata fusion approaches to advance forest characterization. For example, good results have been achieved combining hyperspectral remote sensing data with radar (Koch, 2010), with LiDAR (Clark et al., 2011), with hyperspatial texture (Clark et al., 2005; Ploton et al., 2017), and with long-wave infrared in the 8–12.4 μm range (Harrison et al., 2018) as all have helped map forests with greater detail (e.g., map forest species or types rather than just forest vs. nonforest) and accuracy. Asner et al. (2015) also showed that sparse and highly variable field data when integrated with hyperspectral and LiDAR data provided accurate estimates of forest canopy biochemistry.

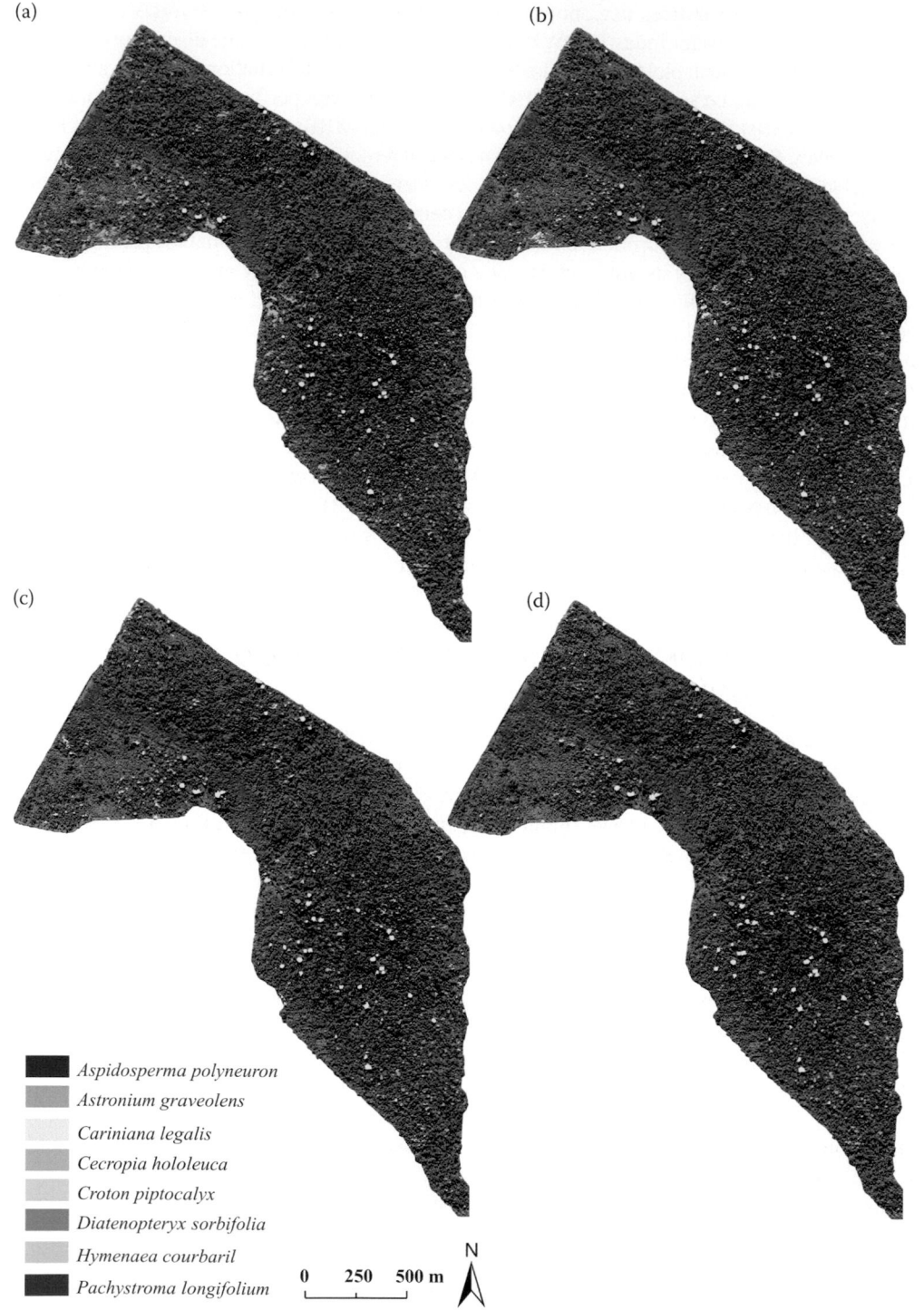

(a) (b)

(c) (d)

■ Aspidosperma polyneuron
■ Astronium graveolens
□ Cariniana legalis
■ Cecropia hololeuca
■ Croton piptocalyx
■ Diatenopteryx sorbifolia
■ Hymenaea courbaril
■ Pachystroma longifolium

0   250   500 m

N

FIGURE 11.11 Maps of species that were produced by (a) linear discriminant analysis (LDA), (b) linear support vector machines (L-SVMs), (c) radial basis function support vector machines (RBF-SVMs), and (d) random forest (RF), each using all the hyperspectral bands. A dark true color composite (R = 639 nm, G = 548 nm, B = 460 nm) of the hyperspectral image is inserted in the background of each map to enhance visualization. (From Ferreira, M.P. et al. 2016. *Remote Sensing of Environment*, 179:66–78.)

In Chapter 7, Dr. Matthew L. Clark presented studies of tropical forest vegetation using remote sensing at leaf and canopy scales. Current literature is limited to laboratory leaf-level studies of biophysical and biochemical properties of forest vegetation using hyperspectral remote sensing. These leaf-level studies quantify variables such as chlorophyll *a* and *b*, pigments (e.g., carotenoids, anthocyanins), and several other variables such as lignin, cellulose, water, and nitrogen. Each of these is best studied using hyperspectral narrow bands (HNBs) from specific portions of the spectrum (e.g., chlorophyll absorption at 680 nm). In this fashion they discuss the extensive literature that is currently available, and the associated difficulties of moving from leaf-level studies in controlled environment to canopy-scale studies involving complex environments. The current, best understanding of canopy-scale studies is through radiative transfer (RT) models. At canopy scales, spectral reflectivity in specific HNBs and hyperspectral vegetation indices (HVIs) are related to quantities such as leaf area index (LAI), biomass, and other quantities (e.g., pigments, Chl, N) as judged using methods such as partial least-squares (PLS), and end-member analysis unmixing the spectral data of hyperspectral pixels into green vegetation, nonphotosynthetic vegetation, and shaded areas and shadow components. However, mapping forest species over broad spatial scales involving imaging spectroscopy data from sensors such as Hyperion, the upcoming HyspIRI, and other similar spaceborne or airborne sensors over large areas is complex given the richness and diversity of forest types and species. Here, Dr. Clark shows the strength of combining hyperspatial data with hyperspectral data to map forest canopy crowns, and to separate species types. He suggests a subpixel multiple-end-member spectral mixture analysis (MESMA) as a potential approach to separating and mapping fractional abundance of species within pixels. He shows that tree crowns can be accurately mapped when combining the power of hyperspectral data with hyperspatial data. Finally, Dr. Clark presents various classification schemes to delineate and map forest tree species. These methods include spectral angle mapper (SAM), MESMA, maximum likelihood (ML), linear discriminant analysis (LDA), decision trees (DTs), and several others. Application of these methods are preceded by feature selection methods to eliminate redundant bands through the use of principal component analysis (PCA), wavelet transforms, stepwise band selection, and HVIs. He rightly recommends that future studies must consider integrating additional spectral variables such as the contributions of bark, epiphytes, and lianas, as well as temporal variables like older leaves and leaf on/leaf off conditions along with currently considered biophysical, biochemical, and structural variables.

11.8   CHARACTERISTICS OF TROPICAL TREE SPECIES IN HYPERSPECTRAL AND MULTISPECTRAL DATA

Remote sensing is widely used in forest studies including land use/land cover classification (LULC), tree crown mapping, tree species identification and mapping, tree height measurements, diameter at breast height (DBH) estimates, detecting selective logging, mapping forest categories (e.g., primary and secondary forests), quantifying regrowth rates, and forest biomass and carbon modeling and mapping. Currently, multispectral, hyperspectral, hyperspatial, and LiDAR data are the four main types of remote sensing data used in forest studies. Multispectral data are acquired typically in 4–10 broad spectral bands in the 400–14,000 nm spectral range with each band in 10–30 m spatial resolution. Such data are gathered from satellites such as Landsat, Satellite Pour l'Observation de la Terre (SPOT), and the Indian Remote Sensing (IRS) satellites which are all widely used due to their free or cheap cost of data acquisition. But these data have limitations in accurately estimating complex forest characteristics such as forest tree species, biomass, and tree height. In contrast, hyperspectral data provide much improved accuracies (Dalponte et al., 2009). Thenkabail et al. (2004a,b) studied tropical forest characteristics using Earth Observing-1 (EO-1) Hyperion data and compared the results with three broad-band datasets from hyperspatial IKONOS, multispectral Landsat Enhanced Thematic Mapper Plus (ETM+), and multispectral EO-1 Advanced Land Imager (ALI). A comparison of data from these four distinct sensors acquired over the Congo rainforests at about the same time are shown in Figure 11.12, and demonstrates the advances one can make through

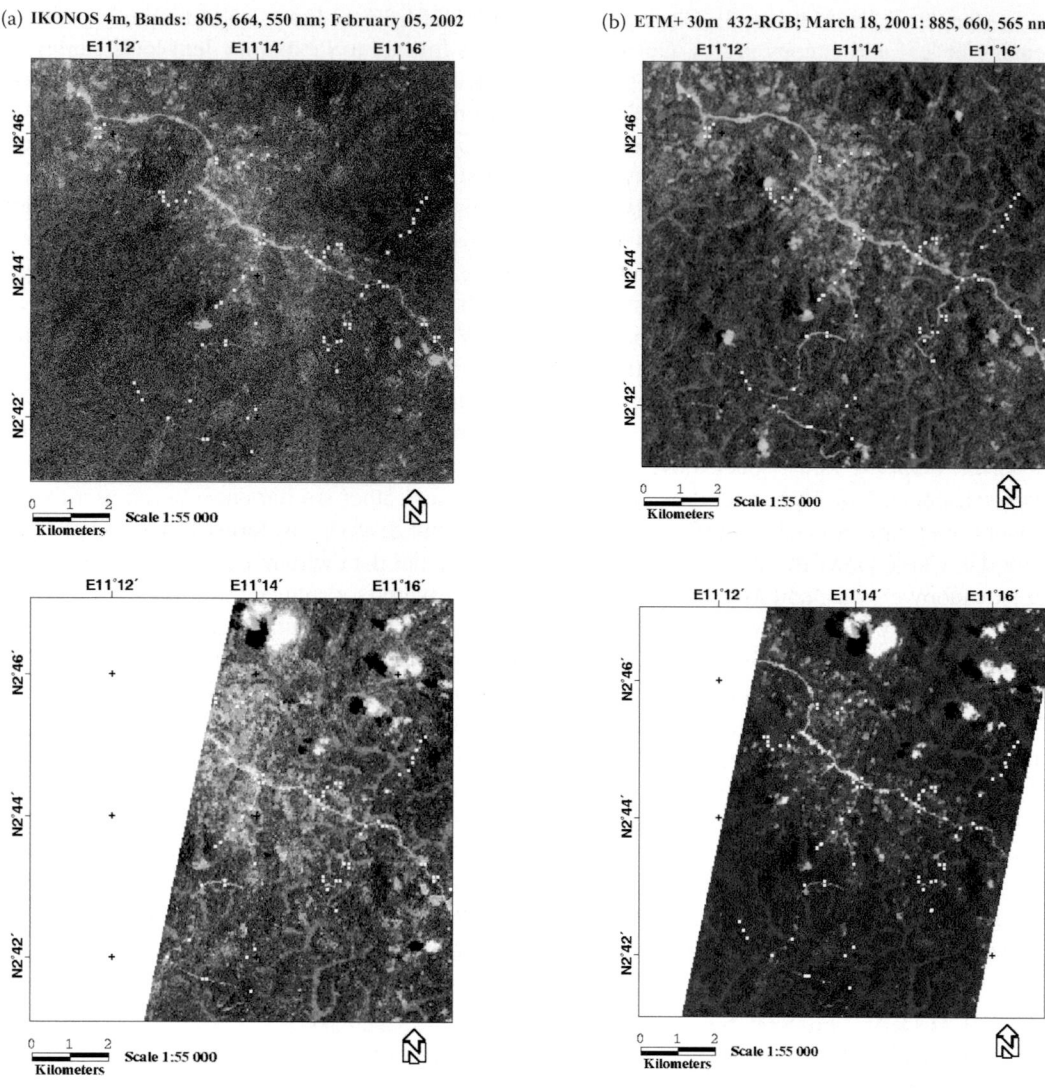

(a) IKONOS 4m, Bands: 805, 664, 550 nm; February 05, 2002

(b) ETM+ 30m 432-RGB; March 18, 2001: 885, 660, 565 nm

(c) Advanced Land Imager (ALI) 30m bands: 865, 660, 565 nm; March 21, 2002

(d) Hyperion 30m 804, 661, 550 RGB; March 21, 2002

FIGURE 11.12 Comparison of hyperspectral data with multispectral data for tropical forests. Satellite remote sensing data of a tropical forest area in Southern Cameroons was obtained from four sensors on dates very close to one another: (a) Hyperspatial IKONOS, (b) multispectral Landsat ETM+, (c) multispectral EO-1 ALI, and (d) hyperspectral EO-1 Hyperion. (From Thenkabail, P.S. et al. 2004a. *Remote Sensing of Environment*, 90:23–43.)

improved classification accuracies in forest class mapping as well as improved quantitative biomass modeling using Hyperion data relative to other datasets (Figure 11.12). This advance through use of hyperspectral Hyperion data was due to fine spectral detail (e.g., Figure 11.13) that supplied enhanced information of forests relative to the broad bandwidths of other sensors (Figure 11.12). Light detection and ranging (LiDAR) provided information on tree height (Dalponte et al., 2012) and stem volume (Magnussen et al., 2010) not easily gathered using hyperspectral or hyperspatial or multispectral data. However, the best results were obtained by fusion of data involving one or more hyperspectral, hyperspatial, and LiDAR sensor sources, and was especially true when mapping numerous complex forest characteristics (Dalponte et al., 2012). Hyperspectral data provided subtle variations in forest species throughout the electromagnetic spectrum in narrow wavebands (Figure 11.13) and thus

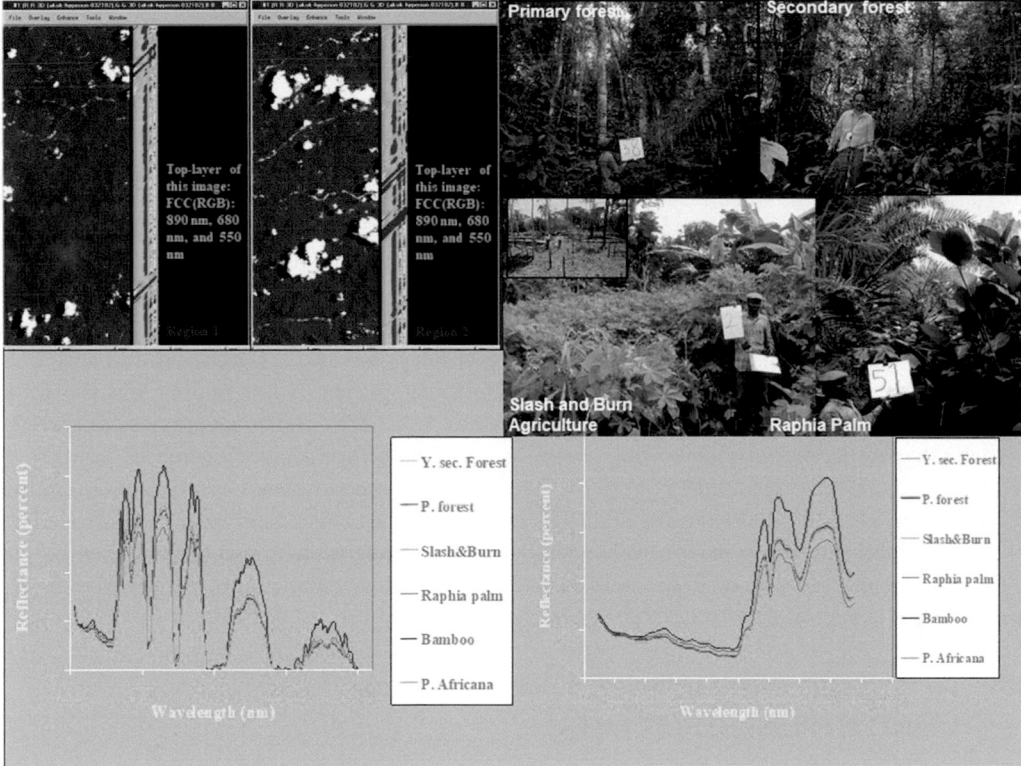

FIGURE 11.13 Hyperspectral Earth Observing-1 Hyperion data over African rainforest to study various vegetation characteristics. Typical spectra of six tropical vegetations in Southern Cameroons are shown.

helped improve accuracy in quantification of forest biophysical and biochemical properties. However, hyperspectral data work best when they are acquired with hyperspatial (submeter to 5-m) capabilities so that the hyperspectral signatures are purely one species (e.g., the hyperspectral signature of a single tree species within a pixel, rather than that of multiple tree species in a pixel). Hyperspatial data also provided textural information for mapping tree crowns and structural features that enabled greater accuracies in forest classification due to purity of signatures. LiDAR provided a powerful mechanism for determining tree height and DBH and indirectly tree species identification, which enhanced estimates of biomass and carbon of individual trees as well as entire forests. Liu and Wu (2018) used high-density LiDAR data for individual tree delineation and Airborne Imaging Spectrometer for Applications (AISA) hyperspectral imagery for pure crown-scale spectra extraction. Alonzo et al. (2014) showed that the accuracies in identifying and mapping small and morphologically unique trees such as the tall palm species *Washingtonia robusta* increased from 29% using hyperspectral bands alone to 71% with hyperspectral and LiDAR data fusion. Tree structural parameters such as tree DBH and crown area were best estimated using hyperspatial imagery (Greenberg et al., 2005). Indeed from the literature it is quite obvious that a combination of hyperspectral, and/or hyperspatial, and/or LiDAR data will provide the best estimates of complex forest characteristics such as tree species detection and mapping, and for other characteristics like forest biomass estimation and carbon sequestration estimation.

In Chapter 8, Dr. Matheus Pinheiro Ferreira and coauthors gathered hyperspectral data of tropical forests in Ghana and Brazil from two airborne instruments: one gathering data in the 400–700 nm range having 122 hyperspectral narrow bands (HNBs) with 4 nm spectral bandwidths and 1 m spatial resolution, and another gathering data in the 970–2500 nm range having 235 hyperspectral narrow bands (HNBs) with 6 nm spectral bandwidths and 1 m spatial resolution. So, the data were not only hyperspectral, but also hyperspatial. They gathered these hyperspectral data in hyperspatial mode for eight species in Brazil and six species in Africa after manually delineating the individual tree crown boundaries accurately. This ensured that the hyperspectral signatures were pure for each species and they obtained data from a large number of pixels (sample size). They also adopted the continuum removal method for spectral feature analysis through feature depth using the United States Geological Survey's (USGS) Processing Routines in Interactive Data Language (IDL) for the Spectroscopic Measurements (PRISM) software. Multiple-end-member Spectral Mixture Analysis (MESMA) was used to determine the within crown variability of trees. Subpixel fractions of green vegetation (GV), nonphotosynthetic vegetation (NPV), and shade of each tree crown were analyzed using MESMA. They related the hyperspectral characteristics to chemical and structural properties of tree species and found strong relationships. For example, they found significant positive relationships between forest species' chlorophyll and leaf water absorption feature depth with hyperspectral data for most species. A key finding was the ability of hyperspectral data to classify tree species with about 15%–17% higher accuracies compared to simulated hyperspatial Worldview-3 data. For example, the average accuracy using 15 bands in the visible and short wave-infrared (VSWIR; 450–2500 nm) from simulated Worldview-3 data was 69.7% relative to 85.1% attained using 30 HNBs in VSWIR. They also demonstrated increased tree species separability using hyperspectral data alone when compared with that of multispectral data alone. Further, increases in accuracy were attained when texture data from hyperspatial data were combined with spectral data from hyperspectral sensors.

11.9 DETECTING AND MAPPING INVASIVE SPECIES USING HYPERSPECTRAL DATA

Exotic species invasion in any place has serious ecological, environmental, and economic impacts (Yang and Everitt, 2010). For example, water hyacinth (*Eichhornia crassipes*) spreads aggressively in lakes and other surface water reservoirs, impacting *zooplankton* and *phytoplankton* productivity in freshwater ecosystems, modifying surface water clarity, and causing hypoxia or a decrease in the concentration of oxygen and related nutrients, such as nitrogen, phosphorous and heavy metals (Thamaga and Dube, 2018). Indeed, losses through invasive species can run into billions of dollars through irreparable damage to native species by causing enormous long-term or permanent damage to local ecosystems. So mapping them, especially early detection (Hestir et al., 2008) as well as their spread over time, is required for proper management. Hyperspatial (sub-meter to ≤5 m) and hyperspectral (≤10 nm) data are powerful tools to detect and delineate invasive species that are often found in patches over large areas (as they spread) in the landscape and not as contiguous cover over large areas. However, hyperspatial and hyperspectral data have their own limitations. Hyperspatial data may introduce significant intra-species spectral variability and result in reduced mapping accuracy, while hyperspectral data are commonly limited to smaller areas, are costly and computationally expensive (Kganyago et al., 2018). Further, regional coverage of such imagery over large areas is limited. Thereby, Kganyago et al. (2018) recommended that the imagery should be routinely acquired over very large areas as done with the Landsat Operational Land Imager (Landsat OLI) with 30-m non-thermal data and Satellite Pour l'Observation de la Terre 6 (SPOT-6) with 1.5 m panchromatic and 6 m multispectral data as they provide excellent opportunities to map invasive alien species. They demonstrated this by mapping the invasive species *Parthenium hysterophorus* in the African Savannas with 83%–86% accuracies. Amaral et al. (2015) used imaging spectroscopy data for mapping invasive plant species *Dendrocalamus sp.* (bamboo) and *Pinus elliottii L.* (slash pine) in a Brazilian neotropical landscape within the tropical Brazilian savanna biome and determined that near-infrared (NIR) and shortwave-infrared (SWIR) hyperspectral data were crucial for identifying and achieving good accuracies. Hestir et al. (2008) mapped three wetland invasive species using the 128-band airborne HyMap hyperspectral data. Their study established that using hyperspectral data perennial pepperweed and water hyacinth

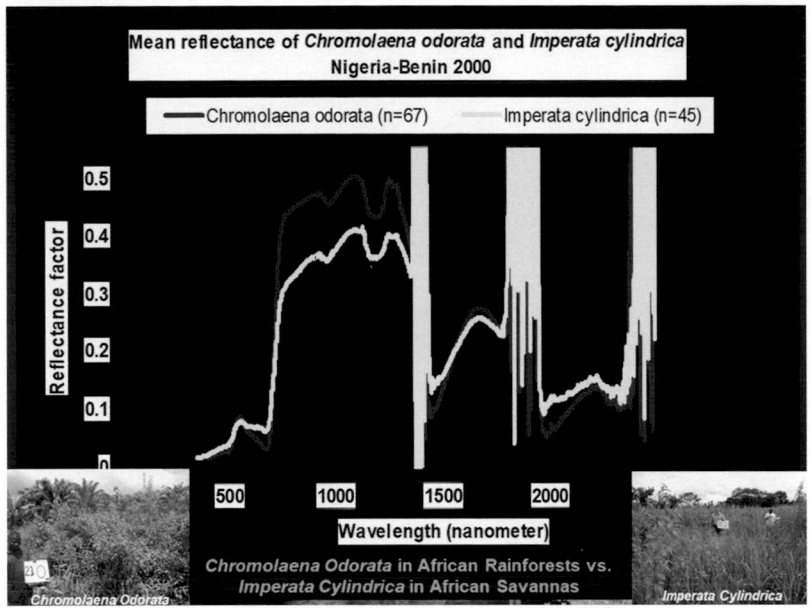

FIGURE 11.14   Two invasive species in Africa. *Chromolaena odorata* is an invasive species found in African rainforests and *Imperata cylindrica* is an invasive species found throughout the African savannas.

were mapped with modest accuracies whereas submerged aquatic vegetation was mapped with high accuracies. Thenkabail et al. (2004a,b) found distinct characteristics of invasive species in the African continent (Figure 11.14), where *Chromolena odorata* was an invasive species found in rainforests and *Imperata cylindrica* was an invasive species found throughout the savannas (Figure 11.14).

In Chapter 9, Dr. Ruiliang Pu discussed invasive plant species detection and mapping in the context of their occurrence in multiple ecosystems such as aquatic systems populated by grasses, shrubs, weeds, and trees. Four basic approaches to delineating and mapping invasive species are addressed and include (1) identifying spectral separability of invasive species from native species; (2) adequate spectral, spatial, and temporal resolutions; (3) methods and approaches used; and (4) uniqueness of invasive species in their biophysical and/or biochemical characteristics, phenologies, and plant structure (e.g., planophile vs. erectophile) relative to native plants. For example, they point out that the invasive species *tamarisk* was detected through classification of the imagery, whereas another invasive species, cheatgrass, was identified via changes in color and distinct seasonality compared to native plants (Noujdina and Ustin, 2008; Pu et al., 2008a,b). The methods discussed by Pu to detect and map invasive species using hyperspectral data included derivative analysis, transformations, spectral matching techniques, hyperspectral vegetation indices (HVIs), spectral mixture analysis, and hyperspectral image classifications. Dr. Pu provides detailed discussions on each of these. Derivative spectra reduce signal-to-noise ratio (SNR) relative to raw hyperspectral data. Spectral matching techniques help match the reference or laboratory spectra to similar spectra from class spectra derived during classification of images (Thenkabail et al., 2007). HVIs help reduce data volume, overcome data redundancies, and help focus on specific biophysical and biochemical quantities studied (Thenkabail et al., 2015). Transformations such as principal component analysis (PCA) also reduce data volumes as well as help to overcome limitations of SNR. Spectral mixture analysis (SMA) was addressed as linear spectral mixture model analysis and mixture tuned matched filtering (MTMF). SMAs help "unmix" proportions of the pixel spectra into distinct end-members: photosynthetically active green vegetation, nonphotosynthetic vegetation, and shade.

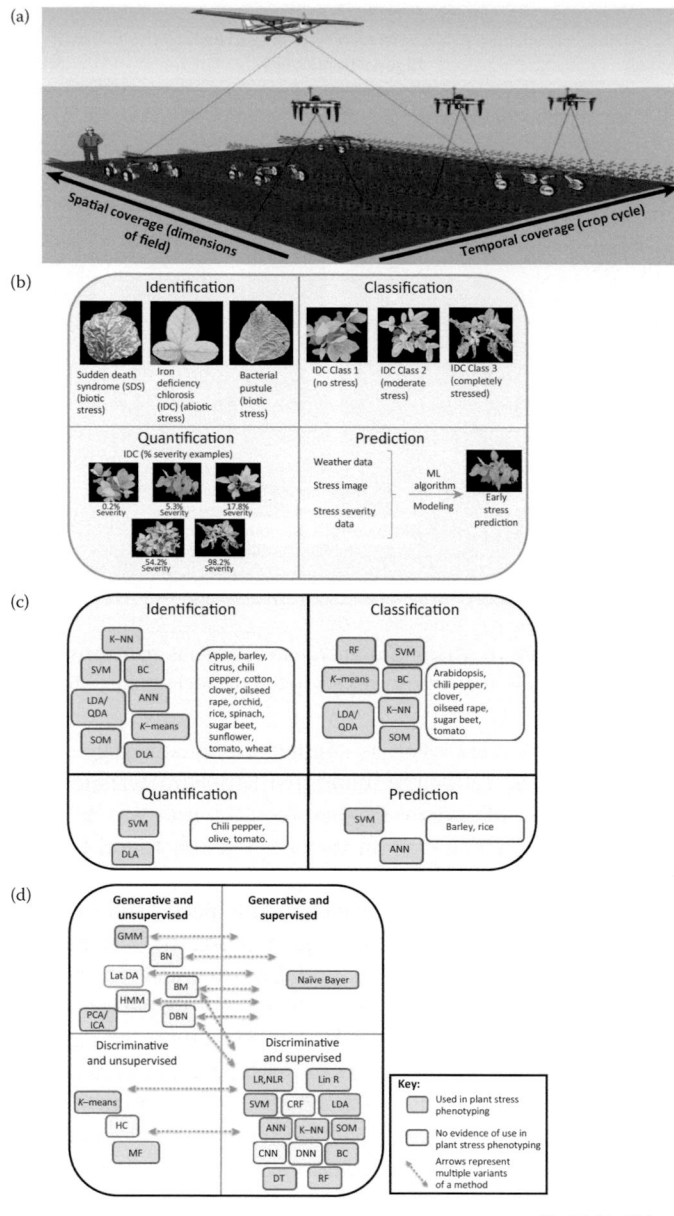

FIGURE 11.15 Machine learning (ML) tools for high-throughput stress phenotyping (a) High-throughput stress phenotyping in a soybean field at various growth stages and at different heights using aircraft, unmanned aerial vehicles, and unmanned ground vehicles. (b) Identification, classification, quantification, and prediction (ICQP) of plant diseases in soybean. (c) ML algorithms used in ICQP of plant stresses. (d) Classification of ML algorithms into generative and discriminative. Abbreviations: ANN, artificial neural network; BC, Bayes classifier; BN, Bayesian network; BM, Boltzmann machine; CRF, conditional random field; CNN, convolutional neural network; DT, decision tree; DNN, deep neural network; GMM, Gaussian mixture models; GP, Gaussian process; HMM, hidden Markov model; HC, hierarchical clustering; ICA, independent component analysis; K-MC, K-means clustering; K-NN, k-nearest neighbor classifier; Lat DA, latent Dirichlet allocation; LDA, linear discriminant analysis; Lin R, linear regression; LR, logistic regression; MF, matrix factorization; NB, naïve Bayes; NLR, nonlinear regression; PCA, principal component analysis; RF, random forests; SOM, self-organizing map; SVM, support vector machine; UAV, unmanned aerial vehicle; UGV, unmanned ground vehicle. (From Singh, A. et al. 2016. *Trends in Plant Science*, 21(2):110–124.)

11.10   VISIBLE, NEAR INFRARED, AND THERMAL SPECTRAL RADIANCE ON-BOARD UAVS FOR HIGH-THROUGHPUT PHENOTYPING OF PLANT BREEDING TRIALS

Robotic and sensor technologies enable us to collect a wide range of plant characteristics in a nondestructive mode either through field-based or unmanned aerial vehicle (UAV)-based or mobile platform-based high-throughput plant phenotyping (HTPP). Remote measurement of plant characteristics (e.g., height, biophysical, biochemical, and structural features) is becoming increasingly popular, especially from high-throughput ground-based or UAV-based platforms. Phenotyping technology can advance plant breeding and requires the throughput of plant screening in the field such as for early-season detection of plant diseases, disease diagnosis, plant height estimation, canopy temperature measurement, normalized difference vegetation index (NDVI) calculations, reflectance measurements, and a host of other plant-related remote measurements (e.g., growth pattern, yield) leading to acceleration of rate of genetic gain in crops (Bai et al., 2016; Shakoor et al., 2017). A reliable high-throughput precision phenotyping capability is of great importance in accelerating genetic gain within breeding programs such as in (1) increasing the size of the breeding program to enable higher selection intensity, (2) enhancing the accuracy of selection (higher repeatability), (3) ensuring adequate genetic variation, (4) accelerating the breeding cycles, and (5) improving decision support tools (Araus et al., 2018; Hu et al., 2018). Hu et al. (2018) measured the sorghum plant height based images captured by a UAV-mounted RGB camera before emergence and at near maturity to generate digital surface models (DSMs). Due to the very nature of the high-throughput phenotyping, large volumes of crop data were gathered on all sorts of crop variables from many sensors. Big-data challenges were handled through mechanisms and methods such as machine learning tools like support vector machines (SVMs) and artificial neural networks (ANNs) (Singh et al., 2016; Figure 11.15).

In Chapter 10, Dr. Scott Chapman and coauthors discussed plot-level phenotypes, termed "traits," such as plant cover, height, biomass, and yield, which were measured through UAV-based HTPP. They also highlighted the important role played in advancing plant breeding using HTPP. For example, they point out that through the advent of high-throughput genomic DNA sequencing, plant breeding is transitioning to a new model. Since HTPP leads to collecting various plant traits repeatedly in close proximity (at a very high spatial resolution of a few cm²), it thus leads to big-data management and processing challenges as well as opportunities. Various data processing methods and techniques are discussed in detail in Chapter 10, whose unique feature is estimations of a wide range of plant biophysical (e.g., plant height, biomass, leaf area index), biochemical (e.g., fraction of photosynthetically active radiation), and physiological (e.g., head counting) characteristics using HTPP along with appropriate methods and techniques for a number of different crops.

ACKNOWLEDGMENT

We would like to thank Dr. Isabella Mariotto, Dr. Michael Marshall, and Dr. Itiya Aneece for their review, insights, and inputs.

REFERENCES

Allen, R.G., Tasumi, M., Trezza, R. 2007. Satellite-based energy balance for mapping evapotranspiration with internalized calibration (METRIC)—Model. *Journal of Irrigation and Drainage Engineering*, 133(4):380–394.

Alonzo, M., Bookhagen, B., Roberts, D.A. 2014. Urban tree species mapping using hyperspectral and lidar data, Remote Sensing of Environment, 148:70–83, ISSN 0034-4257, https://doi.org/10.1016/j.rse.2014.03.018.

Alonzo, M., McFadden, J.P., Nowak, D.J., Roberts, D.A. 2016. Mapping urban forest structure and function using hyperspectral imagery and lidar data. *Urban Forestry & Urban Greening*, 17:135–147, ISSN 1618-8667, https://doi.org/10.1016/j.ufug.2016.04.003.

Alvarado, S.T., Fornazari, T., Cóstola, A., Morellato, L.P.C., Silva, T.S.F. 2017. Drivers of fire occurrence in a mountainous Brazilian cerrado savanna: Tracking long-term fire regimes using remote sensing. *Ecological Indicators*, 78:270–281, ISSN 1470-160X, https://doi.org/10.1016/j.ecolind.2017.02.037.

Amador, E.S., Bandfield, J.L., Thomas, N.H. 2018. A search for minerals associated with serpentinization across Mars using CRISM spectral data. *Icarus*, 311:113–134, ISSN 0019-1035, https://doi.org/10.1016/j.icarus.2018.03.021.

Amaral, C.H., Roberts, D.A., Almeida, T.I.R., Filho, C.R.S. 2015. Mapping invasive species and spectral mixture relationships with neotropical woody formations in southeastern Brazil. *ISPRS Journal of Photogrammetry and Remote Sensing*, 108:80–93, ISSN 0924-2716, https://doi.org/10.1016/j.isprsjprs.2015.06.009.

Anderson, M.C. et al. 2011. Mapping daily evapotranspiration at field to continental scales using geostationary and polar orbiting satellite imagery. *Hyrol Earth Systems Science*, 15:223–239.

Apan, A., Held, A., Phinn, S., Markley, J. 2004. Detecting sugarcane "orange rust" disease using EO-1 Hyperion hyperspectral imagery. *International Journal of Remote Sensing*, 25(2):489–498.

Araus, J.L., Kefauver, S.C., Zaman-Allah, M., Olsen, M.S., Cairns, J.E. 2018. Translating high-throughput phenotyping into genetic gain. *Trends in Plant Science*, ISSN 1360-1385, https://doi.org/10.1016/j.tplants.2018.02.001.

Asner, G., Nepstad, D., Cardinot, G., Ray, D. 2004. Drought stress and carbon uptake in an Amazon forest measured with spaceborne imaging spectroscopy. *Proceedings of the National Academy of Sciences of the United States of America*, 101(16):6039–6044.

Asner, G.P., Martin, R.E., Anderson, C.B., Knapp, D.E. 2015. Quantifying forest canopy traits: Imaging spectroscopy versus field survey. *Remote Sensing of Environment*, 158:15–27, ISSN 0034-4257, https://doi.org/10.1016/j.rse.2014.11.011.

Atzberger C. 2013. Advances in remote sensing of agriculture: Context description, existing operational monitoring systems and major information needs. *Remote Sensing*, 5(2):949–981.

Baccini, A., Walker, W., Carvalho, L., Farina, M., Sulla-Menashe, D., Houghton, R.A. 2017. Tropical forests are a net carbon source based on aboveground measurements of gain and loss. *Science*, 358.

Bagley, J.E., Davis, S.C., Georgescu, M., Hussain, M.Z., Miller, J., Nesbitt, S.W., VanLoocke, A., Bernacchi, C.J. 2014. The biophysical link between climate, water, and vegetation in bioenergy agro-ecosystems. *Biomass and Bioenergy*, 71:187–201, ISSN 0961-9534, https://doi.org/10.1016/j.biombioe.2014.10.007.

Bai, G., Ge, Y., Hussain, W., Baenziger, P.S., Graef, G. 2016. A multi-sensor system for high throughput field phenotyping in soybean and wheat breeding. *Computers and Electronics in Agriculture*, 128:181–192, ISSN 0168-1699, https://doi.org/10.1016/j.compag.2016.08.021.

Baker, V.R., Hamilton, C.W., Burr, D.M., Gulick, V.C., Komatsu, G., Luo, W., Rice, J.W., Rodriguez, J.A.P. 2015. Fluvial geomorphology on Earth-like planetary surfaces: A review. *Geomorphology*, 245:149–182, ISSN 0169-555X, https://doi.org/10.1016/j.geomorph.2015.05.002.

Bastiaanssen, W., Menenti, M., Feddes, R., Holtslag, A. 1998a. A remote sensing surface energy balance algorithm for land (SEBAL). 1. Formulation. *Journal of Hydrology*, 212:198–212.

Bastiaanssen, W., Pelgrum, H., Wang, J., Ma, Y., Moreno, J., Roerink, G., Van der Wal, T. 1998b. A remote sensing surface energy balance algorithm for land (SEBAL).: Part 2: Validation. *Journal of Hydrology*, 212:213–229.

Bates, B.C., Kundzewicz, Z.W., Wu, S., Palutikof, J.P. Eds. 2008. Climate Change and Water. Technical Paper of the *Intergovernmental Panel on Climate Change*, IPCC Secretariat, Geneva, 210 p.

Becker, B.L., Lusch, D.P., Qi, J. 2007. A classification-based assessment of the optimal spectral and spatial resolutions for Great Lakes coastal wetland imagery. *Remote Sensing of Environment*, 108(1):111–120.

Behling, R., Bochow, M., Foerster, S., Roessner, S., Kaufmann, H. 2015. Automated GIS-based derivation of urban ecological indicators using hyperspectral remote sensing and height information. *Ecological Indicators*, 48:218–234, ISSN 1470-160X, https://doi.org/10.1016/j.ecolind.2014.08.003.

Bendig, J., Yu, K., Aasen, H., Bolten, A., Bennertz, S., Broscheit, J., Gnyp, M.L., Bareth, G. 2015. Combining UAV-based plant height from crop surface models, visible, and near infrared vegetation indices for biomass monitoring in barley. *International Journal of Applied Earth Observation and Geoinformation*, 39:79–87, ISSN 0303-2434, https://doi.org/10.1016/j.jag.2015.02.012.

Biard, F., Baret, F. 1997. Crop residue estimation using multiband reflectance. *Remote Sensing of Environment*, 59:530–536.

Biggs, T.W., Petropoulos, G.P., Velpuri, N.M., Marshall, M., Glenn, E.P., Nagler, P., Messina, A. 2015. Remote Sensing of Actual Evapotranspiration from Croplands. In: *Remote Sensing of Water Resources, Disasters, and Urban Studies*. CRC Press, pp. 59–99.

Biggs, T.W., Marshall, M., Messina, A. 2016. Mapping daily and seasonal evapotranspiration from irrigated crops using global climate grids and satellite imagery: Automation and methods comparison. *Water Resources Research*, 52(9): http://doi.org/10.1002/2016WR019107.

Biradar, C.M. et al. 2009. A global map of rainfed cropland areas (GMRCA) at the end of last millennium using remote sensing. *International Journal of Applied Earth Observation and Geoinformation*, 11(2):114–129. doi:10.1016/j.jag.2008.11.002. January, 2009.

Biradar, C.M., Thenkabail, P.S., Platonov, A., Xiangming, X., Geerken, R., Vithanage, J., Turral, H., Noojipady, P. 2008. Water productivity mapping methods using remote sensing. *Journal of Applied Remote Sensing*, 2:023544.

Brown, L., Chen, J.M., Leblanc, S.G., Cihlar, J. 2000. A shortwave infrared modification to the simple ratio for LAI retrieval in boreal forests: An image and model analysis. *Remote Sensing Environment*, 71(1):16–25.

Brown, S., Iverson, L.R., Prasad, A. 2001. *Geographical Distribution of Biomass Carbon in Tropical Southeast Asian Forests: A Database*. ORNL/CDIAC-119, NDP-068. Carbon Dioxide Information Analysis Center, U.S. Department of Energy, Oak Ridge National Laboratory, Oak Ridge, Tennessee, USA.

Brown, M.E. 2015. Satellite remote sensing in agriculture and food security assessment. *Procedia Environmental Sciences*, 29:307, ISSN 1878-0296, https://doi.org/10.1016/j.proenv.2015.07.278.

Cai, X.L. et al. 2009. Water productivity mapping methods and protocols using remote sensing data of various resolutions to support "more crop per drop." *Journal of Applied Remote Sensing*, 3(1):033557. doi:10.1117/1.3257643.

Campbell, T., Fearns, P. 2018. Simple remote sensing detection of Corymbia calophylla flowers using common 3-band imaging sensors. *Remote Sensing Applications: Society and Environment*, ISSN 2352-9385, https://doi.org/10.1016/j.rsase.2018.04.009.

Carvalho, S., Schlerf, M., van der Putten, W.H., Skidmore, A.K. 2013. Hyperspectral reflectance of leaves and flowers of an outbreak species discriminates season and successional stage of vegetation. *International Journal of Applied Earth Observation and Geoinformation*, 24:32–41, ISSN 0303-2434, https://doi.org/10.1016/j.jag.2013.01.005.

Chander, G., Markham, B.L., Helder, D.L. 2009. Summary of current radiometric calibration coefficients for Landsat MSS, TM, ETM+, and EO-1 ALI sensors. *Remote Sensing of Environment*, 113(5):893–903, ISSN 0034-4257, https://doi.org/10.1016/j.rse.2009.01.007.

Chang, A., Jung, J., Maeda, M.M., Landivar, J. 2017. Crop height monitoring with digital imagery from unmanned aerial system (UAS). *Computers and Electronics in Agriculture*, 141:232–237, ISSN 0168-1699, https://doi.org/10.1016/j.compag.2017.07.008.

Chen, J.M., Menges, C.H., Leblanc, S.G. 2005. Global mapping of foliage clumping index using multi-angular satellite data. *Remote Sensing of Environment*, 97(4):447–457, ISSN 0034-4257, https://doi.org/10.1016/j.rse.2005.05.003.

Chen, J., Shen, M., Zhu, X., Tang, Y. 2009. Indicator of flower status derived from *in situ* hyperspectral measurement in an alpine meadow on the Tibetan Plateau. *Ecological Indicators*, 9(4):818–823, ISSN 1470-160X, https://doi.org/10.1016/j.ecolind.2008.09.009.

Chen, S., Zhang, F., Ning, J., Liu, X., Zhang, Z., Yang, S. 2015. Predicting the anthocyanin content of wine grapes by NIR hyperspectral imaging. *Food Chemistry*, 172:788–793, ISSN 0308-8146, https://doi.org/10.1016/j.foodchem.2014.09.119. (http://www.sciencedirect.com/science/article/pii/S0308814614015040)

Chen, F., Wang, K., de Voorde, T.V., Tang, T.F. 2017. Mapping urban land cover from high spatial resolution hyperspectral data: An approach based on simultaneously unmixing similar pixels with jointly sparse spectral mixture analysis. *Remote Sensing of Environment*, 196:324–342, ISSN 0034-4257, https://doi.org/10.1016/j.rse.2017.05.014.

Chowdhury, E.H., Hassan, Q.K. 2015. Operational perspective of remote sensing-based forest fire danger forecasting systems. *ISPRS Journal of Photogrammetry and Remote Sensing*, 104:224–236, ISSN 0924-2716, https://doi.org/10.1016/j.isprsjprs.2014.03.011.

Christian, B., Joshi, N., Saini, M., Mehta, N., Goroshi, S., Nidamanuri, R.R., Thenkabail, P., Desai, A.R., Krishnayya, N.S.R. 2015. Seasonal variations in phenology and productivity of a tropical dry deciduous forest from MODIS and Hyperion. *Agricultural and Forest Meteorology*, 214-215:91–105, ISSN 0168-1923, https://doi.org/10.1016/j.agrformet.2015.08.246.

Clark, M.L., Roberts, D.A., Clark, D.B. 2005. Hyperspectral discrimination of tropical rain forest tree species at leaf to crown scales. *Remote Sensing of Environment*, 96(3–4):375–398, ISSN 0034-4257, https://doi.org/10.1016/j.rse.2005.03.009.

Clark, M.L., Roberts, D.A., Ewel, J.J., Clark, D.B. 2011. Estimation of tropical rain forest aboveground biomass with small-footprint lidar and hyperspectral sensors. *Remote Sensing of Environment*, 115(11):2931–2942, ISSN 0034-4257, https://doi.org/10.1016/j.rse.2010.08.029.

Clark, M.L., Buck-Diaz, J., Evens, J. 2018. Mapping of forest alliances with simulated multi-seasonal hyperspectral satellite imagery. *Remote Sensing of Environment*, 210:490–507, ISSN 0034-4257, https://doi.org/10.1016/j.rse.2018.03.021.

Cleemput, E.V., Vanierschot, L., Fernández-Castilla, B., Honnay, O., Somers, B. 2018. The functional characterization of grass- and shrubland ecosystems using hyperspectral remote sensing: Trends, accuracy and moderating variables. *Remote Sensing of Environment*, 209:747–763, ISSN 0034-4257, https://doi.org/10.1016/j.rse.2018.02.030.

Corgne, S., Hubert-Moy, L., Betbeder, J. 2016. Chapter 6: Monitoring of Agricultural Landscapes Using Remote Sensing Data. In: *Land Surface Remote Sensing in Agriculture and Forest*. Eds. N. Baghdadi and M. Zribi, Elsevier, 221–247, ISBN 9781785481031, https://doi.org/10.1016/B978-1-78548-103-1.50006-6.

Corti, M., Gallina, P.M., Cavalli, D., Cabassi, G. 2017. Hyperspectral imaging of spinach canopy under combined water and nitrogen stress to estimate biomass, water, and nitrogen content. *Biosystems Engineering*, 158:38–50, ISSN 1537-5110, https://doi.org/10.1016/j.biosystemseng.2017.03.006.

Costa dos Santos, J.F., Nunes Romeiro, J.M., de Assis, J.B., Pereira Torres, F.T., Gleriani, J.M. 2018. Potentials and limitations of remote fire monitoring in protected areas. *Science of The Total Environment*, 616–617:1347–1355, ISSN 0048-9697, https://doi.org/10.1016/j.scitotenv.2017.10.182.

Curran, P.J. 2001. Environmental and Ecological Statistics. *Environmental and Ecological Statistics* 8:331–344, https://doi.org/10.1023/A:1012730418844

Dalponte, M., Bruzzone, L., Gianelle, D. 2012. Tree species classification in the Southern Alps based on the fusion of very high geometrical resolution multispectral/hyperspectral images and LiDAR data. *Remote Sensing of Environment*, 123:258–270, ISSN 0034-4257, https://doi.org/10.1016/j.rse.2012.03.013.

Dalponte, M., Bruzzone, L., Vescovo, L., Gianelle, D. 2009. The role of spectral resolution and classifier complexity in the analysis of hyperspectral images of forest areas. *Remote Sensing of Environment*, 113:2345–2355.

Dash, J., Curran, P.J. 2004. The MERIS terrestrial chlorophyll index. *International Journal of Remote Sensing*, 25(23):5403–5413.

Daughtry, C.S.T., Walthall, C.L., Kim, M.S., Brown de Colstoun, E., McMurtrey J.E., 2000. Estimating corn leaf chlorophyll status from leaf and canopy reflectance. *Remote Sensing of Environment,* 74:229–239.

Daughtry, C.S.T., Hunt Jr., E.R., Doraiswamy, P.C., McMurtrey III, J.E. 2005. Remote sensing the spatial distribution of crop residues. *Agronomy Journal*, 97:864–871.

Daughtry, C.S.T., Doraiswamy, P.C., Hunt, E.R., Stern, A.J., McMurtrey, J.E., Prueger, J.H. 2006. Remote sensing of crop residue cover and soil tillage intensity. *Soil and Tillage Research*, 91(1–2):101–108, ISSN 0167-1987, https://doi.org/10.1016/j.still.2005.11.013.

Davidson, E.A. et al. 2012. The Amazon basin in transition. *Nature*, 481(7381):321–328. doi: 10.1038/nature10717.

Deel, L.N., McNeil, B.E., Curtis, P.G., Serbin, S.P., Singh, A., Eshleman, K.N., Townsend, P.A. 2012. Relationship of a Landsat cumulative disturbance index to canopy nitrogen and forest structure. *Remote Sensing of Environment*, 118:40–49, ISSN 0034-4257, https://doi.org/10.1016/j.rse.2011.10.026.

DeFries, R.S., Houghton, R.A., Hansen, M.C., Field, C.B., Skole, D., Townshend, J. 2004. Carbon emissions from tropical deforestation and regrowth based on satellite observations for the 1980s and 1990s. *Proceedings of the National Academy of Sciences USA*, 2002 October 29; 99(22):14256–14261.

de Moura, Y.M. et al. 2017. Spectral analysis of amazon canopy phenology during the dry season using a tower hyperspectral camera and modis observations. *ISPRS Journal of Photogrammetry and Remote Sensing*, 131:52–64, ISSN 0924-2716, https://doi.org/10.1016/j.isprsjprs.2017.07.006.

Dennison, P.E., Matheson, D.S. 2011. Comparison of fire temperature and fractional area modeled from SWIR, MIR, and TIR multispectral and SWIR hyperspectral airborne data. *Remote Sensing of Environment*, 115:876–886.

Dennison, P.E., Charoensiri, K., Roberts, D.A., Peterson, S.H., Green, R.O. 2006. Wildfire temperature and land cover modeling using hyperspectral data. *Remote Sensing of Environment*, 100:212–222.

Denny, E.G. et al. 2014. Standardized phenology monitoring methods to track plant and animal activity for science and resource management applications. *International Journal of Biometeorology*, 58(4):591–601.

Dixon, R.K., Solomon, A.M., Brown, S., Houghton, R.A., Trexier, M.C., Wisniewski, J. 1994. Carbon pools and flux of global forest ecosystems. *Science*, 263(5144):185–190.

Djomo, A.N., Chimi, C.D. 2017. Tree allometric equations for estimation of above, below and total biomass in a tropical moist forest: Case study with application to remote sensing. *Forest Ecology and Management*, 391:184–193, ISSN 0378-1127, https://doi.org/10.1016/j.foreco.2017.02.022.

Eitel, J.U.H., Magney, T.S., Vierling, L.A., Dittmar, G. 2014. Assessment of crop foliar nitrogen using a novel dual-wavelength laser system and implications for conducting laser-based plant physiology. *ISPRS Journal of Photogrammetry and Remote Sensing*, 97:229–240, ISSN 0924-2716, https://doi.org/10.1016/j.isprsjprs.2014.09.009.

Enclona, E.A., Thenkabail, P.S., Celis, D., Diekman, J. 2004. Within-field wheat yield prediction from IKONOS data: A new matrix approach. *International Journal of Remote Sensing*, 25(2):377–388.

FAO. 2008. Food and Agriculture Organization (Content source); Jim Kundell (Topic Editor). 2008. Water profile of Ghana. In: *Encyclopedia of Earth*. Eds. C.J. Cleveland, Environmental Information Coalition, National Council for Science and the Environment, Washington, DC.

Fava, F., Colombo, R., Bocchi, S., Meroni, M., Sitzia, M., Fois, N., Zucca, C. 2009. Identification of hyperspectral vegetation indices for Mediterranean pasture characterization. *International Journal of Applied Earth Observation and Geoinformation*, 11(4):233–243, ISSN 0303-2434, https://doi.org/10.1016/j.jag.2009.02.003.

Fensholt, R. Sandholt, I., 2003. Derivation of a shortwave infrared water stress index from MODIS near- and shortwave infrared data in a semiarid environment. *Remote Sensing of Environment*, 87:111–121, http://dx.doi.org/10.1016/j.rse.2003.07.002

Fernandes, R., Butson, C., Leblanc, S., Latifovic, R. 2003. Landsat-5 TM and Landsat-7 ETM+ based accuracy assessment of leaf area index products for Canada derived from SPOT-4 VEGETATION data. *Canadian Journal of Remote Sensing*, 29(2):241–258.

Ferreira, M.E., Ferreira, L.G., Sano, E.E., Shimabukuro, Y.E. 2007. Spectral linear mixture modelling approaches for land cover mapping of tropical savanna areas in Brazil. *International Journal of Remote Sensing*, 28:413–429.

Ferreira, M.P., Zortea, M., Zanotta, D.C., Shimabukuro, Y.E., de Souza Filho, C.R. 2016. Mapping tree species in tropical seasonal semi-deciduous forests with hyperspectral and multispectral data. *Remote Sensing of Environment*, 179:66–78, ISSN 0034-4257.

Frampton, W.J., Dash, J., Watmough, G., Milton, E.J. 2013. Evaluating the capabilities of Sentinel-2 for quantitative estimation of biophysical variables in vegetation. *ISPRS Journal of Photogrammetry and Remote Sensing*, 82:83–92, ISSN 0924-2716, https://doi.org/10.1016/j.isprsjprs.2013.04.007.

Galvão, L.S., Pereira Filho, W., Abdon, M.M., Novo, E.M.M.L., Silva, J.S.V., Ponzoni, F.J. 2003. Spectral reflectance characterization of shallow lakes from the Brazilian Pantanal wetlands with field and airborne hyperspectral data. *International Journal of Remote Sensing*, 24:21, 4093–4112. doi: 10.1080/0143116031000070382.

Galvão, L.S., Formaggio, A.R., Tisot, D.A. 2005. Discrimination of sugarcane varieties in Southeastern Brazil with EO-1 Hyperion data. *Remote Sensing of Environment*, 94(4):523–534.

Gamon, J., Peñuelas, J., Field, C. 1992. A narrow-waveband spectral index that tracks diurnal changes in photosynthetic efficiency. *Remote Sensing of Environment*, 41(1):35–44.

Gao, B.C. 1996. NDWI—A normalized difference water index for remote sensing of vegetation liquid water from space. *Remote Sensing of Environment*, 58(3):257-266, ISSN 0034-4257, https://doi.org/10.1016/S0034-4257(96)00067-3.

GCOS. 2010. Ensuring the Availability of Global Observations for Climate. http://www.wmo.int/pages/prog/gcos/index.php?name=Publications (accessed September 18, 2018).

Ghulam, A., Porton, I., Freeman, K. 2014. Detecting subcanopy invasive plant species in tropical rainforest by integrating optical and microwave (InSAR/PolInSAR) remote sensing data, and a decision tree algorithm. *ISPRS Journal of Photogrammetry and Remote Sensing*, 88:174–192, ISSN 0924-2716, https://doi.org/10.1016/j.isprsjprs.2013.12.007.

Gilmore, M.S., Wilson E.H., Barrett, N., Civco, D.L., Prisloe, S., Hurd, J.D., Chadwick, C. 2008. Integrating multi-temporal spectral and structural information to map wetland vegetation in a lower Connecticut River tidal marsh. *Remote Sensing of Environment*, 112(11):4048–4060.

Gitelson, A.A., Kaufman, Y.J., Merzlyak, M.N. 1996. Use of a green channel in remote sensing of global vegetation from EOS-MODIS. *Remote Sensing of Environment*, 58:289–298.

Gitelson, A.A., Kaufman, Y.J., Stark, R., Rundquist, D. 2002. Novel algorithms for remote estimation of vegetation fraction. *Remote Sensing of Environment*, 80(1):76–87.

Gitelson, A.A., Gritz, Y., Merzlyak, M.N. 2003a. Relationships between leaf chlorophyll content and spectral reflectance and algorithms for non-destructive chlorophyll assessment in higher plant leaves. *Journal of Plant Physiology*, 160:271–282.

Gitelson, A.A., Viña, A., Arkebauer, T.J., Rundquist, D.C., Keydan, G., Leavitt, B. 2003b. Remote estimation of leaf area index and green leaf biomass in maize canopies. *Geophysical Research Letters*, 30(5):1248.

Gitelson, A., Keydan, G., Merzlyak, M. 2006. Three-band model for noninvasive estimation of chlorophyll, carotenoids, and anthocyanin contents in higher plant leaves. *Geophysical Research Letters*, 33.

Gitelson, A.A., Gamon, J.A., Solovchenko, A. 2017. Multiple drivers of seasonal change in PRI: Implications for photosynthesis. 2 stand level. *Remote Sensing of Environment*, 190:198–206, ISSN 0034-4257, https://doi.org/10.1016/j.rse.2016.12.015.

Gitelson, A., Solovchenko, A. 2018. Non-invasive quantification of foliar pigments: Possibilities and limitations of reflectance- and absorbance-based approaches. *Journal of Photochemistry and Photobiology B: Biology*, 178:537–544, ISSN 1011-1344, https://doi.org/10.1016/j.jphotobiol.2017.11.023.

Gorroño, J., Banks, A.C., Fox, N.P., Underwood, C. 2017. Radiometric inter-sensor cross-calibration uncertainty using a traceable high accuracy reference hyperspectral imager. *ISPRS Journal of Photogrammetry and Remote Sensing*, 130:393–417, ISSN 0924-2716, https://doi.org/10.1016/j.isprsjprs.2017.07.002.

Goswami, J.N., Annadurai, M. 2008. Chandrayaan-1 mission to the moon. *Acta Astronautica*, 63(11–12):1215–1220, ISSN 0094-5765.

Goward, S.N., Chander, G., Pagnutti, M., Marx, A., Ryan, R., Thomas, N., Tetrault, R. 2012. Complementarity of ResourceSat-1 AWiFS and Landsat TM/ETM+ sensors. *Remote Sensing of Environment*, 123:41–56, ISSN 0034-4257, https://doi.org/10.1016/j.rse.2012.03.002.

Greenberg, J.A., Dobrowski, S.Z., Ustin, S.L. 2005. Shadow allometry: Estimating tree structural parameters using hyperspatial image analysis. *Remote Sensing of Environment*, 97(1):15–25, ISSN 0034-4257, https://doi.org/10.1016/j.rse.2005.02.015.

Gross, J., Heumann, B. 2014. Can flowers provide better spectral discrimination between herbaceous wetland species than leaves? *Remote Sensing Letters*, 5(10):892–901. doi:10.1080/2150704X.2014.973077.

Gu, Y.W., Li, S., Gao, W., Wei, H. 2015. Hyperspectral estimation of the cadmium content in leaves of Brassica rapa chinesis based on spectral parameters. *Acta Ecologica Sinica*, 35(13):4445–4453.

Haboudane, D., Miller, J.R., Pattey, E., Zarco-Tejada, P.J., Strachan, I.B. 2004. Hyperspectral vegetation indices and novel algorithms for predicting green LAI of crop canopies: Modeling and validation in the context of precision agriculture. *Remote Sensing of Environment*, 90:337–352.

Haboudane, D., Miller, J.R., Tremblay, N., Zarco-Tejada, P.J., Dextraze, L. 2002. Integrated narrow-band vegetation indices for prediction of crop chlorophyll content for application to precision agriculture. *Remote Sensing of Environment*, 81:416–426.

Hansen, D., Uphoff, N., Lal, R. 2002. *Food Security and Environmental Quality in the Developing World*. CRC Press, Boca Raton, FL.

Hansen, M.C. et al. 2013. High-resolution global maps of 21st-century forest cover change. *Science*, 342(15 November):850–53. Data available on-line from: http://earthenginepartners.appspot.com/science-2013-global-forest.

Harrison, D., Rivard, B., Sánchez-Azofeifa, A. 2018. Classification of tree species based on longwave hyperspectral data from leaves, a case study for a tropical dry forest. *International Journal of Applied Earth Observation and Geoinformation*, 66:93–105, ISSN 0303-2434, https://doi.org/10.1016/j.jag.2017.11.009.

Harvey, K.R., Hill, G.J. 2001. Vegetation mapping of a tropical freshwater swamp in the Northern Territory, Australia: A comparison of aerial photography, Landsat TM and SPOT satellite imagery. *International Journal of Remote Sensing*, 22:2911–2925. doi: 10.1080/01431160119174.

He, L., Zhang, H.Y., Zhang, Y.S., Song, X., Feng, W., Kang, G.Z., Wang, C.Y., Guo, T.C. 2016. Estimating canopy leaf nitrogen concentration in winter wheat based on multi-angular hyperspectral remote sensing. *European Journal of Agronomy*, 73:170–185, ISSN 1161-0301, https://doi.org/10.1016/j.eja.2015.11.017.

Heiskanen, J., Rautiainen, M., Stenberg, P., Mõttus, P., Vesanto, V.H. 2013. Sensitivity of narrowband vegetation indices to boreal forest LAI, reflectance seasonality and species composition. *ISPRS Journal of Photogrammetry and Remote Sensing*, 78:1–14, ISSN 0924-2716, https://doi.org/10.1016/j.isprsjprs.2013.01.001.

Hernández-Clemente, R., Navarro-Cerrillo, R.M., Suárez, F., Morales, L., Zarco-Tejada, P.J. 2011. Assessing structural effects on PRI for stress detection in conifer forests. *Remote Sensing of Environment*, 115:2360–2375.

Hestir, E.L., Khanna, S., Andrew, M.E., Santos, M.J., Viers, J.H., Greenberg, J.A., Rajapakse, S.S., Ustin, S.L. 2008. Identification of invasive vegetation using hyperspectral remote sensing in the California Delta ecosystem. *Remote Sensing of Environment*, 112(11):4034–4047, ISSN 0034-4257, https://doi.org/10.1016/j.rse.2008.01.022.

Hirano, A., Madden, M., Welch, R. 2003. Hyperspectral image data for mapping wetland vegetation. *BioOne Journal*, 23(2).

Hopping, K.A., Yeh, E.T., Gaerrang, Harris, R.B. 2018. Linking people, pixels, and pastures: A multi-method, interdisciplinary investigation of how rangeland management affects vegetation on the Tibetan Plateau. *Applied Geography*, 94:147–162, ISSN 0143-6228, https://doi.org/10.1016/j.apgeog.2018.03.013.

Houghton, R.A. 2005. Aboveground forest biomass and the global carbon balance. *Global Change Biology*, 11(6):945–958.

Houghton, R.A. 2007. Balancing the global carbon budget. *Annual Review of Earth and Planatery Sciences*, 35:313–347.

Houghton, R.A. 2008. Carbon Flux to the Atmosphere from Land-Use Changes: 1850–2005. In: *TRENDS: A Compendium of Data on Global Change.* Carbon Dioxide Information Analysis Center, Oak Ridge National Laboratory, U.S. Department of Energy, Oak Ridge, Tenn., USA.

Houghton, R.A., Birdsey, R.A. Nassikas, A., McGlinchey, D. 2017. *Forests and Land Use: Undervalued Assets for Global Climate Stabilization. Policy Brief.* Falmouth: Woods Hole Research Center.

Hu, P., Chapman, S.C., Wang, X., Potgieter, A., Duan, T., Jordan, D., Guo, Y., Zheng, B. 2018. Estimation of plant height using a high throughput phenotyping platform based on unmanned aerial vehicle and self-calibration: Example for sorghum breeding. *European Journal of Agronomy*, 95:24–32, ISSN 1161-0301, https://doi.org/10.1016/j.eja.2018.02.004.

Hu, Y. 2011. Vegetation Stress Level Monitoring in Mine Area Based on HJ-1 Hyperspectral Data. *Master's thesis*, Qingdao: Shandong University of Science and Technology, 57 p.

Huete, A. 1988. A soil-adjusted vegetation index (SAVI). *Remote Sensing of Environment*, 25(3):295–309.

Inoue, Y., Sakaiya, E., Zhu, Y., Takahashi, W. 2012. Diagnostic mapping of canopy nitrogen content in rice based on hyperspectral measurements. *Remote Sensing of Environment*, 126:210–221, ISSN 0034-4257, https://doi.org/10.1016/j.rse.2012.08.026.

IPCC. 2007. Climate Change 2007: Synthesis Report. In: *Contribution of Working Groups I, II and III to the Fourth Assessment Report of the Intergovernmental Panel on Climate Change.* Eds. Core Writing Team, R.K. Pachauri, and A. Reisinger, IPCC, Geneva, Switzerland, 104 p.

Islam, M.A., Thenkabail, P.S., Kulawardhana, R.W., Alankara, R., Gunasinghe, S., Edussriya, C., Gunawardana, A. 2008. Semi-automated methods for mapping wetlands using Landsat ETM+ and SRTM data. *International Journal of Remote Sensing*, 29(24):7077–7106.

Jacquemoud, S., Ustin, S.L., Verdebout, J., Schmuck, G., Andreoli, G., Hosgood, B. 1996. Estimating leaf biochemistry using the PROSPECT leaf optical properties model. *Remote Sensing of Environment*, 56:194–202.

Jafari, R., Lewis, M.M. 2012. Arid land characterisation with EO-1 Hyperion hyperspectral data. *International Journal of Applied Earth Observation and Geoinformation*, 19:298–307, ISSN 0303-2434, https://doi.org/10.1016/j.jag.2012.06.001.

Jakimow, B., Griffiths, P., van der Linden, S., Hostert, P. 2018. Mapping pasture management in the Brazilian Amazon from dense Landsat time series. *Remote Sensing of Environment*, 205:453–468, ISSN 0034-4257, https://doi.org/10.1016/j.rse.2017.10.009.

Jensen, J.R., Rutchey, K., Koch, M.S., Narumalani, S. 2002. Inland wetland change detection in the Everglades Water Conservation area 2A using a time series of normalized remotely sensed data. *Photogrammetric Engineering and Remote Sensing*, 61(2):199–209.

Jones, K., Lanthier, Y., Voet, P.V.D., Valkengoed, E.V., Taylor, D., Fernández-Prieto, D. 2009. Monitoring and assessment of wetlands using Earth Observation: The GlobWetland project. *Journal of Environmental Management*, 90(7): 2154–2169. doi:10.1016/j.jenvman.2007.07.037.

Jordan, C.F. 1969. Derivation of leaf-area index from quality of light on the forest floor. *Ecology*, 50:663–666.

Kayitakire, F., Hamel, C., Defourny, P. 2006. Retrieving forest structure variables based on image texture analysis and IKONOS-2 imagery. *Remote Sensing of Environment*, 102(3–4):390–401, ISSN 0034-4257, https://doi.org/10.1016/j.rse.2006.02.022.

Kganyago, M., Odindi, J., Adjorlolo, C., Mhangara, P. 2018. Evaluating the capability of Landsat 8 OLI and SPOT 6 for discriminating invasive alien species in the African Savanna landscape. *International Journal of Applied Earth Observation and Geoinformation*, 67:10–19, ISSN 0303-2434, https://doi.org/10.1016/j.jag.2017.12.008.

Khanal, K., Fulton, J., Shearer, S. 2017. An overview of current and potential applications of thermal remote sensing in precision agriculture. *Computers and Electronics in Agriculture*, 139:22–32, ISSN 0168-1699, https://doi.org/10.1016/j.compag.2017.05.001.

Kira, O., Linker, R., Gitelson, A. 2015. Non-destructive estimation of foliar chlorophyll and carotenoid contents: Focus on informative spectral bands. *International Journal of Applied Earth Observation and Geoinformation*, 38:251–260, ISSN 0303-2434, https://doi.org/10.1016/j.jag.2015.01.003.

Koch, B. 2010. Status and future of laser scanning, synthetic aperture radar and hyperspectral remote sensing data for forest biomass assessment. *ISPRS Journal of Photogrammetry and Remote Sensing*, 65(6):581–590, ISSN 0924-2716, https://doi.org/10.1016/j.isprsjprs.2010.09.001.

Kokaly, R.F., Clark, R.N. 1999. Spectroscopic determination of leaf biochemistry using band-depth analysis of absorption features and stepwise multiple linear regression. *Remote Sensing of Environment*, 67(3):267–287.

Kokaly, R.F., Asner, G.P., Ollinger, S.V., Martin, M.E., Wessman, C.A. 2009. Characterizing canopy biochemistry from imaging spectroscopy and its application to ecosystem studies. *Remote Sensing of Environment*, 113(S1):S78–S91.

Kulawardhana, R.W., Thenkabail, P.S., Vithanage, J., Biradar, C., Islam, Md.A., Gunasinghe, S., Alankara, R. 2007. Evaluation of the wetland mapping methods using Landsat ETM+ and SRTM data. *Journal of Spatial Hydrology*, 7(2):62–96, ISSN: 1530-4736.

Lausch, A., Salbach, C., Schmidt, A., Doktor, D., Merbach, I., Pause, M. 2015. Deriving phenology of barley with imaging hyperspectral remote sensing. *Ecological Modelling*, 295:123–135, ISSN 0304-3800, https://doi.org/10.1016/j.ecolmodel.2014.10.001.

le Maire, G., François, C., Soudani, K., Berveiller, D., Pontailler, D., Bréda, N., Genet, H., Davi, H., Dufrêne, E. 2008. Calibration and validation of hyperspectral indices for the estimation of broadleaved forest leaf chlorophyll content, leaf mass per area, leaf area index and leaf canopy biomass. *Remote Sensing of Environment*, 112(10):3846–3864.

Le Quéré, C. et al. 2016. Global carbon budget. *Earth Systems Science Data*, 8:605–649.

Lewis, S.L. et al. 2009. *Nature*, 457:1003–1006. doi:10.1038/nature07771.

Li, L., Wang, S., Ren, T., Wei, Q., Ming, J., Li, J., Li, X., Cong, R., Lu, J. 2018. Ability of models with effective wavelengths to monitor nitrogen and phosphorus status of winter oilseed rape leaves using *in situ* canopy spectroscopy. *Field Crops Research*, 215:173–186, ISSN 0378-4290, https://doi.org/10.1016/j.fcr.2017.10.018.

Li, M., Liu, X., Liu, M. 2010a. Fuzzy neural network model for predicting stress levels in rice fields polluted with heavy metals using hyperspectral data. *Acta Scientiae Circumstantiae*, 30(10):2108–2115.

Li, N., Lue, J., Altermann, W. 2010b. Applications of spectral analysis to monitoring of heavy metal-induced contamination in vegetation. *Spectroscopy and Spectral Analysis*, 30:2508–2511.

Liu, Y., Chen, H., Wu, G., Wu, X. 2010c. Feasibility of estimating heavy metal concentrations in Phragmites Australis using laboratory-based hyperspectral data—A case study along Le'an River, China. *International Journal of Applied Earth Observation and Geoinformation*, 12(Supplement 2):S166–S170, ISSN 0303-2434, https://doi.org/10.1016/j.jag.2010.01.003.

Liu, H., Wu, C. 2018. Crown-level tree species classification from AISA hyperspectral imagery using an innovative pixel-weighting approach. *International Journal of Applied Earth Observation and Geoinformation*, 68:298–307, ISSN 0303-2434, https://doi.org/10.1016/j.jag.2017.12.001.

Lopez, R., Lyon, J., Lyon, L., Lopez, D. 2017. *Wetland Landscape Characterization: Techniques and Applications for GIS Mapping, Remote Sensing, and Image Analysis*. CRC Press, 308 p.

Lopez-Sanchez, J.M., Vicente-Guijalba, F., Erten, E., Campos-Taberner, M., Garcia-Haroc, F.J. 2017. Retrieval of vegetation height in rice fields using polarimetric SAR interferometry with TanDEM-X data. *Remote Sensing of Environment*, 192:30–44.

Lu, D. 2006. The potential and challenge of remote sensing based biomass estimation. *International Journal of Remote Sensing*, 27:1297–1328.

Lunetta, R.S., Balogh, M.E., Merchant, J.W. 1999. Application of multi-temporal Landsat 5 TM imagery for wetland identification. *Photogrammetric Engineering and Remote Sensing*, 65:1303–1310.

Lyon, J. 2001. *Wetland Landscape Characterization: Techniques and Applications for GIS Mapping, Remote Sensing, and Image Analysis*. CRC Press, p. 135.

Magney, T.S., Vierling, L.A., Eitel, J.U.H., Huggins, D.R., Garrity, S.R. 2016. Response of high frequency Photochemical Reflectance Index (PRI) measurements to environmental conditions in wheat. *Remote Sensing of Environment*, 173:84–97, ISSN 0034-4257, https://doi.org/10.1016/j.rse.2015.11.013.

Magnussen, S., Næsset, E., Gobakken, T. 2010. Reliability of LiDAR derived predictors of forest inventory attributes: A case study with Norway spruce. *Remote Sensing of Environment*, 114(4):700–712.

Malhi, Y., Wright, J. 2004. Spatial patterns and recent trends in the climate of tropical rainforest regions. *Philos Trans R Soc Lond B Biol Sci.*, 359(1443):311–329.

Mariotto I., Gutschick V., Clason D.L. 2011. Mapping evapotranspiration from ASTER data through GIS spatial integration of vegetation and terrain features. *Phtotogrammetric Engineering and Remote Sensing*, 77(5):483–493.

Mariotto, I., Thenkabail, P.S., Huete, H., Slonecker, T., Platonov, A. 2013. Hyperspectral versus multispectral crop- biophysical modeling and type discrimination for the HyspIRI mission. *Remote Sensing of Environment*, 139:291–305. IP-049224.

Marshall, M.T., Thenkabail, P.S. 2014. Biomass modeling of four leading World crops using hyperspectral narrowbands in support of HyspIRI mission. *Photogrammetric Engineering and Remote Sensing*, 80(4):757–772. IP-052043.

Marshall, M.T., Thenkabail, P.S. 2015a. Advantage of hyperspectral EO-1 Hyperion over multispectral IKONOS, GeoEye-1, WorldView-2, Landsat ETM+, and MODIS vegetation indices in crop biomass estimation. *ISPRS Journal of Photogrammetry and Remote Sensing*, 108:205–218. http://dx.doi.org/10.1016/j.isprsjprs.2015.08.001. IP-060745.

Marshall M.T., Thenkabail P.S. 2015b. Developing *in situ* non-destructive estimates of crop biomass to address issues of scale in remote sensing. *Remote Sensing*, 7(1):808–835. doi:10.3390/rs70100808. IP-060652.

Marshall, M.T., Thenkabail, P.S., Biggs, T., Post, K. 2016. Hyperspectral narrowband and multispectral broadband indices for remote sensing of crop evapotranspiration and its components (transpiration and soil evaporation). *Agricultural and Forest Meteorology*, 218–219:122–134. IP-065032.

Matheson, D.S., Dennison, P.E. 2012. Evaluating the effects of spatial resolution on hyperspectral fire detection and temperature retrieval. *Remote Sensing of Environment*, 124:780–792, ISSN 0034-4257, https://doi.org/10.1016/j.rse.2012.06.026.

McNairn, H., Protz, R. 1993. Mapping corn residue cover on agricultural fields in Oxford County, Ontario, using thematic mapper. *Canadian Journal of Remote Sensing*, 19:152–159.

Melesse, A.M., Weng, Q., Thenkabail, P., Senay, G. 2007. Remote sensing sensors and applications in environmental resources mapping and modelling. *Special Issue of Remote Sensing of Natural Resources and the Environment Sensors Journal*, 7:3209–3241. http://www.mdpi.org/sensors/papers/s7123209.pdf.

Merzlyak, J.R., Gitelson, A.A., Chivkunova, O.B., Rakitin, V.Y. 1999. Non-destructive optical detection of pigment changes during leaf senescence and fruit ripening. *Physiologia Plantarum*, 106(1):135–141.

Merzlyak, M.N., Chivkunova, O.B., Solovchenko, A.E., Naqvi, K.R. 2008. Light absorption by anthocyanins in juvenile, stressed, and senescing leaves. *Journal of Experimental Botany*, 59(14):3903–3911.

Middleton, E.M., Huemmrich, K.F., Landis, D.R., Black, T.A., Barr, A.G., McCaughey, J.H. 2016. Photosynthetic efficiency of northern forest ecosystems using a MODIS-derived photochemical reflectance index (PRI). *Remote Sensing of Environment*, 187:345–366, ISSN 0034-4257, https://doi.org/10.1016/j.rse.2016.10.021.

Mitsch, W.J., Gosselink, J.G. 2007. *Wetlands*. John Wiley and Sons, 582 p.

Mulla, D.J. 2013. Twenty five years of remote sensing in precision agriculture: Key advances and remaining knowledge gaps. *Biosystems Engineering*, 114(4):358–371, ISSN 1537-5110, https://doi.org/10.1016/j.biosystemseng.2012.08.009.

Nagler, P.L., Inoue, Y., Glenn, E.P., Russ, A.L., Daughtry, C.S.T. 2003. Cellulose absorption index (CAI) to quantify mixed soil-plant litter scenes. *Remote Sensing of Environment*, 87(2–3):310–325, ISSN 0034-4257, https://doi.org/10.1016/j.rse.2003.06.001.

Näsi, R., Honkavaara, E., Blomqvist, M., Lyytikäinen-Saarenmaa, P., Hakala, T., Viljanen, N., Kantola, T., Holopainen, M. 2018. Remote sensing of bark beetle damage in urban forests at individual tree level using a novel hyperspectral camera from UAV and aircraft. *Urban Forestry & Urban Greening*, 30:72–83, ISSN 1618-8667, https://doi.org/10.1016/j.ufug.2018.01.010.

Nepstad, D.C., Stickler, C.M., Soares-Filho, B., Merry, F. 2008. Interactions among Amazon land use, forests, and climate: Prospects for a near-term forest tipping point. *Philosophical Transactions Royal Society*, 363:1737–1746.

Nielsen, E.M., Prince, S.D., Koeln, G.T. 2008. Wetland change mapping for the U.S. mid-Atlantic region using an outlier detection technique. *Remote Sensing of Environment*, 112:4061–4074.

Noujdina, N.V., Ustin, S.L. 2008. Mapping downy brome (*Bromus tectorum*) using multidate AVIRIS data. *Weed Science*, 56:173–179.

Ozesmi, S.L., Bauer, M.E. 2002. Satellite remote sensing of wetlands. *Wetlands Ecology and Management*, 10:381–402.

Panda, S.S., Rao, M.N., Thenkabail, P.S., Fitzerald, J.E. 2015. Remote Sensing Systems – Platforms and Sensors: Aerial, Satellites, UAVs, Optical, Radar, and LiDAR, Chapter 1. In: Thenkabail, P.S., (Editor-in-Chief), 2015. *"Remote Sensing Handbook" (Volume I): Remotely sensed data characterization, classification, and accuracies.* ISBN 9781482217865 - CAT# K22125. Taylor and Francis Inc./CRC Press, Boca Raton, London, New York. pp. 3–60. IP-060641.

Pasqualotto, N., Delegido, J., Wittenberghe, S.V., Verrelst, J., Rivera, J.P., Moreno, J. 2018. Retrieval of canopy water content of different crop types with two new hyperspectral indices: Water absorption area index and depth water index. *International Journal of Applied Earth Observation and Geoinformation*, 67:69–78, ISSN 0303-2434, https://doi.org/10.1016/j.jag.2018.01.002.

Pasquarella, V.J., Holden, C.E., Woodcock, C.E. 2018. Improved mapping of forest type using spectral-temporal Landsat features. *Remote Sensing of Environment*, 210:193–207, ISSN 0034-4257, https://doi.org/10.1016/j.rse.2018.02.064.

Pastor-Guzman, J., Dash, J., Atkinson, P.M. 2018. Remote sensing of mangrove forest phenology and its environmental drivers. *Remote Sensing of Environment*, 205:71–84, ISSN 0034-4257, https://doi.org/10.1016/j.rse.2017.11.009.

Pearlman, S., Barry, P.S., Segal, C.C., Shepanski, J., Beiso, D., Carman, S.L. 2003. Hyperion, a space-based imaging spectrometer. *IEEE Transactions on Geoscience and Remote Sensing*, 41(6):1160–1173.

Pelley, J. 2008. Can wetland restoration cool the planet? *Environmental Science & Technology*, 42(24):8994.

Peng, D., Wu, C., Zhang, X., Yu, L., Huete, A.R., Wang, F., Luo, S., Liu, X., Zhang, H. 2018. Scaling up spring phenology derived from remote sensing images. *Agricultural and Forest Meteorology*, 256–257:207–219, ISSN 0168-1923.

Peñuelas, J., Filella, I., Biel, C., Serrano, L., Save, R. 1993. The reflectance at the 950–970 nm region as an indicator of plant water status. *International Journal of Remote Sensing*, 14:887–1905.

Peñuelas, J., Llusia, J., Pinol, J., Filella, I. 1997. Photochemical reflectance index and leaf photosynthetic radiation-use-efficiency assessment in Mediterranean trees. *International Journal of Remote Sensing*, 18:2863–2868.

Phillips, O.L. et al. 1998. Changes in the carbon balance of tropical forests: Evidence from long-term plots. *Science*, 282:439–442.

Pignatti, S., Cavalli, R.M., Cuomo, V., Fusilli, L., Pascucci, S., Poscolieri, M., Santini, F. 2009. Evaluating Hyperion capability for land cover mapping in a fragmented ecosystem: Pollino National Park, Italy. *Remote Sensing of Environment*, 113(3):622–634, ISSN 0034-4257, https://doi.org/10.1016/j.rse.2008.11.006.

Pinter, P.J., Ritchie, J.C., Hartfield, J.L., Hart, G.F. 2003. The agricultural research service's remote sensing program: An example of interagency collaboration. *Photogrammetric Engineering and Remote Sensing*, 69(6):615–618.

Platonov, A. et al. 2008. Water productivity mapping (WPM) using Landsat ETM+ data for the irrigated croplands of the Syrdarya river Basin in Central Asia. *Sensors Journal*, 8(12):8156–8180. doi:10.3390/s8128156. http://www.mdpi.com/1424-8220/8/12/8156/pdf.

Ploton, P. et al. 2017. Toward a general tropical forest biomass prediction model from very high resolution optical satellite images. *Remote Sensing of Environment*, 200:140–153, ISSN 0034-4257, https://doi.org/10.1016/j.rse.2017.08.001.

Pu, R., Bell, S., Meyer, C., Baggett, L., Zhao, Y. 2012. Mapping and assessing seagrass along the western coast of Florida using Landsat TM and EO-1 ALI/Hyperion imagery. *Estuarine, Coastal and Shelf Science*, 115:234–245, ISSN 0272-7714, https://doi.org/10.1016/j.ecss.2012.09.006.

Pu, R., Gong, P., Biging, G.S., Larrieu, M.R. 2003. Extraction of red edge optical parameters from Hyperion data for estimation of forest leaf area index. *IEEE Transaction on Geosciences and Remote Sensing*, 41(4):916–921.

Pu, R., Gong, P., Tian, Y., Miao X., Carruthers, R. 2008a. Invasive species change detection using artificial neural networks and CASI hyperspectral imagery. *Environmental Monitoring and Assessment*, 140:15–32.

Pu, R., Gong, P., Yu, Q. 2008b. Comparative analysis of EO-1 ALI and Hyperion, and Landsat ETM+ data for mapping forest crown closure and leaf area index. *Sensors*, 8(6):3744–3766.

Pullanagari, R.R., Kereszturi, G., Yule, I.J. 2016. Mapping of macro and micro nutrients of mixed pastures using airborne AisaFENIX hyperspectral imagery. *ISPRS Journal of Photogrammetry and Remote Sensing*, 117:1–10, ISSN 0924-2716, https://doi.org/10.1016/j.isprsjprs.2016.03.010.

Ramirez, F.J.R., Rafael, M., Navarro-Cerrillo, M., Varo-Martínez, A., Quero, J.L., Doerr, S., Hernández-Clemente, R. 2018. Determination of forest fuels characteristics in mortality-affected Pinus forests using integrated hyperspectral and ALS data. *International Journal of Applied Earth Observation and Geoinformation*, 68:157–167, ISSN 0303-2434, https://doi.org/10.1016/j.jag.2018.01.003.

Ramoelo, M.A., Cho, R., Mathieu, S., Madonsela, R., van de Kerchove, Z., Kaszta, E., Wolff, E. 2015. Monitoring grass nutrients and biomass as indicators of rangeland quality and quantity using random forest modelling and WorldView-2 data. *International Journal of Applied Earth Observation and Geoinformation*, 43:43–54, ISSN 0303-2434, https://doi.org/10.1016/j.jag.2014.12.010.

Ramsar. 2004. *The Ramsar Convention Manual: A Guide to the Convention on Wetlands (Ramsar, Iran, 1971)*, 3rd ed. RAMSAR Convention Secretariat, Gland, Switzerland. http://www.ramsar.org/lib/lib_manual2004e.htm.

Ramsey, III, E., Nelson, G., Sapkota, S. 1998. Classifying coastal resources by integrating optical and radar imagery and color infrared photography. *Mangroves and Salt Marshes*, 2(2):109–119.

Ramsey, III, E., Lu, Z., Rangoonwala, A., Rykhus, R. 2006. Multiple baseline radar interferometry applied to coastal landscape classification and change analyses. *GIS Science and Remote Sensing*, 43(4):283–309.

Rebelo, A.J., Somers, B., Esler, K.J., Meire, P. 2018. Can wetland plant functional groups be spectrally discriminated? *Remote Sensing of Environment*, 210:25–34, ISSN 0034-4257, https://doi.org/10.1016/j.rse.2018.02.031.

Ren, H.-Y., Zhuang, D., Singh, A.N., Pan, J., Qiu, D., Shi, R. 2009. Estimation of As and Cu contamination in agricultural soils around a mining area by reflectance spectroscopy; a case study. *Pedosphere*, 19(6):719–726.

Rights and Resources Initiative. 2018. Uncertainty and Opportunity: The status of forest carbon rights and governance frameworks in over half of the world's tropical forests, https://rightsandresources.org/wp-content/uploads/2018/03/EN_Status-of-Forest-Carbon-Rights_RRI_Mar-2018.pdf

Roberts, D.A., Ustin, S., Ogunjemiyo, S., Greenberg, J., Dobrowski, S., Chen, J., Hinckley, T. 2004. Spectral and structural measures of northwest forest vegetation at leaf to landscape scales. *Ecosystems*, 7(5):545–562.

Roberts, D.A., Quattrochi, D.A., Hulley, G.C., Hook, S.J., Green, R.O. 2012. Synergies between VSWIR and TIR data for the urban environment: An evaluation of the potential for the Hyperspectral infrared imager (HyspIRI) decadal survey mission. *Remote Sensing of Environment*, 117:83–101, ISSN 0034-4257, https://doi.org/10.1016/j.rse.2011.07.021.

Rossini, M. et al. 2013. Assessing canopy PRI from airborne imagery to map water stress in maize. *ISPRS Journal of Photogrammetry and Remote Sensing*, 86:168–177, ISSN 0924-2716, https://doi.org/10.1016/j.isprsjprs.2013.10.002.

Rosso, P.H., Pushnik, J.C., Lay, M., Ustin, S.L. 2005. Reflectance properties and physiological responses of *Salicornia virginica* to heavy metal and petroleum contamination. *Environmental Pollution*, 137(2):241–252.

Rouse, J.W., Haas, R.H., Schell, J.A., Deering, D.W., Harlan, J.C. 1974. *Monitoring the vernal advancements and retro gradation of natural vegetation*. NASA/GSFC Final Report Greenbelt, MD, USA. 371 p.

Salmon, J.M., Friedl, M.A., Frolking, S., Wisser, D., Douglas, E.M. 2015. Global rain-fed, irrigated, and paddy croplands: A new high resolution map derived from remote sensing, crop inventories and climate data. *International Journal of Applied Earth Observation and Geoinformation*, 38:321–334, ISSN 0303-2434, https://doi.org/10.1016/j.jag.2015.01.014.

Schowengerdt, R.A. 2007. *Remote Sensing: Models and Methods for Image Processing*. Academic Press (Elsevier), San Diego, California, USA. pp. 509.

Senay, G.B., Bohma, S., Singh, R.K., Gowda, P.H., Velpuri, N.M., Alemu, H., Verdin, J.P. 2013. Operational evapotranspiration mapping using remote sensing and weather datasets: A new parameterization for the SSEB approach. *JAWRA Journal of the American Water Resources Association*, 49(3): Paper No. JAWRA-12-0097-P.

Serbin, G., Hunt, E.R., Daughtry, C.S.T., McCarty, G., Doraiswamy, P. 2009. An improved ASTER index for remote sensing of crop residue. *Remote Sensing*, 1:971–991.

Shakoor, N., Lee, S., Mockler, T.C. 2017. High throughput phenotyping to accelerate crop breeding and monitoring of diseases in the field. *Current Opinion in Plant Biology*, 38:184–192, ISSN 1369-5266, https://doi.org/10.1016/j.pbi.2017.05.006.

Shi, T., Liu, H., Chen, Y., Wang, J., Wu, G. 2016. Estimation of arsenic in agricultural soils using hyperspectral vegetation indices of rice. *Journal of Hazardous Materials*, 308:243–252, ISSN 0304-3894, https://doi.org/10.1016/j.jhazmat.2016.01.022.

Singh, A., Ganapathysubramanian, B., Singh, A.K., Sarkar, S. 2016. Machine learning for high-throughput stress phenotyping in plants. *Trends in Plant Science*, 21(2):110–124, ISSN 1360-1385, https://doi.org/10.1016/j.tplants.2015.10.015.

Smith, R., Adams, M., Maier, S., Craig, R., Kristina, A., Maling, I. 2007. Estimating the area of stubble burning from the number of active fires detected by satellite. *Remote Sensing of Environment*, 109(1):95–106, ISSN 0034-4257, https://doi.org/10.1016/j.rse.2006.12.011.

Stagakis, S., Vanikiotis, T., Sykioti, O. 2016. Estimating forest species abundance through linear unmixing of CHRIS/PROBA imagery. *ISPRS Journal of Photogrammetry and Remote Sensing*, 119:79–89, ISSN 0924-2716, https://doi.org/10.1016/j.isprsjprs.2016.05.013.

Sullivan, D.G., Truman, C.C., Schomberg, H.H., Endale, D.M., Strickland, T.C. 2006. Evaluating techniques for determining tillage regime in the Southeastern Coastal Plain and Piedmont. *Agronomy Journal*, 98:1236–1246.

Teixeira, Antônio de C., Hernandez, F.B.T., Scherer-Warren, M., Andrade, R.G., Leivas, J.F., Victoria, D.C., Bolfe, E.L., Thenkabail, P.S., Franco, R.A.M. 2015. Water Productivity Studies From Earth Observation Data: Characterization, Modeling and Mapping Water Use and Water productivity, Chapter 4. In: *"Remote Sensing Handbook" (Volume III): Remote Sensing of Water Resources, Disasters, and Urban Studies*. Ed. P.S. Thenkabail, (Editor-in-Chief), 2015. Taylor and Francis Inc./CRC Press, Boca Raton, London, New York. ISBN 9781482217919-CAT# K22128. pp. 101–128. IP-058357.

Teluguntla, P., Thenkabail, P.S., Oliphant, A., Xiong, J., Gumma, M.K. 2018. A 30-m Landsat-derived cropland extent product of Australia and China using random forest machine learning algorithm on Google Earth Engine cloud computing platform. *International Journal of Photogrammetry and Remote Sensing (ISPRS) Journal of Photogrammetry and Remote Sensing.* In press.

Teluguntla, P., Thenkabail, P.S., Xiong, J., Gumma, M.K., Congalton, R.G., Oliphant, A., Poehnelt, J., Yadav, K., Rao, M., Massey, R. 2017. Spectral matching techniques (SMTs) and automated cropland classification algorithms (ACCAs) for mapping croplands of Australia using MODIS 250-m time-series (2000–2015) data. *International Journal of Digital Earth.* doi:10.1080/17538947.2016.1267269. IP-074181, http://dx.doi.org/10.1080/17538947.2016.1267269.

Teluguntla, P. et al. 2015. Global Cropland Area Database (GCAD) derived from Remote Sensing in Support of Food Security in the Twenty-first Century: Current Achievements and Future Possibilities, Chapter 6. In: *"Remote Sensing Handbook" Volume II: Land Resources: Monitoring, Modeling, and Mapping: Advances over Last 50 Years and a Vision for the Future.* Ed. P.S. Thenkabail, (Editor-in-Chief), 2015. Taylor and Francis Inc./CRC Press, Boca Raton, London, New York. pp. 800+.

Thamaga, K.H., Dube, T. 2018. Remote sensing of invasive water hyacinth (Eichhornia crassipes): A review on applications and challenges. *Remote Sensing Applications: Society and Environment,* 10:36–46, ISSN 2352-9385, https://doi.org/10.1016/j.rsase.2018.02.005.

Thenkabail, P.S., Smith, R.B., De-Pauw, E. 2000a. Hyperspectral vegetation indices for determining agricultural crop characteristics. *Remote Sensing of Environment,* 71:158–182.

Thenkabail, P.S., Smith, R.B., Pauw, E.D. 2000b. Hyperspectral vegetation indices and their relationships with agricultural crop characteristics. *Remote Sensing of Environment,* 71(2):158–182, ISSN 0034-4257, https://doi.org/10.1016/S0034-4257(99)00067-X.

Thenkabail, P.S., Smith, R.B., De-Pauw, E. 2002. Evaluation of narrowband and broadband vegetation indices for determining optimal hyperspectral wavebands for agricultural crop characterization. *Photogrammetric Engineering and Remote Sensing,* 68(6):607–621.

Thenkabail, P.S., Hall, J., Lin, T., Ashton, M.S., Harris, D., Enclona, E.A. 2003. Detecting floristic structure and pattern across topographic and moisture gradients in a mixed species Central African forest using IKONOS and Landsat-7 ETM+ images. *International Journal of Applied Earth Observation and Geoinformation,* 4: 255–270.

Thenkabail, P.S. 2003. Biophysical and yield information for precision farming from near-real-time and historical Landsat TM images. *International Journal of Remote Sensing,* 24(14):2879–2904. doi:10.1080/01431160710155974.

Thenkabail, P.S. 2004. Inter-sensor relationships between IKONOS and Landsat-7 ETM+ NDVI data in three ecoregions of Africa. *International Journal of Remote Sensing,* 25(2):389–408.

Thenkabail, P.S. 2010. Biophysical and yield information for precision farming from near-real-time and historical Landsat TM images. *International Journal of Remote Sensing,* 24(14):2879–2904. doi:10.1080/01431160710155974.

Thenkabail, P.S., Enclona, E.A., Ashton, M.S., Legg, C., Jean De Dieu, M. 2004a. Hyperion, IKONOS, ALI, and ETM+ sensors in the study of African rainforests. *Remote Sensing of Environment,* 90:23–43.

Thenkabail, P.S., Enclona, E.A., Ashton, M.S., Van Der Meer, V. 2004b. Accuracy assessments of hyperspectral waveband performance for vegetation analysis applications. *Remote Sensing of Environment,* 91(2–3):354–376.

Thenkabail, P.S., GangadharaRao, P., Biggs, T., Krishna, M., Turral, H. 2007. Spectral matching techniques to determine historical land use/land cover (LULC) and irrigated areas using time-series AVHRR pathfinder datasets in the Krishna River Basin, India. *Photogrammetric Engineering and Remote Sensing,* 73(9):1029–1040.

Thenkabail, P.S. Biradar C.M., Noojipady, P., Dheeravath, V., Li, Y.J., Velpuri, M., Gumma, M., Reddy, G.P.O., Turral, H., Cai, X. L., Vithanage, J., Schull, M., Dutta, R. 2009a. Global irrigated area map (GIAM), derived from remote sensing, for the end of the last millennium. *International Journal of Remote Sensing,* 30(14):3679–3733. July 20, 2009.

Thenkabail, P.S., Lyon, J.G., Turral, H., Biradar, C.M. 2009b. *Remote Sensing of Global Croplands for Food Security.* CRC Press-Taylor & Francis Group, Boca Raton, FL. p. 556.

Thenkabail, P.S. Knox, J.W., Ozdogan, M., Gumma, M.K., Congalton, R.G., Wu, Z., Milesi, C., Finkral, A., Marshall, M., Mariotto, I., You, S. Giri, C., Nagler, P. 2012. Assessing future risks to agricultural productivity, water resources and food security: How can remote sensing help? *Photogrammetric Engineering and Remote Sensing,* August 2012 Special Issue on Global Croplands: Highlight Article, 78(8):773–782. IP-035587.

Thenkabail, P.S., Wu, Z. 2012. An automated cropland classification Algorithm (ACCA) for Tajikistan by combining Landsat, MODIS, and secondary data. *Remote Sensing*, 4(10):2890–2918. (65%). Download the paper @ this link: http://www.mdpi.com/2072-4292/4/10/2890. ACCA algorithm at this link: http://www.sciencebase.gov/catalog/folder/4f79f1b7e4b0009bd827f548. IP-035313.

Thenkabail, P.S., Mariotto, I., Gumma, M.K., Middleton, E.M., Landis, D.R., Huemmrich, F.K. 2013. Selection of hyperspectral narrowbands (HNBs) and composition of hyperspectral twoband vegetation indices (HVIs) for biophysical characterization and discrimination of crop types using field reflectance and Hyperion/EO-1 data. *IEEE Journal of Selected Topics in Applied Earth Observations and Remote Sensing*, 6(2):427–439. doi:10.1109/JSTARS.2013.2252601. (80%). IP-037139.

Thenkabail, P.S., Enclona, E.A., Ashton, M.S., Legg, C., Dieu, M.J.D. 2014a. Hyperion, IKONOS, ALI, and ETM+ sensors in the study of African rainforests. *Remote Sensing of Environment*, 90(1):23–43, ISSN 0034-4257, https://doi.org/10.1016/j.rse.2003.11.018.

Thenkabail, P.S., Gumma, M.K., Teluguntla, P., Mohammed, I.A. 2014b. Hyperspectral remote sensing of vegetation and agricultural crops. Highlight article. *Photogrammetric Engineering and Remote Sensing*, 80(4):697–709. IP-052042.

Thenkabail, P.S. 2015. Hyperspectral Remote Sensing for Terrestrial Applications, Chapter 9. In: *"Remote Sensing Handbook" (Volume II): Land Resources Monitoring, Modeling, and Mapping with Remote Sensing*. Ed. P.S. Thenkabail, (Editor-in-Chief), 2015. Taylor and Francis Inc./CRC Press, Boca Raton, London, New York. ISBN 9781482217957—CAT# K22130. pp. 201–236. IP-0606312.

Tiner, R.W. 2009. Global distribution of wetlands. *Encyclopedia of Inland Waters*, 526–530.

Toniol, A.C., Galvão, L.S., Ponzoni, F.J., Sano, E.E., de Jesus Amore, D. 2017. Potential of hyperspectral metrics and classifiers for mapping Brazilian savannas in the rainy and dry seasons. *Remote Sensing Applications: Society and Environment*, 8:20–29, ISSN 2352-9385, https://doi.org/10.1016/j.rsase.2017.07.004.

Trifonov, G.M., Zhizhin, M.N., Melnikov, D.V., Poyda, A.A. 2017. VIIRS nightfire remote sensing volcanoes. *Procedia Computer Science*, 119:307–314, ISSN 1877-0509, https://doi.org/10.1016/j.procs.2017.11.189.

Tucker, C.J. 1979. Red and photographic infrared linear combinations for monitoring vegetation. *Remote Sensing of Environment*, 8(2):127-150, ISSN 0034-4257, https://doi.org/10.1016/0034-4257(79)90013-0.

Turpic, K.R., Klemas, V.V., Byrd, K., Kelly, M., Jo, Y.H. 2015. Prospective HyspIRI global observations of tidal wetlands. *Remote Sensing of Environment*, 167:206–217, ISSN 0034-4257, https://doi.org/10.1016/j.rse.2015.05.008.

UNFCCC, 2008 UNFCCC (United Nations Framework Convention on Climate Change). 2008. *Report of the Conference of the Parties on its Thirteenth Session, Held in Bali from 3 to 15 December 2007.* Addendum, Part 2. Document FCCC/CP/2007/6/Add.1. UNFCCC, Bonn, Germany.

Ustin, S.L., Gitelson, A.A., Jacquemoud, S., Schaepman, M., Asner, G.P., Gamon, J.A., Zarco-Tejada, P. 2009. Retrieval of foliar information about plant pigment systems from high resolution spectroscopy. *Remote Sensing of Environment*, 113(S1):S67–S77, ISSN 0034-4257, https://doi.org/10.1016/j.rse.2008.10.019.

Vadrevu, K.P., Ellicott, E., Badarinath, K.V.S., Vermote, E. 2011. MODIS derived fire characteristics and aerosol optical depth variations during the agricultural residue burning season, north India. *Environmental Pollution*, 159(6):1560–1569, ISSN 0269-7491, https://doi.org/10.1016/j.envpol.2011.03.001.

Van Deventer, A.P., Ward, A.D., Gowda, P.H., Lyon, J.G. 1997. Using Thematic Mapper data to identify contrasting soil plains and tillage practices. *Photogrammetric Engineering and Remote Sensing*, 63:87–93.

van Leeuwen, W.J.D., Orr, B.J., Marsh, S.E., Herrmann, S.M. 2006. Multi-sensor NDVI data continuity: Uncertainties and implications for vegetation monitoring applications. *Remote Sensing of Environment*, 100(1):67–81, ISSN 0034-4257, https://doi.org/10.1016/j.rse.2005.10.002.

Vogelmann, J.E., Rock, B.N., Moss, D.M., 1993. Red edge spectral measurements from sugar maple leaves. *International Journal of Remote Sensing*, 14:(8):1563-1575. doi: 10.1080/01431169308953986.

Wagner, W., Blöschl, G., Pampaloni, P., Calvet, J.-C., Bizzarri, B., Wigneron, J.-P., Kerr, Y. 2007. Operational readiness of microwave remote sensing of soil moisture for hydrologic applications. *Nordic Hydrology*, 38(1):1–20.

Wang, P., Liu, X., Huang, F. 2010. Retrieval model for subtle variation of contamination stressed maize chlorophyll using hyperspectral data. *Spectroscopy and Spectral Analysis*, 30:197–201.

Wang, W., Yao, X., Yao, S.F., Tian, Y.C., Liu, X.J., Ni, J., Cao, W.X., Zhu, Y. 2012. Estimating leaf nitrogen concentration with three-band vegetation indices in rice and wheat. *Field Crops Research*, 129:90–98, ISSN 0378-4290, https://doi.org/10.1016/j.fcr.2012.01.014.

Wang, Y., Tian, F., Huang, Y., Wang, J., Wei, C. 2015a. Monitoring coal fires in Datong coalfield using multi-source remote sensing data. *Transactions of Nonferrous Metals Society of China*, 25(10):3421–3428, ISSN 1003-6326, https://doi.org/10.1016/S1003-6326(15)63977-2.

Wang, Z., Skidmore, A.K., Wang, T., Darvishzadeh, R., Hearne, J. 2015b. Applicability of the PROSPECT model for estimating protein and cellulose+lignin in fresh leaves. *Remote Sensing of Environment*, 168:205–218, ISSN 0034-4257, https://doi.org/10.1016/j.rse.2015.07.007.

Wang, Z., Skidmore, A.K., Darvishzadeh, R., Wang, T. 2018. Mapping forest canopy nitrogen content by inversion of coupled leaf-canopy radiative transfer models from airborne hyperspectral imagery. *Agricultural and Forest Meteorology*, 253–254:247–260, ISSN 0168-1923.

Warda. 2006. *Medium Term Plan 2007–2009. Charting the Future of Rice in Africa.* Africa Rice Center (WARDA), Cotonou, Republic of Benin.

Weber, D., Schaepman-Strub, G., Ecker, K. 2018. Predicting habitat quality of protected dry grasslands using Landsat NDVI phenology. *Ecological Indicators*, 91:447–460, ISSN 1470-160X, https://doi.org/10.1016/j.ecolind.2018.03.081.

Wheeler, D., Guzder-Williams, B., Petersen, R., Thau, D. 2018. Rapid MODIS-based detection of tree cover loss. *International Journal of Applied Earth Observation and Geoinformation*, 69:78–87, ISSN 0303-2434, https://doi.org/10.1016/j.jag.2018.02.007.

White, K., Pontius, J., Schaberg, P. 2014. Remote sensing of spring phenology in northeastern forests: A comparison of methods, field metrics and sources of uncertainty. *Remote Sensing of Environment*, 148:97–107, ISSN 0034-4257, https://doi.org/10.1016/j.rse.2014.03.017.

Wolter, P.T., Townsend, P.A., Sturtevant, B.R. 2009. Estimation of forest structural parameters using 5 and 10-meter SPOT-5 satellite data. *Remote Sensing of Environment*, 113(9):2019–2036, ISSN 0034-4257, https://doi.org/10.1016/j.rse.2009.05.009.

Workie, T.G., Debella, H.J. 2018. Climate change and its effects on vegetation phenology across ecoregions of Ethiopia. *Global Ecology and Conservation*, 13: Article e00366, ISSN 2351-9894, https://doi.org/10.1016/j.gecco.2017.e00366.

Wu, C., Huang, W., Yang, Q., Xie, Q. 2015. Improved estimation of light use efficiency by removal of canopy structural effect from the photochemical reflectance index (PRI). *Agriculture, Ecosystems & Environment*, 199:333–338, ISSN 0167-8809, https://doi.org/10.1016/j.agee.2014.10.017.

Xiong, J., Thenkabail, P.S., Gumma, M.K., Teluguntla, P., Poehnelt, J., Congalton, R.G., Yadav, K., Thau, D. 2017a. Automated cropland mapping of continental Africa using Google Earth Engine cloud computing. *ISPRS Journal of Photogrammetry and Remote Sensing*, 126:225–244, http://www.sciencedirect.com/science/article/pii/S0924271616301575.

Xiong, J., Thenkabail, P.S., Tilton, J.C., Gumma, M.K., Teluguntla, T., Oliphant, A., Congalton, R.G., Yadav, K., Gorelick, N. 2017b. Nominal 30-m cropland extent map of continental Africa by integrating pixel-based and object-based algorithms using Sentinel-2 and Landsat-8 data on Google Earth Engine. *Remote Sensing*, 9(10):1065. doi:10.3390/rs9101065, http://www.mdpi.com/2072-4292/9/10/1065.

Yang, C., Everitt, J.H. 2010. Mapping three invasive weeds using airborne hyperspectral imagery. *Ecological Informatics*, 5(5):429–439, ISSN 1574-9541, https://doi.org/10.1016/j.ecoinf.2010.03.002.

Yang, Z., Wu, W.B., Di, L., Üstündağ, B. 2017. Remote sensing for agricultural applications. *Journal of Integrative Agriculture*, 16(2):239–241, ISSN 2095-3119, https://doi.org/10.1016/S2095-3119(16)61549-6.

Yao, X., Zhu, Y., Tian, Y.C., Feng, W., Cao, W.X. 2010. Exploring hyperspectral bands and estimation indices for leaf nitrogen accumulation in wheat. *International Journal of Applied Earth Observation and Geoinformation*, 12(2):89–100, ISSN 0303-2434, https://doi.org/10.1016/j.jag.2009.11.008.

Zarco-Tejada, P.J., Guillén-Climent, M.L., Hernández-Clemente, R., Catalina, A., González, M.R., Martín, P. 2013. Estimating leaf carotenoid content in vineyards using high resolution hyperspectral imagery acquired from an unmanned aerial vehicle (UAV). *Agricultural and Forest Meteorology*, 171–172:281–294, ISSN 0168-1923, https://doi.org/10.1016/j.agrformet.2012.12.013.

Zarco-Tejada, P.J., Miller, J.R., Noland, T.L., Mohammed, G.H., Sampson, P.H. 2001. Scaling-up and model inversion methods with narrowband optical indices for chlorophyll content estimation in closed forest canopies with hyperspectral data. *IEEE Transactions on Geoscience and Remote Sensing*, 39(7):1491–1507.

Zarco-Tejada, P.J., Ustin, S.L., 2001. Modeling canopy water content for carbon estimates from MODIS data at land EOS validation sites. In: *Proceedings of the IEEE 2001 International Geoscience and Remote Sensing Symposium (IGARSS '01)*, Sydney, Australia, 9–13 July, vol. 1, pp. 342–344.

Zhang, Q. et al. 2017. Improving the ability of the photochemical reflectance index to track canopy light use efficiency through differentiating sunlit and shaded leaves. *Remote Sensing of Environment*, 194:1–15, ISSN 0034-4257, https://doi.org/10.1016/j.rse.2017.03.012.

Zhao, B. et al. 2018. Exploring new spectral bands and vegetation indices for estimating nitrogen nutrition index of summer maize. *European Journal of Agronomy*, 93:113–125, ISSN 1161-0301, https://doi.org/10.1016/j.eja.2017.12.006.

Zheng, B., Campbell, J.B., de Beurs, K.M. 2012. Remote sensing of crop residue cover using multi-temporal Landsat imagery. *Remote Sensing of Environment*, 117:177–183, ISSN 0034-4257, https://doi.org/10.1016/j.rse.2011.09.016.

Zhou, X., Huang, W., Kong, W., Ye, H., Luo, J., Chen, P. 2016. Remote estimation of canopy nitrogen content in winter wheat using airborne hyperspectral reflectance measurements. *Advances in Space Research*, 58(9):1627–1637, ISSN 0273-1177, https://doi.org/10.1016/j.asr.2016.06.034.

Zhuang, Q., Melack, J.M., Zimov, S., Sakha, C., Walter, K.M., Butenhoff, C.L., Khalil, A.K. 2009. Global methane emissions from wetlands, rice paddies, and lakes. *Eos*, 90(5):37–44.

Index